UNDERSTANDING SOILS OF MOUNTAINOUS LANDSCAPES

UNDERSTANDING SOILS OF MOUNTAINOUS LANDSCAPES

Sustainable Use of Soil Ecosystem Services and Management

Edited by

RAHUL BHADOURIA
*Department of Environmental Studies, Delhi College of Arts and Commerce, University of Delhi,
New Delhi, India*

SHIPRA SINGH
School of Environmental Sciences, Jawaharlal Nehru University, New Delhi, India

SACHCHIDANAND TRIPATHI
Department of Botany, Deen Dayal Upadhyaya College, University of Delhi, New Delhi, India

PARDEEP SINGH
Department of Environmental Studies, PGDAV College, University of Delhi, New Delhi, India

ELSEVIER

Elsevier
Radarweg 29, PO Box 211, 1000 AE Amsterdam, Netherlands
The Boulevard, Langford Lane, Kidlington, Oxford OX5 1GB, United Kingdom
50 Hampshire Street, 5th Floor, Cambridge, MA 02139, United States

Notices
Knowledge and best practice in this field are constantly changing. As new research and experience broaden our understanding, changes in research methods, professional practices, or medical treatment may become necessary.

Practitioners and researchers must always rely on their own experience and knowledge in evaluating and using any information, methods, compounds, or experiments described herein. In using such information or methods they should be mindful of their own safety and the safety of others, including parties for whom they have a professional responsibility.

To the fullest extent of the law, neither the Publisher nor the authors, contributors, or editors, assume any liability for any injury and/or damage to persons or property as a matter of products liability, negligence or otherwise, or from any use or operation of any methods, products, instructions, or ideas contained in the material herein.

ISBN: 978-0-323-95925-4

For Information on all Elsevier publications
visit our website at https://www.elsevier.com/books-and-journals

Publisher: Joseph P. Hayton
Acquisitions Editor: Jessica Mack
Editorial Project Manager: Ali Afzal-Khan
Production Project Manager: Bharatwaj Varatharajan
Cover Designer: Matthew Limbert

Typeset by MPS Limited, Chennai, India

Contents

2

Soil microbial processes and nutrient dynamics

3. Integrated remedial and management strategies for sustaining mountainous soil 43

Surbhi Sharma, Neeru Bala, Priyanka Sharma, Joat Singh, Shalini Bahel and Jatinder Kaur Katnoria

4. Microbial perspectives for the agricultural soil health management in mountain forests under climatic stress 59

Soumya Sephalika Swain, Yasaswinee Rout, Phani Bhusan Sahoo and Shubhransu Nayak

5. Soil biological processes of mountainous landscapes: a holistic view 91

Bhawna Tyagi, Simran Takkar and Prabhat Kumar

6. Soil nutrient dynamics under selected tree species explains the soil fertility and restoration potential in a semi-arid forest of the Aravalli Mountain range 115

Shikha Prasad and Ratul Baishya

7. Soil nutrient dynamics under mountainous landscape: issues and challenges 131

S. Sivaranjani and Vijender Pal Panwar

3

Soil physicochemical parameters

4

Land use and land cover change

5

Plant functional traits and ecological sustainability

List of contributors

Nazish Abid Department of Architecture and Interior Design, College of Engineering, University of Bahrain, Bahrain

Pavithra Acharya (G.M) Department of Environmental Science, University of Mysore, Manasagangotri, Karnataka, India

Mustaqeem Ahmad Department of Environment Studies, Panjab University, Chandigarh, India

Mushtaq Ahmad Dar Department of Botany, Panjab University, Chandigarh, India

P.V. Annie Gladys Department of English, Nesamony Memorial Christian College, Marthandam, Kanyakumari, Tamil Nadu, India

Shalini Bahel Department of Electronics Technology, Guru Nanak Dev University, Amritsar, Punjab, India

Ratul Baishya Ecology & Ecosystem Research Laboratory, Department of Botany, University of Delhi, New Delhi, India

Roshan M. Bajracharya Department of Environmental Science and Engineering, School of Science, Kathmandu University, Dhulikhel, Nepal

Neeru Bala Department of Botanical and Environmental Sciences, Guru Nanak Dev University, Amritsar, Punjab, India

Sunny Bansal School of Architecture, Anant National University, Ahmedabad, Gujarat, India

Vidhu Bansal Department of Architecture and Regional Planning, IIT Kharagpur, Kharagpur, West Bengal, India; Anant National University, Ahmedabad, Gujarat, India

Anushree Baruah Department of Botany, The Assam Royal Global University, Guwahati, Assam, India

Daizy R. Batish Department of Botany, Panjab University, Chandigarh, India

Rahul Bhadouria Department of Environmental Studies, Delhi College of Arts and Commerce, University of Delhi, New Delhi, India

Abhishek Kumar Chaubey School of Environmental Sciences, Jawaharlal Nehru University, New Delhi, Delhi, India

Mitrajit Deb Department of Zoology, The Assam Royal Global University, Guwahati, Assam, India

Ipsa Gupta Department of Botany, Panjab University, Chandigarh, India

Sharmila Jagadisan School of Architecture, Vellore Institute of Technology, Vellore, Tamil Nadu, India

Solomon Jeeva Department of Botany, Scott Christian College, Nagercoil, Kanyakumari, Tamil Nadu, India

K.S. Karthika (Kavukattu) ICAR-National Bureau of Soil Survey and Land Use Planning, Hebbal, Bengaluru, Karnataka, India

Jatinder Kaur Katnoria Department of Botanical and Environmental Sciences, Guru Nanak Dev University, Amritsar, Punjab, India

Shalinder Kaur Department of Botany, Panjab University, Chandigarh, India

Solomon Kiruba Department of Zoology, Madras Christian College, Tambaram, Chennai, Tamil Nadu, India

Anil K.S. Kumar ICAR-National Bureau of Soil Survey and Land Use Planning, Hebbal, Bengaluru, Karnataka, India

Prabhat Kumar School of Environmental Sciences, Jawaharlal Nehru University, New Delhi, India

Sachin Kumar Department of Geography, Government Degree College, Shahpur, Kangra, Himachal Pradesh, India

M. Lalitha ICAR-National Bureau of Soil Survey and Land Use Planning, Hebbal, Bengaluru, Karnataka, India

P.A. Lubina Kerala Forest Research Institute, Peechi, Thrissur, Kerala, India

Priyanka Mahajan Department of Botany, Panjab University, Chandigarh, India

Hardik Manek National Institute of Technology, Durgapur, West Bengal, India

P. Maria Antony ATREE's Agasthyamalai Community Conservation Centre (ACCC), Manimutharu, Tamil Nadu, India

Akhilendra Kumar Mishra Department of Geography, Maharaja Bijli Pasi Government Post Graduate College, Ashiyana, Lucknow, Uttar Pradesh, India

Ravi Namasivaya Institute of Wood Science and Technology, Malleswaram, Bengaluru, Karnataka, India

Shubhransu Nayak Odisha Biodiversity Board, Forest, Environment and Climate Change Department, Government of Odisha, Regional Plant Resource Centre Campus, Bhubaneswar, Odisha, India

Vijender Pal Panwar Soil Science Discipline, Forest Ecology and Climate Change Division, Forest Research Institute, Dehradun, Uttarakhand, India

Shivangi Singh Parmar RCG School of Infrastructure Design and Management, IIT Kharagpur, Kharagpur, West Bengal, India; School of Architecture, Vellore Institute of Technology (VIT), Vellore, Tamil Nadu, India

Shikha Prasad Ecology & Ecosystem Research Laboratory, Department of Botany, University of Delhi, New Delhi, India

Brahmacharimayum Preetiva School of Environmental Sciences, Jawaharlal Nehru University, New Delhi, Delhi, India

Riya Raina Department of Environment Studies, Panjab University, Chandigarh, India

Yasaswinee Rout Odisha Biodiversity Board, Forest, Environment and Climate Change Department, Government of Odisha, Regional Plant Resource Centre Campus, Bhubaneswar, Odisha, India

Phani Bhusan Sahoo Biosystematics Laboratory, PG Department of Botany, Maharaja Sriram Chandra Bhanjadeo University, Sri Ram Chandra Vihar, Baripada, Odisha, India

M.C. Sandhya Department of Plant Biotechnology and Cytogenetic, Institute of Forest Genetics and Tree Breeding, Coimbatore, Tamil Nadu, India

Sarika School of Environmental Science, Jawaharlal Nehru University, New Delhi, Delhi, India; Ashoka Trust for Research in Ecology and the Environment, Royal Enclave, Srirampura, Jakkur, Bengaluru, Karnataka, India

Joy Sen Department of Architecture and Regional Planning, IIT Kharagpur, Kharagpur, West Bengal, India

Priyanka Sharma Department of Botanical and Environmental Sciences, Guru Nanak Dev University, Amritsar, Punjab, India

Surbhi Sharma Department of Botanical and Environmental Sciences, Guru Nanak Dev University, Amritsar, Punjab, India

Amit Shoshta Department of Geography, G B Pant Memorial Government College, Rampur Bushahr, Shimla, Himachal Pradesh, India

Him Lal Shrestha Department of Environmental Science and Engineering, School of Science, Kathmandu University, Dhulikhel, Nepal

Harminder P. Singh Department of Environment Studies, Panjab University, Chandigarh, Punjab, India

Joat Singh Department of Botanical and Environmental Sciences, Guru Nanak Dev University, Amritsar, Punjab, India

Rishikesh Singh Department of Botany, Panjab University, Chandigarh, India

Vipin Kumar Singh Department of Botany, K. S. Saket P.G. College, Ayodhya, Uttar Pradesh, India

Jonathan S. Singsit School of Environmental Sciences, Jawaharlal Nehru University, New Delhi, Delhi, India

Bishal K. Sitaula Department of International Environment and Development Studies, Norwegian University of Life Sciences, Ås, Norway

S. Sivaranjani Soil Science Discipline, Forest Ecology and Climate Change Division, Forest Research Institute, Dehradun, Uttarakhand, India

Kakul Smiti School of Environmental Sciences, Jawaharlal Nehru University, New Delhi, Delhi, India

Pratap Srivastava Department of Botany, University of Allahabad, Prayagraj, Uttar Pradesh, India

Sruthi Subbanna Institute of Wood Science and Technology, Malleswaram, Bengaluru, Karnataka, India

Soumya Sephalika Swain Odisha Biodiversity Board, Forest, Environment and Climate Change Department, Government of Odisha, Regional Plant Resource Centre Campus, Bhubaneswar, Odisha, India

Simran Takkar Life Sciences Department, Shiv Nadar University, Greater Noida, Uttar Pradesh, India

G.N. Tanjina Hasnat Institute of Forestry and Environmental Sciences, University of Chittagong, Chattogram, Bangladesh

Bhawna Tyagi School of Environmental Sciences, Jawaharlal Nehru University, New Delhi, India

Syam Viswanath Kerala Forest Research Institute, Peechi, Thrissur, Kerala, India

About the editors

Dr. Rahul Bhadouria is working as an assistant professor at the Department of Environmental Studies, Delhi College of Arts and Commerce, University of Delhi, New Delhi, India. He obtained his doctorate from the Department of Botany, Banaras Hindu University, Varanasi, India, in 2017. The area of his doctoral research was the performance evaluation of tree seedling growth under dry tropical environment. He has published more than 22 papers, 22 book chapters, and 7 edited books in internationally reputed journals/publishers. His current research areas are "Management of soil C dynamics to mitigate climate change," "a perspective on tree seedling survival and growth attributes in tropical dry forests under the realm of climate change," "plant community assembly, functional diversity and soil attributes along the forest-savanna-grassland continuum in India," and "recovery of degraded mountains in central Himalayas."

Affiliations and Expertise
Department of Environmental Studies, Delhi College of Arts and Commerce, University of Delhi, New Delhi, India; School of Environmental Sciences, Jawaharlal Nehru University, New Delhi, India.

Dr. Shipra Singh completed her MPhil and PhD from the School of Environmental Sciences, Jawaharlal Nehru University, New Delhi, India. She obtained her master's degree in Environmental Sciences from Banaras Hindu University, Varanasi, India. The area of her doctoral research work was to characterize plant functional traits of dominant woody species and soil attributes in the Western Himalayan mountainous landscape. Her research interest focuses on montane ecology and climate change, plant functional traits, ecosystem functions and services. She has published several papers in international journals in the field of forest ecology.

Affiliations and Expertise
School of Environmental Sciences, Jawaharlal Nehru University, New Delhi, India

Dr. Sachchidanand Tripathi is working as an assistant professor at the Department of Botany, Deen Dayal Upadhyaya College, University of Delhi, New Delhi, India. He did his postgraduation and PhD from the Department of Botany, Banaras Hindu University, Varanasi, India. His areas of interest are forest ecology, soil ecology, restoration ecology, aquatic ecology, and eco-physiology. He has published more than 21 research/review papers in peer-reviewed reputed international journals, 5 books with international publishers, and 15 book chapters in books published by Elsevier, Springer, etc. He has also edited conference proceedings. He is interested in collaborating with scholars from the same fields as well as from other fields of interest.

Affiliations and Expertise
Department of Botany, Deen Dayal Upadhyaya College, University of Delhi, New Delhi, India

Dr. Pardeep Singh is working as an assistant professor in the Department of Environmental Studies, PGDAV College, University of Delhi, New Delhi, India. He obtained his master's degree from the Department of Environmental Science, Banaras Hindu University, Varanasi, India. He obtained his doctorate from the Indian Institute of Technology (Banaras Hindu University), Varanasi, India. The area of his research is waste management, global climate change, circular economy, and environmental conservation. He has published more than 75 papers in international journals in the field of waste and environmental management and edited more than 45 books with leading international publishers.

Affiliations and Expertise
Department of Environmental Studies, PGDAV College, University of Delhi, New Delhi, India

Foreword

Mountains cover 24% of the terrestrial land area and are home to the world's major biomes, which provide a variety of ecological services. Mountains are home to a diverse range of biodiversity, including microorganisms, plants, and large animals. Mountains supply freshwater to more than half of the world's population as most of the rivers have their origin from mountains. Mountain ecosystems provide tangible benefits that contribute significantly to the economies in the mountains and catchments; for example, subsistence agriculture foods, grazing animal products, timber, fuel wood, and nontimber items such as wildlife and medicinal plants. Mountains also provide significant nonmaterial benefits to lowlands in the form of scenic beauty, biodiversity, and soil nutrients. At the same time, these ecosystems are ecologically fragile and vulnerable due to a variety of factors such as soil formation, land availability, climate change, natural hazards; a lack of access to markets, education, and health care; ineffective governmental or industrial interventions; the high specialization and interdependence of mountain social and land use systems; and globalization. Natural hazards such as avalanches, landslides, and rock falls are common features of mountain ecosystems; however, in the absence of anthropogenic disturbances, these are regulated to some extent by forests.

Soil is a limited natural resource that requires replenishment over time. Mountain soils have long provided vital ecological services such as food security and nutrition to the 900 million people who live nearby, as well as billions more who live downstream around the world. Even though these soils are naturally unstable, they contain 25% of the biodiversity, including agro-biodiversity, which is critical for locally adapted agricultural crops and livestock.

Mountain soils are extremely vulnerable to climate change, deforestation, unsustainable farming practices, and resource extraction, all of which contribute to soil degradation and disasters such as severe flooding and landslides.

This book contains the most recent practical and theoretical concepts, as well as new challenges, related to mountain soils, specifically soil microbial processes and nutrient dynamics, threats to their cultural heritage, plant functional traits, and ecological sustainability. This book also discusses ecological sustainability, urbanization, and soil development in mountains. This publication includes case studies as well as conservation and management strategies. This book also highlighted some of the major challenges of achieving ecological restoration goals in mountain ecosystems around the world. With such comprehensive coverage of a wide range of topics, this book includes a comprehensive research handbook for soil conservation and mountain management strategies. The book provides a framework for understanding and restoring the soils of mountainous landscapes, thereby increasing their involvement in ecosystem services. The importance of this book stems from its unique organization of chapters covering a wide range of topics, which is supported by numerous examples, experiments, and case studies, as well as numerous techniques that can be used.

This publication was created to commemorate the "United Nations Decade on Ecosystem Restoration (2021−2030)" by editors and contributors from various organizations and regions. I believe the book will be a valuable resource for researchers in the field of environment and ecology as well as experts in town and country planning, policymakers, and architects. While congratulating the editors, contributors, and publishers on the publication of this volume, I express my best wishes for its success.

S.C. Garkoti
School of Environmental Sciences,
Jawaharlal Nehru University,
New Delhi, India

Preface

Mountains provide valuable soil, minerals, water, biological diversity, and recreational opportunities. They are critical for the survival of large global humanity as a complex and integrated ecosystem. Mountain ecosystems cover approximately 25% of the world's total land cover, supporting a significant portion of the world's biodiversity and housing 13% of the world's total population. Mountainous landscapes provide a variety of ecosystem services to people who live there and in surrounding areas. These landscapes support a diverse but dispersed array of habitats. The successional changes that occur in the montane ecosystem over time play an important role in determining soil properties such as moisture content, aeration, nutrient concentration, and microbiota, which in turn determines the intensity, quality, and sustainability of primary production directly utilized by humans. The microbial diversity in the soil ecosystem is astounding—there are more species in 1 g of soil than there are mammals on the planet. In comparison to aboveground diversity, scientific understanding of belowground diversity is dispersed and limited. This is primarily due to the scarcity of techniques for soil monitoring, inventorying, and characterizing the functional significance of various soil types and soil organisms. Furthermore, the continuous changes in soil properties are primarily determined by the type of parent rock material from which soils form.

Soils are thought to be a function of independent factors such as topography, vegetation, climate, and parent rock material in the mountainous ecosystem. Thus understanding soil dynamics and their mechanisms is aided by knowledge of the aforementioned factors. However, the formation of microclimatic conditions at regional and even local scales complicates the explanation of soil dynamics and limits the generalization of soil concepts and theories. On the one hand, increased pH of soil leads to increased physical and chemical weathering, resulting in progressive soil development; on the other hand, steep slopes may cause increased soil erosion and unweathered material deposition. Understanding mountain soils, their composition, chemical and biological properties, and the challenges associated with sustainable development in mountains may aid in addressing these challenges and ensuring highly resilient livelihoods for people living in its vicinity as well as downstream communities.

Mountain soils are particularly vulnerable to increased traditional practices that result in significant land use changes. Traditional practices such as slash and burn, overuse of land, and increased use of pesticides and other harmful chemicals all have a negative impact on soil quality and the environment. Climate change, on the other hand, contributes significantly to its degradation in terms of arability, desertification, and soil health. Other practices, such as tourism, have a significant impact on the soil quality of mountainous landscapes. To conserve montane soil, proper knowledge of land resources is required, with a special emphasis on the assessment of critical zones. Furthermore, various management practices such as contour farming, crop rotation, and conservation tillage increase soil efficiency, promoting soil ecosystem stability. The goal of this book is to describe the key characteristics of mountain soil ecosystems, as well as the threats they face and various strategies for mitigating soil degradation and depletion. The book will contribute to a better understanding of the importance of mountain soils and the roles they play in achieving livelihood and sustainable development on a global scale. The book's goal is to provide readers with a comprehensive understanding of the most recent practical and theoretical concepts, as well as emerging issues in mountain soil ecosystems, in order to help them generate a better understanding and problem-solving approach. It also highlights some of the major challenges that mountain ecosystems face due to tourism, agriculture, and other human-caused activities. This book will benefit undergraduate, postgraduate, university research scholars, teachers, environmental scientists, ecologists, agriculturists, biotechnologists, and early career researchers, particularly those working in soil management and resource management, NGOs and government institutions working on environmental conservation and restoration, and policymakers. The book contains a total of 19 chapters:

Chapter 1 provides a better understanding of soil ecosystem that would help in developing mitigation approaches and improving the soil C-sequestration potential of mountain soils.

Chapter 2 provides a detailed review of the threats faced by mountain soils and the conservation and management practices that might lead to sustainable soil management.

Chapter 3 deals with the wide range of processes that affect the mountain soil and

different management strategies to conserve soil and its ecosystem.

Chapter 4 deals with various possible ways to sustainably manage the soil health of mountain forests. This chapter focuses on the use of microorganisms in agricultural practices and safeguarding agrobiodiversity in mountainous regions.

Chapter 5 provides a detailed view of the microbial diversity, their structure, and function and how they change with altitudinal gradients and other topographic features. This chapter also discusses the application of next-generation sequencing approaches like metagenome and metatranscriptome sequencing for identifying the microbial identity, community structures, and function of mountainous environments

Chapter 6 attempts to analyze the nutrient dynamics of native and exotic tree species across different seasons in the semi-arid forest of the Aravalli mountains. The findings would enable the selection of appropriate tree species for plantation, assisting in the long-term soil restoration in the region.

Chapter 7 deals with nutrient dynamics under soils from various sources in different vegetation covers along an altitudinal gradient in a mountainous landscape.

Chapter 8 discusses the impact of various agricultural management practices on the soil microbiome under the changing climate. This chapter explores the role of the nexus between various management practices and climate change in regulating the soil microbial community structure.

Chapter 9 talks about the application of biochar as an alternative to enhance the fertility of the soil and the effects induced by their application in montane landscapes.

Chapter 10 provides an overview of the role of organic debris in soil formation and succession for global conservation and restoration of tropical rain forests including rock ecosystems.

Chapter 11 discusses the wide range of threats faced by mountain soil ecosystems due to tourism practices and the need to develop sustainable tourism practices to avoid disastrous incidents.

Chapter 12 highlights the trans-relationship of soil and forest in the mountainous biome having a negligible anthropogenic intervention.

Chapter 13 deals with the importance of the alluvial fan which is an important depositional landform across all the global climatic regimes. This chapter aims to examine soil development on alluvial fans in the North-Western Himalayas.

Chapter 14 analyzes the effects of soil properties and land use type on soil organic carbons in three different watersheds of Nepal.

Chapter 15 attempts to understand the urbanization process in the mountain towns of Ladakh region in India. This chapter highlights the risks and vulnerability of the mountain soil ecosystem and strives to propose indigenous knowledge systems of building construction based on the soil.

Chapter 16 investigates human-induced stresses which affect the terrain in physical and socioeconomic aspects in two different mountain systems located in the northern and southern parts of India.

Chapter 17 attempts of using Artificial-Neuron-Network-based Cellular Automata Model for simulating changes in land use and land cover and the model can also be used to analyze land-use changes in mountainous areas.

Chapter 18 addresses the functional traits of plants and their link to soil functions, especially in mountainous soil. This chapter explores relatedness of plant traits, soil function and its ecosystem services by delving into the interaction between plant functional traits and carbon and nutrient dynamics, biodiversity pool, and other aspects of land as well as mountainous soil ecosystem.

Chapter 19 elaborates socioeconomic and ecological sustainability of agroforestry systems. This chapter also highlights the impact of agroforestry practices on soil resources and different challenges and opportunities in terms of mountain ecosystems.

Rahul Bhadouria
Shipra Singh
Sachchidanand Tripathi
Pardeep Singh

Soils of mountainous landscapes: introduction

Mountain soils and climate change: importance, threats and mitigation measures

Rishikesh Singh[1], Ipsa Gupta[1], Riya Raina[2], Priyanka Mahajan[1], Pratap Srivastava[3], Vipin Kumar Singh[4] and Daizy R. Batish[1]

[1]Department of Botany, Panjab University, Chandigarh, India [2]Department of Environment Studies, Panjab University, Chandigarh, India [3]Department of Botany, University of Allahabad, Prayagraj, Uttar Pradesh, India [4]Department of Botany, K.S. Saket P.G. College, Ayodhya, Uttar Pradesh, India

1.1 Introduction

Mountains are complex ecosystems with a wide variety of abiotic and biotic components (Messerli and Ives, 1997; Rapp and Silman, 2012; Beniston and Fox, 2015). Mountains are found in all the continents of the Earth (Beniston and Fox, 2015). Most mountainous systems (except those found in Antarctica) are inhabited by humans and other living organisms, and they serve as a source of food for various organisms (Ives, 1992; Körner et al., 2011, 2017; Beniston and Fox, 2015). Significant altitudinal differences observed at mountains, such as the Andes, Alps, Himalayas, and Rockies, cause variation in microclimatic conditions over short distances that is typically observed over large latitudinal ranges in flat systems. Because of the short-distance climatic variability, there is a greater diversity of plant and animal species, community composition, and habitat conditions along the mountain systems (Beniston and Fox, 2015; Tito et al., 2020). Mountains, for example, cover about 12% of the Earth's terrestrial land surface and support more than 85% of the world's amphibian, bird, and mammal species, as well as roughly one-third of the plant communities (Körner et al., 2011; Rahbek et al., 2019a). Within a short altitudinal variation of ~2800 m (i.e., from submontane forests at 800 m asl to the treelines at 3625 m asl) in the Manu National Park, Peru, Farfan-Rios et al. (2015) observed more than 1000 species of lianas, palms, and trees. Mountain ecosystems not only reflect high biodiversity but also show a high level of endemism (Rahbek et al., 2019a). Mountainous ecosystems, in addition to their rich biodiversity, play an important role in climate regulation (Beniston and Fox, 2015). Mountain regions are now attracting significant economic investments in the industries of hydropower, agriculture, tourism, transportation, and communication (Fig. 1.1). Thus,

FIGURE 1.1   Importance of mountain ecosystems.

mountainous ecosystems are among the most vulnerable ecosystems in terms of the occurrence of extreme events, and they are constantly at odds due to various economic development and environmental conservation issues (Beniston and Fox, 2015).

Soil is the topmost surface layer of the Earth formed by the weathering of rocks caused by precipitation, temperature, and wind. It is made up of solids (organic matter and minerals), microorganisms, and gases that are occasionally replaced by water. It provides nutrients and water to plants and trees, as well as habitat for a variety of small animals, insects, and microorganisms (FAO, 2015). For example, over 10,000 types of microorganisms (e.g., bacteria and fungi) reside in the alpine soils (Hagedorn et al., 2019). Soil is critical in mountain regions because it provides a variety of ecosystem services and functions such as agricultural development, food security, C-sequestration, climate change mitigation, and so on (FAO, 2015). Soil development in mountain regions can take centuries to millennia, depending on the climatic and ecosystem processes that occur beneath the surface (Hagedorn et al., 2019). The lower temperature in mountain (alpine) regions limits microbial activity, resulting in the slow and shallow development of mountain soils. Mountain soils are generally acidic with low fertility when compared to lowland soils, but there is significant heterogeneity in soil properties depending on the exposure and steepness of the mountain slopes (FAO, 2015). Temperature-mediated freeze-thaw cycles, for example, reduce soil aggregate stability and water retention capacity in cold mountain regions, resulting in poor fertility. Because of various extreme weather events, the soils on different mountain slopes are more prone to disturbance/erosion and deposition to flatter areas (FAO, 2015). These movements of mountain soil regulate the vegetation type, nutrient transport, microclimatic conditions, etc., particularly in cold regions (Callaghan et al., 2011). Variation in soil properties, such as organic matter content and moisture content, has similar effects on permafrost, infiltration, groundwater recharge, and runoff processes as variation in climate, hydrology, and vegetation types (Osterkamp et al., 2009). Overall, the interrelationships between climate-soil-vegetation in mountain ecosystems are complex, and soil responses to climate and vegetation changes vary across spatiotemporal scales (Rodriguez-Iturbe, 2000; Rasouli et al., 2019).

According to the United States National Oceanic and Atmospheric Administration (NOAA), the majority of the years since 2005 have been among the warmest on record in the last 140 years (IAEA, 2022). It is now widely acknowledged that global climate change is occurring, and high altitude areas, such as alpine regions, are experiencing the fastest warming conditions (Hagedorn et al., 2019). Mountain ecosystems are subjected to extreme climatic variability, which is exacerbated by land-use changes (Hagedorn et al., 2010). For example, the Himalayas are expected to experience a more than threefold increase in mean global temperature and erratic rainfall patterns, influencing various ecosystem functions and processes such as C and other nutrient cycling (Hagedorn et al., 2010). Furthermore, warming in mountain regions will cause altitudinal shifts in plants and animals (MacLean and Beissinger, 2017; Fadrique et al., 2018; Freeman et al., 2018). For example, a $2°C-4°C$ rise in the air temperature may result in $300-600$ m altitudinal migration of plants. Soils play an important role in this migration because they serve as a repository for the nutrients and water needed for plant growth (Hagedorn et al., 2019). However, the rate of migration remains out of sync with the rate of climatic variability, resulting in an unexpected species loss in the near future (Fadrique et al., 2018; Tito et al., 2020). Because of changing climatic conditions, large amounts of C stored in mountain soils (as humus) are expected to be released as CO_2

(Hagedorn et al., 2019). To observe the changes in different ecosystem processes with respect to climate change, several controlled-environment (e.g., by manipulating temperature and CO_2 concentrations) level experiments such as by using infrared heaters, soil heating cables, glasshouses and open-top chambers have been carried out in different regions (Bokhorst et al., 2011; Elmendorf et al., 2015; Kimball, 2016; Wang et al., 2017). However, such experiments are limited to manipulating a few parameters (e.g., temperature and CO_2 conditions), while others (e.g., wind speed, soil moisture, relative humidity, radiation, biotic interactions, and so on) are not. Mountain ecosystems, on the other hand, with higher levels of climatic heterogeneity along altitudes, serve as a natural laboratory system for gaining a clear understanding of the direct and indirect effects of climate change and identifying different mitigation strategies (Tito et al., 2020). Furthermore, it will aid in the comprehension of various ecological processes and functions such as new biotic interactions and processes, factors influencing species assemblages in newer climatic conditions, interactions between abiotic and biotic components for species establishment and migration, and so on (Tito et al., 2020).

Rapid changes in climatic conditions in mountain regions are major sources of concern for local communities, environmentalists, policymakers, and scientists because they affect not only the mountain regions but also the associated lowland areas and the people who live there (Freppaz and Williams, 2015). The hydrological features of mountain systems provide significant benefits to local communities as well as people living in flatter areas (Beniston and Fox, 2015; Rasouli et al., 2019). Climate change is expected to reduce the average soil organic C (SOC) stock in various mountain ecosystems, but the changes will be elevation-dependent (Liu et al., 2021). For example, in a modeling study on SOC dynamics in the middle Qilian Mountains, Liu et al. (2021) predicted a decrease in the SOC stock in the elevation range <3100 to >3900 m whereas above and below (upto 2900 m) these elevation ranges, a decrease was predicted under different Representative Concentration Pathway (RCP) such as RCP2.6, RCP4.5 and RCP8.5 scenarios. Furthermore, soil type, increased nitrogen deposition, CO_2 enrichment, climate extremes, and ecological restoration practices all influence SOC dynamics (Li et al., 2021a). Climate change is expected to cause increased glacier retreat, which will result in the accumulation of C in Glacic Cryosols (soil near glaciers) due to an increase in vegetation (and litter) components (Liu et al., 2021). However, it will also lead to decrease in C locked in the permafrost (Hagedorn et al., 2019). There is also a major concern for the release of greenhouse gases locked in the mountain soils which will further modulate the Earth's surface temperature (Freppaz and Williams, 2015). There are a few questions related to the mountain soils and climate change on which recent researches are being carried out (Hagedorn et al., 2010), viz., how the climate warming will influence the labile C pools?, how the change in microbial community composition influence the soil properties and nutrient dynamics?, how the freeze-and-thaw cycles as well as annual warming and cooling temperatures will influence the soil nutrients and C stocks?, how extreme weather events (e.g., avalanches, landslides, etc.) influence nutrient cycling and vegetation components?, and how the land abandonment will affect the soil functioning? Mountain ecosystems, which are a sensitive and reliable indicator of climate change, provide a natural setting for addressing such complex research questions at various spatiotemporal scales (Hagedorn et al., 2010). This chapter provides a brief overview of the importance of mountain soils, variations in different soil properties along mountain ecosystems, major threats and challenges to mountain soils, and mitigation and adaptation measures used for mountain soil prevention and conservation.

1.2 Importance of mountain soils

Soils are considered as the basis for the healthy food production (FAO, 2015). Mountain soils support natural vegetation, biodiversity, agricultural crops, a source of water for major rivers, food security, and other ecosystem services that support human life on Earth (Beniston and Fox, 2015; FAO, 2015; Feng et al., 2019). Small-scale variations in climatic conditions, plant species, soil properties, and productivity serve as a natural laboratory for understanding the processes at work in the soil system (Hagedorn et al., 2010). Moreover, mountains play important role in hydrological cycles and act as "water towers" for a major population, globally (Viviroli et al., 2011; Beniston and Stoffel, 2013; Fort, 2015). For example, the Himalayas and the Hindu Kush region provide freshwater to 800 million people living in the surrounding area (Wangdi et al., 2017). According to estimates, mountain soils help to ensure food security for 900 million people who live in the mountains and billions more who live downstream (FAO, 2015). Mountain soils aid in climate change mitigation and disaster reduction by managing water and carbon, increasing resilience to floods and droughts, and so on (Fig. 1.2). A large amount of C accumulates in alpine and higher altitude areas, with approximately 90% of the C stored within the surface soil layers, and alpine soils may play a role in the changes in vegetation caused by climate change (Hagedorn et al., 2019). Though mountain soils are fragile in nature, they support 25% of biodiversity and serve as critical gene pools for locally adapted crops and livestock (FAO, 2015). The alpine ecosystem has locked 90% of the C that indirectly regulates the climate, and if it is leaked as greenhouse gases, it will fuel climate change even more (Hagedorn et al., 2019). Even minor changes in temperature and precipitation can cause significant variations in SOC stock in the alpine regions (Stockmann et al., 2013; Ding et al., 2016; Liu et al., 2021). Thus, variations in soil properties in mountain regions with warming conditions provide insight into the effects of climate change (Hagedorn et al., 2019).

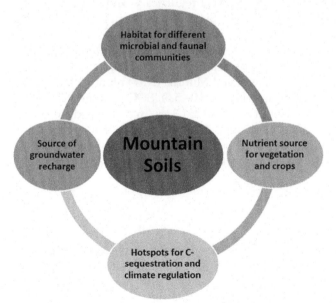

FIGURE 1.2   Multiple benefits of mountain soils for the sustenance of human livelihoods.

In terms of C-sequestration potential, mountain permafrost is critical, containing 66.3 Pg of SOC and accounting for $\sim$4.5% of global C pools, whereas Arctic soils contain 496 Pg of SOC and accounting for $\sim$33% of global C pools (Bockheim and Munroe, 2014). Permafrost is a kind of condition in which the material remains below 0°C for two or more years successively (van Everdingen, 1998). As per the report of the "Mapping and Modeling of Mountain Permafrost" working group of the International Permafrost Association, $\sim$4.88 million km^2 area is occupied by the mountain permafrost (Bockheim and Munroe, 2014). The spatial distribution of C in mountain soils exhibits significant heterogeneity. The C content will be higher near the vegetation, and altitudinal variability will be observed. With warming climatic conditions, the treeline is expected to shift upward, directly increasing soil C-storage potential via litter fall and indirectly decreasing C-storage by exposing previously stored C (e.g., permafrost) and releasing it as CO_2 via enhanced microbial activities (Hagedorn et al., 2019). Hagedorn explained how climatic warming will induce soil microbial communities and their activities, which will enhance nutrient cycling and provide suitable grounds for plant growth in a six-year model experiment on the warming of soils using heating cables at Davos (Hagedorn et al., 2019). Thus, mountain soils are of high ecological importance and play a critical role in the climate change scenario.

1.3 Variation in soil properties in the mountain regions

Various natural and anthropogenic factors (such as elevation, slope, soil type, roofing, snow cover, freeze-thaw cycle, etc.) operate at different levels in mountain regions, resulting in variations in soil properties. The thickness of snow cover, for example, is an important driver of soil nutrient and water dynamics; thus, the frost-and-thaw cycle regulates C and N dynamics as well as soil structure and stability (Hagedorn et al., 2010). Similarly, the composition of the litter regulates microbial activity and nutrient cycling in mountain soils. The presence of phenolics in the litter favors fungal communities, whereas P availability favors bacterial activities, resulting in N-mineralization throughout the decomposition phase (Bárta et al., 2010). Based on a global database of 41 sites and 312 pedons, Bockheim and Munroe (2014) reported that alpine soils with permafrost generally have strong acidic pH (i.e. 5.0−5.5), SOC concentrated in the upper 30−40 cm layer with average profile density of 15.2 ± 1.3 kg/m^2, intermediate cation exchange capacity (20−25 cmolc/kg) and base saturation (44%−85%), and fall within isotic mineral class. Thus, the effects of various major factors such as elevation and slope, as well as soil type/order, on some of the major biophysical properties of mountain soils are described in the sub-sections that follow (Fig. 1.3).

1.3.1 Effect of elevation and slope

Plant species distribution, diversity of herbivores, frugivorous birds, seed predators, and soil biophysical properties were found to vary significantly along elevation gradients in mountain regions, according to studies (Nottingham et al., 2018; Bender et al., 2019; Hargreaves et al., 2019; Tito et al., 2020). For example, a peak of plant species richness was

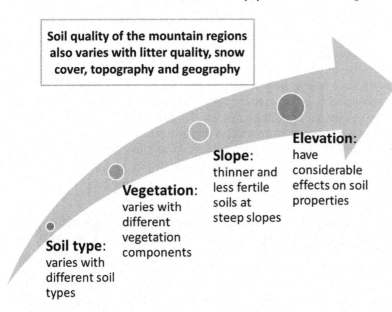

Soil quality of the mountain regions also varies with litter quality, snow cover, topography and geography

Elevation: have considerable effects on soil properties

Slope: thinner and less fertile soils at steep slopes

Vegetation: varies with different vegetation components

Soil type: varies with different soil types

FIGURE 1.3 Impact of different factors on soil biophysical properties in the mountain regions.

observed ~1400 m elevation (Girardin et al., 2010; Tito et al., 2020), and an intense decrease in SOC stocks in mid-elevation zones (~3000−3500 m asl), less variation in lower elevation zones and an accumulation in the higher elevation zones (>4000 m asl) in different mountain regions of the world have been reported during different time periods (Ding et al., 2017; Chen et al., 2019; Liu et al., 2021). Though soil physicochemical properties varied in mountain regions, there was no consistent pattern in their variation across regions (Rahbek et al., 2019b). Local characteristics such as geographic location, topography, and vegetation influence heterogeniety in soil properties. In the Peruvian Andes, for example, an increase in SOC content was observed as elevation increased (Tito et al., 2018), Hawaiian mountains (Townsend et al., 1995), Mount Rainier in the western United States (Ettinger and Hillerislambers, 2017), whereas a decrease in SOC content was observed along with the increasing elevation in the Himalayas (Sheikh et al., 2009). High concentrations of P and K contents are reported at the middle elevations (~3500 m asl) whereas a strong increase in Ca content and a decrease in Mg content with higher elevation was observed in the southwestern Peruvian Andes (Tito et al., 2018). A decrease in N content along the elevation was observed in the Andes (Zimmermann et al., 2010) whereas no trend in N and P content along the elevation was observed in the Mount Rainier (Ettinger and Hillerislambers, 2017) and arid Tianshan Mountain in northwest China (Xu et al., 2019). In general, most of the tropical mountain soils are younger (with N-limitation) at higher elevation whereas older (with P-limitation) at lower elevations (Vitousek and Farrington, 1997). Some studies also found that altitude increased the contribution of particulate and labile C pools, particularly in grassland and treeline regions, but there was no significant increase in total SOC storage in some European mountain regions (Sjögersten et al., 2003; Leifeld et al., 2009). Smaller aggregation and lower aggregate stability can also be attributed to lower SOC stability in higher altitude soils (Hagedorn et al., 2010).

Mountain slopes, like elevation, play an important role in regulating soil properties. Slopes play an important role in regulating soil biophysical properties by influencing small-scale climate gradients (Blagodatskaya et al., 2010). The soils on the steeper mountain slopes are thinner and less fertile than that of the flatter slopes (FAO, 2015). In a mountain valley region, north-east facing slopes with colder and wetter areas reveal fungal dominance and thus higher soil organic matter (SOM) content in top soils than southwest facing slopes in alpine regions (Blagodatskaya et al., 2010). Thus, future research should focus on variation in plant and soil properties along the elevation gradient in order to better understand the impact of climate change in mountain regions (Tito et al., 2020).

1.3.2 Effect of soil type/order

Different types of soils occur in different mountain regions depending on the age of the mountains. Specific soil physicochemical properties are associated with the type of soil formation (Bockheim and Munroe, 2014). Cambisolization in high altitude alpine regions, podsolization in subalpine regions, gleization in most mountain ranges, particularly in bedrock depressions, and paludization in Appalachian Mountain alpine zones are just a few of the soil formation processes that occur in various mountain regions (Bockheim and Munroe, 2014). These processes lead to the formation of different types of soils having varying SOC density. For example, SOC density varied ($15.2 \pm 1.3 \, kg/m^2$ average and ranged from <1.0 to $88.3 \, kg/m^2$) with the type/order of soil in the alpine regions (Bockheim and Munroe, 2014) as follows: Histosols ($52 \, kg/m^2$) > Spodosols ($27.2 \pm 4.5 \, kg/m^2$) > Alfisols ($27 \pm 11 \, kg/m^2$) > Gelisols ($20 \pm 6 \, kg/m^2$) > Mollisols ($18.6 \pm 11 \, kg/m^2$) > Inceptisols ($13.5 \pm 0.86 \, kg/m^2$) > Entisols ($6.2 \pm 1.1 \, kg/m^2$). Inceptisols (Cambisols) are young soils commonly found in alpine regions where pedogenesis is hampered by slow weathering and decomposition processes (Bockheim and Munroe, 2014). The variation in different soil properties along mountainous gradients has been described in different specific chapters of the book, so it is only briefly discussed in this introductory chapter.

1.4 Threats and challenges to the mountain soils

Mountain ecosystems are distinguished by their complex topography, fragility, slow soil formation, increased climatic and vegetation variability, short growing season, and sharp seasonal contrasts (Körner, 2013; Winkler et al., 2016). Several global-scale changes, such as land-use change, deforestation, livestock overgrazing, marginal soil cultivation, urbanization, and climate change, are having a significant impact on mountain ecosystems (FAO, 2015; Fort, 2015). In addition to increasing population and poor infrastructure, the increased frequency of natural hazards such as floods, rock fall and debris flows, landslides, soil erosion, avalanches, and glacial outbursts are major threats observed in mountain regions, affecting overall ecosystem functioning (Hagedorn et al., 2010; FAO, 2015). These threats have an impact on mountain biodiversity and water resources, as well as the livelihood and food security of the local population and other nearby regions (Lutz and Immerzeel, 2013; Kohler et al., 2014; FAO, 2015; Fort, 2015). A detailed view on the threats and challenges to the mountain soils has been elaborated in the following sub-sections (Fig. 1.4).

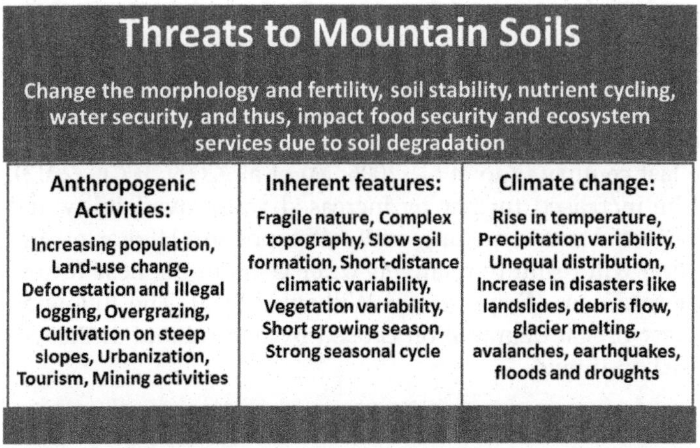

Threats to Mountain Soils

Change the morphology and fertility, soil stability, nutrient cycling, water security, and thus, impact food security and ecosystem services due to soil degradation

Anthropogenic Activities:	Inherent features:	Climate change:
Increasing population, Land-use change, Deforestation and illegal logging, Overgrazing, Cultivation on steep slopes, Urbanization, Tourism, Mining activities	Fragile nature, Complex topography, Slow soil formation, Short-distance climatic variability, Vegetation variability, Short growing season, Strong seasonal cycle	Rise in temperature, Precipitation variability, Unequal distribution, Increase in disasters like landslides, debris flow, glacier melting, avalanches, earthquakes, floods and droughts

FIGURE 1.4   Potential threats to the mountain soils induced by the inherent, anthropogenic and climate change-mediated factors.

1.4.1 Climate change and extreme weather events

According to global climate change projections, the impact of climate change will differ across regions. More warming, for example, will be observed at land, high latitudes, and elevations (Auer et al., 2007; Gobiet et al., 2014). Furthermore, mountains are regarded as one of the most unique regions for determining the impact of climate change, with glacier retreat and permafrost being the most visible effects of climate change (IPCC, 2014; Freppaz and Williams, 2015; Rogora et al., 2018). Climate change, for example, is affecting mountain ecosystems by increasing the frequency and intensity of extreme weather events such as heavy rainfall, glacier melt, droughts, avalanches, and mudslides (Beniston and Fox, 2015; FAO, 2015) resulting in habitat degradation, deterioration in freshwater quality, biodiversity loss, and overall ecosystem services (Stoll et al., 2015; Huss et al., 2017). The major issues encountered in several mountain regions are soil erosion and decreasing slope stability due to complex orographic conditions, climate change-induced extreme events, earthquakes, and so on (Beniston and Fox, 2015). These activities caused several problems for mountain soils, including water erosion and contamination, sediment flow, loss of SOM, biodiversity loss, and nutrient mining, reducing soil productivity and ecosystem services (FAO, 2015). Glacier melting and shrinking is now being reported all over the world, particularly in the mountains of South America, North America, Africa, and Asia. Mount Kilimanjaro, Africa's most important glacier, is melting rapidly and is expected to shrink in the coming decades (Freppaz and Williams, 2015). Changes in permafrost density caused by temperature and/or precipitation may have a negative impact on the surrounding population because they will have a significant impact on hydrological features, discharges to springs and rivers, and water availability (Haeberli, 2013; Fort, 2015).

Snow cover has the potential to influence mountain soil development by regulating the temperature, moisture, and duration of growing conditions, and thus the nutrient cycling. The physical and biological weathering of mountain soils is caused by increased frequency of freeze-thaw cycles, altered nutrient (C and N) dynamics, poor soil particle stability, and increased erodibility after snow melting and deglaciation (Freppaz and Williams, 2015). Thus, climate change-mediated warming and the resulting increase in snow melting are of greater concern, particularly in temperate mountain regions of the Northern Hemisphere (Freppaz and Williams, 2015). The effects of climate change on the mountain ecosystem varied according to temperature and precipitation patterns, biogeographical region, and ecological domain (Zemp et al., 2009; Müller et al., 2010). For example, the Himalayas have experienced a consistent warming trend (three times greater increase in temperature than the global average) over the last century (Yao et al., 2006; Xu et al., 2009; IPCC, 2013). The evapotranspiration rate will be increased further by increased water availability and higher temperatures caused by potential snow melting and soil exposure. Higher evapotranspiration and lower precipitation will result in relatively drier years in the near future, with dire consequences for water resources (Freppaz and Williams, 2015). The following subsection provides a brief overview of soil degradation caused by natural and anthropogenic activities in changing climate conditions.

1.4.2 Soil degradation

Soil degradation is accelerating, owing primarily to the conversion of fertile lands to other land uses and other anthropogenic activities. Every year, approximately 12 million ha of soil are lost, according to estimates (FAO, 2015). Soil is a delicate resource that requires time to regenerate after degradation. Climate change, deforestation, resource extraction and intensive farming practices, and unsustainable land management are the primary causes of mountain soil degradation. Soil degradation in already vulnerable mountain ecosystems not only reduces soil fertility but also causes desertification, flooding, landslides, and a variety of socioeconomic issues (Beniston and Fox, 2015; FAO, 2015). The main causes of soil degradation in mountain regions are precipitation (frequency and duration) and snow movement (avalanches) (Freppaz and Williams, 2015). Higher precipitation during extreme events, for example, may exacerbate soil erosion via debris flow, slope failure, and landslides. Though there is no concrete relationship between precipitation and landslides at the global scale, increased precipitation may cause soil weakening and thus the occurrence of landslides or mass soil movement (Beniston and Fox, 2015). Similarly, avalanche movement carried a significant portion of soil to the bottom or towards running water systems along its path, altering soil morphology and fertility (Freppaz and Williams, 2015). Snow cover changes may have a significant impact on hydrology, biogeochemical processes, and plant diversity, structure, and composition (Gobiet et al., 2014; Magnani et al., 2017; Rogora et al., 2018).

In addition to various edaphic and climatic factors, the presence of various livestock such as yaks may cause damage to the shrub vegetation, which provides instant shade to the soil and aids in the maintenance of soil temperature. Large livestock browsing habits and shrub removal may result in increased soil temperature and relatively higher microbial decomposition (Liu et al., 2021). Mountain soils contain a significant amount of labile C pools, and their exposure to changing climatic scenarios may result in increased decomposition and C-loss (Hagedorn et al., 2010). The Long-Term Ecosystem Research (LTER) Network was established to gain a better understanding of the impact of climate change on mountain ecosystems. The LTER Network is an international monitoring network that collects long-term (periodic) high-quality data to assess the impact of global change on various ecological processes (Rogora et al., 2018). The data are available from the LTER Network (http://www.lter-europe.net/). Aside from other factors, intensive slash-and-burn agriculture is a major threat to mountain soils, particularly steep soils. Due to increasing population, frequent or short-(fallow) length slash-and-burn practices may result in nutrient depletion due to loss of vegetation and biodiversity, a significant increase in soil erosion, and soil quality deterioration (FAO, 2015). The following subsection provides a brief overview of mountain soil erosion.

1.4.3 Soil erosion

Soil erosion is one of the most detrimental phenomena responsible for soil degradation. As per the estimates, natural (geological) erosional processes result in $\sim 2\,t/ha/year$ soil loss (Nearing et al., 2017) which can further increase upto $100\,t/ha/year$ due to various anthropogenic stressors (Julien, 2010). Soil erosion negatively affects soil fertility, food productivity, water storage and scarcity, biodiversity, and overall environmental quality, globally (Lal, 2014; Li and Fang, 2016; Blake et al., 2018). The major regulating factors of soil erosion by water are geology, topography (slope), land use and management (cultivation and conservation) practices, vegetation cover, unreasonable logging, precipitation, snow movements, overgrazing, and soil types (mostly lithosols with a higher percentage of impermeable rocks) (Beniston and Fox, 2015; Panagos et al., 2015; Misthos et al., 2019). Because of their slow soil formation and frequent soil displacement, convex hillslope parts (e.g., summits and shoulders) are the most vulnerable to soil erosion caused by climate change (Misthos et al., 2019). Changes in precipitation amount, pattern, and frequency caused by climate change have a direct impact on soil erosion, whereas loss of SOC content due to decomposition and poor soil structural development have an indirect impact on soil erosion (Li and Fang, 2016; Duulatov et al., 2019; Chapman et al., 2021). Changes in relative humidity, wind and cloud patterns, extreme events, and so on, as a result of a changing climate scenario, affect soil erosion in addition to temperature and precipitation (Misthos et al., 2019). In addition to these factors, forest fires in mountainous areas exacerbate soil erosion rates (Kosmas et al., 2006).

Overall, there are several threats to mountain soils that affect the physicochemical properties and overall ecosystem services of mountainous regions, either directly or indirectly. Various chapters of the book provide a detailed examination of the threats to mountain soils.

1.5 Mitigation measures for the protection of mountain soils

Mountain soils, particularly those at higher elevations, are critical because they are major sources of water, biodiversity, and regulate global atmospheric and cryospheric systems (Bockheim and Munroe, 2014). High-elevation mountain soils store a significant amount of SOC in permafrost, which may be released back into the atmosphere as a result of climate change (Kabala and Zapart, 2012; Schuur et al., 2013; Zollinger et al., 2013). As stated in previous sections, soil development is a slower process than vegetation development (Innes, 1991). As a result, appropriate long-term strategies for mitigating soil degradation in mountainous regions are required under changing climate scenarios (Liu et al., 2021). Evaluations of SOC distribution and spatiotemporal dynamics are critical factors for appropriate management of mountain soils under changing climate scenarios (Liu et al., 2021). The dynamics of soil C in mountain regions under changing climate scenarios are critical. There have been a few studies on SOC dynamics, particularly depth-wise C distribution in alpine soils and changing precipitation and temperature patterns in different mountain regions (Li et al., 2021a). Temperature and precipitation interactions may shape ecosystem functioning in a climate change scenario.

The C-gain and -loss under a changing climate scenario will vary with the ecosystems, as soil warming will increase C-loss due to CO_2 release, while precipitation may increase vegetation productivity (Hagedorn et al., 2010). For example, the Qilian Mountains are experiencing cold-to-warm (0.3°C/decade) and dry-to-wet (15.4 mm/decade) transformations in the past four decades (Yang et al., 2021). Temperature increases may promote SOC decomposition by microorganisms (Wang et al., 2019; Jakšić et al., 2021), whereas increased precipitation may boost vegetation productivity and C-sequestration, particularly in the mountains' mid- and low-elevation zones. Under different RCPs scenarios in the Qilian Mountains, the subalpine meadow may experience an intense C-loss (0.76−3.64 kg/m^2) whereas the alpine desert may show an accumulation of C (0.76−1.06 kg/m^2) in the soils (Liu et al., 2021). Moreover, the soil having higher initial SOC stock may show considerable C loss under the warming conditions at different mountainous ecosystems (Crowther et al., 2016; Prietzel et al., 2016; Liu et al., 2021). However, the overall ecosystem productivity of the Qilian Mountains at different elevation zones will be guided by the soil water availability (Zhu et al., 2019; Liu et al., 2021). Moreover, depth-wise distribution of SOC may vary with the changing climate conditions, as SOC at the 20 cm depth will be more responsive to precipitation whereas SOC at 100 cm depth will be highly influenced by the enhanced microbial activities induced by the increased temperature and precipitation conditions in the alpine regions (Li et al., 2021a). SOC at 20 cm may remain constant due to balanced input-output conditions, but enhanced microbial activities may lead to higher SOC decomposition and reduction in the SOC stock upto 100 cm depth in the Qilian Mountains. After a few years of deglaciation, high elevation soils have been shown to have significant microbial activities, which are responsible for SOC decomposition under these extreme conditions (Freppaz and Williams, 2015). Atmospheric deposition of inorganic and organic nutrients may also serve as a nutrient source for microbial activities in these barren areas, assisting in soil formation. Overall, SOC loss and accumulation with future changing climate scenarios vary with depth and elevation zones; thus, mitigation strategies should include these factors as the focal point (Liu et al., 2021).

People living in mountain regions act as custodians of mountain ecosystems by incorporating traditional knowledge and indigenous practices for sustainable resource management (soil, water, vegetation) (FAO, 2015). Over the centuries, mountain people have developed a number of solutions and practices that aid in ecosystem resilience by incorporating new knowledge systems. Terrace farming and agroforestry systems are the best examples of this type of evolution in mountain ecosystems around the world (FAO, 2015). Terraces and similar structures could be used to limit erosion and soil degradation in extremely steep landscapes. Traditional slash-and-burn agricultural practices used in various hilly areas aid in improving short-term soil fertility by releasing nutrients quickly to support crops. It is a

FIGURE 1.5 Potential measures for the conservation of mountain soils under the changing climate scenario.

Mitigation measures for mountain soil conservation

❑ Agroforestry

❑ Terrace farming at steep slopes

❑ Roofing

❑ Shrub canopy and vegetation conservation

❑ Sustainable slash-and-burn agricultural practices

❑ Reducing cultivation activities at fragile landscapes

❑ Reducing mining, illegal construction, overgrazing, overexploitation of natural resources and logging activities

sustainable practice if enough fallow time is allowed between two cycles of slash-and-burn, which aids in the natural recovery of vegetation and soil conditions, particularly biological activities (FAO, 2015). Furthermore, measures like the shelter effect of native shrub canopy (e.g., *Caragana jubata*, *Salix gilashanica*, and *Potentilla fruticosa*) are responsible for lowering soil temperature in mid-elevation zones and could be used to reduce SOC loss caused by increased temperature (Zhu et al., 2019). Reducing cultivation activities and encouraging the conversion of cultivated land to grasslands may help to improve the SOC stock in mountain regions even more (Liu et al., 2021). In addition to various management strategies, there is a need to slow the pace of various anthropogenic stressors to mountain ecosystems, such as coal mining, illegal construction, overgrazing, resource overexploitation, logging activities, and so on (Li et al., 2021b). Overall, there is a need to investigate the causes and consequences of mountain soils under current climate change scenarios, as well as to develop appropriate site-specific mitigation measures for threats to sustainable mountain ecosystems. A brief illustrative description of different mitigation measures for the conservation of mountain soils has been represented in Fig. 1.5.

1.6 Conclusion and future prospects

Mountain soils are critical to the survival of a variety of people and organisms. They provide several ecosystem services to mountain people, such as food security, water availability, and livelihood options; help to reduce the frequency and occurrence of extreme weather events; and act as hotspots of soil C-sequestration. Variations in temperature and precipitation caused by climate change are inducing various mechanisms that have direct and indirect effects on mountain soils. Extreme weather events such as flooding, debris flow, landslides, and avalanches cause soil movement and degradation, whereas increased microbial activity in deglaciated higher elevation soils causes stored soil C loss to the atmosphere. In addition to C-loss, evidence of C-gains due to increased vegetation productivity has been found in several mountainous ecosystems. In general, there is a need to investigate the interaction of climate change and mountain soils in order to develop future climate change mitigation strategies.

A better understanding of mountain soils will aid in better forecasting of extreme weather events, resulting in prudent policy formulation and the safety of mountain people and resources on regional and national scales (Freppaz and Williams, 2015). According to Freppaz and Williams (2015), for understanding and managing the mountain regions in the climate change scenarios, the priorities for the future research should be on exploring the:

1. impacts on land degradation by emphasizing slope stability, soil erosion, sediment transport, and distribution,
2. impacts on the livelihoods of the people who live there by emphasizing their food security, agricultural sustainability, water availability, energy security, and transportation activities,
3. impacts on hydrological features by drawing attention to the issue of water cycling, runoff, and river discharges,
4. impacts on soil nutrient dynamics by emphasizing SOC dynamics, greenhouse gas emissions, and feedback mechanisms, and
5. impacts on the cryosphere by emphasizing issues such as snow cover, permafrost, glacier dynamics, and the land-water-ecosystem quality nexus.

Acknowledgments

RS gratefully acknowledge the financial support as National Post-Doctoral Fellowship (NPDF) Scheme from the Science and Engineering Research Board (SERB), New Delhi, India (Grant Number: PDF/2020/001607). IG is thankful to University Grants Commission (UGC), New Delhi, India for providing Senior Research Fellowship. PM is thankful to UGC, New Delhi, India for providing research fellowship as UGC-PDFWM Scheme.

References

Auer, I., Böhm, R., Jurkovic, A., Lipa, W., Orlik, A., Potzmann, R., et al., 2007. HISTALP—historical instrumental climatological surface time series of the Greater Alpine Region. International Journal of Climatology 27, 17–46. Available from: https://doi.org/10.1002/joc.1377.

Bárta, J., Applová, M., Vaněk, D., et al., 2010. Effect of available P and phenolics on mineral N release in acidified spruce forest: connection with lignin-degrading enzymes and bacterial and fungal communities. Biogeochemistry 97, 71–87. Available from: https://doi.org/10.1007/s10533-009-9363-3.

Bender, I.M.A., Kissling, W.D., Böhning-Gaese, K., Hensen, I., Kühn, I., Nowak, L., et al., 2019. Projected impacts of climate change on functional diversity of frugivorous birds along a tropical elevational gradient. Scientific Reports 9, 17708. Available from: https://doi.org/10.1038/s41598-019-53409-6.

Beniston, M., Stoffel, M., 2013. Assessing the impacts of climatic change on mountain water resources. The Science of the Total Environment 493, 1129–1137. Available from: https://doi.org/10.1016/j.scitotenv.2013.11.122.

Beniston, M., Fox, D.G., 2015. Impacts of climate change on mountain regions. Chapter 5. https://climatechange.lta.org/wp-content/uploads/cct/2015/03/SAR_Chapter-5_IPCC_Mountains.pdf (accessed 16.03.2022).

Blagodatskaya, E., Dannemann, M., Gasche, R., et al., 2010. Microclimate and forest management alter fungal-tobacterial ratio and N_2O-emission during rewetting in the forest floor and mineral soil. Biogeochemistry 97 (1), 55–70. Available from: https://doi.org/10.1007/s10533-009-9310-3.

Blake, W.H., Rabinovich, A., Wynants, M., Kelly, C., Nasseri, M., Ngondya, I., et al., 2018. Soil erosion in East Africa: an interdisciplinary approach to realising pastoral land management change. Environmental Research Letters 13 (12), 124014.

Bockheim, J.G., Munroe, J.S., 2014. Organic carbon pools and genesis of alpine soils with permafrost: a review. Arctic, Antarctic, and Alpine Research 46 (4), 987–1006.

Bokhorst, S., Bjerke, J.W., Street, L.E., Callaghan, T.V., Phoenix, G.K., 2011. Impacts of multiple extreme winter warming events on sub-Arctic heathland: phenology, reproduction, growth, and CO_2 flux responses. Global Change Biology 17, 2817–2830. Available from: https://doi.org/10.1111/j.1365-2486.2011.02424.x.

Callaghan, T.V., Johansson, M., Brown, R.D., Groisman, P.Y., Labba, N., Radionov, V., et al., 2011. Multiple effects of changes in Arctic snow cover. Ambio 40, 32–45.

Chapman, S., Birch, C.E., Galdos, M.V., Pope, E., Davie, J., Bradshaw, C., et al., 2021. Assessing the impact of climate change on soil erosion in East Africa using a convection-permitting climate model. Environmental Research Letters 16 (8), 084006.

Chen, D.Y., Xue, M.Y., Duan, X.W., Feng, D.T., Huang, Y., Rong, L., 2019. Changes in topsoil organic carbon from 1986 to 2010 in a mountainous plateau region in Southwest China. Land Degradation & Development 31, 734–747.

Crowther, T.W., Todd-Brown, K.E.O., Rowe, C.W., Wieder, W.R., Carey, J.C., Machmuller, M.B., et al., 2016. Quantifying global soil carbon losses in response to warming. Nature 540, 104–108.

Ding, J.Z., Li, F., Yang, G.B., Chen, L.Y., Zhang, B.B., Liu, L., et al., 2016. The permafrost carbon inventory on the Tibetan Plateau: a new evaluation using deep sediment cores. Global Change Biology 22, 2688–2701.

Ding, J.Z., Chen, L.Y., Ji, C.J., Hugelius, G., Li, Y.N., Liu, L., et al., 2017. Decadal soil carbon accumulation across Tibetan permafrost regions. Nature Geoscience 10, 420–424.

Duulatov, E., Chen, X., Amanambu, A.C., Ochege, F.U., Orozbaev, R., Issanova, G., et al., 2019. Projected rainfall erosivity over Central Asia based on CMIP5 climate models. Water 11, 897.

Elmendorf, S.C., Henry, G.H.R., Hollister, R.D., Fosaa, A.M., Gould, W.A., Hermanutz, L., et al., 2015. Corrections for Elmendorf et al., experiment, monitoring, and gradient methods used to infer climate change effects on plant communities yield consistent patterns. Proceedings of the National Academy of Sciences of the United States of America 112, E4156. Available from: https://doi.org/10.1073/pnas.1511529112.

Ettinger, A., Hillerislambers, J., 2017. Competition and facilitation may lead to asymmetric range shift dynamics with climate change. Global Change Biology 23, 1–13. Available from: https://doi.org/10.1111/gcb.13649.

Fadrique, B., Báez, S., Duque, Á., Malizia, A., Blundo, C., Carilla, J., et al., 2018. Widespread but heterogeneous responses of Andean forests to climate change. Nature 564, 207–212. Available from: https://doi.org/10.1038/s41586-018-0715-9.

Farfan-Rios, W., Garcia-Cabrera, K., Salinas, N., Raurau-Quisiyupanqui, M.N., Silman, M.R., 2015. Lista anotada de árboles y afines en los bosques montanos del sureste peruano: la importancia de seguir recolectando. Revista Peruana de Biología 22, 145–174. Available from: https://doi.org/10.15381/rpb.v22i2.11351.

Feng, Q., Yang, L.S., Deo, R.C., AghaKouchak, A., Adamowski, J.F., Stone, R., et al., 2019. Domino effect of climate change over two millennia in ancient China's Hexi Corridor. Nature Sustainability 2, 957–961.

Food and Agriculture Organization (FAO), 2015. Understanding Mountain Soils: A Contribution from Mountain Areas to the International Year of Soils 2015. By Romeo, R., Vita, A., Manuelli, S., Zanini, E., Freppaz, M., Stanchi, S. In collaboration with the Mountain Partnership Secretariat, The Global Soil Partnership and the University of Turin, Rome, Italy.

Fort, M., 2015. Impact of climate change on mountain environment dynamics. An introduction. Journal of Alpine Research Revue de Géographie Alpine (103–2), . Available from: https://doi.org/10.4000/rga.2877.

Freeman, B.G., Scholer, M.N., Ruiz-Gutierrez, V., Fitzpatrick, J.W., 2018. Climate change causes upslope shifts and mountaintop extirpations in a tropical bird community. Proceedings of the National Academy of Sciences of the United States of America 115, 11982–11987. Available from: https://doi.org/10.1073/pnas.1804224115.

Freppaz, M., Williams, M.W., 2015. Mountain soils and climate change. Understanding Mountain Soils: A Contribution from Mountain Areas to the International Year of Soils 2015. FAO, Rome, p. 2015, ISBN 978-92-5-108804-3.

1. Soils of mountainous landscapes: introduction

Girardin, C.A.J., Malhi, Y., Aragão, L.E.O.C., Mamani, M., Huaraca Huasco, W., Durand, L., et al., 2010. Net primary productivity allocation and cycling of carbon along a tropical forest elevational transect in the Peruvian Andes. Global Change Biology 16, 3176–3192. Available from: https://doi.org/10.1111/j.1365-2486.2010.02235.x.

Gobiet, A., Kotlarski, S., Beniston, M., Heinrich, G., Rajczak, J., Stoffel, M., 2014. 21st century climate change in the European Alps—a review. The Science of the Total Environment 493, 1138–1151.

Haeberli, W., 2013. Mountain permafrost – research frontiers and a special long-term challenge. Cold Regions Science and Technology 96, 71–76.

Hagedorn, F., Mulder, J., Jandl, R., 2010. Mountain soils under a changing climate and land-use. Biogeochemistry 97 (1), 1–5.

Hagedorn, F., Gavazov, K., Alexander, J.M., 2019. Above-and belowground linkages shape responses of mountain vegetation to climate change. Science (New York, N.Y.) 365 (6458), 1119–1123.

Hargreaves, A.L., Suárez, E., Mehltreter, K., Myers-Smith, I., Vanderplank, S.E., Slinn, H.L., et al., 2019. Seed predation increases from the arctic to the equator and from high to low elevations. Science Advances 5, . Available from: https://doi.org/10.1126/sciadv.aau4403. eaau4403.

Huss, M., Bookhagen, B., Huggel, C., Jacobsen, D., Bradley, R.S., Clague, J.J., et al., 2017. Toward mountains without permanent snow and ice. Earth's Future 5 (5), 418–435. Available from: https://doi.org/10.1002/2016EF000514:.

Innes, J., 1991. High-altitude and high-latitude tree growth in relation to past, present and future global climate change. Holocene 1, 168–173.

Intergovernmental Panel on Climate Change (IPCC), 2013. Climate Change 2013: The Physical Science Basis. Contribution of Working Group I to the Fifth Assessment Report of the Intergovernmental Panel on Climate Change. Cambridge University Press, Cambridge, United Kingdom, and New York.

Intergovernmental Panel on Climate Change (IPCC), 2014. Climate Change 2014: Synthesis Report. In: Pachauri, R., Meyer, L. (Eds.), Contribution of Working Groups I, II and III to the Fifth Assessment Report of the Intergovernmental Panel on Climate Change. IPCC, Geneva, Switzerland. ISBN: 978-92-9169-143-2.

International Atomic Energy Agency (IAEA), 2022. Understanding the effects of climate change on soil and water resources in mountainous regions. https://www.iaea.org/sites/default/files/19/11/climate_change_web.pdf (accessed 19.03.2022).

Ives, J.D., 1992. In: Stone, R. (Ed.), Preface, The State of the World's Mountains. Zed Books, London, UK, pp. xiii–xvi.

Jakšíc, S., Ninkov, J., Milíc, S., Vasin, J., Zivanov, M., Jaksic, D., et al., 2021. Influence of slope gradient and aspect on soil organic carbon content in the region of Nis, Serbia. Sustainability 13, 8332.

Julien, P.Y., 2010. Erosion and Sedimentation. Cambridge University Press, United Kingdom.

Kabala, C., Zapart, J., 2012. Initial soil development and carbon accumulation on moraines of the rapidly retreating Werenskiold Glacier, SW Spitsbergen, Svalbard archipelago. Geoderma 175–176, 9–20.

Kimball, B.A., 2016. Crop responses to elevated CO_2 and interactions with H_2O, N, and temperature. Current Opinion in Plant Biology 31, 36–43. Available from: https://doi.org/10.1016/j.pbi.2016.03.006.

Mountains and climate change: a global concern. In: Kohler, T., Wehrli, A., Jurek, M. (Eds.), Sustainable Mountain Development Series. Centre for Development and Environment (CDE), Swiss Agency for Development and Cooperation (SDC) and Geographica Bernensia, Bern, Switzerland, 136 pp.

Körner, C., 2013. Alpine ecosystems. In: second ed. Levin, S.A. (Ed.), Encyclopedia of Biodiversity, Vol. 1. Academic Press, Amsterdam, The Netherlands, pp. 148–157.

Körner, C., Paulsen, J., Spehn, E.M., 2011. A definition of mountains and their bioclimatic belts for global comparisons of biodiversity data. Alpine Botany 121, 73–78. Available from: https://doi.org/10.1007/s00035-011-0094-4.

Körner, C., Jetz, W., Paulsen, J., Payne, D., Rudmann-Maurer, K., Spehn, E.M., 2017. A global inventory of mountains for bio-geographical applications. Alpine Botany 127, 1–15. Available from: https://doi.org/10.1007/s00035-016-0182-6.

Kosmas, C., Danalatos, N., Kosma, D., Kosmopoulou, P., 2006. In: Boardman, J., Poesen, J. (Eds.), Soil Erosion in Europe. John Wiley & Sons, Chichester, pp. 279–288.

Lal, R., 2014. Desertification and soil erosion. Global Environmental Change. Springer, Dordrecht, pp. 369–378.

Leifeld, J., Zimmermann, M., Fuhrer, J., et al., 2009. Storage and turnover of carbon in grassland soils along an elevation gradient in the Swiss Alps. Global Change Biology 15, 668–679.

Li, Z., Fang, H., 2016. Impacts of climate change on water erosion: a review. Earth-Science Reviews 163, 94–117.

Li, H.W., Wu, Y.P., Chen, J., Zhao, F.B., Wang, F., Sun, Y.Z., et al., 2021a. Responses of soil organic carbon to climate change in the Qilian Mountains and its future projection. Journal of Hydrology 596, 126110.

Li, Z.X., Feng, Q., Li, Z.J., Wang, X.F., Gui, J., Zhang, B.J., et al., 2021b. Reversing conflict between humans and the environment—the experience in the Qilian Mountains. Renewable and Sustainable Energy Reviews 148, 111333.

Liu, W., Zhu, M., Li, Y., Zhang, J., Yang, L., Zhang, C., 2021. Assessing soil organic carbon stock dynamics under future climate change scenarios in the middle Qilian Mountains. Forests 12, 1698. Available from: https://doi.org/10.3390/f12121698.

Lutz, A.F., Immerzeel, W.W., 2013. Water availability analysis for the upper Indus, Ganges, Brahmaputra, Salween and Mekong river basins. Final Report to ICIMOD, September 2013. Future Water Report, No. 127. Wageningen, The Netherlands, Future Water.

MacLean, S.A., Beissinger, S.R., 2017. Species' traits as predictors of range shifts under contemporary climate change: a review and meta-analysis. Global Change Biology 23, 4094–4105. Available from: https://doi.org/10.1111/gcb.13736.

Magnani, A., Viglietti, D., Godone, D., Williams, M.W., Balestrini, R., Freppaz, M., 2017. Interannual variability of soil N and C forms in response to snow-cover duration and pedoclimatic conditions in alpine tundra, northwest Italy. Arctic, Antarctic, and Alpine Research 49 (2), 227–242.

Messerli, B., Ives, J. (Eds.), 1997. Mountains of the Word. A Global Priority. Parthenon Publishing, New York, 496 p.

Misthos, L.M., Papada, L., Panagiotopoulos, G., Gakis, N., Kaliampakos, D., 2019. Estimating climate change-based soil loss using erosion models and UAV imagery in the Metsovo Mountain Region. In: 4th Joint International Symposium on Deformation Monitoring (JISDM). 15–17 May 2019, Athens, Greece.

Müller, F., Baessler, C., Schubert, H., Klotz, S. (Eds.), 2010. Long-Term Ecological Research: Between Theory and Application. Springer.

Nearing, M.A., Xie, Y., Liu, B., Ye, Y., 2017. Natural and anthropogenic rates of soil erosion. International Soil and Water Conservation Research 5 (2), 77–84.

Nottingham, A.T., Fierer, N., Turner, B.L., Whitaker, J., Ostle, N.J., McNamara, N.P., et al., 2018. Microbes follow Humboldt: temperature drives plant and soil microbial diversity patterns from the Amazon to the Andes. Ecology 99, 2455–2466. Available from: https://doi.org/10.1002/ecy.2482.

Osterkamp, T., Jorgenson, M., Schuur, E., Shur, Y., Kanevskiy, M., Vogel, J., et al., 2009. Physical and ecological changes associated with warming permafrost and thermokarst in interior Alaska. Permafrost and Periglacial Processes 20, 235–256.

Panagos, P., Borrelli, P., Poesen, J., Ballabio, C., Lugato, E., Meusburger, K., et al., 2015. The new assessment of soil loss by water erosion in Europe. Environmental Science & Policy 54, 438–447.

Prietzel, J., Zimmermann, L., Schubert, A., Christophel, D., 2016. Organic matter losses in German Alps forest soils since the 1970s most likely caused by warming. Nature Geoscience 9, 543–550.

Rahbek, C., Borregaard, M.K., Antonelli, A., Colwell, R.K., Holt, B.G., Nogues-Bravo, D., et al., 2019a. Building mountain biodiversity: geological and evolutionary processes. Science (New York, N.Y.) 365, 1114–1119. Available from: https://doi.org/10.1126/science.aax0151.

Rahbek, C., Borregaard, M.K., Colwell, R.K., Dalsgaard, B., Holt, B.G., Morueta-Holme, N., et al., 2019b. Humboldt's enigma: what causes global patterns of mountain biodiversity? Science (New York, N.Y.) 365, 1108–1113. Available from: https://doi.org/10.1126/science.aax0149.

Rapp, J., Silman, M., 2012. Diurnal, seasonal, and altitudinal trends in microclimate across a tropical montane cloud forest. Climate Research 55, 17–32. Available from: https://doi.org/10.3354/cr01127.

Rasouli, K., Pomeroy, J.W., Whitfield, P.H., 2019. Are the effects of vegetation and soil changes as important as climate change impacts on hydrological processes. Hydrology and Earth System Sciences 23 (12), 4933–4954.

Rodriguez-Iturbe, I., 2000. Ecohydrology: a hydrologic perspective of climate-soil-vegetation dynamics. Water Resources Research 36, 3–9.

Rogora, M., Frate, L., Carranza, M.L., Freppaz, M., Stanisci, A., Bertani, I., et al., 2018. Assessment of climate change effects on mountain ecosystems through a cross-site analysis in the Alps and Apennines. The Science of the Total Environment 624, 1429–1442.

Schuur, E.A.G., Abbott, B.W., Bowden, W.B., Brovkin, V., Camill, P., Canadell, J.G., et al., 2013. Expert assessment of vulnerability of permafrost carbon to climate change. Climatic Change 119, 359–374.

Sheikh, M.A., Kumar, M., Bussmann, R.W., 2009. Altitudinal variation in soil organic carbon stock in coniferous subtropical and broadleaf temperate forests in Garhwal himalaya. Carbon Balance and Management 4, 6. Available from: https://doi.org/10.1186/1750-0680-4-6.

Sjögersten, S., Turner, B.L., Mathieu, N., et al., 2003. Soil organic matter biochemistry and potential susceptibility to climate change across the forest-tundra ecotone in the Fennoscandian mountains. Global Change Biology 9, 759–772.

Stockmann, U., Adams, M.A., Crawford, J.W., Field, D.J., Henakaarchchi, N., Jenkins, M., et al., 2013. The knowns, known unknowns and unknowns of sequestration of soil organic carbon. Agriculture, Ecosystems & Environment 164, 80–99.

Stoll, S., Frenzel, M., Burkhard, B., Adamescu, M., Augustaitis, A., Baeßler, C., et al., 2015. Assessment of ecosystem integrity and service gradients across Europe using the LTER Europe network. Ecological Modelling 295, 75–87. Available from: https://doi.org/10.1016/j.ecolmodel.2014.06.019.

Tito, R., Vasconcelos, H.L., Feeley, K.J., 2018. Global climate change increases risk of crop yield losses and food insecurity in the tropical Andes. Global Change Biology 24, e592–e602. Available from: https://doi.org/10.1111/gcb.13959.

Tito, R., Vasconcelos, H.L., Feeley, K.J., 2020. Mountain ecosystems as natural laboratories for climate change experiments. Frontiers in Forests and Global Change 3, 38. Available from: https://doi.org/10.3389/ffgc.2020.00038.

Townsend, A.R., Vitousek, P.M., Trumbore, S.E., 1995. Soil organic matter dynamics along gradients in temperature and land use on the Island of Hawaii. Ecology 76, 721–733. Available from: https://doi.org/10.2307/1939339.

van Everdingen, R. (Ed.), 1998. Multi-Language Glossary of Permafrost and Related Ground-Ice Terms. National Snow and Ice Data Center, Boulder, CO.

Vitousek, P.M., Farrington, H., 1997. Nutrient limitation and soil development: experimental test of a biogeochemical theory. Biogeochemistry 37, 63–75.

Viviroli, D., Archer, D.R., Buytaert, W., Fowler, H.J., Greenwood, G.B., Hamlet, A.F., et al., 2011. Climate change and mountain water resources: overview and recommendations for research, management and policy. Hydrology and Earth System Sciences 15, 471–504. Available from: https://doi.org/10.5194/hess-15-471-2011.

Wang, P., Limpens, J., Mommer, L., van Ruijven, J., Nauta, A.L., Berendse, F., et al., 2017. Above- and below-ground responses of four tundra plant functional types to deep soil heating and surface soil fertilization. Journal of Ecology 105, 947–957. Available from: https://doi.org/10.1111/1365-2745.12718.

Wang, S., Zhuang, Q.L., Yang, Z.J., Yu, N., Jin, X.X., 2019. Temporal and spatial changes of soil organic carbon stocks in the forest area of northeastern China. Forests 10, 1023.

Wangdi, N., Om, K., Thinley, C., Drukpa, D., Dorji, T., Darabant, A., et al., 2017. Climate change in remote mountain regions: a throughfall-exclusion experiment to simulate monsoon failure in the Himalayas. Mountain Research and Development 37 (3), 294–309.

Winkler, M., Lamprecht, A., Steinbauer, K., Hülber, K., Theurillat, J.-P., Breiner, F., et al., 2016. The rich sides of mountain summits – a pan-European view on aspect preferences of alpine plants. Journal of Biogeography 43, 2261–2273. Available from: https://doi.org/10.1111/jbi.12835.

Xu, J., Grumbine, R.E., Shrestha, A., Eriksson, M., Yang, X., Wang, Y., et al., 2009. The melting Himalayas: cascading effects of climate change on water, biodiversity, and livelihoods. Conservation Biology: The Journal of the Society for Conservation Biology 23 (3), 520–530.

Xu, Z., Chang, Y., Li, L., Luo, Q., Xu, Z., Li, X., et al., 2019. Climatic and topographic variables control soil nitrogen, phosphorus, and nitrogen: phosphorus ratios in a *Picea schrenkiana* forest of the Tianshan Mountains. PLoS One 14, e0211839. Available from: https://doi.org/10.1371/journal.pone.0211839.

Yang, L.S., Feng, Q., Adamowski, J.F., Alizadeh, M.R., Yin, Z.L., Wen, X.H., et al., 2021. The role of climate change and vegetation greening on the variation of terrestrial evapotranspiration in northwest China's Qilian Mountains. The Science of the Total Environment 759, 143532.

Yao, T., Guo, X., Thompson, L., Duan, K., Wang, N., Pu, J., et al., 2006. $d^{18}O$ record and temperature change over the past 100 years in ice cores on the Tibetan Plateau. Science China Series D 49 (1), 1–9.

Zemp, M., Hoelzle, M., Haeberli, W., 2009. Six decades of glacier mass balance observations — a review of the worldwide monitoring network. Annals of Glaciology 50, 101—111.

Zhu, M., Feng, Q., Qin, Y.Y., Cao, J.J., Zhang, M.X., Liu, W., et al., 2019. The role of topography in shaping the spatial patterns of soil organic carbon. Catena 176, 296—305.

Zimmermann, M., Meir, P., Bird, M.I., Malhi, Y., Ccahuana, A.J.Q., 2010. Temporal variation and climate dependence of soil respiration and its components along a 3000 m altitudinal tropical forest gradient. Global Biogeochemical Cycles 24, 1—13. Available from: https://doi.org/10.1029/2010GB003787.

Zollinger, B., Alewell, C., Kneisel, C., Meusburger, K., Gärtner, H., Brandová, D., et al., 2013. Effect of permafrost on the formation of soil organic carbon pools and their physical-chemical properties in the eastern Swiss Alps. Catena 110, 70—85.

Threats to mountainous soils: conservation and management strategies

Akhilendra Kumar Mishra

Department of Geography, Maharaja Bijli Pasi Government Post Graduate College, Ashiyana, Lucknow, Uttar Pradesh, India

O U T L I N E

2.1 Introduction

Soil is a limited resource that plays an important role in ecosystem services, human survival, environmental protection, and agricultural productivity (Kumar et al., 2021). Soil is vital to life on Earth because it provides various types of nutrients, water, minerals, and other elements to various flora and fauna. It has the ability to retain, absorb, and filter water, as well as control floods and droughts. Because soil contains more carbon than the atmosphere and plants, it also plays an important role in carbon storage, reducing emissions, and preventing climate change. Soil is a natural substance made up of solids, liquids, and gases that exist on the land's surface and take up space. Weathering of parent rocks, transfer of weathered materials, and deposition all contribute to the formation of soil layers. Has the ability to support the transformation of energy and matter in the natural environment or the rooted plan (USDA, 1999). Soil is the foundation of agriculture, providing nutrition to crops, flora, and fauna as well as food security for 7 billion people around the world. The population is expected to reach 9 billion by 2030, putting additional strain on the soil. At the moment, the urban population accounts for 55% of the total, implying that this population is also dependent on agriculture performed by the rural population for food grains. Agriculture is one of the oldest methods of altering natural resources through which humans interact with nature for crop production, livestock production, and overall well-being (Dale et al., 2013; Singh et al., 2019). Weathering of rocks, transfer of weathered materials, and deposition all contribute to the formation of soil. Temperature, rainfall, and wind, among other factors, play an important role in this work. At the same time, mineral gases, water, organic matter, and microorganisms play important roles in its development; without them, the soil does not develop fully (Srivastava et al., 2018). Soil contains non-consolidated minerals, organic matter, and other micronutrients that act as a natural medium for plant growth and development. In soil formation and development, parent rocks, climate, flora-fauna, relief, and time all play a role. Mountains were estimated to house 15% of the world's population in 2017 (FAO, 2019). Mountainous soil is beneficial not only to those who live in the mountains, but also to those who live downstream. The soil of the hilly region provides food, fodder, medicinal plants, and other wood and non-timber forest products, which are the main source of income for the people who live there. Mountainous areas are water recharge areas in groundwater that are used for a variety of purposes by a portion of the world's population (Stanchi et al., 2021).

2.2 Mountainous soil

Because most mountainous areas have low temperatures, biological activity is slow. Soil formation and development are also slow in such areas, and the soil is shallow. Mountainous soils are generally defined as being weakly developed, skeletal, thin layer, acidic, nitrogen deficient, and relatively less fertile. They typically become less fertile and less developed as elevation increases. Soil concentration decreases during the freeze-thaw cycle in cold regions, as does their stability, fertility, and water retention. Mountainous soils are primarily found on hill slopes and are formed by the deposition of organic matter from forests. These soils are most commonly found at elevations ranging from 900 to 1800 m. These soils form as slopes on stony, unstable surfaces. Rock fragments that have been physically decomposed by frost action can occur in large quantities in soil but do not undergo many chemical changes. These are also referred to as screen soils (Majid, 2007). "The relationship between people of mountains and soil is fundamental. Unscientific farming and forest management practices increase the pressure on these critical ecosystems, accelerating erosion, decreasing fertility, and triggering natural hazards. Although mountains are characterized by high vulnerability to natural hazards, the risks can be somewhat lower when the areas are low density settled. However, when mountain areas are high-density populated the impact of natural disasters can be very disruptive with consequences on people living in the lowlands" (Romeo et al., 2015). Improving the health of mountainous soil can provide significant benefits such as increased agricultural productivity and other ecosystem services (Eze et al., 2021).

2.3 Major threats to mountainous soils

Along with the advancement of human civilization, humans have made significant advances in technologies for basic needs such as food production (Jiao et al., 2012). Because of the world's growing population, there has been a significant shift in land use for agricultural expansion (Montgomery, 2007; Tarolli and Sofia, 2016, 2020). Water is one of the major causes of soil erosion, and it has a negative impact on ecosystem services, agricultural production, drinking water, and carbon stocks (Panagos et al., 2015; Tarolli and Straffelini, 2020). Unscientific grape cultivation on steep slopes of hills and mountains also contributes to soil erosion (Prosdocimi et al., 2016; Tarolli and Straffelini, 2020). Major threads to mountainous soil are given in Fig. 2.1.

2.3.1 Decreasing fertility

Soil fertility is the ability of soil to support agricultural and forestry plant growth. Plant nutrients are stored in the upper layers of the soil before being recycled. The amount available for the plant to use is reduced as a result of these layers being removed by erosion. The continuous use of chemical fertilizers, as well as erosion, reduce soil fertility. We are unable to replenish soil fertility artificially in the same amount that it is destroyed by agricultural activities, resulting in a decrease in soil fertility. Because of the continued use of chemical fertilizers, this problem has gotten worse. Hilly areas have a thin, immature, and fragile layer of soil, which complicates matters even more.

2.3.2 Soil erosion

Soil erosion is defined as a faster rate of soil removal than natural replacement. Soil erosion is the process of eroding soil particles, rock fragments, soil aggregates, and organic matter from their original location and transporting them to another location using various

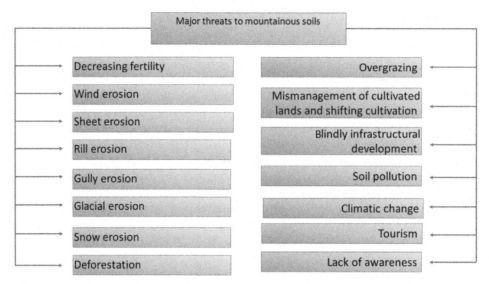

FIGURE 2.1   Major threats to mountainous soils.

processes (Poesen, 2018; Spalevic et al., 2017a,b). Soil erosion may be divided majorly into the following types:

2.3.2.1 Wind erosion

The wind is a major cause of soil erosion. The wind blows away the dry, unconsolidated material. It's a major issue at high altitudes in arid and semi-arid climates. Windborne erosion causes soil fertility to deteriorate, agricultural productivity to decline, snow cover to decrease, and water content to decrease (Clow et al., 2016; Duniway et al., 2019; Joyce et al., 2013; Painter et al., 2010).

2.3.2.2 Sheet erosion

It is the removal of surface debris in general. It occurs as a result of overland flow on a low gradient. Slopes, usually at slow speeds and over long distances. It is common due to the humidity.

2.3.2.3 Rill erosion

The upper layer of the surface is usually washed away by the action of running water, forming many small channels with depths of only a few inches, and most of which water only flows during the rainy season. Rills, on the other hand, are more localized and grow in humid climates.

2.3.2.4 Gully erosion

Large-scale deforestation causes the reduction of forest areas, which weakens the soil in the absence of tree roots and encourages soil erosion, e.g., Tennessee Valley, United States

(Majid, 2007). Alley erosion is a common type of water-induced erosion that results in the loss of fertile soil and soil moisture (Mokarram and Zarei, 2021).

2.3.2.5 Glacial erosion

Glacial erosion is the most common phenomenon where the average temperature is below 0°C. Glacial erosion is caused by the slow movement of a large mass of ice. Measures to prevent vegetation and other forms of erosion have little effect when compared to the immense force generated by the massive mass of ice. Another distinguishing feature of glacial erosion is that the soil is only damaged at the ice's edges, in new ice channels, and by meltwater.

2.3.2.6 Snow erosion

Glacial regions are associated with glacial erosion. The glacial zone is defined as an area with continuous snow cover. There is a distinction to be made between glacial erosion and snow erosion. Avalanche channels erode where there is a lot of ice pressure and velocity. Soil erosion is also caused by the very slow movement of snow, particularly on slopes with a high friction action on the top surface of the soil. There is an increase in erosion caused by ice pressure in areas where waterlogging and runoff cooperate.

2.3.3 Deforestation

Forests provide essential ecosystem services such as soil erosion control, ecosystem stability, and climate and energy flow regulation. The reduction in forest cover is primarily due to agricultural expansion, frequent forest fires, increased transportation, increased urbanization, rapid industrialization, and other factors (Zachar, 2011). The most powerful cause of land-use change is deforestation. Deforestation causes a cascade of side effects that disrupt the ecological balance. Soil erosion, decreased soil fertility, increased greenhouse gas emissions, decreased habitat, increased animal migration, decreased biodiversity, increased landslides and avalanches, increased frequency of floods and droughts are all consequences of deforestation. Furthermore, the main side effects are increased sediment deposition in reservoirs, wetlands, lakes, and river pools, severe eutrophication, resource pollution, climate change, and socioeconomic off-site effects (Abd Elbasit et al., 2013; FAO, 2015; Indarto and Mutaqin, 2016; Gharibreza et al., 2020; Gharibreza and Ashraf, 2014). Forest destruction alters soil quality and can have significant local and global consequences for ecological water processes and heat budgets (Osterkamp et al., 2009; Rasouli et al., 2019). Furthermore, the short and long-term socioeconomic impacts of deforestation have harmed forest-based income (Gharibreza et al., 2020).

2.3.4 Overgrazing

"Herds of cattle and sheep are often cantered on the same area for too long in many livestock farms. The result of this process comes as overgrazing, repeated trampling, and soil removal during traffic. Thinning of grass reduces the protective cover and increases soil erosion, especially on hillsides. Overgrazing reduces soil organic matter content, soil structure becomes poor and increases water and wind erosion. Grazing alters the cover, species composition, and structure of dryland plant and biological crust communities, although these effects depend on the timing, intensity, as well as duration of grazing, and the evolutionary history and resilience of the plants and soils to herbivory and associated trampling" (Mack and Thompson, 1982; Fleischner, 1994; Webb and Strong, 2011; Aubault et al., 2015; Duniway et al., 2019). Overgrazing not only alters the structure of the vegetation, but it also causes soil erosion and a decrease in fertility and carbon storage (Aryal et al., 2015; Sharma et al., 2014; Wen et al., 2013; Wang et al., 2019; Wen et al., 2013).

2.3.5 Mismanagement of cultivated lands

Soil erosion is primarily caused by the spread of agriculture on sloping, shallow, and marginal lands. Soil deterioration is caused by unscientific agriculture and tillage, increased traffic, shifting cultivation, increased use of chemical fertilizers, irrigation from low-quality water, and a lack of vegetation cover. When the soil is damaged, it takes a long time to heal. As a result of increased population pressure on agriculture in some hilly areas, farmers are being forced to use more hill land for crop production.

In many developing countries, rising population pressures have increased food demand, and in some regions, as the global climate warms, agricultural spread in the mountains is expanding. This has resulted in two types of serious issues. To begin, many forests on steep slopes were cut down and converted into agricultural areas to meet local food demands, and many other forests were heavily utilized by the collection of fuelwood, fodder, and forest produce to support agriculture. Soil erosion has been accelerated by the large-scale conversion of hill forests into cropland for the production of cash crops. Extensive biomass extraction has hampered forests' ability to recycle carbon and nutrients, resulting in a decline in tree productive capacity and soil loss.

The impact of Jhum farming has long been a topic of debate among academics. According to some researchers, it causes deforestation, river sedimentation, and soil erosion. Other researchers have discovered that shifting cultivation improves soil conditions and soil health for cultivating a variety of crops. It is an important multi-cropping agroforestry system for hilly areas (Romeo et al., 2015). Some key factors for mountain soil hazards can be identified. It is associated with land waste. The overuse of farm machinery has weakened the hilly areas. Simultaneously, soil erosion has increased (Tarolli and Straffelini, 2020; Torquati et al., 2015).

In many countries, soil erosion and other erosion related to land resources have emerged as a major spatial problem (Fistikoglu and Harmancioglu, 2002; Hoyos, 2005; Pandey and Rizvi, 2009; Prasannakumar et al., 2012). Because of the enormous environmental impacts on ecology, soil, agriculture, and water, environmentalists and scientists are particularly interested in the study of soil erosion and soil degradation in mountainous areas (Khaledi Darvishan et al., 2017; Spalevic et al., 2017a,b). As erosion increases, infertile alluvial deposits form on the fertile soil of small alluvials near the main aqueduct, resulting in infertile alluvial deposits. The fertile land is buried in the deposition, reducing the fertility of the surrounding area. Erosion is observed on all slopes, but it is more pronounced due to vegetation removal or on steep slopes without vegetation (Spalevic, 2011; Spalevic et al., 2017a,b).

2.3.6 Blindly infrastructural development

Since the last century, the development and accessibility of transportation have increased widespread forest exploitation in the European Alps, resulting in severe soil erosion. Because of the increased production of minerals and energy sources, increased storage of forest produce, meeting the growing demand for food and recreational activities, and the increased construction of infrastructure in the world's mountains, many problems have negatively affected the hilly soil. The rapid development of transportation routes to mines and forests in the Rocky Mountains, Andes, and Himalayas has resulted in not only soil erosion but also increased sediment deposition and flood outbreaks in low-lying areas. Uncontrolled mining operations are also detrimental to the hilly environment, resulting in widespread soil degradation and deterioration of the hill ecology (Wang et al., 2019).

2.3.7 Tourism

Rapid urbanization has put enormous strain on the resources of mountainous regions, just as it has in other areas, and infrastructure development has been encouraged to extend the service life of summer and winter tourism. This type of movement at tourist destinations has been shown to be detrimental to the development and health of the soil. Tourists in mountainous areas try to balance a more snow-covered surface with weather conditions that amplify erosive forces and soil loss. Rising summer tourism has resulted in a rapid expansion of trail construction for trekking and mountain biking, causing new disturbances that are harmful to runoff processes, soil quality, and development. Furthermore, as urban development expands to support mountain tourism, it weakens the mountain surface, increasing surface runoff and exacerbates the problem of erosion. The uncontrolled disposal of solid waste generated by the tourism industry, as well as over-exploitation of soil, vegetation, and other local resources, is causing the ecosystem to degrade. Rapid infrastructure development required for sustainable tourism can have a negative impact on local aesthetic and cultural properties. At the same time, it may reduce tourism's future income potential (Reinfeld, 2003; Oli and Zomer, 2011; Wang et al., 2019). There has been an increased demand for sub-alpine wood for the lodge's construction, as well as extensive use of forests and shrubs for fuel used in the lodge by tourists visiting the alpine areas (Byers, 2014; Wang et al., 2019).

2.3.8 Soil pollution

Air pollution, mining activities, tourism, industrialization, and transportation are all likely to pollute the mountainous soil. Air pollution is commonly found in hilly valleys during heavy traffic for large mining operations. They cause acidic rain, which is especially harmful to the area's mountainous soils and flora. Heavy industrial developments in some parts of Eastern Europe during the postwar period caused long-term damage to forests and soils in mountainous areas. The study's findings revealed that not only have mines contributed to the serious problem of heavy metal pollution, but also the highly toxic and non-toxic risk of soil heavy metal pollution poses a serious threat to the public, particularly children and people living nearby. Heavy metal emissions from traffic activities are a significant source of pollution for the agricultural land ecosystem along roads

TABLE 2.1   Increase in CO_2 in last 140 years (WMO, 2021).

Year	Increase in CO_2 (ppm)
1880	280
1980	315
2000	370
2020	413.2

Source: WMO, 2021. WMO Air Quality and Climate Bulletin, No. 1.

and railways (Zhang et al., 2012). On the Tibetan plateau, the main heavy metals in the soil are manganese (Mn) and chromium (Cr). The natural intensity of arsenic (As) is also about 20 mg/kg, which is very high (Sheng et al., 2012; Wang et al., 2019).

As a result, there is an urgent need for mountainous soil conservation and management. Various conservation measures for mountainous soils have been developed, which will be elaborated in the following sections of the chapter.

2.3.9 Climate change

Greenhouse impact and global warming are the main reasons for climate change. Global warming is a consequence of increasing greenhouse gases like CO_2, CH_4, N_2O, CFCs etc. in the atmosphere. Since the industrial revolution, there has been a rapid increase in the combustion of fossil fuels, which has increased greenhouse gas emissions; however, deforestation has reduced CO_2 sink, complicating the problem. The amount of carbon dioxide in the atmosphere in 1880 was 280 ppm, which increased to 413.2 ppm in 2020 (Table 2.1). Preindustrialization, an increase of about 49% has been recorded in the amount of CO_2 between 1750 and 2020. According to the IPCC (2021) report, the global temperature is projected to increase by about 1.5°C over the next 40 years (IPCC, 2021). As a result of rising temperatures, the moisture content of mountainous soils has decreased. It also has an impact on the composition and structure of mountain soils.

2.4 Conservation and management of mountainous soils

Conservation and management are primarily concerned with resource allocation. Conservation of resources is a component of resource management. The survey, evaluation, use, and conservation of a resource are all examples of management. To make strategic use of resources in order to meet the needs of the current generation while also providing for the needs of future generations. It entails attempting to save resources from degradation as well as restoring resources that have become degraded. The concept of sustainable development is emphasized in this, as is the use of such methods of use that there is minimal degradation to the environment.

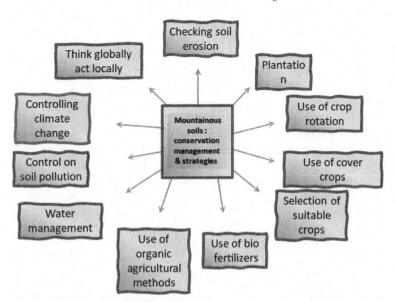

FIGURE 2.2 Mountainous soils. Conservation management and strategies—at a glance.

There are a number of methods/management practices that have been evolved for the conservation of mountainous soils. Some of the major management practices have been presented below (Fig. 2.2).

2.4.1 Checking soil erosion

It is estimated that degradation is currently affecting 65% of the world's soil resources. The primary cause of soil degradation is erosion. Every year, 75 billion tonnes of fertile soil worth approximately $400 billion are lost (Romeo et al., 2015).

Soil erosion is a serious issue that affects the entire world. Only by doing the groundwork at the ground level will the soil of the hill slopes be protected by effective erosion control (Toy et al., 2002). We conclude by discussing possible management procedures for reducing wind erosion and preserving areas with air erosion, as well as the dissemination of technical knowledge on other major erosion control measures (Duniway et al., 2019).

To control mountain soil erosion, plantation, controlled grazing, crop rotation, crop cover, water management, small dams, terrace farming, contour tillage, scientific and organic farming, and other methods should be used. This chapter expands on the topics mentioned above.

2.4.2 Plantation

Large changes in the snow and runoff system may cause changes in the vegetation, soil, and climate, or they may interact with one another (Rasouli et al., 2019). The forests contribute significantly to hill agriculture by providing food, fuelwood, fodder, construction material, and timber, among other things. They keep the soil healthy by providing cover and highly developed root systems, protecting it from erosion and landslides, increasing soil fertility, improving the ability of the soil to hold water, and assisting in the maintenance of normal air and soil temperature. Sustainable forest management and the restoration of degraded forests are critical components of developing watershed management systems. At the same time, they aid in providing a sporadic water supply on the downstream side. For centuries, traditional agroforestry techniques have been used successfully in mountainous areas.

Forests significantly contribute to food production by improving soil fertility for crop production. The hill people of the Himalayan region rely more on agriculture and forest-based resources for a living. Forests help to improve soil fertility. Efforts are being made to restore soil stability through natural and human-assisted reforestation. In the river valleys, an elaborate agroforestry system is being developed. Soil conservation through agroforestry is an important method in the Sikkim Himalayas (Mishra et al., 2019). Many crop species and plants have adapted to grow on mountainous soils, which serve many important functions, including erosion control.

2.4.3 Pastoralism

Pastoralism is a traditional but enduring way of life in many mountain and highland areas. It benefits hill people by making flexible use of pasture, water, salt, and forest resources. This arrangement creates a balance between the dry and wet seasons, or between summer and winter grazing lands. To deal with harsh environments, shepherds also use high genetic diversity and diversity in herd structure management. Shepherds are creating a wide range of economic and social benefits from areas with low biomass productivity through herd management in such areas that are not suitable for intensive harvesting. Along with providing food and fuel, this type of traditional method contributes significantly to climate change mitigation, flood and erosion control, and nutrient cycling. Ancient human populations and their livestock have also made significant contributions to understanding how they altered the local environment over time (Ventresca Miller et al., 2020).

2.4.4 Contouring

Farming on sloping land is one of the major causes of rapid erosion in areas where soil and water conservation measures have not been widely implemented. Contour farming is required to protect weak slopes from accelerated erosion. Breaking the order of the slope in areas with high rainfall necessitates the construction of permanent vegetation contour dams along the divergence trenches. These measures reduce erosion by lowering runoff volume and velocity. Contour ridge systems are used all over the world to promote grain production and control soil erosion (Barton et al., 2004; Fang et al., 2020; An et al., 2021). Thus, finer topography and ridge geometry are the two main factors influencing seepage and nutrients (Liu et al., 2016; An et al., 2021).

2.4.5 Terracing

Bench terraces are an effective solution for sustainable crop production on steep slopes in hilly areas, but their development and maintenance are costly. Terraced systems can help in mountainous soil conservation and agricultural development on a large scale if hill people are given adequate financial support and technical training. This has the potential to significantly improve the landscape of the hilly region. This system is now recognized as an important agricultural heritage system all over the world. Terracing systems, for example, protect the soil in the Andes Mountains. Contributes to the formation of deep, fertile soil. Increases water storage and promotes better irrigation water use by expanding cultivated areas and ensuring that sunlight is available for a longer period of time during the day. They also promote soil conservation by assisting in the protection of agricultural areas from potential landslides (Romeo et al., 2015).

2.4.6 Use of crop rotation and selection of suitable crops

Agriculture production, as well as soil and environmental health, must be considered for human civilization. Crop rotation, intercropping, and crop diversification can all be used to conserve soil in large areas (Sharma et al., 2017). Crop rotation can be beneficial in the world's major mountainous regions, such as the Himalayas, Alps, Rocky Mountains, and Andes, among others. Crop rotation, intercropping methods, and crop diversification soil management techniques can all have a significant impact on soil productivity (Roper et al., 2017).

The effects of unscientific human activity on mountain soils all over the world have resulted in rapid degradation. Hill soil nutrients have depleted. Organic farming methods can be used to address these issues, which will benefit the produced crop products' human health while also adapting to the local ecology (Thapa, 1996).

According to estimates from the United Nations Food and Agriculture Organization, approximately 45% of the world's hilly areas are currently unsuitable for agricultural and animal husbandry. As a result, the mountain people, who are mostly farmers and herders, have had to resort to methods that harm the environment in order to survive. They have adopted complex and diverse farming systems on cropland, pastures, and forests, employing various types of soil at various altitudes and seasons of the year. They are, for example, growing more sun-seeking plants on the hottest slopes and bringing the animals to graze on high summer pastures after the snow melts. Many of the world's most important food crops, such as corn and potatoes, are grown on a large scale in mountainous areas, which has a negative impact on the ecology. In hilly areas, the number of domestic animals has also increased. This has exacerbated problems such as erosion in hilly areas. Native livestock breeds such as yak, mountain goats, and sheep are well adapted to harsh mountainous conditions, with the ability to withstand hot and cold, wet and dry conditions.

2.4.7 Use of cover crops

The continuous cultivation of the same crop has a negative impact on an area's soil fertility. Simultaneously, soil erosion is encouraged. Crop rotation, intercropping, crop diversification, and cover cropping can all be used to address this issue (Chavarria et al., 2016).

The cover cropping method is used to prevent soil erosion by air and water (Langdale et al., 1991).

2.4.8 Sustainable and organic mixed farming

The calculation of soil nutrient balance was measured as the net difference between entry and exit of nutrients into the soil, with the soil being assumed as a "black box" (Smaling and Stoorvogel, 1990; Van Beek et al., 2016). Subsidies have encouraged the agricultural sector in many parts of the world to adopt the best agricultural management methods. However, farmers in many parts of the world continue to be denied such facilities; sustainable development cannot be achieved unless all of those farmers are reached. It is not only the responsibility of local farmers to keep productive agricultural land fertile and prevent soil nutrient loss, but also of government institutions and other local communities (Neitsch et al., 2011).

Locals in many mountainous regions around the world practice organic farming because they lack access to expensive investments. These farmers use biological methods to maintain soil fertility and control pests and diseases with natural enemies. Pest management entails raising insects that eat pests that harm crops and can be done organically to avoid environmental damage. Hilly areas can be developed sustainably by combining agriculture and animal husbandry and emphasizing a holistic mixed agricultural approach.

Priority is given to balancing soil health and agricultural production in sustainable agriculture management. To maintain crop yield and soil health on sloping land, a different approach can be used, such as minimum soil disturbances, crop cover, and crop rotation. It should continue to rotate the main food crops with crops that fertilize the soil and fodder crops to encourage deep root growth, re-establish nitrogen and organic matter, and optimize access to moisture and nutrients (Romeo et al., 2015). Most Indian states in the Himalayan region, as well as the majority of foreign countries sharing the Himalayan borders, have launched an innovative planned organic farming to obtain quality products in their region in order to strengthen the health and economy for better livelihood and sustainable development (Das et al., 2016; Mandal et al., 2006; Singh et al., 2021). Sikkim is India's first organic state, where organic farming has benefited both the hill soil and the local ecology from an environmental standpoint. Some other hill states in India have also adopted this method, which should be promoted on a larger scale (Mandal et al., 2006; Tewari et al., 2017; Singh et al., 2021).

2.4.9 Water management

The check dam is one of many effective soil and water conservation methods used around the world. Check dams have been used in many parts of the world for centuries. People have benefited from check dams in land reforms and agricultural production over the last several hundred years (Abbasi et al., 2019). Water management in hilly areas can enrich soil. That is, by retaining the optimal amount of water in the soil structure through water management, erosion can be avoided. At the moment, small dams can be built to make this work more sustainable. Furthermore, check dams have been shown to be a useful mechanism for controlling soil erosion and flooding on a large scale, both theoretically and practically (Abbasi et al., 2019).

2.4.10 Controlling soil pollution

A study of remote ecosystems in the world's mountainous regions reveals that these areas are also not immune to atmospheric pollution deposition. Mountainous areas, such as the Himalayas, trap atmospheric contaminants from cold condensation and increased atmospheric deposition due to convergence, resulting in a situation of loaded metal pollution, which can lead to a decline in the quality of soil, air, and water bodies, affecting the health of all living organisms (Fernández et al., 2018). The control of heavy metals such as As, Pb, Cd, Ni, Cu, Hg, and Zn has been prioritized as a result of the assessment of pollution and health risk by the appropriate government agencies. In addition, mines such as lead-zinc, manganese, tungsten, and others are prioritized as control mine categories. So that the hilly soil of such areas can be protected from heavy metal pollution (Hong-gui et al., 2012).

2.4.11 Climate change and mountainous soils

Climate change has also had an adverse effect on mountainous areas. This is common in many mountainous areas around the world. As a result, climate change will have a direct impact on the dynamics of snow cover and biodiversity in the eastern Himalayas. Because of future high temperatures, potential evaporation will likely increase, which will be harmful to the hilly soil. A relatively dry year with more evaporation than precipitation has a negative impact on water resources and drought-related problems. Dry soil erosion is exacerbated by strong winds in semi-arid and arid climates Temperature rises, changes in rainfall patterns, monsoon onset, duration, and intensity, and annual and monthly variations in streamflow in the Himalayan region in recent decades all indicate that the region is a climate change hotspot (Kumar et al., 2021; Szabo et al., 2016). Climate change impacts should be considered in conjunction with socioeconomic issues in the context of long-term development (Maikhuri et al., 2003). As per Paris Agreement, 2015, the goal is to keep global warming below 2°C and, if possible, below 1.5°C by the end of the 21st century.

2.4.12 Awareness

When working with mountain soils, it's important to remember that most people don't know much about them because most soil survey programs are focused on more fertile arable land. As a result, there is a significant lack of integrated ecosystem management. Awareness campaigns should be carried out among the hill people. Hill people's and others' education courses should include hill soil conservation and management so that the next generation can play an important role in hill soil conservation and management with better knowledge, practice, and training from childhood.

2.5 Conclusion

By nature, mountainous soil is a very soft and thin layer. The soil of that area is the foundation of the most important activities such as forest, animal husbandry, agriculture, herb collection, and so on, for the difficult livelihood of the people in mountainous areas. To address the soil crisis caused by rapid population growth and indiscriminate development in mountainous areas, the people of that region, as well as the country's national, provincial, and local governments, will need to take serious soil conservation measures on the ground. To prevent soil erosion in such mountainous areas, measures such as plantation, terrace farming, crop rotation, crop cover, small dams, promotion of traditional local methods, use of organic agriculture, and sustainable agriculture, among others, should be implemented. It must be incorporated into daily life. Local governments should implement the measures proposed by the United Nations and the Food and Agriculture Organization.

References

Abbasi, N.A., Xu, X., Lucas-Borja, M.E., Dang, W., Liu, B., 2019. The use of check dams in watershed management projects: examples from around the world. Science of the Total Environment 676, 683−691.

Abd Elbasit, M.A., Huang, J., Ojha, C.S.P., Yasuda, H., Adam, E.O., 2013. Spatiotemporal changes of rainfall erosivity in Loess Plateau, China. International Scholarly Research Notices 2013.

An, J., Geng, J., Yang, H., Song, H., Wang, B., 2021. Effect of ridge height, row grade, and field slope on nutrient losses in runoff in contour ridge systems under seepage with rainfall condition. International Journal of Environmental Research and Public Health 18 (4), 2022.

Aryal, D.R., De Jong, B.H., Ochoa-Gaona, S., Mendoza-Vega, J., Esparza-Olguin, L., 2015. Successional and seasonal variation in litterfall and associated nutrient transfer in semi-evergreen tropical forests of SE Mexico. Nutrient Cycling In Agroecosystems 103 (1), 45−60.

Aubault, H., Webb, N.P., Strong, C.L., McTainsh, G.H., Leys, J.F., Scanlan, J.C., 2015. Grazing impacts on the susceptibility of rangelands to wind erosion: the effects of stocking rate, stocking strategy and land condition. Aeolian Research 17, 89−99.

Barton, A.P., Fullen, M.A., Mitchell, D.J., Hocking, T.J., Liu, L., Bo, Z.W., et al., 2004. Effects of soil conservation measures on erosion rates and crop productivity on subtropical Ultisols in Yunnan Province, China. Agriculture, Ecosystems & Environment 104 (2), 343−357.

Byers, A., 2014. Contemporary human impacts on subalpine and alpine ecosystems of the Hinku Valley, Makalu-Barun National Park and Buffer Zone, Nepal. Himalaya 33, 25−41.

Chavarria, D.N., Verdenelli, R.A., Serri, D.L., Restovich, S.B., Andriulo, A.E., Meriles, J.M., et al., 2016. Effect of cover crops on microbial community structure and related enzyme activities and macronutrient availability. European Journal of Soil Biology 76, 74−82.

Clow, D.W., Williams, M.W., Schuster, P.F., 2016. Increasing aeolian dust deposition to snowpacks in the Rocky Mountains inferred from snowpack, wet deposition, and aerosol chemistry. Atmospheric Environment 146, 183−194.

Dale, V.H., Kline, K.L., Kaffka, S.R., Langeveld, J.W.A., 2013. A landscape perspective on sustainability of agricultural systems. Landscape Ecology 28 (6), 1111−1123.

Das, A., Ramkrushna, G.I., Makdoh, B., Sarkar, D., Layek, J., Mandal, S., et al., 2016. Managing soils of the lower Himalayas, Encyclopedia of Soil Science, third ed. CRC Press, p. 10.

Duniway, M.C., Pfennigwerth, A.A., Fick, S.E., Nauman, T.W., Belnap, J., Barger, N.N., 2019. Wind erosion and dust from US drylands: a review of causes, consequences, and solutions in a changing world. Ecosphere 10 (3), e02650.

Eze, S., Dougill, A.J., Banwart, S.A., Sallu, S.M., Smith, H.E., Tripathi, H.G., et al., 2021. Farmers' indicators of soil health in the African highlands. Catena 203, 105336.

Fang, H., Gu, X., Jiang, T., Yang, J., Li, Y., Huang, P., et al., 2020. An optimized model for simulating grain-filling of maize and regulating nitrogen application rates under different film mulching and nitrogen fertilizer regimes on the Loess Plateau, China. Soil and Tillage Research 199, 104546.

FAO, 2015. Food and Agriculture Organization of the United Nations.

Fernández, S., Cotos-Yáñez, T., Roca-Pardiñas, J., Ordóñez, C., 2018. Geographically weighted principal components analysis to assess diffuse pollution sources of soil heavy metal: application to rough mountain areas in Northwest Spain. Geoderma 311, 120–129.

Fistikoglu, O., Harmancioglu, N.B., 2002. Integration of GIS with USLE in assessment of soil erosion. Water Resources Management 16 (6), 447–467.

Fleischner, T.L., 1994. Ecological costs of livestock grazing in western North America. Conservation Biology 8 (3), 629–644.

Gharibreza, M., Ashraf, M.A., 2014. Applied limnology. Springer Japan, Tokyo.

Gharibreza, M., Zaman, M., Porto, P., Fulajtar, E., Parsaei, L., Eisaei, H., 2020. Assessment of deforestation impact on soil erosion in loess formation using 137Cs method (case study: Golestan Province, Iran). International Soil and Water Conservation Research 8 (4), 393–405.

Hong-gui, D., Teng-Feng, G., Ming-Hui, L., Xu, D., 2012. Comprehensive assessment model on heavy metal pollution in soil. International Journal of Electrochemical Science 7 (6), 5286–5296.

Hoyos, N., 2005. Spatial modeling of soil erosion potential in a tropical watershed of the Colombian Andes. Catena 63 (1), 85–108.

Indarto, J., Mutaqin, D.J., 2016. An overview of theoretical and empirical studies on deforestation.

IPCC, 2021. Climate Change 2021: The Physical Science Basis. Contribution of Working Group I to the Sixth Assessment Report of the Intergovernmental Panel on Climate Change.

Jiao, Y., Li, X., Liang, L., Takeuchi, K., Okuro, T., Zhang, D., et al., 2012. Indigenous ecological knowledge and natural resource management in the cultural landscape of China's Hani Terraces. Ecological Research 27 (2), 247–263.

Joyce, L.A., Briske, D.D., Brown, J.R., Polley, H.W., McCarl, B.A., Bailey, D.W., 2013. Climate change and North American rangelands: assessment of mitigation and adaptation strategies. Rangeland Ecology & Management 66 (5), 512–528.

Khaledi Darvishan, A., Behzadfar, M., Spalevic, V., Kalonde, P., Ouallali, A., el Mouatassime, S., 2017. Calculation of sediment yield in the S2-1 watershed of the Shirindareh River Basin, Iran. Transcultural Studies 63 (3).

Kumar, A., Pramanik, M., Chaudhary, S., Negi, M.S., 2021. Land evaluation for sustainable development of Himalayan agriculture using RS-GIS in conjunction with analytic hierarchy process and frequency ratio. Journal of the Saudi Society of Agricultural Sciences 20 (1), 1–17.

Langdale, G.W., Blevins, R.L., Karlen, D.L., McCool, D.K., Nearing, M.A., Skidmore, E.L., et al., 1991. Cover crop effects on soil erosion by wind and water. Cover Crops for Clean Water. Soil and Water Conservation Society, pp. 15–22.

Liu, L., Liu, Q.J., Yu, X.X., 2016. The influences of row grade, ridge height and field slope on the seepage hydraulics of row sideslopes in contour ridge systems. Catena 147, 686–694.

Mack, R.N., Thompson, J.N., 1982. Evolution in steppe with few large, hooved mammals. The American Naturalist 119 (6), 757–773.

Maikhuri, R.K., Rao, K.S., Patnaik, S., Saxena, K.G., Ramakrishnan, P.S., 2003. Assessment of vulnerability of forests, meadows and mountain ecosystems due to climate change. ENVIS Bulletin 11 (2), 1–9.

Majid, H., 2007. Fundamentals of Physical Geography. Rawat Publications.

Mandal, S., Mohanty, S., Datta, K.K., Tripathi, A.K., Hore, D.K., Verma, M.R., 2006. Internalising Meghalaya towards organic agriculture: issues and priorities. 14th Annual Conference of the Agricultural Economics Research Association 19, 1–18.

Mishra, P., Rai, A., Rai, S., 2019. Agronomic measures in traditional soil and water conservation practices in the Sikkim Himalaya, India. American Research Journal of Agriculture 5, 1–16.

Montgomery, D.R., 2007. Soil erosion and agricultural sustainability. Proceedings of the National Academy of Sciences 104 (33), 13268–13272.

Mokarram, M., Zarei, A.R., 2021. Determining prone areas to gully erosion and the impact of land use change on it by using multiple-criteria decision-making algorithm in arid and semi-arid regions. Geoderma 403, 115379.

Neitsch, S.L., Arnold, J.G., Kiniry, J.R., Williams, J.R., 2011. Soil and Water Assessment Tool Theoretical Documentation Version 2009. Texas Water Resources Institute.

Oli, K.P., Zomer, R., 2011. Kailash Sacred Landscape Conservation Initiative: Feasibility Assessment Report. International Centre for Integrated Mountain Development (ICIMOD).

Osterkamp, T.E., Jorgenson, M.T., Schuur, E.A.G., Shur, Y.L., Kanevskiy, M.Z., Vogel, J.G., et al., 2009. Physical and ecological changes associated with warming permafrost and thermokarst in interior Alaska. Permafrost and Periglacial Processes 20 (3), 235–256.

Painter, T.H., Deems, J.S., Belnap, J., Hamlet, A.F., Landry, C.C., Udall, B., 2010. Response of Colorado River run-off to dust radiative forcing in snow. Proceedings of the National Academy of Sciences 107 (40), 17125–17130.

Panagos, P., Borrelli, P., Poesen, J., Ballabio, C., Lugato, E., Meusburger, K., et al., 2015. The new assessment of soil loss by water erosion in Europe. Environmental Science & Policy 54, 438–447.

Pandey, K.B., Rizvi, S.I., 2009. Protective effect of resveratrol on formation of membrane protein carbonyls and lipid peroxidation in erythrocytes subjected to oxidative stress. Applied Physiology, Nutrition, and Metabolism 34 (6), 1093–1097.

Poesen, J., 2018. Soil erosion in the Anthropocene: research needs. Earth Surface Processes and Landforms 43 (1), 64–84.

Prasannakumar, V., Vijith, H., Abinod, S., Geetha, N.J.G.F., 2012. Estimation of soil erosion risk within a small mountainous sub-watershed in Kerala, India, using Revised Universal Soil Loss Equation (RUSLE) and geo-information technology. Geoscience Frontiers 3 (2), 209–215.

Prosdocimi, M., Cerdà, A., Tarolli, P., 2016. Soil water erosion on Mediterranean vineyards: a review. Catena 141, 1–21.

Rasouli, K., Pomeroy, J.W., Whitfield, P.H., 2019. Are the effects of vegetation and soil changes as important as climate change impacts on hydrological processes? Hydrology and Earth System Sciences 23 (12), 4933–4954.

Reinfeld, M.A., 2003. Tourism and the politics of cultural preservation: a case study of Bhutan. Journal of Public and International Affairs-Princeton 14, 125–143.

Romeo, R., Vita, A., Manuelli, S., Zanini, E., Freppaz, M., Stanchi, S., 2015. Understanding Mountain Soils: A Contribution from Mountain Areas to the International Year of Soils 2015. FAO, Rome.

Roper, W.R., Osmond, D.L., Heitman, J.L., Wagger, M.G., Reberg-Horton, S.C., 2017. Soil health indicators do not differentiate among agronomic management systems in North Carolina soils. Soil Science Society of America Journal 81 (4), 828–843.

Sharma, C.M., Mishra, A.K., Prakash, O., Dimri, S., Baluni, P., 2014. Assessment of forest structure and woody plant regeneration on ridge tops at upper Bhagirathi basin in Garhwal Himalaya. Tropical Plant Research 1 (3), 62–71.

Sharma, N.K., Singh, R.J., Mandal, D., Kumar, A., Alam, N.M., Keesstra, S., 2017. Increasing farmer's income and reducing soil erosion using intercropping in rainfed maize-wheat rotation of Himalaya, India. Agriculture, Ecosystems & Environment 247, 43–53.

Sheng, J., Wang, X., Gong, P., Tian, L., Yao, T., 2012. Heavy metals of the Tibetan top soils. Environmental Science and Pollution Research 19 (8), 3362–3370.

Singh, C., Chauhan, N., Upadhyay, S.K., Singh, R., Rani, A., 2021. The Himalayan natural resources: challenges and conservation for sustainable development. Journal of Pharmacognosy Phytochemistry 10 (1), 1643–1648.

Singh, R., Srivastava, P., Singh, P., Upadhyay, S., Raghubanshi, A.S., 2019. Human overpopulation and food security: challenges for the agriculture sustainability. Urban Agriculture and Food Systems: Breakthroughs in Research and Practice. IGI Global, pp. 439–467.

Smaling, E.M.A., Stoorvogel, J.J., 1990. Assessment of Soil Nutrient Depletion in Sub-Saharan Africa: 1983–2000. Main Report. SC-DLO (Report/Winand Staring Centre 28), Wageningen.

Spalevic, V., 2011. Impact of Land Use on Runoff and Soil Erosion in Polimlje (Doctoral dissertation, Doctoral thesis). Faculty of Agriculture of the University of Belgrade, Serbia.

Spalevic, V., Lakicevic, M., Radanovic, D., Billi, P., Barovic, G., Vujacic, D., et al., 2017a. Ecological-economic (Eco-Eco) modelling in the River Basins of Mountainous Regions: impact of land cover changes on sediment yield in the Velicka Rijeka, Montenegro. Notulae Botanicae Horti Agrobotanici Cluj-Napoca 45 (2), 602–610.

Spalevic, V., Radanovic, D., Skataric, G., Billi, P., Barovic, G., Curovic, M., et al., 2017b. Ecological-Economic (Eco-Eco) modelling in the mountainous river basins: impact of land cover changes on soil erosion. Agriculture & Forestry/Poljoprivreda i Sumarstvo 63 (4).

Srivastava, P., Singh, R., Bhadouria, R., Singh, P., Tripathi, S., Singh, H., et al., 2018. Physical and biological processes controlling soil C dynamics, Sustainable Agriculture Reviews, 33. Springer, Cham, pp. 171–202.

Stanchi, S., D'Amico, M.E., Pintaldi, E., Colombo, N., Romeo, R., Freppaz, M., 2021. Mountain soils.

Szabo, S., Gácsi, Z., Balazs, B., 2016. Specific features of NDVI, NDWI and MNDWI as reflected in land cover categories. Landscape & Environment 10 (3–4), 194–202.

1. Soils of mountainous landscapes: introduction

Tarolli, P., Sofia, G., 2016. Human topographic signatures and derived geomorphic processes across landscapes. Geomorphology 255, 140−161.

Tarolli, P., Straffelini, E., 2020. Agriculture in hilly and mountainous landscapes: threats, monitoring and sustainable management. Geography and Sustainability 1 (1), 70−76.

Tewari, V.P.V.P., Verma, R.K., Von Gadow, K., 2017. Climate change effects in the Western Himalayan ecosystems of India: evidence and strategies. Forest Ecosystems 4 (1), 1−9.

Thapa, G.B., 1996. Land use, land management and environment in a subsistence mountain economy in Nepal. Agriculture, Ecosystems & Environment 57 (1), 57−71.

Torquati, B., Giacchè, G., Venanzi, S., 2015. Economic analysis of the traditional cultural vineyard landscapes in Italy. Journal of rural studies 39, 122−132.

Toy, T.J., Foster, G.R., Renard, K.G., 2002. Soil Erosion: Processes, Prediction, Measurement, and Control. John Wiley & Sons.

USDA, 1999. United States Department of Agriculture.

Van Beek, C.L., Elias, E., Yihenew, G.S., Heesmans, H., Tsegaye, A., Feyisa, H., et al., 2016. Soil nutrient balances under diverse agro-ecological settings in Ethiopia. Nutrient Cycling in Agroecosystems 106 (3), 257−274.

Ventresca Miller, A.R., Spengler, R., Haruda, A., Miller, B., Wilkin, S., Robinson, S., et al., 2020. Ecosystem engineering among ancient pastoralists in northern Central Asia. Frontiers in Earth Science 8, 168.

Wang, Y., Wu, N., Kunze, C., Long, R., Perlik, M., 2019. Drivers of change to mountain sustainability in the Hindu Kush Himalaya. The Hindu Kush Himalaya Assessment. Springer, Cham, pp. 17−56.

Webb, N.P., Strong, C.L., 2011. Soil erodibility dynamics and its representation for wind erosion and dust emission models. Aeolian Research 3 (2), 165−179.

WMO, 2021. WMO Air Quality and Climate Bulletin, No. 1.

Wen, L., Dong, S., Li, Y., Wang, X., Li, X., Shi, J., Dong, Q., 2013. The impact of land degradation on the C pools in alpine grasslands of the Qinghai-Tibet Plateau. Plant and Soil 368 (1), 329−340.

Zachar, D., 2011. Soil Erosion. Elsevier, pp. 33−34.

Zhang, F., Yan, X., Zeng, C., Zhang, M., Shrestha, S., Devkota, L.P., et al., 2012. Influence of traffic activity on heavy metal concentrations of roadside farmland soil in mountainous areas. International Journal of Environmental Research and Public Health 9 (5), 1715−1731.

Further reading

Blanco, H., Lal, R., 2008. Principles of Soil Conservation and Management, vol. 167169. Springer, New York.

COP 21, 2015. The Paris Agreement.

Gholami, L., Banasik, K., Sadeghi, S.H., Darvishan, A.K., Hejduk, L., 2014. Effectiveness of straw mulch on infiltration, splash erosion, runoff, and sediment in laboratory conditions. Journal of Water and Land Development .

Gholami, R., Moradzadeh, A., Rasouli, V., Hanachi, J., 2014. Practical application of failure criteria in determining safe mud weight windows in drilling operations. Journal of Rock Mechanics and Geotechnical Engineering 6 (1), 13−25.

Grimalt, J.O., Van Drooge, B.L., Ribes, A., Vilanova, R.M., Fernandez, P., Appleby, P., 2004. Persistent organochlorine compounds in soils and sediments of European high altitude mountain lakes. Chemosphere 54 (10), 1549−1561.

Jaiswal, D.K., 2015. Current need of organic farming for enhancing sustainable agriculture. Journal of Cleaner Production 102, 545e547.

Li, Z., Ma, Z., van der Kuijp, T.J., Yuan, Z., Huang, L., 2014. A review of soil heavy metal pollution from mines in China: pollution and health risk assessment. Science of the Total Environment 468, 843−853.

Smaling, E.M.A., Stoorvogel, J.J., Windmeijer, P.N., 1993. Calculating soil nutrient balances in Africa at different scales. Fertilizer Research 35 (3), 237−250.

Song, L., Li, L., Zhang, C.J., Huang, L., Guo, J.S., Zhu, B., et al., 2021. CRWS-mountain project: coordinate remediation techniques and devices for water-soil pollution in mountain areas in China. Journal of Mountain Science 18 (9), 2441−2446.

Soil microbial processes and nutrient dynamics

CHAPTER

3

Integrated remedial and management strategies for sustaining mountainous soil

Surbhi Sharma[1], Neeru Bala[1], Priyanka Sharma[1], Joat Singh[1], Shalini Bahel[2] and Jatinder Kaur Katnoria[1]

[1]Department of Botanical and Environmental Sciences, Guru Nanak Dev University, Amritsar, Punjab, India [2]Department of Electronics Technology, Guru Nanak Dev University, Amritsar, Punjab, India

3.1 Introduction

Mountains cover 24% of the earth's physical area and are home to 12% of the world's population, with another 14% living nearby (Sharma et al., 2010). Mountainous habitats are fragile ecosystems with diverse structures and functions that are abundant in biodiversity, water, soil, and minerals. Furthermore, they are important sites of traditional ecological knowledge as well as a vital center of biological and cultural diversity, and they have a large impact on the climate on a variety of scales (Jing-Yun et al., 2004). Climate change, on the other hand, is causing widespread elevational changes in mountains, raising the risk of species extinction (Elsen et al., 2020). Agriculture, tourism, and natural resource use are all sources of income for mountain communities. Half of the world's biodiversity hotspots are found in mountain regions. A shift in the intensity of human land use, the introduction of new crops or techniques, and rising population densities can all have a significant impact on these systems (Skeldon, 1985). Mountain soils are highly susceptible to change and can react quickly to environmental changes. Chemical and physical weathering, as well as mineral transformation, contribute to progressive soil development in hilly areas, whereas unweathered materials are eroded and deposited, resulting in deterioration of mountainous soils (Egli and Poulenard, 2016). Soil is a major resource for crop and vegetable production, and it aids in climate change adaptation by participating in biogeochemical cycles such as the carbon, water, and nitrogen cycles (Singh and Bakshi, 2010). The role of soil in improving flood and drought resistance is an important factor in water management. Despite being considered unstable, mountain soil supports 25% of terrestrial biodiversity and serves as a vital gene pool for locally adapted crops and animals. Mountain soils vary in composition depending on the source rock and are formed as a result of mechanical weathering and organic matter deposition (Dwevedi et al., 2017). Mountain soils can be found at lower elevations as well as higher elevations with enough rainfall. Mountain soils at higher elevations are typically shallow and deficient in nutrients, particularly nitrogen, which plants require. Peat accumulates in various mountain locations due to damp climate and frosted conditions, resulting in the formation of moist and acidic soils. *Tephra*, or eruptive ash, contributes to soil fertility in volcanic areas. Soil erosion and associated deterioration caused by agricultural practices and flash floods reduce soil fertility (Prasannakumar et al., 2012). The entry of eroded soil into bodies of water is a worldwide environmental concern because it reduces water quality by causing sedimentation and increases the likelihood of flooding (Zhou, 2008). Flooding, on the other hand, causes soil erosion, which leads to a loss of soil fertility as well as a slew of other negative environmental consequences, putting long-term agricultural production at risk (Prasannakumar et al., 2012). High rates of erosion have been well documented in mountainous areas, which are caused by natural factors such as the steepness of the terrain combined with excessive rain (Cao et al., 2010; János et al., 2013). According to one study, unsuitable land use or agricultural management practices such as continuous tillage with no soil cover, overgrazing, and poor land management exacerbated erosion processes (Al-Wadaey and Ziadat, 2014).

Mountain soils have historically generated a variety of critical ecosystem services that ensure food security and nutrition for millions of mountain people worldwide, as well as billions who live downstream. Mountains are a valuable legacy threatened by climate change and human activity. Mountain ecosystems are also extremely vulnerable, facing both natural and man-made threats. Various changes, such as changes in land use patterns and climate, have an unprecedented impact on mountainous soils, making mountainous people's livelihoods and food security vulnerable to these changes (Romeo et al., 2015). Human activities such as mining, inadequate infrastructure, and tourism cause disasters such as floods, landslides, debris flows, and glacial lake outbursts in most mountain regions. Climate change is exacerbating the risk effect by increasing the frequency of extreme events such as heavy rain, droughts, and glacier melting (Van, 2006). Mountains are among the most affected by climate change and provide some of the most visible evidence, such as glacier loss. In reality, mountainous soil is under severe threat and requires immediate conservation and management. Furthermore, soil conservation and management provide a number of agronomic, environmental, and economic benefits. Given the preceding discussion, the primary goal of this chapter is to provide a framework for the various threats and conservation strategies for mountainous soil.

3.2 Threats to mountainous soil

Mountainous soils play an important role in the functioning and conservation of unique ecosystems, as well as in mountain range hydrology. Thus, changes in alpine soil characteristics caused by land use can raise concerns about their fertility and associated services. Soil degradation in mountainous areas is a global problem. Mountain soils' fragile characteristics, such as slow soil formation, delayed pedogenesis, and sharp and steep slopes, contribute to degrading processes such as water erosion, topsoil truncation, organic content loss, chemical and physical quality degradation, and desertification (Fort, 2015; Egli et al., 2012). When marginal regions are abandoned, natural hazards become even more prevalent. It has an impact not only on the soil ecosystem, but also on the native and endemic flora and fauna of that specific hill/alpine (Dax et al., 2021). Some of the most important threats to mountainous soil are as follows.

3.2.1 Natural disasters

Natural disasters can be extremely disruptive, especially when mountain areas are densely populated with people living in the lowlands; thus, subsistence farming in high-altitude regions necessitates special care for soil management in such difficult alpine environments (Romeo et al., 2015). The increasing frequency of landslides along mountain slopes is another threat to alpine ecology (Ganie et al., 2019). Landslides have a large impact on the natural population and cause habitat fragmentation (Dar and Naqshi, 2002). Furthermore, landslides play an important role in making the soil more vulnerable due to changes in the physio-chemical characteristics of the soil, which can make the natural habitat of endemic species unfavorable and lead to extinction (Tali et al., 2015). Avalanches exacerbate the vulnerability of mountain ecosystems. Massive, stand-replacing disturbances are relatively rare in European Alps forests due to the high degree of landscape fragmentation and widespread man-made management, such as active fire and avalanche suppression (Brotons et al., 2013; Kulakowski et al., 2006; Vacchiano et al., 2015).

3.2.2 Climate change

The impact of climate change on mountain soils may have off-site consequences in addition to localized events. The Intergovernmental Panel on Climate Change (IPCC) highlighted the high susceptibility of mountain areas to climate change in its report (Pachauri et al., 2014). Mountain and soil dwellers are thought to share a vital and dynamic bond. Unsustainable farming and forest management practices put even more strain on this delicate and vulnerable environment, hastening erosion, reducing fertility, and increasing the risk of natural disasters. Despite the fact that mountainous ecosystems are particularly vulnerable to natural disasters, the risks are mitigated when the areas are sparsely populated (Gobiet et al., 2014). Climate change is a global issue, but it has a particularly serious and troubling impact on mountains (Beniston, 2003). It has resulted in warming and weather extremes that are becoming more severe in high elevation areas. The rate of warming increases with altitude, just as it does at high latitudes, where ice loss is much faster than predicted (Lutz and Immerzeel, 2013). As a result, high mountain regions experience faster temperature oscillations and significantly greater daily temperature differences than lowland areas Williams et al. (1998) conducted a long-term snow-fence experiment at the Niwot Ridge Long-Term Ecological Research (NWT) site in the Colorado Front Range of the Rocky Mountains, U.S.A., to investigate the effects of climate change on alpine ecology and biogeochemical cycles. Many studies have also been conducted and reported that climate change caused variations in snow duration, depth, and extent, resulting in significant fluctuations in the carbon and nitrogen contents of alpine ecosystem soils (Watson and Haeberli, 2004; Schickhoff, 2011).

3.2.3 Anthropogenic threats

Changes in land use have had a significant impact on vulnerable ecosystems. Farming practices on steep slopes, for example, frequently result in forest loss and can cause landslides and increased surface run-off. Overgrazing has also resulted in land degradation, the formation of bogs, and the extinction of plant species. Excessive grazing in hilly/alpine regions has resulted in the formation of boggy areas, a reduction in plant diversity, and increased soil erosion susceptibility (Ganie and Tali, 2013). One of the most serious threats to soil is water-induced soil erosion (surface runoff and drop erosivity), which has a significant impact on ecosystem services, agricultural production, drinking water, and carbon stocks (Panagos et al., 2015). All of this posed a risk to residents of mountainous areas as well as native wildlife (Tarolli and Straffelini, 2020). Because of the increased livestock on alpine meadows, animal species have migrated further into the high mountains, increasing the risk of predators preying on domestic cattle. The rapid movement of people and resources into and out of the mountains exacerbated their problems. The operations pollute even inaccessible mountain areas at high altitudes. As a result of all the threats, the *Agenda 21* chapter on sustainable mountain development recognized that mountain ecosystems were rapidly changing, and that proper management of mountain resources and socioeconomic development of affected people required immediate attention. The United Nations General Assembly also declared 2002 to be the International Year of Mountains in order to raise awareness about the importance of long-term mountain development (Cocean, 2015).

3.3 Management strategies for conservation of mountainous soil

There is widespread agreement that soil degradation in mountainous areas is a worldwide issue. Mountain soils are naturally prone to degradation processes such as water erosion, deterioration of physical and chemical quality, and desertification. Loss of original soil profile integrity, disruption of a prior pedogenic route due to site disturbance, slow pedogenesis, and steep slopes were some of the variables that favored soil profile erosion and limited restoration success (Moorhead, 2015). Topsoil loss, limited organic matter inputs, and harsh weather, which alters soil biodiversity and organic material turnover, are all factors that contribute to mountainous soil degradation. Overgrazing, deforestation, monocropping, up and down-slope plowing, and urbanization all contribute to severe soil degradation, which accelerates erosion and decreases soil fertility, increasing the risk of natural disasters (Griffiths et al., 2009). Mountainous areas require a combination of biological and structural soil and water conservation techniques due to steep slopes, varied terrain, and regularly alternating dry and wet spells in order to develop a protective vegetative cover and minimize downward landslides caused by splash, rill, and gully erosion. Because of the wide range of processes that affect mountain soils, as well as their potential on-site and off-site consequences, hazard and risk assessment are critical. Officials working to protect the mountainous terrain have put in place specific measures to protect the soil from various degradation processes. The following are some mountainous soil conservation methods.

3.3.1 Contouring

Contour cultivation is a well-known method for controlling mountainous soil erosion and reducing runoff. Contour farming is a tillage technique that reduces soil erosion while increasing crop yield. In contour farming, tillage, planting, and other farming operations are performed on the contour of the field slope. Cropping on sloping terrain is a major source of rapid erosion unless soil and water conservation practices are widely implemented. On a moderate slope, this method works well. On gentle slopes, contour farming is used (i.e., planting, plowing and weeding across rather than down the slope). In areas where rainfall is heavier, permanent vegetated contour bunds or ridges with retention or diversion ditches are required to break up the slope, increase rainwater collection and infiltration, reduce runoff volume and velocity, and thus avoid erosion (Traoré et al., 2004; Liu et al., 2016). Contour farming is used on gentle slopes to reduce runoff and soil erosion. This method can also increase crop yield in dry and semi-arid environments by retaining soil moisture. The results showed that contour farming reduced yearly runoff by 10% when compared to cultivation perpendicular to the slope. When compared to cultivation and planting on the slope, cultivation and planting along contour lines reduced soil and water losses by 49.5% and 32%, respectively (Farahani et al., 2016).

3.3.2 Terracing

Terracing is an ancient method of water and soil protection. Terracing is the process of constructing a canal and a bank or a single terrace wall, such as an earthen ridge or a stone wall (Hanway and Laflen, 1974; Morgan, 2009). The most important soil conservation strategy is the maintenance of a permanent soil cover, which must be used in conjunction with terracing. It is also the most widely used method of soil conservation on the planet. As seen in Figs. 3.1 and 3.2, terracing lessens slope steepness and divides the slope

FIGURE 3.1 Terraces in Rajouri region (Jammu and Kashmir, India).

FIGURE 3.2 Terraces in Doda region (Jammu and Kashmir, India).

into tiny, softly sloping sections. Progressive or bench terraces are required for long-term farming on steep slopes, which are costly and time-consuming to build and maintain. However, as evidenced by world-class terracing systems, several of which have been designated as Globally Important Agricultural Heritage Systems (GIAHS). Terracing systems, for example, conserve soil in the Andes by allowing deeper soil development and the expansion of farmed areas, as well as ensuring that sunlight is available for longer periods of time during the day, increasing water storage and allowing more efficient irrigation water usage. The use of rocks in the construction of terrace faces that support mountain slopes protects agricultural areas from future landslides (Dorren and Imeson, 2005).

During a strong storm, a significant amount of precipitation falls at the soil surface and, depending on the soil type, infiltrates the water, while the remainder flows in the form of runoffs, which accumulate in natural depressions and flow downhill to natural deposition zones. Increased runoff causes increased velocity and volume. In sandy soils, the critical runoff velocity is reported as 5 m/s while in clay soils, it is 8 m/s, at the time when soil particles have become disengaged from soil aggregates and begin to get transported over the surface (International Institute of Tropical Agriculture, 2000, Dorren and Rey, 2004).

3.3.3 Enhancement of biodiversity

The use of native plants in the mountainous agricultural region benefits biodiversity and prevents soil erosion (Wall et al., 2015). Soil biota is critical for soil quality and aids in the prevention of deterioration and desertification. Soil biota account for a significant portion of global terrestrial biodiversity and play critical ecological roles in processes such as biomass breakdown, nutrient cycling, CO_2 mitigation, and disease suppression, among others. Soil aggregates are naturally occurring clusters or clumps of soil mass in which the forces holding the particles together are significantly greater than the forces between adjacent aggregates (Lynch and Bragg, 1985). Plant-based materials that enter the soil as particulates undergo biological and chemical transformations as well as direct stabilization. Particulate organic matter is colonized by a microbial population and absorbs mineral particles and metal ions at the same time. Plant debris that is labile is consumed by the microbial population and eventually enters the soil organic matter pool as microbial products. As a result, increasing the activity and species diversity of soil fauna and flora (micro, meso, and macro) is critical for restoring and improving soil quality and reducing the risk of soil deterioration. People have long recognized the importance of macro-organisms like earthworms and termites in soil quality restoration. As a result, soil degradation risks can be reduced by implementing land use and management strategies that improve soil biological processes, as well as selectively inoculating soils with beneficial organisms (Lavelle et al., 1992; Schonbeck, 2017).

3.3.4 Enhancement of soil resilience

Soil resilience is defined as a soil's inherent ability to recover from deterioration and revert to a new equilibrium that is comparable to its previous state. Another definition of soil resilience is a system's ability to restore its "functional and structural integrity" (Seybold et al., 1999; Dorren and Imeson, 2005; Blanco and Lal, 2008). Soils are affected by a variety of degradative processes, including erosion, compaction, salinization, and acidification. However, depending on the severity and duration of degradative processes, as well as the strength of restorative mechanisms, most soils have an inherent ability to resist exogenous and endogenous disturbances and recover. The ability of a soil to rebound from disturbances is an important and fundamental property. To put it another way, a soil has an inherent ability to regenerate itself, which, when combined with good management, has the potential to reverse soil deterioration (Blanco and Lal, 2008). Soil can establish a self-regenerating system to combat degradative processes by using restorative management systems. Practices that degrade the soil must be systematically matched with practices that improve the soil's resilience. Adoption of practices that increase the input of soil organic matter is a significant way to improve soil resilience (Srivastava et al., 2016). Organic matter helps to improve soil pore structure, water infiltration, compaction, runoff, and erosion. Improvements in micro porosity and pore structure are critical to the soil's water retention and transmission properties. When pressures are released, high levels of soil organic matter act as a sponge, reducing soil compressibility while increasing resilience (Connolly, 1998; Schaeffer et al., 2016). Soil restoration necessitates changes in farming practices, land use, and human attitudes (Srivastava et al., 2016). Conservation tillage is one strategy for improving the resilience of moderately degraded soils. This change in management initiates soil recovery, allowing for faster regeneration of soil properties. Soils with less tillage are more resilient due to higher soil organic matter content and soil organism activity. Long-term conservation tillage restores degraded soils by reducing soil disturbance and increasing residue input. However, the benefits of no-till practices for improving soil resilience may vary depending on soil degradation, soil type, and climate. No-till management may not be sufficient to restore severely degraded soils (Blanco and Lal, 2008). Some of the factors influencing soil resilience is given in Table 3.1. Building strong soil management capabilities, advocating for comprehensive policies and governance, and participating in soil research and soil information systems are all critical for long-term agricultural systems that benefit everyone.

3.3.5 Assessment of soil erosion hazards

Soil erosion can be a slow, unnoticed process, or it can happen quickly, resulting in significant topsoil loss. Reduced agricultural productivity, poor surface water quality, and clogged drainage systems could all be the result of soil loss in croplands. Soil erosion reduces farmland productivity and pollutes nearby waterways and lakes. Erosion is the separation, movement, and deposition of soil caused by water, wind, or tillage. Scientific management of soil, water, and other resources on watersheds is critical for preventing erosion and rapid siltation in bodies of water, but it is difficult to implement on large and difficult watersheds. Remote sensing applications and geographic information systems (GIS) can help with soil, water, and other resource management (Yadav and Singh, 2015). Large amounts of data collected via remote sensing techniques can be effectively managed and exploited with the help of GIS. Soil erosion is caused by natural and man-made factors interacting in complex ways. Because such variables change over time and space, assessing soil erosion becomes even more difficult (Phinzi and Ngetar, 2019). Recent advances in geographic information technology have supplemented existing models and provided effective methods of monitoring, analyzing, and managing earth resources. The use of digital elevation models (DEMs), remote sensing data, and GIS can successfully enable the rapid and comprehensive evaluation of mountainous soil hazards (Arnous and Green, 2011; Metternicht et al., 2005).

3.3.6 Development of soil erosion models

Soil erosion models have made it possible to study soil erosion with great precision, particularly for conservation purposes. Many erosion models, such as the Universal Soil Loss Equation/Revised/Modified Universal Soil Loss Equation (USLE/RUSLE/MULSE) and the Water Erosion Prediction Project, are used to measure soil erosion and develop effective erosion management strategies (WEPP) (Maqsoom et al., 2020). These models are elaborated in the following sections.

3.3.6.1 *Universal soil loss equation*

A model has been developed by (Wischmeier and Smith, 1978). It was presented for sheet and rill erosion based on a large amount of experimental data from agricultural plots. The equation was created using single agricultural plots and is only applicable to

TABLE 3.1 Factors influencing soil resilience.

S. No.	Elements that influence soil resilience and restoration		References
1.	Enhancing the biomass of microorganisms	Organic matter in the soil and soil microorganisms are inextricably linked. Microbial biomass is the living component of soil organic matter, and it can act as a significant source or sink for soil carbon and nutrients, as well as influence the amount of organic carbon and nitrogen retained in soil organic matter (SOM). Most nutrient release processes are catalyzed by microorganisms. Microbial biomass is higher in systems with high organic matter inputs and accessible soil organic matter, enhancing the soil's restorative capability.	Landgraf and Klose (2002) Jiang-shan et al. (2005) Fuke et al. (2021)
2.	Improving aeration	Soil structure is a physical property of soil that influences soil water balance, gas flow, plant growth and development, and, eventually, plant production. Biochar has gained international attention as a soil amendment capable of improving soil structure while also increasing soil resilience. The ability of gas to pass through the pore space of the soil is measured by soil air permeability. Soil pores found within (intra) and between (inter) aggregates act as water and air flow channels in the soil. The movement of water and gas in the soil profile is influenced not only by the number and size of pores, but also by pore connectivity and tortuosity, which are closely related to the geometric properties of soil pore structure. As a result, by increasing soil porosity and introducing soil amendments, soil quality can be improved.	Niu et al. (2012) Amoakwah et al. (2017)
3.	Creating a favorable hydrological balance	Overuse of water resources by a growing population has resulted in an unbalance in the hydrological cycle, resulting in a drop in water table level, which is the primary cause of soil hydrological desertification. Industries dump their waste effluent into the seas, where it eventually returns to the land via the hydrological cycle, degrading soil quality and resilience. As a result, excessive use of water resources should be avoided to the greatest extent possible in order to prevent soil deterioration and improve soil resilience.	Loch et al. (2014) Loucks and Van Beek (2017)
4.	Improving soil biodiversity	Biodiversity loss is the single most serious threat to the protection and long-term use of mountainous soil. Soil organisms provide a variety of important ecosystem services to the environment, such as soil formation, organic matter decomposition (which affects soil fertility and plant growth), water infiltration and retention, pollutant degradation, and pest biocontrol. As a result, soil biodiversity and ecological services are important for global health. Despite these critical services, however, there is little specific knowledge about soil biota.	Asefa et al. (2003) Jeffery and Gardi (2010) Geisen et al. (2019)

(*Continued*)

2. Soil microbial processes and nutrient dynamics

TABLE 3.1   (Continued)

S. No.	Elements that influence soil resilience and restoration		References
5.	Promoting elemental cycling and establishing a healthy elemental balance	Soil carbon (C) sequestration is the process by which atmospheric CO_2 is transferred into the soil of a land unit via its plants. Increased biodiversity and improved elemental recycling are two benefits of soil C sequestration. Land use, soil management, and farming practices all influence soil organic C (SOC). Soil management strategies such as conservation agriculture must be used to increase SOC concentration and provide a positive C budget in order to restore soil quality.	Lal et al. (2015) Lal et al. (2015)

areas of up to one hectare (ha). The USLE equation takes into account slope length (L factor), steepness (S factor), climate (R factor), soils (K factor), cropping (C factor), and management (P factor). The goal of this model was to predict the average yearly soil movement from a specific field plot under specific land use and management conditions. The model is only valid for 20-year average data, not for individual storms.

The USLE for estimating average annual soil erosion is;

A = RKLSCP

where, A = average annual soil loss in t/a (tons per acre), R = rainfall erosivity index, K = soil erodibility factor, LS = topographic factor-L is for slope length and S is for slope, C = cropping factor, P = conservation practice factor.

3.3.6.2 Revised universal soil loss equation

RUSLE is an erosion model that predicts long-term average annual soil loss (A) from specific field slopes in certain cropping and management systems, as well as from rangeland. RUSLE's utility and validity for this purpose have been validated by widespread use (Renard and Ferreira, 1993). It also applies to non-agricultural settings such as construction sites. The changes made to the USLE to produce the RUSLE are listed below.

- Computerizing the algorithms to aid in calculations.
- New rainfall-runoff erosivity term (R) in the Western US, based on more than 1200 auge locations.
- Some changes and additions for the Eastern United States, including corrections for high R-factor areas with flat slopes to account for splash erosion caused by raindrops falling on ponded water.
- Development of a seasonally variable soil erodibility term (K).
- A novel method for calculating the cover management term (C), with sub-factors representing prior land use, crop canopy, surface cover, and surface roughness.
- New slope length and steepness (LS) algorithms reflecting rill to inter-rill erosion ratio.
- The capacity to calculate LS products for the slopes of varying shapes
- New conservation practices value (P) for rangelands, strip crop rotations, contour factor values and subsurface drainage.

3.3.6.3 Modified universal soil loss equation

The model was created to estimate the sediment load produced by each storm, taking into account not only rainfall erosivity but also runoff volume. The MUSLE can estimate soil loss from a single event, but it, as well as the USLE and RUSLE, cannot estimate sediment detachment, entrainment, transport, deposition, and redistribution within the watershed and are thus of limited use (Williams, 1975). MUSLE is a modification of USLE which was modified by Williams (1975) and his analysis revealed that using the product of the volume of runoff and peak discharge for an event yielded more accurate sediment yield predictions, especially for large events, than the USLE with the R factor ().

The MUSLE is expressed as

$$Y = 11.8 \left(Q_p V_q \right)^{0.56} KLSCP$$

where, Y = Sediment yield from storm (tonnes), Q = Peak runoff rate (m^3/s), V_q = Volume of storm runoff (m^3)

- All other factors K, (LS), C and P have the same meaning as in USLE.
- The values of Q_p and V_q can be obtained by appropriate runoff models.
- In this model V_q is considered to represent the detachment process and Q_p is the sediment transport.
- It is a sediment yield model and does not need a separate estimation of sediment delivery ratio and applies to individual storms.
- It also improves the accuracy of sediment yield prediction. It has the advantage of being able to model daily, monthly, and annual sediment yields of a watershed by combining appropriate hydrological models with MUSLE.

3.3.6.4 Water erosion prediction project model

This model is a physically-based, distributed parameter, single-event simulation erosion prediction model developed by Flanagan and Nearing (1995). The model includes processes such as erosion, sediment transport, and deposition across the landscape and in channels via a transport equation. WEPP was originally designed for agricultural modeling, but it is now widely used in forestry, fisheries, rangeland, and mining studies, as well as climate research. WEPP has the ability to change the monthly precipitation amounts, the monthly number of wet days (for storm "intensification"), and the monthly maximum and minimum temperatures. WEPP does not simulate gully erosion or mass wasting. WEPP's primary limitation is the small area it models (<2.6 km^2).

3.4 Conclusion

This study explains how to conserve and manage mountainous soil to a greater extent by improving soil resilience capacity and employing various techniques such as terracing and contouring. An increasing number of reports of anthropogenic hazards, as well as the frightening rate of climate change, highlighted the urgent need to investigate remedial procedures and strategies to preserve mountainous soil. As a result, research into existing geospatial technologies for erosion modeling is becoming increasingly important. This is reflected in the growing number of studies that model RUSLE erosion using remote sensing and GIS. Despite the challenges, the current work is likely to advance understanding of the role of geospatial technology in determining individual RUSLE characteristics. Future research, on the other hand, must focus on the error evaluation of RUSLE parameters obtained through remote sensing.

Conflict of interest

The authors state that there are no conflicts of interest in the publication of this chapter.

References

Al-Wadaey, A., Ziadat, F., 2014. A participatory GIS approach to identify critical land degradation areas and prioritize soil conservation for mountainous olive groves (case study). Journal of Mountain Science 11 (3), 782−791.

Amoakwah, E., Frimpong, K.A., Okae-Anti, D., Arthur, E., 2017. Soil water retention, air flow and pore structure characteristics after corn cob biochar application to a tropical sandy loam. Geoderma 307, 189−197.

Arnous, M.O., Green, D.R., 2011. GIS and remote sensing as tools for conducting geo-hazards risk assessment along Gulf of Aqaba coastal zone, Egypt. Journal of Coastal Conservation 15 (4), 457−475.

Asefa, D.T., Oba, G., Weladji, R.B., Colman, J.E., 2003. An assessment of restoration of biodiversity in degraded high mountain grazing lands in northern Ethiopia. Land Degradation & Development 14 (1), 25−38.

Beniston, M., 2003. Climatic change in mountain regions: a review of possible impacts. Climate Variability and Change in High Elevation Regions: Past, Present and Future. Springer, pp. 5−31.

Blanco, H., Lal, R., 2008. Principles of Soil Conservation and Management. Springer, New York, 167169.

Brotons, L., Aquilué, N., De Cáceres, M., Fortin, M.J., Fall, A., 2013. How fire history, fire suppression practices and climate change affect wildfire regimes in Mediterranean landscapes. PLoS One 8 (5), e62392.

Cao, S., Lee, K.T., Ho, J., Liu, X., Huang, E., Yang, K., 2010. Analysis of runoff in ungauged mountain watersheds in Sichuan, China using kinematic-wave-based GIUH model. Journal of Mountain Science 7 (2), 157−166.

Cocean, G., 2015. The importance of addressing anthropogenic threats in the assessment of karst geosites in the apuseni mountains (Romania). GeoJournal of Tourism and Geosites 16 (2), 198−205.

Connolly, R.D., 1998. Modelling effects of soil structure on the water balance of soil−crop systems: a review. Soil and Tillage Research 48 (1−2), 1−19.

Dar, G.H., Naqshi, A.R., 2002. Plant resources of Kashmir: diversity, utilization and conservation. Natural Resources of Western Himalaya. *Valley Book House, Srinagar, Kashmir*, pp. 109−122.

Dax, T., Schroll, K., Machold, I., Derszniak-Noirjean, M., Schuh, B., Gaupp-Berghausen, M., 2021. Land Abandonment in Mountain Areas of the EU: An Inevitable Side Effect of Farming Modernization and Neglected Threat to Sustainable Land Use. Land 10 (6), 591.

Dorren, L., Rey, F., 2004. In: Boix-Fayons, C., Imeson, A. (Eds.), A Review of the Effect of Terracing on Erosion. *Briefing Papers of the 2nd SCAPE Workshop*, pp. 97−108.

Dorren, L.K.A., Imeson, A.C., 2005. Soil erosion and the adaptive cycle metaphor. Land Degradation & Development 16 (6), 509−516.

Dwevedi, A., Kumar, P., Kumar, P., Kumar, Y., Sharma, Y.K., Kayastha, A.M., 2017. Soil sensors: detailed insight into research updates, significance and future prospects. New Pesticides and Soil Sensors. Academic Press, pp. 561−594.

Egli, M., Favilli, F., Krebs, R., Pichler, B., Dahms, D., 2012. Soil organic carbon and nitrogen accumulation rates in cold and alpine environments over 1 Ma. Geoderma 183−184, 109−123.

Egli, M., Poulenard, J., 2016. Soils of mountainous landscapes. International Encyclopedia of Geography: People, the Earth, Environment and Technology. Wiley, pp. 1−10.

Elsen, P.R., Monahan, W.B., Merenlender, A.M., 2020. Topography and human pressure in mountain ranges alter expected species responses to climate change. Nature Communications 11 (1), 1—10.

Farahani, S.S., Soheili, F.F., Asoodar, M.A., 2016. Effects of contour farming on runoff and soil erosion reduction: a review study. Agriculture 101, 44089—44093.

Flanagan, D.C., Nearing, M.A., 1995. USDA-Water Erosion Prediction Project: hillslope profile and watershed model documentation. National Soil Erosion Research Laboratory (NSERL) Report 10, 1—123.

Fort, M., 2015. Impact of climate change on mountain environment dynamics. An introduction. Journal of Alpine Research | Revue de géographie alpine 103-2.

Fuke, P., Kumar, M., Sawarkar, A.D., Pandey, A., Singh, L., 2021. Role of microbial diversity to influence the growth and environmental remediation capacity of bamboo: a review. Industrial Crops and Products 167, 113567.

Ganie, A.H., Tali, B.A., 2013. Vanishing medicinal plants of Kashmir Himalaya. http://www.GreaterKashmir.com/news/gk-magazine/vanishing-medicinalplants-of-kashmir-himalaya/153598.html.

Ganie, A.H., Tali, B.A., Khuroo, A.A., Reshi, Z.A., Nawchoo, I.A., 2019. Impact assessment of anthropogenic threats to high-valued medicinal plants of Kashmir Himalaya, India. Journal for Nature Conservation 50, 125715.

Geisen, S., Wall, D.H., van der Putten, W.H., 2019. Challenges and opportunities for soil biodiversity in the anthropocene. Current Biology 29 (19), R1036—R1044.

Gobiet, A., Kotlarski, S., Beniston, M., Heinrich, G., Rajczak, J., Stoffel, M., 2014. 21st century climate change in the European Alps—a review. Science of the Total Environment 493, 1138—1151.

Griffiths, R.P., Madritch, M.D., Swanson, A.K., 2009. The effects of topography on forest soil characteristics in the Oregon Cascade Mountains (USA): Implications for the effects of climate change on soil properties. Forest Ecology and Management 257 (1), 1—7.

Hanway, J.J., Laflen, J.M., 1974. Plant Nutrient Losses from Tile-Outlet Terraces. American Society of Agronomy, Crop Science Society of America, and Soil Science Society of America, pp. 351—356, 3, 4.

János, M., Klaudia, K., Mária, S., Andrea, K.B., András, K., László, M., et al., 2013. Hazards and landscape changes (degradations) on Hungarian karst mountains due to natural and human effects. Journal of Mountain Science 10 (1), 16—28.

Jeffery, S., Gardi, C., 2010. Soil biodiversity under threat — a review. Acta Societatis Zoologicae Bohemicae 74 (1—2), 7—12.

Jiang-shan, Z., Jian-fen, G., Guang-shui, C., Wei, Q., 2005. Soil microbial biomass and its controls. Journal of Forestry Research 16 (4), 327—330.

Jing-Yun, F.A.N.G., Ze-Hao, S.H.E.N., Hai-Ting, C.U.I., 2004. Ecological characteristics of mountains and research issues of mountain ecology. Biodiversity Science 12 (1), 10—19.

Kulakowski, D., Rixen, C., Bebi, P., 2006. Changes in forest structure and in the relative importance of climatic stress as a result of suppression of avalanche disturbances. Forest Ecology and Management 223 (1—3), 66—74.

Lal, R., Negassa, W., Lorenz, K., 2015. Carbon sequestration in soil. Current Opinion in Environmental Sustainability 15, 79—86.

Landgraf, D., Klose, S., 2002. Mobile and readily available C and N fractions and their relationship to microbial biomass and selected enzyme activities in a sandy soil under different management systems. Journal of Plant Nutrition and Soil Science 165 (1), 9—16.

Lavelle, P., Blanchart, E., Martin, A., Spain, A.V., Martin, S., 1992. Impact of soil fauna on the properties of soils in the humid tropics. Myths and Science of Soils of the Tropics 29, 157—185.

Liu, Q.J., An, J., Zhang, G.H., Wu, X.Y., 2016. The effect of row grade and length on soil erosion from concentrated flow in furrows of contouring ridge systems. Soil and Tillage Research 160, 92—100.

Loch, A., Adamson, D., Mallawaarachchi, T., 2014. Role of hydrology and economics in water management policy under increasing uncertainty. Journal of Hydrology 518, 5—16.

Loucks, D.P., Van Beek, E., 2017. Water Resource Systems Planning and Management. Springer International Publishing AG, Switzerland.

Lutz, A.F., Immerzeel, W.W., 2013. Water availability analysis for the upper Indus, Ganges, Brahmaputra, Salween and Mekong river basins. Final Report to International Centre for Integrated Mountain Development (ICIMOD). Future Water Report, 127.

Lynch, J.M., Bragg, E., 1985. Microorganisms and soil aggregate stability. Advances in Soil Science. Springer, New York, pp. 133—171.

Maqsoom, A., Aslam, B., Hassan, U., Kazmi, Z.A., Sodangi, M., Tufail, R.F., et al., 2020. Geospatial assessment of soil erosion intensity and sediment yield using the revised universal soil loss equation (RUSLE) model. ISPRS International Journal of Geo-Information 9 (6), 356.

Metternicht, G., Hurni, L., Gogu, R., 2005. Remote sensing of landslides: an analysis of the potential contribution to geo-spatial systems for hazard assessment in mountainous environments. Remote Sensing of Environment 98 (2–3), 284–303.

Moorhead, K.K., 2015. A pedogenic view of ecosystem restoration. Ecological Restoration 33 (4), 341–351.

Morgan, R.P.C., 2009. Soil Erosion and Conservation. John Wiley and Sons.

Niu, W., Guo, Q., Zhou, X., Helmers, M.J., 2012. Effect of aeration and soil water redistribution on the air permeability under subsurface drip irrigation. Soil Science Society of America Journal 76 (3), 815–820.

Pachauri, R.K., Allen, M.R., Barros, V.R., Broome, J., Cramer, W., Christ, R., et al., 2014. Climate change 2014: synthesis report. Contribution of Working Groups I, II and III to the fifth assessment report of the Intergovernmental Panel on Climate Change. IPCC, p. 151.

Panagos, P., Borrelli, P., Poesen, J., Ballabio, C., Lugato, E., Meusburger, K., et al., 2015. The new assessment of soil loss by water erosion in Europe. Environmental Science & Policy 54, 438–447.

Phinzi, K., Ngetar, N.S., 2019. The assessment of water-borne erosion at catchment level using GIS-based RUSLE and remote sensing: a review. International Soil and Water Conservation Research 7 (1), 27–46.

Prasannakumar, V., Vijith, H., Abinod, S., Geetha, N.J.G.F., 2012. Estimation of soil erosion risk within a small mountainous sub-watershed in Kerala, India, using Revised Universal Soil Loss Equation (RUSLE) and geo-information technology. Geoscience Frontiers 3 (2), 209–215.

Renard, K.G., Ferreira, V.A., 1993. RUSLE model description and database sensitivity.

Romeo, R., Vita, A., Manuelli, S., Zanini, E., Freppaz, M., Stanchi, S., 2015. Understanding Mountain Soils: A Contribution from Mountain Areas to the International Year of Soils 2015. FAO, Rome.

Schaeffer, A., Amelung, W., Hollert, H., Kaestner, M., Kandeler, E., Kruse, J., et al., 2016. The impact of chemical pollution on the resilience of soils under multiple stresses: a conceptual framework for future research. Science of the Total Environment 568, 1076–1085.

Schickhoff, U., 2011. In: Millington, A., Blumler, M., Schickhoff, U. (Eds.), Dynamics of Mountain Ecosystems. *Handbook of Biogeography*, pp. 313–337.

Schonbeck, M., 2017. Soil Health and Organic Farming. *Organic Farming Research Foundation*, Santa Cruz, CA.

Seybold, C.A., Herrick, J.E., Brejda, J.J., 1999. Soil resilience: a fundamental component of soil quality. Soil Science 164 (4), 224–234.

Sharma, E., Chettri, N., Oli, K.P., 2010. Mountain biodiversity conservation and management: a paradigm shift in policies and practices in the Hindu Kush-Himalayas. Ecological Research 25 (5), 909–923.

Singh, S., Bakshi, B.R., 2010. May. Enhancing the reliability of C and N accounting in economic activities: integration of bio-geochemical cycle with Eco-LCA. Proceedings of the 2010 IEEE International Symposium on Sustainable Systems and Technology. IEEE, pp. 1–6.

Skeldon, R., 1985. Population pressure, mobility, and socio-economic change in mountainous environments: regions of refuge in comparative perspective. Mountain Research and Development 233–250.

Srivastava, P., Singh, R., Tripathi, S., Raghubanshi, A.S., 2016. An urgent need for sustainable thinking in agriculture—an Indian scenario. Ecological Indicators 67, 611–622.

Tali, B.A., Ganie, A.H., Nawchoo, I.A., Wani, A.A., Reshi, Z.A., 2015. Assessment of threat status of selected endemic medicinal plants using IUCN regional guidelines: a case study from Kashmir Himalaya. Journal for Nature Conservation 23, 80–89.

Tarolli, P., Straffelini, E., 2020. Agriculture in hilly and mountainous landscapes: threats, monitoring and sustainable management. Geography and Sustainability 1 (1), 70–76.

Traoré, K.B., Gigou, J.S., Coulibaly, H., Doumbia, M.D., 2004. Contoured ridge-tillage increases cereal yields and carbon sequestration. 13th International Soil Conservation Organisation Conference:"Conserving Soil and Water for Society: Sharing Solutions". ISCO.

Vacchiano, G., Maggioni, M., Perseghin, G., Motta, R., 2015. Effect of avalanche frequency on forest ecosystem services in a spruce–fir mountain forest. Cold Regions Science and Technology 115, 9–21.

Van Aalst, M.K., 2006. The impacts of climate change on the risk of natural disasters. Disasters 30 (1), 5–18.

Wall, D.H., Nielsen, U.N., Six, J., 2015. Soil biodiversity and human health. Nature 528 (7580), 69–76.

Watson, R.T., Haeberli, W., 2004. Environmental threats, mitigation strategies and high-mountain areas. AMBIO: A Journal of the Human Environment 33 (sp13), 2–10.

Williams, J.R., 1975. Sediment routing for agricultural watersheds 1. JAWRA Journal of the American Water Resources Association 11 (5), 965–974.

Williams, M.W., Brooks, P.D., Seastedt, T., 1998. Nitrogen and carbon soil dynamics in response to climate change in a high-elevation ecosystem in the Rocky Mountains, USA. Arctic and Alpine Research 30 (1), 26–30.

Yadav, N., Singh, K.K., 2015. A Review Approach to USLE and Its Derivatives.

Zhou, T.F., 2008. Advances on petrogenesis and metallogeny study of the mineralization belt of the Middle and Lower Reaches of the Yangtze River area. Acta Petrolei Sinica 24, 1666–1678.

Microbial perspectives for the agricultural soil health management in mountain forests under climatic stress

Soumya Sephalika Swain[1], Yasaswinee Rout[1], Phani Bhusan Sahoo[2] and Shubhransu Nayak[1]

[1]Odisha Biodiversity Board, Forest, Environment and Climate Change Department, Government of Odisha, Regional Plant Resource Centre Campus, Bhubaneswar, Odisha, India
[2]Biosystematics Laboratory, PG Department of Botany, Maharaja Sriram Chandra Bhanjadeo University, Sri Ram Chandra Vihar, Baripada, Odisha, India

O U T L I N E

4.1 Introduction: Indian mountain forests and agriculture

Mountains are a major source of biodiversity and scenic beauty. It offers a natural habitat for wild flora and fauna. The undisturbed mountain forest is a repository for many wild crop plants as well as other biological diversity (https://www.fao.org/3/ca8642en/ca8642en.pdf). The majority of the people who live in the mountain region are impoverished and rely entirely on the mountain forest for a living. Forest dwelling farmers (mostly tribals) generally practice natural agriculture and primarily cultivate landraces and wild crops, which have potential genetic diversity and an abundant microbial population, in order to cope with the various abiotic and biotic stresses caused by global climate change (Biswas, 2003). The microbial diversity associated with crop species, soil, water, and natural biofertilizers has been demonstrated to be distinct and abundant, thereby maintaining agricultural soil health (Tilak et al., 2005).

Food insecurity affects one out of every eight people worldwide, as well as one out of every three people in rural mountain areas. It indicates that approximately 300 million people living in mountain areas face food insecurity, with approximately half of them suffering from chronic hunger (Li et al., 2019). India has the world's second largest population. According to the report "The State of Food Security and Nutrition, 2020," 189.2 million people in India are undernourished, accounting for 14% of the total population, and 20% of children under the age of five are wasting, which means having a low weight for their height. As a result, agricultural product production must increase to meet rising demand. To make matters worse, climate change has emerged as a major threat to agriculture and the forest ecosystem. Growing human population, degradation of natural habitats, introduction of high yielding varieties, and erosion of plant genetic diversity pose significant challenges to food security and environmental sustainability in the 21st century (https://www.fao.org/3/i6583e/i6583e.pdf). As a result, proper development and management of agrodiversity, particularly in mountain and hilly regions, is an urgent need. To address this major challenge, it is critical to comprehend the potential of natural agricultural practices practiced by mountain forest dwellers, as well as the scenario of microbial diversity associated with forest agroecosystems (Singh and Singh, 2017).

The Eastern Ghats, Western Ghats, Aravalli Ranges, Vindhaya Ranges, Satpura Range, Purvanchal Range, and the Himalayas are the seven mountain ranges found in India. These mountain areas have a diverse floristic and faunal diversity, which helps to strengthen agricultural diversity and local people's livelihoods. The state of Odisha is located on the east coast of the Deccan peninsula. Odisha State has a tribal population of more than 20% and is endowed with a diverse array of floral and faunal diversity. It has a variety of natural ecosystems, including agricultural habitats, mountain and hilly forests, fresh and saline wetlands, and open grasslands. Forests cover approximately 39.31% of the State's geographical area (Forest Survey of India, 2019), and it harbors four out of the sixteen major forest types occurring in India (Champion and Seth, 1968). With one "National Park" (another proposed), one "Biosphere Reserve," two "Tiger Reserves," and 19 "Wildlife Sanctuaries," approximately 10.37% of the total forest area and 5.19% of the total geographical area of the State have been protected. Since ancient times, tribal farmers in these areas have practiced mostly organic agriculture. Several crops have been cultivated in the diversified agro-climatic zones of the State which include cereals like rice, wheat, small millets, finger millets, foxtail millets, niger, etc. oil seeds like castor, linseed, mustard, sunflower, groundnut, cashew, pulses like arhar, black gram, green gram, lentils, jute, sugarcane, coconut and turmeric are of great significance (Panda and Palita, 2021). Several varieties of turmeric grown in Kandhamal district, various types of millets in Koraput, and maize varieties in Gajapati district have been identified as having special importance in both national and international markets. Furthermore, tribal farmers in Odisha cultivate a variety of tuber and root crops such as Elephant Foot Yam (EFY), Taro, Sweet potato, and Yam varieties, which serve as their primary source of income (Lenka et al., 2012). Odisha is thought to be the "secondary center of origin of cultivated rice," with approximately 15,000 traditional rice varieties out of 50,000 found worldwide (Manikanta et al., 2019). Years of tribal farmers' efforts in the conservation of 130 indigenous rice cultivars were finally recognized in 2012, when the Koraput district was designated as a "Globally Important Agricultural Heritage System (GIAHS)" by the Food and Agricultural Organization (FAO) (http://www.fao.org/giahs/giahsaroundtheworld/designated-sites/asia-and-the%20pacific/en).

With this in mind, an attempt was made in this chapter to discuss agricultural ecosystems and soil health in mountain/hill forests in the context of India and Odisha. The tribals' natural agricultural practices have been addressed, as has the microbial diversity associated with these forest agroecosystems, and their potential roles in soil health management have also been highlighted.

4.2 Impacts of climate change on agro-diversity in mountain forests

According to the Intergovernmental Panel for Climate Change (IPCC) (2007), Climate change is caused by increased concentrations of greenhouse gases in the atmosphere, such as carbon dioxide (CO_2), methane (CH_4), nitrous oxide (N_2O), and halocarbons. Greenhouse gases influence the absorbency, scattering, and emission of radiation in the atmosphere. It is also believed that climate change in mountain areas gradually affects agro-diversity in low-lying areas (Fort, 2015). Climate change has had a significant impact on agro-diversity in mountain regions in recent years (Negi et al., 2012). The people of the mountains are rapidly confronted with the consequences of climate change. They are adapting their land use systems for agriculture in response to climate variability. However, the time may soon come when the consequences will be out of control, affecting both local people and agriculture. Due to severe climate change, important wild relative crop plants and wild animals may become extinct. Mountain communities that rely on agriculture and live in a fragile ecosystem may face an immediate risk of crop loss and a decrease in livestock diversity, which could lead to increased malnutrition and hunger (Kohler and Maselli, 2009).

4.3 Importance of agriculture in mountain and hill forests

4.3.1 Mountain forests and the interlinked agro-diversity

The forest situated on about 2500 m or higher elevation above the sea level, regardless of the sharp changes in elevation of slopes within a short distance and slope with 300—2500 m of elevation is called mountain forest (Kapos et al., 2000). There are several hill forests in elevated landscapes below this range. Mountain forest covers approximately 900 million hectares of land on Earth, accounting for 20% of total forest cover (https://www.fao.org/sustainable-forest-management/toolbox/modules/mountainforests/basic-knowledge/en/). The biodiversity in mountain regions is similar to our mother nature's life insurance, with more biological diversity ensuring better environmental security. Mountain forests are very unique in nature, with a variety of topographic conditions, frequent slopes, different microclimatic conditions, and rich macro and micronutrients that support various ecological niches in forest ecosystems (Shimrah et al., 2012). Regardless of natural disasters such as landslides, cyclones, and floods, mountain forests continue to be more diverse in terms of constituents and biodiversity. In addition to these, there are numerous socioeconomic constraints such as a labor shortage, a poor agricultural management system, a lack of markets, and a lack of entrepreneurship (Sharma et al., 2017). Agriculture and forest diversity are inextricably linked and mutually beneficial. Water, pollinators, healthy soil with all micro and macronutrients, and microorganisms are all provided by the forest environment. Agriculture, in turn, ensures food security and aids in the preservation of ecosystem services (Bommarco et al., 2013). Invasive agriculture is one of the major causes of deforestation, but it is becoming unavoidable in order to meet the food needs of forest dwellers. Nonetheless, modern agricultural practices based on low-cost, high-yielding strategies endanger natural agro-biodiversity and ecosystem services. As a result, both agriculture and forest should be managed in a balanced manner using a holistic approach to reduce climate resilience (Erisman et al., 2016).

4.3.2 Agricultural production system and traditional agricultural practices by forest farmers

Over millennia, India has been a treasure trove of intellectual knowledge, spiritual wisdom, and biological wealth (Elango et al., 2020). Odisha, in particular, is a repository of traditional agricultural practices. Agro-diversity in Odisha's hilly regions includes landraces and traditional varieties of crops, wild crops, and so on, which are the major sources of genetic resources, as discussed in Section 4.1. Traditional agricultural practices in India include a variety of methods based on agro-climatic conditions, vegetation, and soil types (Patel et al., 2020). Various traditional agricultural practices reported across India include bun cultivation, bidd cultivation, bamboo drip cultivation, jhum cultivation, jhola (terrace) cultivation, myda system of rice cultivation, mulching, multi cropping, shifting cultivation, kharri method, utera cropping, lahee method, transplanting of crops, double transplanting etc. (Sharma et al., 2020). Aside from these, a variety of plant and animal products are used for disease and pest management, crop enhancement and improvement, and crop storage. These practices not only contributed to the health of the forest ecosystem, but also to the health of the forest soil. Fig. 4.1 depicts the significance of forest landscapes and traditional forest farmer practices. Some of the important agricultural practices used by forest dwellers, particularly tribal populations, are briefly discussed in the following sections.

4.3.2.1 Utilization of wild crops and landraces

Elevated region tribes have learned to use wild plants for food, medicine, and commerce. Wild edible crops, known for their high nutrient content, have played an important role in meeting the daily food needs of rural and tribal communities. Some traditional agricultural crop plants documented in various parts of the State includes indigenous varieties of rice, millets, barley, mays, oil seeds (*Helianthus* sp., *Brassica* sp., etc.), tuber plants (*Dioscorea* sp., *Amorphophallus* sp., *Colocasia* sp., *Zingiber* sp., *Curcuma* sp.) (Fig. 4.4). Tuber plants belonging to the family *Dioscoreaceae* (Wild Yam) are commonly utilized as food supplements among the indigenous population (Panda and Palita, 2021).

4.3.2.2 Insect and pest management

The tribal people use their traditional knowledges to replace synthetic chemicals for crop protection from insect pests. They apply plants parts like Karanj leaf (*Pongamia pinnata*), Neem leaf (*Azadirachta indica*), Arakha leaf (*Calotropis procera*), Karada leaf (*Cleistanthus collinus*) (Meshram, 2010) and also cow urine, cow dung, ash of leaves to manage various insects and pests.

4.3.2.3 Guarding crop fields from wild animals

Agricultural fields in the interior forest lack modern facilities for animal protection. To protect their crop fields from wild animals, the local tribal people build bamboo and straw tree houses. Other traditional practices used by tribals to protect their crop fields include the construction of fences made of bamboo sticks and other plant stems around crop fields and the hanging of polythene bags around crop fields (Observation by Odisha Biodiversity Board).

4.3.2.4 Organic manures

Farmers in forested areas primarily use organic manures rather than costly chemical fertilizers. Cow dung compost, also known as Farm Yard Manure (FYM), is primarily used in crop fields to increase and maintain soil fertility (Qamar and Saif, 2018). Furthermore, the tribal people continue to develop various organic manures using their traditional knowledge. This organic fertilizer's main ingredients are cow dung, urine, flours, fertile soil, and molasses.

FIGURE 4.1   Mountain agriculture and traditional practices in Odisha, India (A) Agricultural landscape of mountain/hill forest in Kuldiha Wildlife Sanctuary, Balasore, (B) Terrace farming in Mahendragiri hills (Proposed Biosphere Reserve), Gajpati, (C) Sabri river, Gupteswar Reserve Forest (proposed Biodiversity Heritage Site), Koraput, (D) Tree house to guard agricultural field, (E) Compost yard in a village of Badrama-Khalasuni Wildlife Sanctuary, Sambalpur, and (F) Traditional storage practice of maize in Koraput.

4.3.2.5 Intercropping and crop rotation

The diversity of agro-ecosystems in mountain forests is thought to be more resistant to anthropogenic and natural threats while also yielding higher production. Crop diversity helps to regulate natural defenses and maintain nutrient balance in the soil, protects against pests, and provides resilience to weather extremes (Scherr and McNeely, 2008; Thrupp, 2002). Rotation with other crops after monocrop cultivation results in nutrient restoration and microbial diversity of crop soil (Gaudin et al., 2015; Patel et al., 2020). Crop rotation is traditionally practiced by forest tribals of Odisha.

4.3.2.6 Traditional storage techniques

Despite the fact that the green revolution increased crop production in India, approximately 10% of agricultural produce is lost after harvest (Dhaliwal and Singh, 2010; Prakash et al., 2016). Tribal farmers in the forest have adopted completely eco-friendly traditional storage structures made of bamboo sticks coated with cow dung to repel pests. Furthermore, they use natural ingredients such as sal resin (jhuna) to smoke during daily rituals to control insects and storage pathogens (Nayak et al., 2020a).

Thus, the agricultural ecosystem in mountain forests as a whole is a blessing for forest dwellers, particularly tribal people. The natural and non-invasive crop management method also ensures that no alien elements are introduced into the healthy forest soil. The effect of such natural means on mountain soil health and soil microbial diversity, as well as vice versa, will be discussed in the following sections.

4.4 Sustenance of soil health by forest agriculture system

4.4.1 Healthy soil and microorganisms

Agriculturally healthy soil refers to its function within a natural and managed ecosystem, which is critical for long-term productivity and environmental sustainability. Excessive use of chemical fertilizers, synthetic pesticides, and chemical plant growth hormones in agricultural practices has been one of the primary causes of agricultural soil health degradation. The assessment of soil health is directly dependent on various indicators, which are broadly classified into three components: physical structure, chemical composition, and biological properties as depicted in Fig. 4.2.

Soil health is largely determined by microbial diversity and biomass, soil respiration, enzyme dynamics, and other parameters, with diversity of various microbial groups playing an important role (Nielsen et al., 2002). Soil microbes, which include millions of bacteria, fungi, protozoa, algae, and other microorganisms, play a wide range of important roles, from cycling macro and micro nutrients to protecting plants from the harmful effects of abiotic and biotic stresses (Bardgett and Van Der Putten, 2014; Nayak et al., 2020b). Microbial diversity of soil depends upon the physical property and the amount of "soil organic matter" (SOM). A schematic representation of soil microorganisms is given in Fig. 4.3.

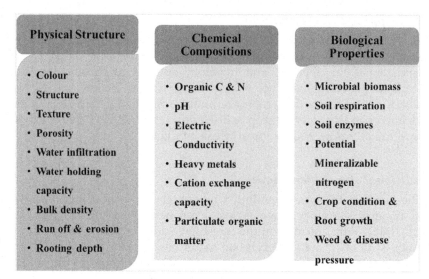

FIGURE 4.2   Physico-chemical and biological components of a healthy soil.

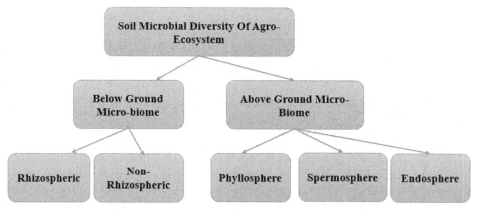

FIGURE 4.3   Schematic representation of various microbe-plant association in crop plants.

4.4.2 Wild crops, landraces and associated microbial diversity

Crop wild relatives are cultivated crops' wild ancestors. The genetic potential of wild crops is greater than that of cultivated crops, allowing them to withstand a variety of biotic and abiotic stresses. Many traditional and indigenous crop plants, as well as their wild relatives, have a higher nutritional value than domesticated crops. Several studies have proved that wild food possesses various micro and macro-nutritional properties (Samson and Pretty, 2006; McMichael et al., 2007; Vincetti et al., 2008) along with essential vitamins. Wild crops are required for natural resource stability, fortification of native agriculture, and food crop production (Slikkerveer, 1994).

TABLE 4.1   Rhizospheric Bacterial and Fungal diversity isolated from wild crops and landraces.

Wild crop	Species	Microbial association	References
Rice	*Oryza rufipogon*	*Rhizobium* sp., *Janthinobacterium* sp., *Pedobacter* sp., *Acidiphilum* sp., *Flavobacterium* sp., *Brevundimonas* sp., *Stenotrophomonas* sp., *Cellvibrio* sp., *Cohnella* sp., *Leadbetterella* sp.	Xu et al. (2020)
		Gibberella sp., *Mortierella* sp., *Arthrinium* sp., *Pyrenochaetopsis* sp., *Pleosporales* sp., *Saccharicola* sp., *Neosetophoma* sp.	
		Glomus sp.	
Barley	*Hordeum spontaneum*	*Paenibacillus polymyxa*, *Bacillus pumilus*, *Bacillus cereus*, *Bacillus megaterium*	Timmusk et al. (2011)
Yam	*Dioscorea wallichi*	*Aspergillus* sp., *Penicillium* sp., *Trichoderma* sp.	Sivakumar et al. (2009)

Wild crop plants cultivated by tribal forest farmers harbor a wide diversity of microorganisms that may directly promote plant growth or indirectly can create conducive environment for plant sustenance (Yadav et al., 2018) and can be broadly classified into three different categories: rhizospheric, endophytic and phyllospheric. Among these, the root microbiome is a major determinant of plant vegetative and reproductive growth, as well as plant health management. Microorganisms complement and supplement host plants by improving nutrient uptake, protecting against pests and pathogens, and assisting in the tolerance of abiotic stresses such as submergence, drought, salt, frost, desiccation, and so on (Berendsen et al., 2012). The volume of soil adhering to the rhizoplane is referred to as the rhizosphere. It is regarded as a diversity hotspot of microorganisms that greatly benefit plants (Brimecombe et al., 2001). The association of soil microbiota with plant roots, as well as their configuration and activity, are solely determined by rhizodeposition or root exudates. Rhizosphere microbial diversity exceeds bulk density, with Rhizobacteria constituting the largest group of all microorganisms (Brimecombe et al., 2001; Bahadur et al., 2017; Verma et al., 2017; Kumar et al., 2017). Table 4.1 shows some of the rhizospheric microorganisms of wild crops.

Endophytic microorganisms have long been thought to be the biochemical factory, producing a diverse range of bioactive natural compounds. Endophytes can influence host plant growth, maturation, flowering, and fruiting through a variety of mechanisms such as atmospheric nitrogen fixation, mineral solubilization, and the production of siderophores, phytohormones, and other secondary metabolites. Because wild crop relatives can withstand abiotic and biotic stresses and can impart better traits to cultivated crops, they should be prioritized for use in organic agriculture in hill forests. Diversity of such potential microorganisms is presented in Table 4.2. Fig. 4.4 represents some important wild crops and landraces cultivated by tribal forest farmers of Odisha.

4.4.3 Importance of macrofungi in nutrient enrichment of mountain soil

Fungi are one of the most abundant and diverse microbial biomasses found in forest soil, mountains, and hills. Fungi are one of the most successful and long-lasting inhabitant

TABLE 4.2 Endophytic bacterial and fungal diversity isolated from wild crops and landraces of agricultural and horticultural crop plants.

Species	Plant part used	Organism	Associated microorganism	Source
Barley				
Hordeum murinum	Seed	**Fungi**	*Cladosporium herbarum, Alternaria alternate, Humicola grisea* sp.	Murphy et al. (2018a)
Hordeum secalinum	Seed		*Penicillium* sp.	Murphy et al. (2018b)
Brassica				
Brassica barrelieri	Seed	**Bacteria**	*Methylobacterium oryzae, Methylobacterium fujisawaense, Sphingomonas paucimobilis, Pseudomonas lactis*	Roodi et al. (2020)
Brassica elongata			*Methylobacterium fujisawaense*	
Brassica gravinae			*Methylobacterium phyllosphaerae*	
Brassica incana			*Methylobacterium phyllosphaerae*	
Brassica juncea			*Methylobacterium fujisawaense, Stenotrophomonas rhizophila, Pseudomonas lactis, Bacillus mycoides, Plantibacter flavus*	
Brassica napus			*Methylobacterium phyllosphaerae, Methylobacterium fujisawaense, Kocuria palustris, Caulobacter mirabilis*	
Brassica nigra			*Methylobacterium fujisawaense, Pseudomonas lactis, Sphingobium yanoikuyae,*	
Brassica rapa			*Methylobacterium fujisawaense, Methylobacterium extorquens, Novosphingobium clariflavum, Novosphingobium resinovorum, Brevundimonas vesicularis*	
Brassica oleracea			*Stenotrophomonas rhizophila, Pseudomonas lactis*	
Maize				
Zea mays ssp. *mays*	Seed	**Bacteria**	*Paenibacillus* sp.	Mousa et al. (2015), Shehata et al. (2016)
	Root		*Burkholderia gladioli*	
Zea diploperennis,	Seed		*Paenibacillus* sp.	
	Root		*Burkholderia gladioli*	
Zea mays ssp. *parviglumis*	Root		*Burkholderia gladioli*	
			Citrobacter sp., *Paenibacillus* sp.	
Peanut				
Arachis duranensis	Root nodule		*Bradyrhizobium* sp., *Agrobacterium* sp., *Burkholderia* sp., *Herbaspirillum* sp.	Chen et al. (2014)
Rice and Landraces				
Oryza alata	Stem	Bacteria	*Azospirillum amazonensis, Flavobacterium gleum*	Elbeltagy et al. (2000)
	Seed		*Pantoea ananas*	
Oryza barthii	Stem		*Herbaspirillum rubrisubalbicans*	
Oryza brachyantha	Stem		*Methylobacterium* sp.	

(Continued)

TABLE 4.2   (Continued)

Species	Plant part used	Organism	Associated microorganism	Source
Oryza breviligulata	Root		*Bradyrhizobium* sp.	Chaintreuil et al. (2000)
Oryza eichingeri	Leaf, Stem, Root		*Burkholderia* sp., *Stenotrophomonas* sp., *Rhizobium* sp., *Paenibacillus* sp.	Banik et al. (2016)
Oryza glandiglumis	Stem		*Azospirillum lipoferum*	Elbeltagy et al. (2001)
Oryza granulata	Leaf		*Pseudomonas* sp.	Koomnok et al. (2007)
	Stem		*Azospirillum* sp., *Herbaspirillum* sp. *Beijerinckia* sp.	
	Root		*Azospirillum* sp., *Herbaspirillum* sp.	
	Root	Fungi	*Exophiala* sp., *Cladophialophora* sp., *Harpophora* sp., *Periconia macrospinosa*, *Ceratobasidium/Rhizoctonia* complex	Yuan et al. (2010a)
			Harpophora oryzae	Yuan et al. (2010b)
Oryza latifolia	Stem	Bacteria	*Methylobacterium* sp.	Elbeltagy et al. (2000)
Oryza longiglumis	Stem		*Methylobacterium* sp.	
Oryza meridionalis	Seed		*Herbaspirillum seropedicae*, *Methylobacterium* sp.	
	Stem		*Methylobacterium* sp.	
Oryza minuta	Stem	Bacteria	*Methylobacterium* sp.	
	Root		*Azoarcus* sp.	Engelhard et al. (2000)
Oryza nivara	Root		*Gallionella* sp.	
	Leaf		*Azospirillum* sp., *Pseudomonas* sp.	Koomnok et al. (2007)
	Stem		*Azospirillum* sp., *Beijerinckia* sp.	
	Root		*Azospirillum* sp., *Herbaspirillum* sp. *Beijerinckia* sp.	
Oryza officinalis	Stem		*Herbaspirillum* sp.	Elbeltagy et al. (2001), Engelhard et al. (2000), Koomnok et al. (2007)
	Root		*Azoarcus* sp., *Sphingomonas paucimobilis*	
Oryza ridleyi	Stem		*Rhodopseudomonas palustris*	
Oryza rufipogon	Stem		*Methylobacterium* sp., *Sphingomonas adheasiva*	
			Azospirillum brasilense, *Enterobacter cancerogenus*, *Herbaspirillum seropedicae*	
			Azospirillum sp., *Herbaspirillum* sp. *Beijerinckia* sp.	
	Leaf	Bacteria	*Pseudomonas* sp., *Azospirillum* sp.	
	Root		*Beijerinckia* sp., *Azospirillum* sp., *Herbaspirillum* sp.	
	Stem, nodal roots		*Citrobacter amalonaticus*, *Klebsiella* sp., *Enterobacter* sp., *Pantoea* sp., *Phytobacter diazotrophicus*	Zhang et al. (2008)
	Stem		*Bacillus cereus*, *Cellulosimicrobium cellulans*, *Stenotrophomonas maltophilia*, *Microbacterium* sp., *Proteus* sp., *Staphylococcus* sp., *Erwinia* sp., *Ochrobactrum* sp., *Enterobacter* sp.	Borah et al. (2018)
Shampakatari Ushapari	Root	Fungi	*Claroideoglomus* sp., *Paraglomus* sp., *Glomus* sp.	Parvin et al. (2021)

(Continued)

2. Soil microbial processes and nutrient dynamics

TABLE 4.2 (Continued)

Species	Plant part used	Organism	Associated microorganism	Source
Strawberry				
Fragaria vesca	Root	Fungi	*Humicola* sp., *Volutella rosea*, *Trichoderma* sp., *Paraphoma* sp., *Dactylonectaria* sp., *Cadophora* sp.	Yokoya et al. (2017)
Wheat				
Aegilops sharonensis	Seed, Stem	Fungi	*Aspergillus* sp., *Chaetomium* sp., *Alternaria infectoria*, *Alternaria chalatospora*, *Chaetomium* sp.	Ofek-Lalzar et al. (2016)
Triticum aestivum landrace	Seed, Seedling	Bacteria	Pseudomonadaceae, Comamonadaceae, Enterobacteraceae	Ozkurt et al. (2020)
Triticum boeoticum			Pseudomonadaceae, Comamonadaceae, Halomonadaceae, Vibrionaceae,	
Triticum dicoccoids			Pseudomonadaceae, Comamonadaceae, Halomonadaceae, Vibrionaceae, Oxalobacteraceae, Bacillaceae.	
	Seed, Stem	Fungi	*Aspergillus* sp., *Chaetomium* sp., *Alternaria infectoria*, *Alternaria chalatospora*, *Cladosporium* sp.	Ofek-Lalzar et al. (2016)
Triticum urartu	Seed, Seedling	Fungi	Halomonadaceae, Vibrionaceae, Bacillacea, Xanthomonadaceae, Pseudomonadaceae,	Ozkurt et al. (2020)
Triticum dichasians	Seeds	Fungi	*Neotyphodium* sp.	Marshall et al. (1999)
Triticum tripsacoides				
Triticum cylindricum			*Acremonium* sp.	
Triticum columnare				
Triticum monococcum				
Triticum neglecta				
Triticum recta				
Triticum triunciale				
Triticum turgidum				
Triticum umbellatum				
Triticum dicoccoides, Aegilops sharonensis	Stem	Fungi	*Alternaria infectoria*, *Cladosporium ramotenellum*, *Stemphylium vesicariam*	Sun et al. (2020)
Triticum aestivum landrace	Seed	Fungi	Pleosporales, Pleosporaceae, Aureobasidiaceae	Ozkurt et al. (2020)
Triticum boeoticum			Trichosphaeriaceae, Chaetomiaceae, Pleosporaceae	
Triticum dicoccoids			Saccharomycetaceae, Aspergillaceae	
Triticum urartu			Trichosphaeriaceae, Chaetomiaceae	
Tuber crops				
Dioscorea zingiberensis	Rhizome	Fungi	*Nectria* sp., *Fusarium redolens*, *Rhizopycnis* sp., *Penicillium* sp., *Alternaria longissima*	Xu et al. (2008)

2. Soil microbial processes and nutrient dynamics

FIGURE 4.4 "Wild Relatives" of crop plants cultivated in Odisha for food and fodder. (A) tribal market for wild crops (B) *Amorphophallus paeoniifolius*, (C) *Curcuma longa*, (D) *Momordica charantia*, (E) *Solanum torvum*, (F) *Solanum violaceum*, (G) *Cajanus cajanifolius*, and (H) *Vigna adenantha*.

of mountain soil due to their high level of plasticity and ability to adapt to different forms. They primarily grow on dead plant materials and, in some cases, are pathogenic to plants. Many fungi species act as bio-absorbents of toxic metals (Baldrian, 2003). Soil fungi are classified into three types based on their role in the ecosystem: biological control agents, ecosystem regulators, and decomposers (Swift, 2005; Gardi and Jeffery, 2009). In mountain forest ecosystems, nearly half of the net aboveground primary production returns to the forest soil as litter (Swift et al., 1979). This decay is carried out by macro and microfungi, as well as a decomposing bacterial community and ground-dwelling lower invertebrates (Kaspari et al., 2008; Nayak and Mukherjee, 2015). Decay can be classified as in-ground or above-ground. Because of the presence of adequate moisture and an abundance of wood-inhabiting microbes, including fungi, in-ground decay occurs quickly, whereas above-ground decay is more difficult due to a lack of moisture and microorganisms on the ground (Goodell et al., 2020). Decomposition of dead wood and litter is critical for nutrient cycling and freeing up the ecosystem's limited physical space. One of the most important decomposers of plant cell wall material is macrofungi. Their mycelia structures are capable of externally digesting their food by secreting a variety of digestive enzymes and other chemicals into the substrate in which they grow. Water is required for both enzyme secretion and the diffusion of food material into the fungus during the external digestion process. Cellulose, a polysaccharide of glucose molecules, is the most common organic compound on the planet, accounting for approximately 50% of plant cell walls (Boer et al., 2005). Lignin, a complex polymer of phenolic compounds, accounts for 18%−35% of the cell wall of wood. Furthermore, the intimate integration of lignin with the holocellulose components makes the wood resistant to degradation (Ten Have and Teunissen, 2001). Lignin, a complex polymer of phenolic compounds, accounts for 18%−35% of the wood cell wall. Furthermore, the way lignin is intimately integrated with the holocellulose components makes the wood resistant to degradation (Goodell et al., 2020). Although it is an essential component of the plant cell, it is frequently regarded as a weak link in the wood because fungi attack hemicelluloses in the early stages of decay (Winandy and Morrell, 1993; Curling et al., 2002). Degrading fungi are classified into different categories based on the appearance of the degraded wood, such as brown rot, white rot, and so on. The most common types of fungi are brown rot fungi. Ascomycota and Basidiomycota are the most common producers of polysaccharide degrading enzyme (Osińska-Jaroszuk et al., 2015). The cellulose degrading capacity of Glomeromycota is limited (Kowalchuk et al., 2002; Talbot et al., 2008) but was found greater in recent studies.

Fungal succession on dead plant material has been studied profoundly. Chapela and Boddy (1988) reported that after drying up of the dead wood to a certain extent, the first fungal colonies present on the low levels of living tree grow actively on the deadwood. Those mostly include soft rot fungal groups belonging to Ascomycota, such as *Xylaria* sp. followed by brown rot fungi and white rot fungi belonging to Basidiomycota. Wood decaying fungi are unique in nature as they can decompose lignified cell walls (Blanchette, 1991). Basidiomycetes, also known as polypores, are the most common wood decaying fungi. Polypores are mostly saprophytic and come in a variety of shapes and sizes that are hard, tough, corky, leathery, or woody. Because dry woods are not conducive to fungal growth, the wood must contain more than 28% moisture. Although Ascomycota are the first to colonize leaf litter, the decaying process is a little faster than in the woods (Snajdr et al., 2011).

One of the most important foods for tribal and indigenous communities is wild edible mushrooms. The nutritional value of wild edible mushrooms includes a high concentration of proteins, vitamins, fiber, and macro-micronutrients. These mushrooms are also important in the decomposition of dead wood and litter, and are thus regarded as solid waste managers in the forest ecosystem. They return carbon, minerals, and nutrients to the forest soil while also improving soil health. Furthermore, for better nutrient uptake, most land plants are associated with these macrofungi. As a result, exploiting these symbioses in the natural environment has high environmental and economic value. Overall, the decaying and degrading activities of these forest macrofungi result in fine nutrient deposition in the soil, which gradually flows to agricultural lands, increasing soil fertility. Some of these profound decomposers of forest soil are represented in Fig. 4.5.

FIGURE 4.5 Decomposer Macrofungi of Forest Ecosystems in Odisha, India (A) *Daldinia concentrica*, (B) *Dacryopinax spathularia*, (C) *Xylaria filliformis*, (D) *Ganoderma* sp. (E) *Microporous xanthopus*, (F) *Pycnoporous cinnabarinus*, (G) *Auricularia auricula-judae*, (H) *Schizophyllum commune*, and (I) *Xylaria* sp.

4.4.4 Significance of natural water streams, rivers, and rain wash in mountain agriculture

Rivers, streams, and seasonal rains are fluid mixtures of water, sediments, aquatic organisms, and diverse vegetation. The shape, width, and course of streams and rivers change over time as a result of soil erosion, sediment deposition, and sediment transport through the river and banks. Throughout, however, the dynamic equilibrium between sediment discharges, river slope, sediment load, and sediment size is maintained (Lane, 1955). In this way, ubiquitous microorganisms have evolved in diverse and fascinating ways in a variety of environments, reflecting their localized niche. Run-off water from flash rains and long-duration rains contains root exudates and surface soil particles, as well as topsoil microbes. These water streams are used for irrigation in the agricultural fields by farmers in the mountains and hills. This process expands the agricultural soil's microbial population and diversity.

4.5 Threats to mountain forest agriculture

Globally, agricultural diversity is under threat due to intensive agricultural practices such as the overuse of chemical fertilizers and synthetic pesticides, the replacement of landraces with High Yielding and hybrid crop varieties, the decline in the number of natural pollinators, urbanization, anthropogenic activities, and so on. In addition, climate change causes the emergence of new virulent pathogen strains and the infestation of aggressive pests, among other things. Crop production is also reduced as a result of the erratic weather patterns (Patterson et al., 1999; Rosenzweig et al., 2001; Bale et al., 2002; Malla, 2008; Khouri et al., 2011; Negi et al., 2012). Land degradation is a major problem in mountain agriculture and, as a result, a source of concern for mountain soil. Natural disasters, as well as anthropogenic activities such as physical degradation, chemical contamination, and interference in biological processes, have all contributed significantly to the degradation process. Some of the major natural issues are fragile geological structures, widespread forest fires, avalanches, dry landslides, and flooding, while deforestation, construction work, unscientific farming, overgrazing, shifting cultivation, timber and fodder production issues, and soil erosion (Fig. 4.6) (Acharya and Kafle, 2009). Climate change

FIGURE 4.6 Various threats to agriculture in mountain agriculture system in Odisha, India. (A) Forest fire in Gupteswar Reserve Forest, (B) Aftermath of forest fire, and (C) Deforestation.

and weather cycle has a large impact on agriculture mostly in developing countries. Global agriculture is largely affected by irregular changes in temperature, erratic rainfall patterns and CO_2 level leading to non-seasonal rainfall, droughts, floods, soil erosion etc. These changes lead to reduction in soil health and the effects are more in tropical regions. Effect of climate change is much lower globally than regional (Kumar et al., 2004).

4.6 Microbial potential for soil health restoration

Soil health is closely related to sustainable agriculture, where the extent of soil microbial diversity and their efficiency in biogeochemical activities determine the level of soil health and functional biodiversity (Dastager et al., 2011). Microbial communities are everywhere, play an active role in biological functioning, and provide numerous ecosystem services. They are, however, sensitive to environmental changes (Bouchez et al., 2016; Liang et al., 2019; Roy et al., 2020).

4.6.1 Organic agriculture and the wide microbial diversity

Farmers in forests, as discussed in previous sections, use natural ingredients for various agricultural operations such as soil preparation, sowing, manuring, irrigation, weeding, pest control, fertility, harvesting, storage, and so on. These practices are very similar to those used in organic agriculture. Traditional agricultural practices have been tampered with or, in large part, dominated by the use of harmful chemical agro products, which is why modern organic farming has developed (Adamchak, 2020). All of these activities have the potential to use microorganisms to create a more sustainable agricultural ecosystem. The key to successful crop production is soil fertility. A farmer may be able to manage the physical structure of the farm land and even its chemical composition directly, but it is the most difficult task to observe, manage, and maintain the biological components of the soil, particularly the microbial diversity, in difficult agroecosystems such as those found in mountains. Management of microbial diversity in agricultural soil could result in the successful cycling of critical nutrients such as carbon and nitrogen, as well as the breakdown or decomposition of crop residues, resulting in increased plant growth (Doran, 2002; Godfray et al., 2010). Several bacterial genera like *Rhizobium* sp., *Azotobacter* sp., *Azospirillum* sp., *Pseudomonas* sp., *Pantoea* sp., and *Bacillus* sp. regarding have the capability to produce siderophores and other "Plant Growth Promoting (PGP)" biochemicals and now have been utilized in organic agriculture (Bertola et al., 2021).

4.6.2 Biofertilizers in organic agriculture in hill forests

Living microorganisms are active components of biofertilizers, which are expected to improve plant growth, yield, soil health, and ecosystem health (Vessey, 2003). The base material consists primarily of agricultural residues and waste, as well as various dead (mostly lower) organisms and living organisms, and it is economical, eco-friendly, efficient, and productive (Sahoo et al., 2013). Agricultural wastes include shredded plant parts such as leaves, stalks, barks, flowers, fruits, leftover vegetables, and green manures, which become a natural source of organic nutrients in the soil and thus the cheapest one that mountain farmers can use to improve agricultural soil fertility (Lim and Matu, 2015). Furthermore, rhizosphere microbial strains with Plant Growth Promotion (PGP) activities such as atmospheric nitrogen fixation, organic phosphate solubilization, and so on can be used in the production of biofertilizer (Sahoo et al., 2013). The most important microbial group used in the production of biofertilizer include fungi, bacteria, cyanobacteria, micro-algae etc.

Microbial biofertilizer is an important component of "Integrated Nutrient Management (INM)," as these are low-cost renewable sources for agricultural sustainability. These add a variety of nutrients to the soil, increasing soil fertility and thus crop yield (Edgerton, 2009). When microbial biofertilizers applied to the soil, they can provide important macronutrients like nitrogen, phosphorous, calcium, potassium, magnesium and sulfur and micronutrients like zinc, copper, iron, boron, and molybdenum. Because it maintains soil fertility and sustainability, biofertilizer has become an essential component of organic farming. The following sections discuss various microorganisms used in biofertilizers and are shown in Table 4.3.

4.6.2.1 *Nitrogen fixers* Rhizobium

These symbiotic organisms having association with legumes only belong to the family Rhizobiaceae and have the ability to fix 50–100 kg of Nitrogen per hectare. They are mostly associated with pulse legumes such as black grams, red grams, chickpeas, peas, lentils, and so on, oil-seed legumes such as groundnut and soybean, and forage legumes such as lucerne and berseem. The colonization of legume plants results in the formation of root nodules, which are lumpy outgrowths where atmospheric nitrogen is fixed and gradually transferred to the plants in inorganic forms (Mishra et al., 2013).

4.6.2.2 *Azospirillum*

The heterotropic, symbiotic Azospirillum species belonging to family Spirilaceae have the ability to fix about 20–40 kg of Nitrogen per hectare in addition to the production of growth regulating biochemicals. Because *Azospirillum* grows and fixes nitrogen on organic acid salts such as malic and aspartic acid, it can form symbiotic relationships with a wide range of plants, particularly those that use the C4-dicarboxylic acid pathway in photosynthesis, also known as the "Hatch and Slack pathway" (Mishra et al., 2013).

4.6.2.3 *Azotobacter*

These free-living, aerobic, heterotrophic bacteria are members of the Azotobacteriaceae family. They can thrive in alkaline to neutral soils and produce anti-fungal substances that inhibit the vegetative and conidial growth of a large number of plant pathogenic fungi, primarily in the rhizosphere, preventing seedling mortality to some extent (Mishra et al., 2013).

4.6.2.4 *Blue Green Algae (Cyanobacteria) and* Azolla

The "Blue Green Algae (BGA)" group which can fix about 20–30 kg of atmospheric Nitrogen per hectare nitrogen is classified into eight families. This autotrophic microbial community produces a variety of growth hormones such as gibberellic acid, auxin, and

TABLE 4.3 Microbial biofertilizers associated with different crops (Debnath et al., 2019).

Type of crops	Crop plants	Microbial biofertilizers	Roles
Cereal grains	Rice	*Bacillus thuringiensis*	Lengthening of shoots
		Cyanobacteria	Enhance production
		Azolla sp.	Grains and straw yields have increased and weed emergence has decreased
	Wheat	*Aspergillus niger*	A better growth rate and higher uptake of P
		Azospirillum brasilense	Improved plant growth, root depth, and fresh weight of roots and shoots, as well as increased nutrient use efficiency.
		Azotobacter sp.	Enhanced grain yield
	Maize	VAM	Plants with increased P concentrations
		Azospirillum brasilense	Growth of plants and improvement of biochemical traits.
Common pulses	Bean, Common bean, faba bean, Peas, Pigeon pea, Cowpea, Fenugreek	*Rhizobium* sp. *Bradyrhizobium* sp.	Improved grain and straw yield, vegetative growth parameters, shoot mineral content, harvest index, agronomic fertilizer efficiency, nodule dry weights, and enhanced seed quality.
	Chickpea	*Pseudomonas* spp.	Nodulation and growth stimulation of plants
Millets	Pearl millet,	*Azotobacter* sp.	An increase in plant height, dry matter accumulation, number of effective tillers, grain per ear, and grain and stover yields were achieved
	Foxtail millet	*Azospirillum lipoferum*	Improvements in seed weight, panicle, shoot and root dry weight, total N content of the shoot, and root as well as grain yields.
	Amaranth	*Bacillus* spp.	Improved nutrient use efficiency
Horticultural Crops	Cauliflower	*Azotobacter* sp.	Enhanced morphological characteristics and yield
Economically important crops	Jatropha	VAM	Reduced salt stress
	Mulberry	*Azotobacter* sp.	Silk filament length, cocoon weight, shell weight, and shell ratio have increased as well increased.

Source: Courtesy: Debnath, S., Rawat, D., Mukherjee, A.K., Adhikary, S. and Kundu, R., 2019. Applications and constraints of plant beneficial microorganisms in agriculture. Biostimulants in Plant Science.

Indole Acetic Acid (IAA). These microbial biofertilizers or biomanures, in addition to enriching the soil with Nitrogen, also add significant amounts of other important macro and micronutrients such as P, K, S, Zn, Fe, and Mb (Mishra et al., 2013).

4.6.2.5 Phosphate solubilizers

Various species of bacteria and fungi, such as *Aspergillus* sp., have been studied and shown to have the ability to solubilize insoluble rock phosphate, inorganic phosphates such as dicalcium phosphate, tricalcium phosphate, hydroxyapatite, and others and convert them to plant available form. For this purpose, both anaerobic and aerobic bacterial strains were used (Mishra et al., 2013; Nayak et al., 2020c).

4.6.2.6 *Vesicular Arbuscular Mycorrhiza (VAM)*

The VAM fungi have the ability to solubilize phosphates, increasing phosphorus availability to plants. Furthermore, VAM strengthens plants' resistance to pathogens, pests, and adverse climatic conditions (Mishra et al., 2013).

4.6.3 Microbial biopesticides; a better choice for mountain agriculture

Biopesticides are agricultural pesticides that contain biocontrol agents such as plants, animals, bacteria, or certain mineral compounds. They are used to manage and prevent pests and pathogens. Natural organisms and/or substances derived from natural materials, such as genes or metabolites, may be used as active ingredients to inhibit pests through specific biological effects rather than broad-spectrum synthetic pesticides (Sporleder and Lacey, 2013). To increase crop production volume and economic value, synthetic chemicals are used globally for plant production and disease management, which violates the "organic farming" framework adopted by many hilly states of India (Trivedi et al., 2012). As a result, farmers all over the world are using biopesticides to combat those pests. Biopesticides have higher target specificity than synthetic chemicals and help beneficial microbes survive and reproduce in treated crops (Sporleder and Lacey, 2013). They are proficient and safe for humans, pose less risk to non-target organisms, and leave no residue in crops. In mountain regions, the possibility of successful cultivation through the use of microbial inoculants as biocontrol agents could be extremely rewarding (Singh et al., 2017). As a result, over the last three decades, microbial pesticides have been developed as "Biological Control Agents (BCA)." Globally, numerous research projects are being carried out to develop biological control measures using new microbial inoculants. Some of the most effective inoculants are listed below.

4.6.3.1 *Bacterial inoculants*

Bacillus thuringiensis (Bt), belonging to family Bacillaceae has been the most experimented and commercially exploited bacterial species. The insecticidal mechanism involves the biosynthesis of bacterial "crystal toxins" (Cry & Cyt) along with "parasporal bodies" during the sporulation phase and some other bacterial toxins and virulence factors, synthesized and released during vegetative growth of the bacterium (VIP) (Jurat-Fuentes and Jackson, 2012). Different strains of bacteria are characterized by the production of a variety of insecticidal proteinaceous toxins and other strain specific insecticidal biomolecules due to the presence of a few toxin gene sequences (Pigott and Ellar, 2007). *Lysinibacillus sphaericus* (formerly known as *Bacillus sphaericus*) acts against blackflies and mosquitoes. The mechanism involves the production of crystal proteins like BiN A & Bin B and mosquitocidal toxin Mtx (Charles et al., 2000). Insecticidal toxins produced by bacterial species belonging to *Paenibacillus* genus showed high molecular homology with cry toxins of *B. thuringiensis*. *Brevibacillus laterosporus* has been found to be having broad spectrum of pesticidal activity. These bacteria possess swollen sporangium with a "canoe" shaped parasporal body and have several virulence factors (Ruiu, 2013; Marche et al., 2017; Marche et al., 2018). Gamma proteobacteria, a heterogenous group, includes several entomopathogenic bacterial species like endosymbionts of insecticidal nematodes *Photorhabdus*, *Xenorhabdus* whereas *Serratia* sp. having toxin mediated insecticidal mechanism (French-Constant and Waterfield, 2006; Hurst et al., 2000). *Yersinia entomophaga*, non-spore bearing species, produce toxin complex Yen-Tc produces both toxin and chitinases (Landsberg et al., 2011). *Pseudomonas entomophila*, holds a toxin secretion system that acts by ingestion (Vodovar et al., 2006).

Chromobacterium subtsugae, a beta propteo bacterium, is a commercially successful strain that possesses insecticidal activity against a no. of species like *Diptera, Hemiptera, Lepidoptera* and *Colepotera* (Martin et al., 2007). Another Betapropteo bacterium, *Burkholderia rinojensis* kill a diverse chewing and sucking type insects and mites by ingestion and contact method. An insecticidal toxin "Macrocyclic lactone derivative" produced by Streptomyces sp., damages the "peripheral nervous system" of insects. Natural and semisynthetic derivatives of Spinosins produced by *Saccharopolyspora spinosa* is a potent and broad-spectrum insecticidal toxin having good commercial use.

4.6.3.2 Fungal inoculants

Fungal pathogens mostly cause diseases in horticultural crops which eventually result in less yield (Khandelwal et al., 2012). *Beauveria bassiana* and *Beauveria brongniartii* strains are now used as active ingredients in different formulations due to their varying virulence against diverse targets. Other fungal species commonly used worldwide due to their insecticidal properties include *Verticillium lecanii, Metarhizium anisopliae, Hirsutella* sp., *Lecanicillium* sp., *Paecilomyces* sp. and *Isaria* sp.

4.6.3.3 Baculoviruses inoculants

Baculoviruses are DNA viruses that belong to the Baculoviridea family and are primarily pathogens of invertebrate pests. The viruses attack the host by forming crystalline "occlusion" bodies that contain infectious particles. Baculoviruses are classified into two types based on their occlusion bodies: (1) nucleopolyhedroviruses (NPVs) and (2) granuloviruses (GVs) (Rohrmann, 2011). Baculoviruses attack the pest orally. In most cases, the infection begins soon after consuming pathogen-contaminated foods. Occlusion bodies release virions known as "Occlusion Derived Viruses (ODVs)," which directly interact with microvillar epithelial cells via their envelope proteins (i.e., PIFs or per os infectivity factors). Infected midgut cell nuclei then produce a second type of virions known as budded viruses (BVs), which ensure the virus' continued spread in the host body.

4.6.3.4 Nematode inoculants

Pathogenic nematodes of insects enter the host body through natural openings such as the oral cavity, anus, and spiracles and release their symbiotic bacteria in the homocoel. Furthermore, bacterial proliferation is accompanied by the release of toxins and virulence factors into the host cell, favoring the creation of an appropriate environment for the nematode.

4.6.4 Microbial supplements derived from composting for mountain soil

Soil microorganisms are always an important part of any agro-ecosystem, especially in mountain soils where population and diversity are critical for ecosystem sustainability (Mutumba et al., 2018). Furthermore, because of their pervasiveness, microbes are preferred for meeting the agricultural needs of the hour. Water availability, fertilizer application accuracy, organic manure accuracy, and an effective microbial population are critical to booming agricultural output. Compost application in agricultural fields, in this view, plays an important role in agricultural production in hilly areas.

According to the "United States Department of Agriculture (USDA)," "Composting is a process that works to speed up the natural decay of organic materials by providing the ideal conditions for detritus-eating organisms to thrive." The richly organic layer "forest floor" of mountain forests also known as the "litter layer" or the "O" horizon which mainly consists of decomposed and dried plant materials like leaves, shed vegetative parts, barks, branches and stems existing in various stages of decomposition above the soil surfaces (Osman, 2013). Humus is a type of forest product that is composed of decomposed plant parts. The microorganisms involved in this process produce various lytic enzymes that break down organic matter to form humus, as well as CO_2, water, and heat. Compost matures with time and temperature, becoming a stable organic matter (Henry and Bergeron, 2005). It is estimated that the microbial population involved in the composting process is 10^6 to 10^9 per gram of compost. Furthermore, the use of decomposed compost in the soil results in the slow and steady release of nutrients into the soil that are available for plant uptake (Duong, 2013). The composting process is divided into three stages performed by mesophilic bacteria (moderate-temperature phase), thermophilic bacteria (thermophilic phase), and mesophilic bacteria (curing phase). Mesophilic bacteria are most active in the final stage of composting, where they collaborate with fungal decomposers and actinomycetes populations to break down partially degraded organic matter. Longer curing stages result in the development of a diverse microbial population to operate the composting process, making the soil healthier and more nutrient-dense (Epstein, 1997). Forest farmers apply these microorganisms via forest compost, or they are naturally added to agricultural soil via runoff rain water, making the soil fertile and healthy.

4.6.4.1 Biological functions of compost for enrichment of soil

Organic substances or biological materials added to agricultural soils have numerous advantages. Some of the physical improvements include increased soil aggregation and decreased soil bulk density, restoration of soil hydraulic conductivity, improvement of soil water holding capacity, and cation exchange capability. Organic substances also buffer abrupt changes in soil pH, creating a favorable environment for the enhancement of microbial diversity and biological and biochemical activities. In this regard, the addition of biologically decomposed compost to agricultural soil can significantly improve its physical texture, increase stability by forming soil aggregation, and thus prevent soil erosion. This binding of soil particles also contributes to increased permeability by increasing and maintaining soil pore size. Furthermore, the nutrient richness and dynamics of the soil can help to stabilize the establishment of vegetation. Matured compost contains virtually all macro and micro nutrients in available forms required for proper plant growth and development, particularly nitrogen. Compost with a pH of 6—8 has buffering capability, which can help the soil withstand sudden changes in soil pH. Compostable substances can also bind with heavy metals and a variety of agrochemicals such as insecticides, fungicides, bactericides, other pesticides, herbicides, weedicides, and other harmful contaminants. Compostable materials can reduce the leachability of these chemicals, preventing their uptake by plants and, ultimately, the food chain (Brady et al., 2008; Henry and Bergeron, 2005; Hudson, 1994; Singer and Munns, 2006). Soil microbial population developed in composting process also help in pesticide degradation, fertilizers and hydrocarbons (Brown et al., 2004; Fogarty and Tuovinen, 1991; Henry and Bergeron, 2005). The microbial diversity associated with composting is presented in Table 4.4.

TABLE 4.4 Important bacteria, actinomycetes and fungal species, their structure and habitat found in compost.

Name of Microorganisms	Shape, Respiration, Gram stain blue (positive) or Red (negative)	Habitat
Bacteria		
Alcaligenes faecalis	Rod-shaped bacteria that are Gram-negative, aerobic, but some strains can be also capable of anaerobic respiration.	Organisms that inhabit the gastrointestinal tracts of vertebrates, decaying materials, dairy products, water, and soil and also are capable of infecting respiratory and gastrointestinal tracts of humans.
Arthrobacter sp.	Gram-Negative rods	Usually observed in aerial surfaces of plants, wastewater sediments, and soils.
(Bacillus) *Brevibacillus brevis*	Rod-shaped, aerobic, Gram-positive bacteria that produce spores.	Usually found in decaying matter, soil, air, and water.
Bacillus sp., (*Bacillus megaterium, Bacillus pumilus, Bacillus sphaericus, Bacillus coagulans, Bacillus circulans, Bacillus licheniformis, Bacillus subtilis*).	*Bacillus*, a rod-shaped genus of Gram-positive, spore-forming bacteria is known to be obligate aerobes (oxygen reliant), or facultative anaerobes.	The anaerobes in this genus can grow in soil, water, and a wide range of other environments. Some species, such as *Bacillus cereus, Bacillus subtilis*, and *Bacillus licheniformis*, infect wounds and cause disease, whereas *Bacillus stearothermophilus* is a thermophile (heat resistant).
Clostridium thermocelium	Spore-producing Gram-positive, anaerobic, and thermophilic rod.	Generally found in plants and animals. In ruminant animals such as cattle and horses, it aids in the breakdown of cellulose.
Escherichia coli and other *Enterobacteriaceae*	Facultative anaerobes, Gram negative rods.	Most are located in the intestinal tracts of humans and animals. Through contaminated water, food, and direct contact, pathogenic *E. coli* can be spread.
Flavobacterium sp.	Gram-negative, rods	Usually found in soil or water.
Pseudomonas sp.	Gram-negative rods, aerobic	The species of *Pseudomonas aeruginosa* are well distributed in water and in plant seeds. *Pseudomonas syringae* is not a human pathogen, but it can infect plants.

(Continued)

TABLE 4.4   (Continued)

Name of Microorganisms	Shape, Respiration, Gram stain blue (positive) or Red (negative)	Habitat
Serratia sp.	Gram-negative, facultatively anaerobic rods	Despite being common in the environment, *S. marcescens* can be pathogenic.
Streptococcus sp. (*Enterococcus*)	Gram positive cocci	These organisms are found in the respiratory tract, alimentary tract, genitourinary tract, the skin of both humans and animals, as well as in soil and feces.
Thermus sp.	Temperatures as high as 70°C (160°F) are ideal for these rod-shaped, gram-negative thermophilic bacteria and also thrive as low as 50°C (120°F) and as high as 80°C (180°F).	It is found in soil, sewage, thermal spring, feces, and meat.
Actinomycetes		
Streptomyces sp.	Gram positive, filamentous	Over two-thirds of the clinically useful antibiotics come from *Streptomyces*.
Frankia sp.	Gram positive	Approximately 15% of the world's naturally fixed nitrogen is produced by species of *Frankia* and their host plants.
Other Actinomycetes like *Micromonospora* sp.	Some are facultatively anaerobic, but most are anaerobic.	As part of the composting process, decomposers digest tough plant tissues, such as cellulose and lignin in bark and stems and chitin in insect exoskeletons.
Fungi		
Aspergillus fumigatus	Fungal mould	Usually found in organic matters like compost, bird droppings, tobacco, and stored foods; this fungus can survive at temperatures as high as 50°C.
Humicola grisea, Humicola insolens, Humicola lanuginosa	Thermophilic Fungal mould	As a common component of compost, it is largely present in soil and plant material.
Trichoderma sp.	Free living fungi	Found in forest and bark or decorticated wood and helps in the production of cellulases and degradation of complex polysaccharides.
Penicillium sp. (*Penicillium dupontii*)	Fast growing fungi	Found in soil, on decaying vegetation or on wood, dried leaves, fresh fruits, and vegetables.

4.6.5 Farm yard manure

In addition to natural composting, the most common natural fertilizer used by people in hill forests is farm yard manure (FYM). FYM is the most common type of organic manure used by farmers in hill forest areas. FYM's composition and functions are similar to those of compost in terms of improving soil physical, chemical, and biological properties and increasing the microbial population in mountain agricultural soil (Qamar and Saif, 2018; Jangir et al., 2019). However, the preparation and application of indigenous FYM is dependent on a variety of factors. The availability of crop-based or forest-based bedding materials is dependent on ecological zone, access to crop residues and fodders, labor availability, and other factors. All the base materials need to be fully and properly decomposed to be an ideal FYM (TECA-FAO: https://www.fao.org/teca/fr/technolo-gies/7524). That is why microbes play such an important role in releasing nutrients from organic matter in a form that plants can absorb. The use of improperly and partially decomposed FYM may increase the population of earthworms known as "farmer's friends." FYM fortified with these micro miniatures aids in soil loosening and supplements the ecosystem's microorganisms as decomposers. FYM's population of essential microorganisms helps to sustain active soil life better than inorganic sources.

The addition of manure as fertilizer increases the diversity of microbial populations in the soil, where they play critical roles in nutrient cycling, making nutrients more accessible to plants and improving soil physical properties. FYM performs several important functions in soil to improve its organic properties. The FYM's bacterial population converts organic matter into plant-available nutrients. FYM mostly consist of microbial population belonging to *Bacillus* sp. *Pseudomonas* sp., *Klebsiella* sp. and some free-living nitrogen-fixing biofertilizers like *Clostridium* sp., have been exploited to produce biofertilizers. *Bacillus* and *Pseudomonas* sp. have been proved to be efficient phosphate-solubilizers (Alfa et al., 2014; Barzee et al., 2019; Maurer et al., 2019). Higher nitrogenous components from animal excreta may suppress disease-causing pathogens and harmful insect pests, which could worsen the poor, nutrient-deficient, and lifeless soil. FYM also aids plants in developing healthy root and rhizosphere ecosystems, which may reduce runoff.

4.7 Microbial prospective for agricultural practices for hill forests

As the world population grows rapidly in the 21st century, agricultural production is being challenged to be more sustainable in order to ensure the availability of sufficient food commodities, fiber plants, and eco-friendly biofuels (FAO, 2017). As policymakers, scientists, and industries work to increase agricultural production, biotechnological exploitation of microorganisms in various agricultural operations could be a useful tool for future sustainability. As a result, microbial strains with precise efficiency and well-established Plant Growth Promotion (PGP) activity must be chosen to be used as an inoculant for the development of biofertilizers and biopesticides in order to achieve consistent, ecological, and economical results under variable field conditions. In various field conditions, a combination of conventional methods and microbial inoculants can result in significant positive improvements in the agricultural ecosystem and the environment as a whole. Global research on microbial bioformulations has resulted in the development of several commercial products with improved field application efficiency. The application of such products results in crop resistance to abiotic and biotic stresses, as well as an extended shelf life. These microbial products are also affordable to farmers in forest areas, particularly in mountain regions, and microbial inoculants have been shown to be competent in comparison to soil microflora (Singh et al., 2016). There have been numerous applications of microbial biopesticide products to control a variety of crop diseases and pests. *Bacillus thuringiensis*, a widely used bacterium, could control diamondback moths and the *Helicoverpa* pathogen on cotton, pigeon-pea, and tomato. Trichoderma-based products had a significant effect on rot and wilt pathogens in various crops. Sugarcane borers were effectively controlled by *Trichogramma*. The NPV could also control *Helicoverpa* on gram (Abhiram et al., 2018).

4.8 Conclusion: community participation with legal back up

Mountain or hill forest ecosystems are vulnerable but extremely productive for agriculture. Contrary to modern agricultural practices, the majority of hill and mountain forest inhabitant farmers still practice organic agriculture by default. This type of agricultural practice boosts the microbial population as well as other soil properties. This service works in reverse, as the increased microbial population makes the soil more fertile, resulting in higher agricultural yields. Several studies have revealed the unique microbial population of hills and forests, which must be properly utilized to maintain soil health. Microbial-based biofertilizers, biopesticides, and other bioinoculants have proven to be unbeatable tools for forest agriculture. Furthermore, these unrivaled indigenous tiny service providers must be legally protected in order to streamline the flow of financial benefits resulting from commercial utilization. The Biological Diversity Act of 2002, enacted in India, grants local people sovereign rights over their own biodiversity and bioresources. This Act empowers local residents to form Biodiversity Management Committees (BMCs) in local government. The BMC is a legal and statutory body that governs access to bioresources within its jurisdiction. The Biological Diversity Act of 2002 allows for the regulation and protection of indigenous useful microorganisms through Intellectual Property Rights. As a result, integrating the participation of local communities with legal support and the use of microbial products in agriculture can be a long-term approach to preserving soil health in hill and mountain forests.

Acknowledgment

The authors are thankful to the Chairman and Member Secretary, Odisha Biodiversity Board, Bhubaneswar, Odisha, India for kind support.

References

Abhiram, P., Sandhya, N., Chaitanya, K.V.M.S., Reddy, S.S., 2018. Microbial pesticides a better choice than chemical pesticides. Journal of Pharmacognosy and Phytochemistry 7 (5), 2897–2899.
Acharya, A.K., Kafle, N., 2009. Land degradation issues in Nepal and its management through agroforestry. Journal of Agriculture and Environment 10, 133–143.

Adamchak, R., 2020. Organic Farming. Encyclopaedia Brinannica, Inc. Available from: https://www.britannica.com/topic/organic-farming.

Alfa, M.I., Adie, D.B., Igboro, S.B., Oranusi, U.S., Dahunsi, S.O., Akali, D.M., 2014. Assessment of biofertilizer quality and health implications of anaerobic digestion effluent of cow dung and chicken droppings. Renewable Energy 63, 681−686.

Bahadur, I., Maurya, B.R., Meena, V.S., Shah, M., Kumar, A., Aeron, A., 2017. Mineral release dynamics of tricalcium phosphate and waste muscovite by mineral-soublizing rhizobacteria isolated from indo-gangetic plains of India. Geomicrobiology Journal 34 (5), 454−466.

Baldrian, P., 2003. Interactions of heavy metals with white-rot fungi. Enzyme and Microbial Technology 32 (1), 78−91.

Bale, J.S., Masters, G.J., Hodkinson, I.D., Awmack, C., Bezemer, T.M., Brown, V.K., et al., 2002. Herbivory in global climate change research: direct effects of rising temperature on insect herbivores. Global Change Biology 8 (1), 1−16.

Banik, A., Mukhopadhaya, S.K., Dangar, T.K., 2016. Characterization of N2-fixing plant growth promoting endophytic and epiphytic bacterial community of Indian cultivated and wild rice (Oryza spp.) genotypes. Planta 243 (3), 799−812.

Bardgett, R.D., Van Der Putten, W.H., 2014. Belowground biodiversity and ecosystem functioning. Nature 515 (7528), 505−511.

Barzee, T.J., Edalati, A., El-Mashad, H., Wang, D., Scow, K., Zhang, R., 2019. Digestate biofertilizers support similar or higher tomato yields and quality than mineral fertilizer in a subsurface drip fertigation system. Frontiers in Sustainable Food Systems 3, 58.

Berendsen, R.L., Pieterse, C.M., Bakker, P.A., 2012. The rhizosphere microbiome and plant health. Trends in Plant Science 17 (8), 478−486.

Bertola, M., Ferrarini, A., Visioli, G., 2021. Improvement of soil microbial diversity through sustainable agricultural practices and its evaluation by-omics approaches: a perspective for the environment, food quality and human safety. Microorganisms 9 (7), 1400.

Biswas, P.K., 2003. Forest, people and livelihoods: the need for participatory management. In: XII World Forestry Congress, Quebec, Canada. http://www.fao.org/docrep/ARTICLE/WFC/XII/0586-C1.HTM (accessed 09.09.2014).

Blanchette, R.A., 1991. Delignification by wood-decay fungi. Annual Review of Phytopathology 29 (1), 381−403.

Boer, W.D., Folman, L.B., Summerbell, R.C., Boddy, L., 2005. Living in a fungal world: impact of fungi on soil bacterial niche development. FEMS Microbiology Reviews 29 (4), 795−811.

Bommarco, R., Kleijn, D., Potts, S.G., 2013. Ecological intensification: harnessing ecosystem services for food security. Trends in Ecology & Evolution 28 (4), 230−238.

Borah, M., Das, S., Baruah, H., Boro, R.C., Barooah, M., 2018. Diversity of culturable endophytic bacteria from wild and cultivated rice showed potential plant growth promoting activities. BioRxiv 310797.

Bouchez, T., Blieux, A.L., Dequiedt, S., Domaizon, I., Dufresne, A., Ferreira, S., et al., 2016. Molecular microbiology methods for environmental diagnosis. Environmental Chemistry Letters 14 (4), 423−441.

Brady, N.C., Weil, R.R., Weil, R.R., 2008. The Nature and Properties of Soils, vol. 13. Prentice Hall, Upper Saddle River, NJ, pp. 662−710.

Brimecombe, M.J., De Lelj, F.A., Lynch, J.M., 2001. The rhizosphere. The effect of root exudates on rhizosphere microbial populations. The Rhizosphere. CRC Press, pp. 95−140.

Brown, S., Chaney, R., Hallfrisch, J., Ryan, J.A., Berti, W.R., 2004. In situ soil treatments to reduce the phyto-and bioavailability of lead, zinc, and cadmium. Journal of Environmental Quality 33 (2), 522−531.

Chaintreuil, C., Giraud, E., Prin, Y., Lorquin, J., Bâ, A., Gillis, M., et al., 2000. Photosynthetic *bradyrhizobia* are natural endophytes of the African wild rice *Oryza breviligulata*. Applied and Environmental Microbiology 66 (12), 5437−5447.

Champion, H.G., Seth, S.K., 1968. A Revised Survey of the Forest Types of India. Manager of Publications.

Chapela, I.H., Boddy, L., 1988. Fungal colonization of attached beech branches: II. Spatial and temporal organization of communities arising from latent invaders in bark and functional sapwood, under different moisture regimes. New Phytologist 110 (1), 47−57.

Charles, J.F., Silva-Filha, M.H., Nielsen-LeRoux, C., 2000. Mode of action of *Bacillus sphaericus* on mosquito larvae: incidence on resistance. Entomopathogenic Bacteria: From Laboratory to Field Application. Springer, Dordrecht, pp. 237−252.

Chen, J.Y., Gu, J., Wang, E.T., Ma, X.X., Kang, S.T., Huang, L.Z., et al., 2014. Wild peanut *Arachis duranensis* are nodulated by diverse and novel *Bradyrhizobium* species in acid soils. Systematic and Applied Microbiology 37 (7), 525–532.

Curling, S.F., Clausen, C.A., Winandy, J.E., 2002. Relationships between mechanical properties, weight loss, and chemical composition of wood during incipient brown-rot decay. Forest Product Journal 52, 34–39.

Dastager, S.G., Li, W.J., Saadoun, I., Miransari, M., 2011. Microbial diversity-sustaining earth and industry. Applied and Environmental Soil Science .

Debnath, S., Rawat, D., Mukherjee, A.K., Adhikary, S., Kundu, R., 2019. Applications and constraints of plant beneficial microorganisms in agriculture. Biostimulants in Plant Science .

Dhaliwal, R.K., Singh, G., 2010. Traditional food grain storage practices of Punjab. Indian Journal of Traditional Knowledge 9 (3), 526–553.

Doran, J.W., 2002. Soil health and global sustainability: translating science into practice. Agriculture, Ecosystems & Environment 88 (2), 119–127.

Duong, T.T.T., 2013. Compost Effects on Soil Properties and Plant Growth (Doctoral dissertation).

Edgerton, M.D., 2009. Increasing crop productivity to meet global needs for feed, food, and fuel. Plant Physiology 149 (1), 7–13.

Elango, K., Sobhana, E., Sujithra, P., Bharath, D., Ahuja, A., 2020. Traditional agricultural practices as a tool for management of insects and nematode pests of crops: an overview. Journal of Entomology and Zoology Studies 8 (3), 237–245.

Elbeltagy, A., Nshioka, K., Suzuki, H., Sato, T., Sato, Y.I., Morisaki, H., et al., 2000. Isolation and characterization of endophytic bacteria from wild and traditionally cultivated rice varieties. Soil Science and Plant Nutrition 46 (3), 617–629.

Elbeltagy, A., Nshioka, K., Sato, T., Suzuki, H., Ye, B., Hamada, T., et al., 2001. Endophytic colonization and in planta nitrogen fixation by a *Herbaspirillum* species isolated from wild rice species. Applied and Environmental Microbiology 67 (11), 5285–5293.

Engelhard, M., Hurek, T., Reinhold-Hurek, B., 2000. Preferential occurrence of diazotrophic endophytes, *Azoarcus* spp., in wild rice species and land races of *Oryza sativain* comparision with modern races. Environmental Microbiology 2 (2), 131–141.

Epstein, E., 1997. The Science of Composting. Technomic Publishing Company, Lancaster, PA, p. 1429.

Erisman, J.W., Eekeren, N.V., Wit, J.D., Koopmans, C., Cuijpers, W., Oerlemans, N., et al., 2016. Agriculture and biodiversity: a better balance benefits both. AIMS Agriculture and Food 1 (2), 157–174.

Fogarty, A.M., Tuovinen, O.H., 1991. Microbiological degradation of pesticides in yard waste composting. Microbiological Reviews 55 (2), 225–233.

Forest Survey of India, 2019. State of the Forest Report. Dehradun.

Fort, M., 2015. Impact of climate change on mountain environment dynamics. An introduction. Journal of Alpine Research. Revue de géographie alpine 103-2.

French-Constant, R., Waterfield, N., 2006. An ABC guide to the bacterial toxin, complexes. Advances in Applied Microbiology 58, 169–183.

Gardi, C., Jeffery, S., 2009. Soil Biodiversity. EUR-OP, Brussels.

Gaudin, A.C., Tolhurst, T.N., Ker, A.P., Janovicek, K., Tortora, C., Martin, R.C., et al., 2015. Increasing crop diversity mitigates weather variations and improves yield stability. Cropping System Diversity and Resilience 2, e0113261.

Godfray, H.C.J., Beddington, J.R., Crute, I.R., Haddad, L., Lawrence, D., Muir, J.F., et al., 2010. Food security: the challenge of feeding 9 billion people. Science (New York, N.Y.) 327 (5967), 812–818.

Goodell, B., Winandy, J.E., Morrell, J.J., 2020. Fungal degradation of wood: emerging data, new insights and changing perceptions. Coatings 10 (12), 1210.

Henry, C., Bergeron, K., 2005. Compost Use in Forest Land Restoration. Karen Bergeron, King County Department of Natural Resources. Available from: https://www.epa.gov/sites/default/files/2019-12/documents/compost-forest-land-restore-2005.pdf.

Hudson, B.D., 1994. Soil organic matter and available water capacity. Journal of Soil and Water Conservation 49 (2), 189–194.

Hurst, M.R., Glare, T.R., Jackson, T.A., Ronson, C.W., 2000. Plasmid-located pathogenicity determinants of *Serratia entomophila*, the causal agent of amber disease of grass grub, show similarity to the insecticidal toxins of *Photorhabdus luminescens*. Journal of Bacteriology 182 (18), 5127–5138.

IPCC, 2007. Climate Change 2007; Synthesis Report. Contribution of working groups I, II and II to the Fourth Assessment report of the Intergovernmental Panle on Climate Change, Geneva, Switzerland, IPCC.

Jangir, C.K., Kumar, S., Meena, R.S., 2019. Significance of soil organic matter to soil quality and evaluation of sustainability. Sustainable Agriculture. Scientific Publisher, Jodhpur, pp. 357–381.

Jurat-Fuentes, J.L., Jackson, T.A., 2012. Bacterial entomopathogens. Insect Pathology. Elsevier, pp. 265–349.

Kapos, V., Rhind, J., Edwards, M., Price, M.F., Ravilious, C., 2000. Developing a Map of the World's Mountain Forests. Forests in Sustainable Mountain Development: A State of Knowledge Report for 2000. Task Force on Forests in Sustainable Mountain Development, pp. 4–19.

Kaspari, M., Garcia, M.N., Harms, K.E., Santana, M., Wright, S.J., Yavitt, J.B., 2008. Multiple nutrients limit litterfall and decomposition in a tropical forest. Ecology Letters 11 (1), 35–43.

Khandelwal, M., Datta, S., Mehta, J., Naruka, R., Makhijani, K., Sharma, G., et al., 2012. Isolation, characterization and biomass production of *Trichoderma viride* using various agro products—a biocontrol agent. Advances in Applied Science Research 3 (6), 3950–3955.

Khouri, N., Shideed, K., Kherallah, M., 2011. Food security: perspectives from the Arab World. Food Security .

Kohler, T., Maselli, D., 2009. Mountains and climate change. From understanding to action. Geographica Bernensia .

Koomnok, C., Teaumroong, N., Rerkasem, B., Lumyong, S., 2007. Diazotroph endophytic bacteria in cultivated and wild rice in Thailand. Science Asia 33, 429–435.

Kowalchuk, G.A., Buma, D.S., de Boer, W., Klinkhamer, P.G., van Veen, J.A., 2002. Effects of above-ground plant species composition and diversity on the diversity of soil-borne microorganisms. Antonie Van Leeuwenhoek 81 (1), 509–520.

Kumar, K.K., Kumar, K.R., Ashrit, R.G., Deshpande, N.R., Hansen, J.W., 2004. Climate impacts on Indian agriculture. International Journal of Climatology: A Journal of the Royal Meteorological Society 24 (11), 1375–1393.

Kumar, A., Maurya, B.R., Raghuwanshi, R., Meena, V.S., Islam, M.T., 2017. Co-inoculation with Enterobacter and Rhizobacteria on yield and nutrient uptake by wheat (*Triticum aestivum* L.) in the alluvial soil under indogangetic plain of India. Journal of Plant Growth Regulation 36 (3), 608–617.

Landsberg, M.J., Jones, S.A., Rothnagel, R., Busby, J.N., Marshall, S.D., Simpson, R.M., et al., 2011. 3D structure of the *Yersinia entomophaga* toxin complex and implications for insecticidal activity. Proceedings of the National Academy of Sciences 108 (51), 20544–20549.

Lane, E.W., 1955. Importance of fluvial morphology in hydraulic engineering. Proceedings. American Society of Civil Engineers 81, *Paper no. 745*.

Lenka, A., Nedunchezhiyan, M., Jata, S.K., Sahoo, B., 2012. Livelihood improvement and nutritional security through tuber crop in Odisha. Editor's Note 50.

Li, X., El Solh, M., Siddique, K., 2019. Mountain Agriculture: Opportunities for Harnessing Zero Hunger in Asia. Food and Agriculture Organization of the United Nations (FAO).

Liang, X., Zhuang, J., Löffler, F.E., Zhang, Y., DeBruyn, J.M., Wilhelm, S.W., et al., 2019. Viral and bacterial community responses to stimulated Fe (III)-bioreduction during simulated subsurface bioremediation. Environmental Microbiology 21 (6), 2043–2055.

Lim, S.F., Matu, S.U., 2015. Utilization of agro-wastes to produce biofertilizer. International Journal of Energy and Environmental Engineering 6 (1), 31–35.

Malla, G., 2008. Climate change and its impact on Nepalese agriculture. Journal of Agriculture and Environment 9, 62–71.

Manikanta, N., Mandal, T.K., Maitra, S., Adhikary, R., 2019. Performance of traditional rice (Oryza sativa L.) varieties under system of rice intensification during Kharif Season in South Odisha conditions. International Journal of Agriculture, Environment and Biotechnology 12 (3), 287–292.

Marche, M.G., Camiolo, S., Porceddu, A., Ruiu, L., 2018. Survey of *Brevibacillus laterosporus* insecticidal protein genes and virulence factors. Journal of Invertebrate Pathology 155, 38–43.

Marche, M.G., Mura, M.E., Falchi, G., Ruiu, L., 2017. Spore surface proteins of *Brevibacillus laterosporus* are involved in insect pathogenesis. Scientific Reports 7 (1), 1–10.

Marshall, D., Tunali, B., Nelson, L.R., 1999. Occurrence of fungal endophytes in species of wild *Triticum*. Crop Science 39 (5), 1507–1512.

Martin, P.A., Gundersen-Rindal, D., Blackburn, M., Buyer, J., 2007. *Chromobacterium subtsugae* sp. nov., a betaproteobacterium toxic to Colorado potato beetle and other insect pests. International Journal of Systematic and Evolutionary Microbiology 57 (5), 993–999.

Maurer, C., Seiler-Petzold, J., Schulz, R., Müller, J., 2019. Short-term nitrogen uptake of barley from differently processed biogas digestate in pot experiments. Energies 12 (4), 696.

McMichael, A.J., Powles, J.W., Butler, C.D., Uauy, R., 2007. Food, livestock production, energy, climate change, and health. The Lancet 370 (9594), 1253–1263.

Meshram, P.B., 2010. Role of some biopesticides in management of some forest insect pests. Journal of Biopesticides 3 (1), 250.

Mishra, D., Rajvir, S., Mishra, U., Kumar, S.S., 2013. Role of bio-fertilizer in organic agriculture: a review. Research Journal of Recent Sciences 2502. ISSN, 2277.

Mousa, W.K., Shearer, C.R., Limay-Rios, V., Zhou, T., Raizada, M.N., 2015. Bacterial endophytes from wild maize suppress *Fusarium graminearum* in modern maize and inhibit mycotoxin accumulation. Frontiers in Plant Science 6, 805.

Murphy, B.R., Hodkinson, T.R., Doohan, F.M., 2018a. Endophytic *Cladosporium* strains from a crop wild relative increase grain yield in barley. Biology and Environment: Proceedings of the Royal Irish Academy 118 (3), 147–156. Royal Irish Academy.

Murphy, B.R., Jadwiszczak, M.J., Soldi, E., Hodkinson, T.R., 2018b. Endophytes from the crop wild relative *Hordeum secalinum* L. improve agronomic traits in unstressed and salt-stressed barley. Cogent Food & Agriculture 4 (1), 1549195.

Mutumba, F.A., Zagal, E., Gerding, M., Castillo-Rosales, D., Paulino, L., Schoebitz, M., 2018. Plant growth promoting rhizobacteria for improved water stress tolerance in wheat genotypes. Journal of Soil Science and Plant Nutrition 18 (4), 1080–1096.

Nayak, S., Mukherjee, A.K., 2015. Management of agricultural wastes using microbial agents. Waste Management: Challenges, Threats and Opportunities 65–91.

Nayak, S., Dhua, U., Samanta, S., Chhotaray, A., 2020a. Management of indoor airborne *Aspergillus flavus* by traditional air purifiers commonly used in India. Journal of Pure and Applied Microbiology 14 (2), 1577–1588. Available from: https://doi.org/10.22207/JPAM.14.2.56.

Nayak, S., Samanta, S., Mukherjee, A.K., 2020b. Beneficial role of *Aspergillus* sp. in agricultural soil and environment. Frontiers in Soil and Environmental Microbiology. CRC Press, pp. 17–36.

Nayak, S., Dhua, U., Samanta, S., 2020c. Antagonistic activity of cowshed *Bacillus* sp. bacteria against aflatoxigenic and sclerotic *Aspergillus flavus*. Journal of Biological Control 34 (1), 52–58.

Negi, G.C.S., Samal, P.K., Kuniyal, J.C., Kothyari, B.P., Sharma, R.K., Dhyani, P.P., 2012. Impact of climate change on the western Himalayan Mountain ecosystems: an overview. Tropical Ecology 53 (3), 345–356.

Nielsen, M.N., Winding, A., Binnerup, S., 2002. Microorganisms as indicators of soil health. National Environmental Research Institute Technical Report.

Ofek-Lalzar, M., Gur, Y., Ben-Moshe, S., Sharon, O., Kosman, E., Mochli, E., et al., 2016. Diversity of fungal endophytes in recent and ancient wheat ancestors *Triticum dicoccoides* and *Aegilops sharonensis*. FEMS Microbiology Ecology 92 (10), fiw152.

Osińska-Jaroszuk, M., Jarosz-Wilkołazka, A., Jaroszuk-Ściseł, J., Szałapata, K., Nowak, A., Jaszek, M., et al., 2015. Extracellular polysaccharides from Ascomycota and Basidiomycota: production conditions, biochemical characteristics, and biological properties. World Journal of Microbiology and Biotechnology 31 (12), 1823–1844.

Osman, K.T., 2013. Organic matter of forest soils. Forest Soils. Springer, Cham, pp. 63–76.

Ozkurt, E., Hassani, M.A., Sesiz, U., Künzel, S., Dagan, T., Ozkan, H., et al., 2020. Seed-derived microbial colonization of wild Emmer and domesticated bread Wheat (*Triticum dicoccoides* and *T. aestivum*) seedlings shows pronounced differences in overall diversity and composition. Mbio 11 (6), e02637-20.

Panda, D., Palita, S.K., 2021. Potential of underutilized wild crops in Koraput, Odisha, India for improving nutritional security and promoting climate resilience. Current Science 120 (6), 00113891.

Parvin, S., Van Geel, M., Ali, M.M., Yeasmin, T., Lievens, B., Honnay, O., 2021. A comparision of the arbuscular mycorrhizal fungal communities among Bangladeshi modern high yielding and traditional rice varieties. Plant and Soil 462 (1), 109–124.

Patel, S.K., Sharma, A., Singh, G.S., 2020. Traditional agricultural practices in India: an approach for environmental sustainability and food security. Energy, Ecology and Environment 5 (4), 253–271.

Patterson, D.T., Westbrook, J.K., Joyce, R.J.V., Lingren, P.D., Rogasik, J., 1999. Weeds, insects, and diseases. Climatic Change 43 (4), 711–727.

Pigott, C.R., Ellar, D.J., 2007. Role of receptors in *Bacillus thuringiensis* crystal toxin activity. Microbiology and Molecular Biology Reviews 71 (2), 255–281.

Prakash, B.G., Raghavendra, K.V., Gowthami, R., Shashank, R., 2016. Indigenous practices for eco-friendly storage of food grains and seeds. Advance in Plants & Agriculture Research 3 (4), 00101.

Qamar, S.U.R., Saif, A., 2018. An overview on microorganisms contribute in increasing soil fertility. Biomedical Journal of Scientific & Technical Research 2 (1), 2131−2132.

Rohrmann, G.F., 2011. Baculovirus molecular biology. National Library of Medicine. National Centre for Biotechnology Information, Bethesda, MD, 188. https://ir.library.oregonstate.edu/concern/defaults/7d279043k.

Roodi, D., Millner, J.P., McGill, C., Johnson, R.D., Jauregui, R., Card, S.D., 2020. *Methylobacterium*, a major component of the culturable bacterial endophyte community of wild Brassica seed. PeerJ—The Journal of Life and Environmental Science 8, e9514.

Rosenzweig, C., Iglesius, A., Yang, X.B., Epstein, P.R., Chivian, E., 2001. *Climate change and extreme weather events-* implications for food production, plant diseases, and pests. Global Change and Human Health 2, 92−104.

Roy, K., Ghosh, D., DeBruyn, J.M., Dasgupta, T., Wommack, K.E., Liang, X., et al., 2020. Temporal dynamics of soil virus and bacterial populations in agricultural and early plant successional soils. Frontiers in Microbiology 11, 1494.

Ruiu, L., 2013. *Brevibacillus laterosporus*, a pathogen of invertebrates and a broad-spectrum antimicrobial species. Insects 4 (3), 476−492.

Sahoo, R.K., Bhardwaj, D., Tuteja, N., 2013. Biofertilizers: a sustainable eco-friendly agricultural approach to crop improvement. Plant Acclimation to Environmental Stress. Springer, New York, pp. 403−432.

Samson, C., Pretty, J., 2006. Environmental and health benefits of hunting lifestyles & diets for the Innu of Labrador. Food Policy 31 (6), 528−553.

Scherr, S.J., McNeely, J.A., 2008. Biodiversity conservation and agricultural sustainability: towards a new paradigm of 'ecoagriculture' landscapes. Philosophical Transactions of Royal Society B. Biological Sciences 363 (1491), 477−494.

Sharma, L., Pradhan, B., Bhutia, K.D., 2017. Farmer's perceived problems and constraints for organic vegetable production in Sikkim. Age (Melbourne, Vic.) 16, 26−66.

Sharma, I.P., Kanta, C., Dwivedi, T., Rani, R., 2020. Indigenous agricultural practices: a supreme key to maintaining biodiversity. Microbiological Advancements for Higher Altitude Agro-Ecosystems and Sustainability 91−112.

Shehata, H.R., Lyons, E.M., Jordan, K.S., Raizada, M.N., 2016. Bacterial endophytes from wild and ancient maize are able to suppress the fungal pathogen *Sclerotinia homoeocarpa*. Journal of Applied Microbiology 120 (3), 756−769.

Shimrah, T., Bharali, S., Rao, K.S., Saxena, K.G., 2012. Cultural landscapes: the basis for linking biodiversity conservation with the sustainable development in West Kameng, Arunachal Pradesh. Cultural Landscape the Basis for Linking Biodiversity Conservation with the Sustainable Development. UNESCO, NIE, New Delhi, pp. 105−147.

Singer, M.J., Munns, D.N., 2006. Soils: An Introduction. Pearson Prentice Hall, Upper Saddle River, NJ.

Singh, R., Singh, G.S., 2017. Traditional agriculture: a climate-smart approach for sustainable food production. Energy, Ecology and Environment 2 (5), 296−316.

Singh, J.S., Koushal, S., Kumar, A., Vimal, S.R., Gupta, V.K., 2016. Book review: microbial inoculants in sustainable agricultural productivity-Vol. II: functional application. Frontiers in Microbiology 7, 2105.

Singh, D.P., Singh, H.B., Prabha, R. (Eds.), 2017. Plant-Microbe Interactions in Agro-Ecological Perspectives. Springer, New Delhi.

Sivakumar, N., Remya, R., Al Bahry, S.A.I.F., 2009. Partial characterization of proteases produced by three fungal isolates from the rhizosphere of wild yam Dioscoreawallichii. Journal of Applied Biological Sciences 3 (3), 71−75.

Slikkerveer, L., 1994. Indigenous agricultural knowledge systems in developing countries: a bibliography. Indigenous Knowledge Systems Research and Development Studies, No. 1. Special Issue: INDAKS Project Report 1 in collaboration with the European Commission DG XII. Leiden Ethnosystems and Development Programme (LEAD), Leiden.

Snajdr, J., Cajthaml, T., Valaskova, V., Merhautova, V., Petrankova, M., Spetz, P., et al., 2011. Transformation of Quercus petraea litter: successive changes in litter chemistry are reflected in differential enzyme activity and changes in the microbial community composition. FEMS Microbiology Ecology 75 (2), 291−303.

Sporleder, M., Lacey, L.A., 2013. Biopesticides. In: Giordanengo, P., Vincent, C., Alyokhin, A. (Eds.), Insect Pests of Potato: Global Perspectives on Biology and Management. Elsevier, Oxford, UK, pp. 463−497.

Sun, X., Kosman, E., Sharon, A., 2020. Stem endophytic mycobiota in wild and domesticated wheat: structural differences and hidden resources for wheat improvement. Journal of Fungi 6 (3), 180.

Swift, M.J., 2005. Human impacts on biodiversity and ecosystem services: an overview. (capítulo 31), The Fungal Community. Its Organization and Role in the Ecosystem, third ed. CRC press, Taylor & Francis Group. NY, EEUU, Boca Raton, FL, p. 936.

Swift, M.J., Heal, O.W., Anderson, J.M., Anderson, J.M., 1979. Decomposition in Terrestrial Ecosystems, vol. 5. *University of California Press.*

Talbot, J.M., Allison, S.D., Treseder, K.K., 2008. Decomposers in disguise: mycorrhizal fungi as regulators of soil C dynamics in ecosystems under global change. Functional Ecology 22 (6), 955–963.

Ten Have, R., Teunissen, P.J., 2001. Oxidative mechanisms involved in lignin degradation by white-rot fungi. Chemical Reviews 101 (11), 3397–3414.

Thrupp, L.A., 2002. Linking agricultural biodiversity and foodsecurity: the valuable role of agrobiodiversity for sustainable agriculture. International Affairs 76 (2), 283–297.

Tilak, K.V.B.R., Ranganayaki, N., Pal, K.K., De, R., Saxena, A.K., Nautiyal, C.S., et al., 2005. Diversity of plant growth and soil health supporting bacteria. Current Science 136–150.

Timmusk, S., Paalme, V., Pavlicek, T., Bergquist, J., Vangala, A., Danilas, T., et al., 2011. Bacterial distribution in the rhizosphere of wild barley under contrasting microclimates. PLoS One 6 (3), e17968.

Trivedi, P., Pandey, A., Palni, L.M.S., 2012. Bacterial inoculants for field applications under mountain ecosystem: present initiatives and future prospects. In: Bacteria in Agrobiology: Plant Probiotics, pp. 15–44.

Verma, P., Yadav, A.N., Kumar, V., Singh, D.P., Saxena, A.K., 2017. Beneficial plant microbes interactions: biodiversity of microbes from diverse extreme environments and its impact for crops improvement. Plant-Microbe Interactions in Agro-Ecological Perspectives. Springer Nature, Singapore, pp. 543–580.

Vessey, J.K., 2003. Plant growth promoting rhizobacteria as biofertilizers. Plant and Soil 255 (2), 571–586.

Vincetti, B., Eyzaguirre, P., Johns, T., 2008. The nutritional role of forest plant foods for rural communities. Human Health & Forests: A Global Overview of Issues, Practice & Policy. Earthsan, pp. 63–96.

Vodovar, N., Vallenet, D., Cruveiller, S., Rouy, Z., Barbe, V., Acosta, C., et al., 2006. Complete genome sequence of the entomopathogenic and metabolically versatile soil bacterium *Pseudomonas entomophila*. Nature Biotechnology 24 (6), 673–679.

Winandy, J.E., Morrell, J.J., 1993. Relationship between incipient decay, strength, and chemical composition of Douglas-fir heartwood. Wood and Fiber Science 25 (3), 278–288.

Xu, L., Zhou, L., Zhao, J., Li, J., Li, X., Wang, J., 2008. Fungal endophytes from *Dioscorea zingiberensis* rhizomes and their antibacterial activity. Letters in Applied Microbiology 46 (1), 68–72.

Xu, S., Chang, J., Chang, C., Tian, L., Li, X., Tian, C., 2020. Rhizospheric microbiomes help Dongxiang common wild rice (*Oryza rufipogon* Griff.) rather than *Leersia hexandra* Swartz survive under cold stress. Archives of Agronomy and Soil Science 1–13.

Yadav, A.N., Verma, P., Kumar, V., Sangwan, P., Mishra, S., Panjir, N., et al., 2018. Biodiversity of the genus *Penicillium* in different habitats. New and Future Developments in Microbial Biotechnology and Bioengineering. Elsevier, pp. 3–18.

Yokoya, K., Postel, S., Fang, R., Sarasan, V., 2017. Endophytic fungal diversity of *Fragaria vesca*, a crop wild relative of strawberry. Along environmental gradients within a small geographical area. Peer J 5, e2860.

Yuan, Z.L., Zhang, C.L., Lin, F.C., Kubicek, C.P., 2010a. Identity, diversity and molecular phylogeny of the endophytic mycobiota in the roots of rare wild rice (*Oryza granulate*) from a nature reserve in Yunnan, China. Applied and Environmental Microbiology 76 (5), 1642–1652.

Yuan, Z.L., Lin, F.C., Zhang, C.L., Kubicek, C.P., 2010b. A new species of *Harpophora* (Magnaporthaceae) recovered from healthy wild rice (*Oryza granulate*) roots, representing a novel member of a beneficial dark septate endophyte. FEMS Microbiology Letters 307 (1), 94–101.

Zhang, G.X., Peng, G.X., Wang, E.T., Yan, H., Yuan, Q.H., Zhang, W., et al., 2008. Diverse endophytic nitrogen-fixing bacteria isolated from wild rice *Oryza rufipogon* and description of *Phytobacter diazotrophicus* gen. nov. sp. nov. Archivs of Microbiology 189 (5), 4.

Soil biological processes of mountainous landscapes: a holistic view

Bhawna Tyagi[1], Simran Takkar[2] and Prabhat Kumar[1]

[1]School of Environmental Sciences, Jawaharlal Nehru University, New Delhi, India
[2]Life Sciences Department, Shiv Nadar University, Greater Noida, Uttar Pradesh, India

OUTLINE

Understanding Soils of Mountainous Landscapes
DOI: https://doi.org/10.1016/B978-0-323-95925-4.00008-X

91

5.1 Introduction

Mountains are known as regions having increased elevation (>200 m) with simultaneous declining total land area (Looby and Martin, 2020). They can be found on every continent in a variety of terrestrial ecosystems. Mountain landscapes cover nearly 25% of the global land area and can be found at all latitudes and longitudes (Donhauser and Frey, 2018). Mountains are an important source of water as they provide 60%−80% of the global fresh water (Barry, 2008). Mountains are also important for economic concerns because they provide significant mineral resources, forest reserves, and hydropower. Mountainous ecosystems are made up of a variety of low-temperature habitats such as soils, bare rocks, snow, permafrost, and glaciers. Because of their limited accessibility, protection from human influence, and incorporation of a wide range of climatic gradients within a small region, mountainous regions are considered major hotspots of diversity (Pauli et al., 2015). Despite being biodiversity hotspots, mountains cover only 12% of the world's terrestrial area outside of Antarctica but support approximately 33% of terrestrial species diversity and nearly half of the planet's biodiversity hotspots (Körner et al., 2011). Mountain ecosystems play a dominant role in global biogeochemical cycles because they are a primary and important source of global terrain complexity. Furthermore, mountains serve as key indicators of climate change because warming occurs at a faster rate at higher elevations than at lower elevations. Mountains are thought to be the most climate-sensitive ecosystems on the planet.

Microbes have dominated many activities in the mountain environment, including soil development, the maintenance of major elemental cycles, the development, growth, and survival of plants locally, and the global balance of greenhouse gases. Temperature and climatic variables such as humidity, for example, have a strong influence on the diversity and activity of microbes in an environment (Classen et al., 2015). Microbes have been shown to be highly sensitive to abiotic environments and external biotic influences such as litter quantity and quality. Certainly, microbial abundance typically peaks in high soil organic matter (SOM) zones, which shift with high plant productivity zones (Xu et al., 2014). The study of diversity in mountains has a long history, and soil microbes in mountains are assumed to be intensely outlined by mountain gradients. The main effect of climate on the structure, diversity, and dynamics of microorganisms is indicated by mountain gradients (Sherman et al., 2012). There is 34% decrease in total microbial diversity with increase in elevation has been seen according to time (Looby and Martin, 2020). The growth of soil microbes is also governed by specialized processes such as decomposition and necrosis for energy. Both processes are much more active on mountain gradients because of large carbon stocks, as opposed to mountain biomass and soil (Donhauser and Frey, 2018). Soil microbes are influenced by mountain climate and its changes, so understanding the impact of climate change on soil microbial diversity in mountain ecosystems is essential.

Because of the difficulties in entry and transportation, high altitude mountains are largely understudied in terms of microbial ecology (Gavazov et al., 2017). As a result, in this chapter, we will concentrate on the alpine vegetation zones, as well as the geographical and pedological features that determine the mountainous soil ecosystems. We will investigate microbial habitats and their microbial structure in relation to climate change. We will also talk about the microbial ecology of mountains and the effects of climate change on mountainous soil. This research will focus on the ongoing experimental methods.

5.2 Environment of mountain ecosystem

The climate and vegetation in the mountains differ from those in the lowlands. Mountain climates differ due to altitude, latitude, and relief (Nie et al., 2018). It is well known that atmospheric temperature decreases by about $0.5°C−0.6°C$ per 100 m of altitude, so climate varies with altitude (He et al., 2021). Whereas mountain relief affects climate in such a way that mountains face the wind path, allowing the air to cool as it rises, resulting in high precipitation on windward mountain slopes in the form of orographic precipitation. On the other hand, as rain shadows form, the relative humidity of the air falls and it warms as it descends leeward slopes, decreasing the likelihood of precipitation and triggering the drier climate zone (Smith and Cleef, 1988). Mountains in the desert receive very little rainfall due to dry air, which inhibits precipitation under any environmental conditions. Latitude also has an impact on climate conditions in mountains. Mountains in equatorial regions have no winter and summer seasons, whereas mountains in temperate regions have distinct clear seasons (Ma et al., 2004). Microclimate differences are also significant in mountain regions due to the diverse characteristics of steep slopes that present different environments due to changes in precipitation and solar energy receipt (Pepin et al., 2015). Mountain slopes facing the equator are significantly warmer than opposite slopes in temperate latitudes. This may have an impact on the vegetation both directly and indirectly because the amount of time snow spends on the surface in spring affects vegetation development, ultimately affecting the land's usefulness for grazing (Nie et al., 2018).

5.2.1 Mountain ecosystems and factors changing with elevation

Various environmental conditions (temperature, solar radiation, and atmospheric pressure) vary rapidly with altitude in mountain ecosystems. Solar radiation changes have an impact on the microbial composition above the treeline in alpine ecosystems where unfiltered solar radiation touches the soil surface. Warm temperatures at low altitude have a direct impact on aboveground and belowground community and ecosystem ecological processes (Sundqvist et al., 2013).

Warmer temperatures at lower elevations cause less accumulation of SOM and the release of more soluble nutrients in the soil matrix due to an increased rate of decomposition; this dynamic favors bacteria over fungi (Kotas et al., 2018). On a broad scale, the amount of nutrients such as nitrogen, potassium, and phosphorus decreases with altitude, resulting in nutrient deficient litterfall in mountain forests due to reduced litter inputs (Armas-Herrera et al., 2020). Mountainous soils have a high organic matter content, and the rate of nitrogen mineralization at higher altitudes is significantly slower. Temperatures have an indirect impact on microbial communities as a result of the direct effect of climate on plant communities. The ecosystems with warmer temperatures at lower elevations have a higher diversity of tree species, photosynthesis, biomass, and productivity. As a result, at low altitudes, the abundance and diversity of leaf litter influence microbial diversity and abundance in soil (Hättenschwiler and Jørgensen, 2010). Similarly, as plant ecological processes change, the amount and quality of carbon stocks decrease with cold temperatures at high elevations, affecting microbial communities due to changes in abiotic and biotic environments (Hättenschwiler and Jørgensen, 2010).

According to research, the accumulation of organic matter in soil at high elevations has a direct impact on the structure and diversity of microbes such as bacterial, fungal, and archaeal, which are positively correlated with soil carbon concentration (Looby and Martin, 2020). At high elevations, the mountainous soil becomes anoxic. At higher elevations, soil depths and other important edaphic factors decrease. Soil pH is important for characterizing bacterial community structure, and it also serves as an environmental filter for soil fungi at small spatial scales. Because of higher organic carbon stocks, mountainous soils become increasingly acidic as elevation increases (Glassman et al., 2017).

5.2.2 Geographical controls of mountain climate

Mountain landscape are widely scattered globally and form a complex topography showing different mountain climate along different regions (Donhauser and Frey, 2018). The main factors of geographical control of mountain climate are shown in Fig. 5.1. The most important factor influencing mountain climate is latitude. Seasonal and diurnal patterns of incoming solar radiation have been observed at various latitudes. Low temperature and less radiation are more prominent in high latitude regions than in low latitude regions (Margesin, 2012).

Altitude is another major factor impacting the mountain climate. Owing to the altitude dependent temperature decrease i.e., decrease in air temperature by 6°C per 1000 m with altitude, alpine regions are mostly covered of low-temperature conquered ecosystems (Schulz et al., 2013). Sharp altitudinal gradients in the mountains are the foundation for determining different climatic settings at a small spatial scale. Furthermore, snow cover has a strong influence on soil temperature because it insulates the ground and reduces variations in air temperature. According to recent studies, mountains are very sensitive to

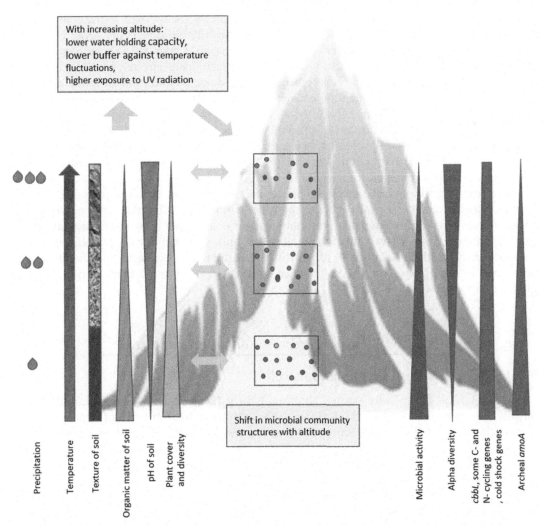

FIGURE 5.1 Diagram showing the shift in microbial community structures, soil properties and plant diversity along with altitude and aspect.

climate change, and some studies have found that high elevations experience more warming than low elevations (Rangwala and Miller, 2012).

The local mountain climate, on the other hand, is primarily influenced by topographic factors such as slope and angle aspects, precipitation, depressions versus ridges, hydrological processes, air flow, and heat transfer (Nagy and Grabherr, 2009). The main climatic feature of mountain landscapes is high precipitation, which increases with altitude in mountains. Because of orographic enhancement, the windward side of mountain regions receives more precipitation than the lowlands. A rain shadow effect, on the other hand, results in less precipitation on the leeward side. According to some studies, the amount of precipitation is expected to increase as global warming continues (Collins et al., 2013). There will be no consistent patterns of precipitation variation in the mountains, as opposed to the arctic, where the extent of precipitation is expected to increase significantly (Stocker et al., 2014). However, it is expected that snowfall will decrease with warming, and snowfall can be converted to rain in both the arctic and high altitude regions, with significant effects on soil temperature (Phillips and Schweizer, 2007). Variable precipitation and soil moisture on elevation gradients have significant direct and unintended consequences for microbial diversity and composition in soil. Microbe habitats become moistened as soil moisture increases, resulting in a decrease in soil oxygen levels and, ultimately, affecting microbe-related aspects such as physiology, abundance, and ecosystem-level microbial mediated practices (Kooch and Noghre, 2020). Because of low evapotranspiration, frequent orographic cloud cover, and shallow soils and waterlogging at cooler temperatures, mountainous ecosystems receive a lot of rain.

The biotic constituents of mountain bionetworks would be severely impacted by the aforementioned climatic variations. Soil microbe activity and diversity are heavily influenced by temperature and moisture. Climate change will further alter the elemental cycles, causing fluctuations in soil nutrition availability and productivity, survival, and biotic component interactions (Classen et al., 2015). Soils are shallow at high altitudes in mountains due to scraping off of the soil by ice formed on high mountains during the last glacial interval (Reyes et al., 2017).

5.3 Soil biological processes of mountainous zone

5.3.1 Characteristics of mountainous soil

Mountains soil is typically young, shallow, poorly formed, acidic, skeletal, and low in fertility. Temperature changes influence microbe activity, which in turn influences soil formation. Because of the varied topography of mountains, soil composition varies greatly (slope, aspect and altitude). Because there is more water available on the north-facing slope, soil development is faster than on the south-facing slope, and it has a high weathering process, microbiological activity, and vegetation development. The processes that influence mountainous soil formation are very dynamic in space and time, and they can be caused by natural hazards such as landslides or by human intervention such as land use change. Human disturbance increases the heterogeneity of soil development in mountainous areas (Egli and Poulenard, 2016).

The fertility of mountainous soil decreases as altitude increases. The harsh climatic conditions and freezing-thawing cycles reduce the stability and water retention capacity of mountainous soil. Plant establishment capacity of mountainous soil decreases with increasing altitude, as does soil potential for organic matter accumulation (Romeo et al., 2015). Despite the harsh conditions, various plant species evolve in mountainous areas and aid in soil erosion prevention. The destruction of vegetative habitats in mountainous environments is caused by hydrological processes that are more active at higher altitudes and have a significant negative impact on soils. Because of the harsh environment, alluvial plains have a higher rate of soil formation than mountain areas (climate and topography). The rate of soil erosion is very high in mountainous regions, indicating that soil formation is slower than soil erosion (Egli and Poulenard, 2016; Romeo et al., 2015).

5.3.2 Soil formation in mountains

Climate, parent material, biological factors, topography, and time are the major factors responsible for soil formation in the mountain ecosystem. Extreme climatic environments fundamentally control soil development in these regions. The nature and rate of weathering are governed by major climatic variables such as temperature and moisture (King et al., 2010). Physical weathering of rocks is favored by cold and arid climates, whereas chemical and biological weathering is favored by warm and moist climates. Mountain areas have complex topography, which affects the soil development process because microclimatic conditions associated with topography have a large impact on the biological, physical, and chemical processes of soil development Because mountain soils are formed primarily by material transported along slopes, slope angle and length have an impact on soil development. The dense vegetation cover that forms a tight root network beneath the soil has the greatest influence on soil stability. The majority of mountain soils are shallow, rocky, infertile, and immature underdeveloped horizons. Entisols, inceptisols, mollisols, histosols, and permafrost are the major soil types found in the mountains (Price and Harden, 2013). Mountains soils are generally deficient in nutrients essential for plant growth, particularly nitrogen. The process of high erosion of loose materials is very common in mountains (Smith and Cleef, 1988). Additionally, due to cool and wet climate, numerous mountain lands gather peat which produces wet and acidic soils.

5.4 Microbial diversity and growth dynamics in mountains

Mountain zone low temperatures and snow cover occurring in most seasons significantly limit the biological, chemical, and physical progressions of mountain soil, affecting the microbial life of mountainous soil (Zumsteg et al., 2013). For a long time, it was assumed that microbes remained silent at subzero temperatures in mountains, but recent research has shown that microbial activity can occur at extremely low temperatures (Panikov et al., 2006). A number of mechanisms of microbial adaptation to low temperatures are now well understood, including adapted cellular compartments that ensure the fluidity of cellular membranes, proper enzyme function at low thermal energy, and prevention of cell death by ice (Nikrad et al., 2016). Because liquid water is required as a solvent for enzymes, membranes, and allowing substrate diffusion, microbial activity in an ice-covered environment is primarily limited to that small unfrozen water with high salts, biofilm materials, and suspended matter (Bakermans, 2008). High-alpine soil microorganisms survive in oligotrophic and severely nutrient-limited conditions, as organic matter declines sharply at higher altitudes due to a lack of vegetation and harsh climatic conditions. However, coarse-textured soils with insufficient plant cover and shallow soils with low water holding capacity offer only marginal protection to microbial community structures from environmental stressors (Rime et al., 2015). Despite these hostile environments, several specially adapted microbes can survive in harsh mountain ecosystems. Microbes in mountain environments frequently form biofilms in order to survive environmental stressors and form biological soil crusts (Hartmann and Frey, 2013).

Climate variables, as well as soil and vegetation properties, are known to be major drivers of microbial community arrangements in mountains. The composition of the soil microbial community in mountains reflects the complexity and heterogeneity of such a harsh environment. Because of the scarcity of studies on methodology and the heterogeneity of mountainous environmental settings, alpha and beta diversity of soil microbes, as well as the implications of microbial diversity for biogeochemical processes and microbial ecosystem services, have received little attention. Soil physicochemical properties (total carbon, total nitrogen, C:N ratio, dissolved organic carbon, pH, nutrients, and moisture), temperature, precipitation, and vegetation type are the primary environmental drivers of bacterial community structure (Yuan et al., 2015). However, the main environmental factors regulating fungal community structures are temperature, SOM level, pH, soil nutrients and plant diversity (Liu et al., 2015). Similarly, the main environmental factors controlling archaeal community structures are pH, soil organic content, clay, cation-exchange capacity, moisture, total nitrogen, potassium, NH_4^+, and NO_3^- (Wang et al., 2015).

Despite the strict environmental settings, mountain soils favor huge microbial diversity (Frey et al., 2016). Microorganisms appear to be more adaptable to extreme environments at higher altitudes and cold environments, and they have distinct distribution patterns when compared to flora and fauna (Shen et al., 2014). Recent studies investigated the bacterial, archaeal or fungal community structure, abundance and phylogenetic relationships (Yue et al., 2015; Liu et al., 2019). Liu et al. (2015) has reported the falling alpha-diversity of arbuscular mycorrhizal fungi along altitude gradient based on phylogenetic marker genes while investigating the microbial diversity on the mountains in China. Similarly, a report suggested the declining bacterial alpha-diversity along altitude gradient in Changbai Mountain (Shen et al., 2015). For archaea, a peak in alpha-diversity at mid-altitudes on Mt. Fuji, Japan was also reported (Singh et al., 2012). Henceforth, the microbial community constructions reflect the alterations in environmental settings related to elevations in mountainous landscapes.

Surveys and studies on the factors influencing microbial community variation along altitudinal gradients are insufficiently studied, with the majority of them focusing on bacterial communities and archaea. On the other hand, surveys related to the study of fungal communities along altitudinal gradients and the factors that influence them are scarce (Zhao et al., 2019). These surveys are primarily concerned with explaining microbial community structure while avoiding microbial dominance measures in altitudinal gradients. To conduct integrative studies, both microbial abundance and composition must be considered (Siles and Margesin, 2016). The majority of recent studies on the effects of climate change on soil have focused on soil warming, precipitation changes, and rising CO_2 levels. Although these studies provide useful statistics in the short term, they provide little information about long-term responses. As a result, it is critical to investigate the role of soil microbes in the context of global warming because it involves temperature increases and rising CO_2 levels in the atmosphere, which may alter the composition of soil microbial communities. As a result, studying the altitudinal gradient will aid in closing this research gap. Furthermore, studies on microbial diversity and richness along elevation gradients in mountainous soils are critical because they will aid in understanding the microbial response of microorganisms to future climate change.

5.4.1 Microbial function in mountainous soil

In mountains, the functional gene dominance and alpha- and beta-diversity among functional genes change with altitude. Along the gradient in mountain environments, a number of different functional communities with stress genes, nitrogen and carbon sequestering genes, and ammonia monooxygenase A (amoA) involved in nitrification, catalyzing oxidation of ammonium nitrite were identified (Čapková et al., 2016). Genes responsible for reduction of nitrate along with the substrate NO_3^- were also identified (Guo et al., 2015). In comparison to other locations, the presence of the Rubisco (ribulose-1,5-bisphosphate carboxylase/oxygenase) gene, which is responsible for CO_2 fixation, and genes responsible for starch and cellulose degradation were reported in low amounts at lower altitude (Liu et al., 2019). The environmental features like soil pH and precipitation best explicate the dominance of bacterial community and archaeal amoA. Yuan et al. (2015) emphasized the importance of nitrogen cycling in mountainous soil, where archaeal amoA quantity decreased with increasing elevation at the surface but persisted at lower depths along a transect of the Tibetan Plateau's mountains. Because CO_2 fixation by autotrophic microbes is important in carbon cycling in mountainous landscapes, the dominance of the cbbL (ribulose bisphosphate carboxylase large chain) gene along the elevation gradient on mountains was also investigated (Guo et al., 2015). It is possible that the bacterial community of cbbL has shifted along the transect due to changes in temperature and water content of soil, nutrients, and vegetation. Rhizobiales, Burkholderiales, and Actinomycetales were discovered to be the most common autotrophic microbes in mountains (Guo et al., 2015).

Microbial communities are responsible for soil development, pioneer plant arrival, and plant growth in high mountain ecosystems, as demonstrated by numerous studies over the last decade (Cartwright et al., 2020). The main colonizers of the common habitations of high altitude mountains are archaea, bacteria, fungi and algae (Ciccazzo et al., 2016). Microbial communities are thought to be the first inhabitants in mountain regions, performing elemental transformation, carbon and nitrogen fixation, pioneer plant growth, and soil fertility maintenance. These practices are important in high mountains to evaluate pedogenic processes and understand the concerns of climate change and rapid glacier melting. Soil microbes play an important role in biogeochemical cycling and are susceptible to environmental changes, which affect soil nutrient cycling. Soil microbes in mountainous forests primarily consume carboxylic acids, amino acids, and carbohydrates, with less consumption of polyamines and polyphenols. Soil microorganisms consume carbon sources in the following order: amino acids > carboxylic acids > polymers > polyphenols in different landscapes: rural natural forest, suburban forest, and urban forest. He et al. (2021) discovered significant differences in the microbial community dynamics of soil in urban-rural forests in the mountains. They also discovered that the microbial communities in urban forests have lower metabolic potential and functional diversity than those in natural forests.

5.4.2 Significance of soil microbial community dynamics in nutrient cycling in high mountain environment

Mountain ranges are found across the world, and their complex topography results in significant climate variation, both between and within regions (Gruber and Haeberli, 2009). Soil microorganisms are primarily responsible for biogeochemical cycles. The geographical distribution of the soil microbial community influences the productivity of soil carbon and nitrogen through interactions with vegetation and a variety of soil parameters. Carbon and nitrogen gradient effects in soil, as essential biogenic elements, could have a significant impact on microbial community diversity. It is difficult to understand the relationship between soil microbes and soil nitrogen and carbon fractions. Carbon and nitrogen sequestration and stability in soil require a thorough understanding of these interactions (Liu et al., 2019). According to the studies, soil microbial community diversity varies across elevational gradients but the same pattern is not followed by the plants and animals (Fierer et al., 2011). The majority of research on soil microorganism pattern diversity is primarily focused on bacterial diversity of mountain ecosystems, while soil archaeal and fungal community studies are scarce, and the available reports require further investigation. The research into the biotic and abiotic factors that explain the diversity of microbial community distribution at mountain elevation is still in its early stages. This is due to the interaction of various environmental parameters and unknown factors related to the soil microbial community (Shen et al., 2020).

Environmental parameters such as temperature, soil physicochemical assets, precipitation, and plant diversity vary with altitude. Changes in these characteristics, as well as their interaction, will have an impact on the geographic distribution patterns of the soil microbial community along an altitudinal gradient, either directly or indirectly. According to recent research, bacteria and fungi respond differently to soil physicochemical properties, causing the bacterial and fungal community structure of soil to shift across an elevation gradient (Shen et al., 2020). The overall distribution patterns of microbial communities in mountainous regions remain unknown. Examining the distribution pattern of soil microbes across an altitudinal gradient suggests relevant information and a useful source for systematically investigating soil microorganism biogeography (Han et al., 2018).

5.5 Response of soil microbes to changing mountain environment

Mountains can be found on every continent and support a wide range of soil microbiota, which are frequently grouped together over short distances. Various environmental conditions change in mountain bionetworks, and these are frequently in opposition to elevation. The general pattern in elevation response of soil microbial community is given in Fig. 5.2. Solar radiations, atmospheric pressure, and temperature are related to environment gradients that changes predictably and rapidly with altitude (Körner, 2007). Because temperature affects the soil microbiota directly and indirectly, decreasing temperature with increasing altitude is an important factor in determining microbial diversity on mountains. Variation in UV-B radiation levels may also affect soil microbe biology, but only in alpine ecosystems above the treeline, where solar radiations can reach the soil's surface. Sundqvist et al. (2013) discovered that lower elevations at warmer temperatures have an impact on the soil microbial ecosystem, both below and above ground. According to evolutionary studies and functional traits, certain fungi prefer warmer temperatures, which has an indirect effect on the community and diversity composition. Warmer temperatures at lower elevations result in a faster rate of decomposition, more soluble nutrients in the soil, and less SOM accumulation, resulting in greater bacterial diversity over fungi (Whitaker et al., 2014; Kotas et al., 2018). Because the concentrations of potassium,

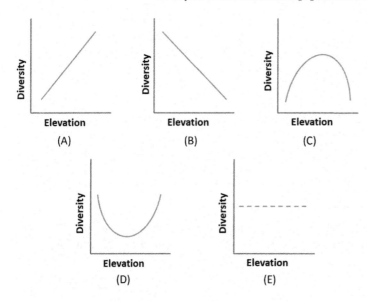

FIGURE 5.2 The general pattern in elevation response of soil microbial community in mountain landscape. (A) Increasing microbial diversity, (B) declining microbial diversity, (C) mid-elevation microbial diversity, (D) U shaped microbial diversity, (E) no trend microbial diversity.

nitrogen, and phosphorus decrease with elevation, mountain forests have fewer available nutrients than lowland forests. Mountain soils have more SOM per unit area, and nitrogen mineralization levels are lower at higher altitudes, whereas tropical mountains are thought to be phosphorus limited, and mountain forests are nitrogen limited (Tanner et al., 1998). Temperature influences microbial diversity by influencing plant communities. Warmer ecosystems at lower elevations have more biomass, photosynthesis, and a higher diversity of tree species (Hättenschwiler et al., 2005; Hättenschwiler and Jørgensen, 2010). Similarly, as the quality and quantity of carbon input decreases and plant ecological processes change at higher elevations and cooler temperatures, the soil microbial composition is disturbed. The cooler temperatures at higher elevations result in poor litter quality and a slow decomposition rate, favoring fungi over bacteria and causing a high accumulation of organic matter. Mountains will eventually become an important global carbon reservoir as a result of this. The accumulation of SOM in high elevation ecosystems determines the diversity and structure of microbes in aboveground taxonomy, as it has been discovered globally that bacterial, fungal, and archaeal communities are positively related to soil carbon levels (Angel et al., 2010; Tedersoo et al., 2014; Delgado-Baquerizo et al., 2016). Variations in soil moisture and precipitation on elevation gradients have an indirect and direct impact on microbial diversity and soil composition, but precipitation has a less consistent global pattern on mountain gradients than temperature gradients. As soil moisture increases, microbial diversity is wetted, resulting in a decrease in soil oxygen levels. These variables have an impact on the abundance, ecosystem level, and physiology of the microbial community. When combined with low transpiration at cooler temperatures, various mountain ecosystems receive high precipitation levels. At higher elevations, shallow mountain soils, orographic cloud cover, anoxic soil conditions, and waterlogging occur, primarily on tropical mountains. Similarly, edaphic factors change with altitude, such as decreasing soil depth at higher altitudes. However, as elevation increases, the pH of the soil becomes acidic due to high organic carbon by-products. Recent research on the diversity pattern of soil microbiota in mountain gradients has focused on bacteria, followed by fungi and archaea (Ciccazzo et al., 2016). The major investigations on the basis of changing mountain gradients with respect to soil microbial diversity has occurred in Asia (47%), majorly in (14%) Tibetan plateau, prevailing in North America (10%), Central America (2%), South America (13%), Oceania (5%). and Europe (23%). In 2017, Peay investigated the pattern of microbial diversity of fungi, bacteria, and archaea (more than one microbial group) with respect to single mountain gradient.

5.6 The impact of climate change on microbes and carbon and nitrogen cycle of mountains

Soil microbes have major role in biogeochemical cycles at local and global levels (Schimel and Schaeffer, 2012). Mountain areas have recently gained significant attention because of their importance in storage of terrestrial carbon (Hagedorn et al., 2010; Chang et al., 2014; Liu et al., 2016). Microbes are essential in the global C-budget equation because they play critical roles in all major element cycles. Mountain biogeochemical cycles would be more severely impacted than lowland biogeochemical cycles due to higher temperatures in mountain regions compared to the global average, similarly to Arctic biogeochemical cycles (Rousk and Bengtson, 2014). As the temperature rises, all biogeochemical reactions can accelerate, increasing photosynthesis and respiration in an ecosystem's input and output. Temperature, moisture, nutrient and substrate availability, soil physicochemical properties, and biotic interaction are all factors that influence microbial reactions. Each of these variables is expected to change in response to temperature changes or to be influenced by temperature changes. As a result, complex interactions and feedbacks between physicochemical and biological components are required for net carbon and nitrogen budgets (Donhauser and Frey, 2018).

In a comprehensive global assessment of warming trials, the warming effect of soil organic carbon (SOC) stocks was found to be strongly dependent on the starting SOC concentration, regardless of temperature or soil features (Crowther et al., 2016). According to the findings, with warming of climate, nival zone soils with low carbon stocks are more likely to become carbon sinks (Ohtsuka et al., 2008). Increasing primary production overcomes increased respiration and increased leaching losses in space for time substitution studies (Guelland et al., 2013). According to a global scale study on N-cycling, increasing atmospheric temperature leads to increased net nitrogen mineralization and nitrification, as well as an increase in soil organic and plant nitrogen pools (Bai et al., 2013). The effects of global warming differed dramatically across habitats (Dawes et al., 2017). The increase in total soil nitrogen leads to an increase in primary production in the extreme oligotrophic alpine region due to a long-term rise in temperature. Chronosequence studies, such as C-cycling, indicate that total soil nitrogen will increase with warming in the long run due to increased primary production in extreme oligotrophic alpine regions. However, accurate, dependable predictions of the amount and timing of carbon and nitrogen cycling variations as a result of warming can only be made by understanding the complex biotic and abiotic interactions that coexist with direct warming effects (Donhauser and Frey, 2018).

In Fig. 5.3, both photosynthesis and respiration dominate the carbon cycle. Carbon in the atmosphere is primarily transported to the soil by "carbon-fixing" autotrophic microorganisms such as photosynthetic plants and photo or chemoautotrophic microorganisms that convert atmospheric CO_2 into SOM. Fixed carbon is then released back into the atmosphere through a variety of processes, the most common of which are fermentation, methanogenesis, and methane oxidation. In another process, "organic carbon-consuming" microorganisms convert SOM into atmospheric carbon by using carbon from plants, animals, and microbes for metabolic activities, storing carbon in biomass, and releasing residual CO_2 back into the atmosphere (Fig. 5.4). Indicates that increasing temperature influences nitrogen cycles. Increase in temperature leads to the increase in microbial

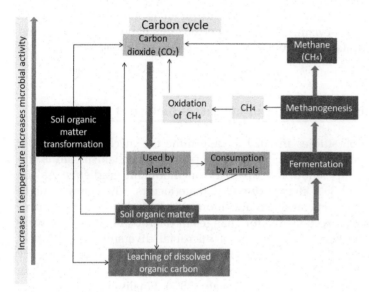

FIGURE 5.3 Carbon cycle in mountainous soil.

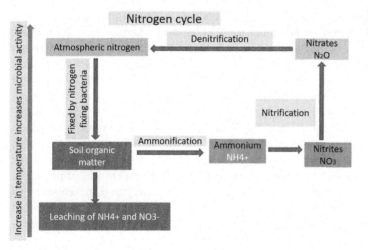

FIGURE 5.4 Nitrogen cycle in mountainous soil.

2. Soil microbial processes and nutrient dynamics

activity. N-cycling reaction which mainly includes nitrogen fixation (symbiotic and non-symbiotic association with plants), nitrification (*nitrosomonas* and nitrobacter) in aerobic conditions and ammonification (*bacillus* and *clostridium* etc.) and denitrification (*thiobacillus* and *micrococcus* etc.) in anaerobic conditions. The effect of global warming increases soil microbial activity in the carbon and nitrogen cycles of mountainous soil. Mountainous soil microorganisms will become more active in all carbon and nitrogen cycle reactions as a result of global warming. Microbial activity will increase SOM mineralization, but SOM uptake by plants and carbon and nitrogen fixing microorganisms may also accelerate at higher temperatures. Because of the high hydrological activity in mountainous environments, a large amount of carbon and nitrogen drains with the soil solution. Organic matter mineralization, methanogenesis, and denitrification emit CO_2, CH_4, and N_2O, which are reduced to N_2 by photosynthesis, methane oxidation, and N_2O reduction to N_2. It is impossible to predict how the balance of distinct C- and N-cycling reactions, and thus greenhouse gas uptake and emission, would change in a mountainous environment in response to climate change.

5.7 Mechanism responsible for changes in mountain biological community

Lobby and Martin reviewed 65 studies in 2020 to determine the mechanisms driving changes in microbial biology, and they discovered that primary mechanism shifts are responsible for soil microbial composition and diversity on mountain gradients. As the primary mechanism, they reported 36% change due to climate factors, 30% change due to edaphic factors, 18% change due to vegetation composition, and 16% change due to nutrients and SOM. The foremost factors responsible for changes in microbial community were not consistent across taxa. For bacteria, key drivers of microbial diversity were edaphic factors (48%) followed by climate (30%). Likewise, for fungi key drivers of community composition were climate (48%), and edaphic factors (24%). For mycorrhizae fungi main drivers across elevation were vegetation (44%) and climate (38%). Finally, for archaea only edaphic factors (77%) were driving changes in the community composition. These findings show how edaphic and climate changes along gradients act as taxon-specific and major determinants of soil microbial community structure. Lobby and Martin were also interested in whether archaea, fungi, and bacteria show coordinated responses across gradients in addition to cross-linked patterns of induvial taxa. In this regard, they discovered and reported that patterns of diversity primarily diverge between taxa along a similar gradient of elevation, as well as the factors responsible for these patterns of diversity. Variations in precipitation and temperature determined the pattern of diversity differently for the induvial class of soil microbial community, according to studies on tropical mountains (Peay et al., 2017; Nottingham et al., 2016). For example, on the Hawai Island along the elevation gradient of 950 m, the factors accompanying the upsurge in richness were different for induvial taxa that is archaea oxidizing ammonia increased with mineralization of nitrogen, increased soil pH and carbon upsurged the bacterial community, and fungi increased with higher soil carbon. Therefore, the non-linear and distinctive patterns on tropical mountain are more prone to have complex effects on diversity of microbes found in rich species ecosystem in future.

5.8 Changing function of microbial diversity with effect of contemporary environment

Higher complexity of topography and climate changes promotes speciation and endemism, increases isolation and stress that leads to higher turnover rates along with elevation gradients (Decaëns, 2010; Steinbauer et al., 2016). Frequent changes in abiotic and biotic factors on mountains acts as trail filters that has been linked to changes in microbial diversity across elevations (Caiafa et al., 2017). The sheer gradients on mountain emphasizes level of nutrients, edaphic and climate sensitivity, and microbial diversity traits. These factors mainly effect function of ecosystems such as nitrogen cycle and decomposition in which soi bacteria and fungi interacts in synergistic manner. This synergistic relationship promotes uptake and fixation of plant nitrogen, elevates stress tolerance of plants (Giauque and Hawkes, 2013), and play role of tailored and synergism in decomposition (Schneider et al., 2012), each of these can be reformed when taxonomic group turnover at various rates on mountain gradients. Mountain gradients role of changing functional roles in microbe diversity is a promising path for improving the understanding in what way the microbe mediated nutrient cycles (nitrogen and carbon) vary with changing elevations in mountains.

5.8.1 Nutrient limitations, mountain conditions, and decomposers

Decomposition rates are lower at higher elevations due to low soil fertility, wet soils, and cool temperatures, which result in poor litter, limited nutrient cycles, and a lower detritus breakdown rate in mountain forest soils. Such cycles are most common in humid tropics mountain forests at elevations with regular orographic cloud cover. The microbial composition and functions of dominant soil taxonomic groups are extremely sensitive to changes in biotic and abiotic conditions that shape decomposition dynamics. The bacterial and fungal decomposition of cellulose is affected by nutrient limitation at high elevations. For example, a large number of genes involved in carbon breakdown, such as exoglucanase and isopullulanase, were discovered at higher elevations on the Tibetan Plateau (Yang et al., 2014). Diversity of microbes also tracks the fertility gradient in mountain ecosystem as reported in China at Changbai Mountains.

5.8.2 Nitrogen cycling and microbial function

The changes caused by microbes in the nitrogen cycle with elevation gradients are less well understood than decomposition, and the majority of this research has been focused on tropical mountains. Nitrogen limitation increases with elevation, which increases the ratio of soil carbon: nitrogen, resulting in dominant fungi communities and a higher demand for nitrogen by microbes in mountain upper ecosystems. There are more studies on how availability of nutrients affects the fixation of nitrogen in mountain ecosystems (Matson et al., 2015) whereas there is less evidence on how nitrogen fixing bacteria diverge with elevations. Elevation responses of fungi symbionts, on the other hand, have been extensively studied, with ectomycorrhizal species increasing with elevation and endomycorrhizal species decreasing. As nitrogen availability decreases with elevation, the functions of microbes in the nitrogen cycle change, as do the activities of the symbiotic microbial community on similar gradients.

5.8.3 Climate change and microbe traits variation

It is thought that following the differential responses of microbial diversity along with its gradient is important for studying the effects of climate change using mountains. Various fungi have characteristics that make them decomposers or stress tolerators. Stress tolerators contribute significantly to the accumulation of SOM in soil by producing recalcitrant compounds, whereas decomposers reduce SOM. As climate conditions change, there may be tradeoffs between microbial communities that move mountains toward carbon sources or sinks (Hotaling et al., 2017). In boreal forests, for example, mycorrhizal necromass is a higher constituent of SOM than plant litter, and parallel driving shifts in microbe biology in mountain regions play an important role in the carbon cycle by changing carbon storage. Microbial diversity tracking along elevation gradients is also important because greenhouse gases such as methane, nitrous oxide, and carbon dioxide are released during microbe-mediated progressions. During decomposition, fungi and bacteria denitrify nitrate and emit carbon dioxide. Moreover, bacteria and archaea are consumers and producers of methane. Tropical forest soils are considered important budgets for greenhouse gases. In 2011, Sousa and coresearchers observed that in Brazilian Atlantic Forest at 900 m gradient elevation greenhouse gases discharge was strongly associated to moisture and soil temperature, signifying that warm climate might support higher nitrogen fluxes and decomposition (Sousa Neto et al., 2011).

5.9 Techniques for identification of soil microbes at different mountain environment

The analysis of soil microbial diversity not only identifies total microbial biomass and community diversity, but it also discovers microbe growth, distribution, interactions, and functions among species. As a result, accurate and dependable molecular techniques are discussed below to investigate how the dynamics and structure of microbes affect the function of mountain ecosystems.

5.9.1 Calorimetric and microscopic based techniques

Epifluorescence microscopy is a technique the microbes existing in the sample are labeled with fluoresce marker (Watteau and Villemin, 2018; Khan et al., 2020). This microscope has a filter that allows the cell to appear in a specific color. Transmission electron microscopes and scanning electron microscopes have also been used to identify bacteria (determines the surface of microbes).

5.9.2 Clone library method

Before the invention of high throughput sequencing, a technique that was widely used was clone library-based analysis. In this, product amplified with PCR from DNA samples are directed to form clones, these are sequenced containing fragments of 16S rRNA genes. Next, for the identification of taxonomic group and phylogenetic analysis, sequenced 16S rRNA genes are compared with databases such as SILVA, ribosomal database project, and greengene. In 2002, Dunbar, showed that soil sample requires forty thousand clones to document richness upto 50% (Dunbar et al., 2002). In 2016, Pascual and co-workers used clone library technique for the determination of bacterial diversity in the rhizosphere region. Similarly, in 2016 Čapková studied the diversity of cyanobacteria on the basis of molecular phylogenesis using 16S rRNA gene sequencing from the high altitudes of Western Himalayas (Čapková et al., 2016). The main benefit of using this technique is that one induvial can handle thousands of clones at once. This technique, however, has the disadvantage of being labor intensive and time consuming.

5.9.3 Polymerase chain reaction techniques

It is a technique for amplifying a sequence of RNA or DNA using primers and then separating the resulting product using gel electrophoresis. Siles and Margesin used quantitative PCR to study microbial diversity (archaea, fungi, and bacteria) and changes in composition and abundance along an altitude gradient in 2016 (Siles and Margesin, 2016).

5.9.3.1 Denaturing gradient gel electrophoresis

It is a technique based on the different fragments of DNA mostly 200–700 bp, denatured in different range of temperatures of 50°C–65°C in the presence of chemicals such as urea. The DNA fragment with more denaturation will have the lesser electrophoretic mobility (Singh et al., 2006).

5.9.3.2 Terminal restriction fragment length polymorphism (TRFLP)

Another PCR-based method for determining the microbial community. Marker genes are linked with fluorescently labeled primers, followed by restriction digestion, separation, and analysis on an automated sequencer. Only label terminal restriction fragments are detected, and their length heterogeneity demonstrates the microbial community complexity as seen on the electropherogram. TRFLP electrophoresis is performed on a sequencing gel, which provides high sensitivity, resolution, and precise size of induvial fragments. Each induvial genetic variant represents a single peak, whereas the height of the peak represents the community's relative abundance.

5.9.3.3 Real time quantitative PCR

The early exponential phase reaction allows for the quantification and detection of PCR amplicons using this technique. This PCR method is more capable of quantifying the number of gene copies present in the sample. It includes fluorescent markers for quantifying the product at the end of each cycle of amplification, and the amount of fluorescence is directly related to the amount of product at the end of each cycle. The initial target DNA concentration is linked to the threshold cycle, which is defined as the number of cycles in which fluorescence rises above the background level. Finally, the DNA target is quantified using a calibration curve that is linked to the threshold cycles.

5.9.4 Hybridization techniques

5.9.4.1 Fluorescence in situ hybridization (FISH)

FISH is an excellent technique for rapid and reliable microorganism identification from soil samples, and it can identify prokaryotic microbes in the absence of cultivation. It entails the hybridization of complementary oligodeoxynucleotides to ribosomal RNA sequences, which include phylogenetic group sequence signatures. In laboratory, FISH is carried out by fixing the whole cells of soil samples containing microbes by paraformaldehyde or ethanol, and their 23S OR 16S rRNA is hybridized through fluorescently labeled probes. The labeled cells are then examined under confocal microscopy. Ma and colleagues determined the microbe-mediated community in different vegetation and altitude gradients in Xinjiang Mountain in 2004 using the FISH (in situ hybridization) staining technique, which provided an excellent model for investigation.

5.9.4.2 Membrane and solution hybridization

Membrane hybridization involves the hybridization of species-specific or group probes with a community of RNA or DNA immobilized on a membrane to produce relative community abundance. Microbe double-stranded DNA is taken via thermal reassociation and dissociation in solution hybridization, which estimates the DNA complexity community.

5.9.5 Next generation sequencing

Conventional DNA sequencing is based on Sanger's 1977 invention of the dideoxy chain termination technique. However, novel DNA sequencing techniques provide high throughput and speed, as they can complete projects in a week that took several years to complete using the Sanger technique. The advantage of this technique is that it can determine the data sequence from single amplified DNA fragments, eliminating the need for DNA fragment clones (Ansorge, 2009). Nonetheless, recent advances in NGS have enabled a variety of available platforms and methods for sequencing with varying capabilities and costs. This technique has been used by various researchers to assess soil microbes (Schöler et al., 2017; Rawat and Joshi, 2019). Pyrosequencing is a new option for cost-effective, high-throughput sequencing. It is based on the detection of pyrophosphate release during DNA synthesis. This technique eliminates the need for cloning, thereby eliminating cloning-related artifacts and the production of aberrant recombinants.

5.10 Conclusions and future perspectives

The current chapter describes the high microbial diversity of mountainous soils and opens up new avenues for microbial resources that could be used in a variety of biotechnological and pharmaceutical applications. To accomplish this, advanced and novel molecular tools such as metagenomics, as well as traditional isolation of cold-adapted microorganisms, can be useful for exploring such microbial assets. On the other hand, isolation methods such as high-throughput isolation chips and advanced sequencing techniques have been developed to isolate microbes from mountainous soil for the purposes of studying their ecological role and utilizing their biotechnological potential. This chapter discusses the climate and soil interactions in mountains, the composition of microbial groups in such conditions, and the impact of global climate change on diversity, diversity, and function of microbes in mountain soils. A deeper understanding of microbial diversity and their interactions in mountains, as well as the impact of climate change on microbial diversity and function in mountains, has been developed through extensive research and studies.

In this study, we propose four major areas of study for microbial diversity and their processes along altitude gradients. To begin, proper effort should be made to identify and quantify the mechanisms controlling microbial abundance, structure, and functions across altitude gradients in the mountains. An integrated model of environmental selection, competition, and spatial assets in constructing bacterial groups along an altitude gradient can be used for this. The second area is the need for a large number of long-term control trials at different altitudes in order to improve our understanding of observational studies and identify the exact mechanisms driving microbial communities. The upcoming research should include, among other things, through fall exclusions, soil transplants, and plant-soil reciprocal transplants. The third area of focus is standardizing the methods for replicating sampling on the same and across mountains at the regional and global levels, as this is an important way to govern if strong, universal patterns and responsible tools exist. Both observational and experimental research are thought to require regulation of high-throughput sequencing techniques in order to allow quantitative analyses of multi-study datasets. The fourth suggested area is to look into elevation diversity patterns using functional traits. This field is thought to benefit from a focus on microbial function rather than microbial diversity. Direct measurements of microbial function should be included in the studies. These statistics will be more useful for ecosystem models that interpret the balances between stress tolerance and enzyme production. With the incorporation of the aforementioned recommendations, it is hoped that future research will improve understanding of the fundamental tools and key environmental characteristics determining the forms of microbial community structure in mountain ecosystems, as well as aid in the prediction of ecosystem-level climate change concerns. It is possible to say that generalized patterns will remain undefinable until the precise taxonomic and functional category are specified and improvements are made to the sequencing methods. Although significant work has been done and this field has advanced rapidly, much more work remains to be done in order to synthesize the biogeography and functional ecology of soil microbes in mountainous ecosystems. Furthermore, biogeographical approaches may be used to reveal the spatial distribution and mechanisms of microbial adaptation to the local environment of mountains. The seasonality and depth-related configurations that shape the microbial ecology of mountains should be studied more closely, as these areas have previously been overlooked. To accomplish this, comprehensive interdisciplinary studies are needed to gain knowledge of microbial ecology at the ecosystem level. Furthermore, investigating the changes in symbiotic or competitive interactions of microbes as a result of environmental conditions and analyzing co-occurrence patterns may provide a systematic understanding of microbial responses to warming. Furthermore, uncultivable techniques such as metagenomic and metatranscriptomic could be successfully used in mountains to identify the important microbial functions that shape the ecosystem of that environment. These approaches may help to uncover the link between microbial functional potential and biogeochemical dynamics.

References

Angel, R., Soares, M.I.M., Ungar, E.D., Gillor, O., 2010. Biogeography of soil archaea and bacteria along a steep precipitation gradient. The ISME Journal 4 (4), 553–563.

Ansorge, W.J., 2009. Next-generation DNA sequencing techniques. New Biotechnology 25 (4), 195–203.

Armas-Herrera, C.M., Badía-Villas, D., Mora, J.L., Gómez, D., 2020. Plant-topsoil relationships underlying subalpine grassland patchiness. Science of the Total Environment 712, 134483.

Bai, E., Li, S., Xu, W., Li, W., Dai, W., Jiang, P., 2013. A meta-analysis of experimental warming effects on terrestrial nitrogen pools and dynamics. New Phytologist 199 (2), 441–451.

Bakermans, C., 2008. Limits for microbial life at subzero temperatures. Psychrophiles: From Biodiversity to Biotechnology. Springer, Berlin, Heidelberg, pp. 17–28.

Barry, R.G., 2008. Mountains and their climatological study. Mountain Weather and Climate 3, 1–23.

Caiafa, M.V., Gómez-Hernández, M., Williams-Linera, G., Ramírez-Cruz, V., 2017. Functional diversity of macromycete communities along an environmental gradient in a Mexican seasonally dry tropical forest. Fungal Ecology 28, 66–75.

Čapková, K., Hauer, T., Řeháková, K., Doležal, J., 2016. Some like it high! Phylogenetic diversity of high-elevation cyanobacterial community from biological soil crusts of western Himalaya. Microbial Ecology 71 (1), 113–123.

Cartwright, J.M., Littlefield, C.E., Michalak, J.L., Lawler, J.J., Dobrowski, S.Z., 2020. Topographic, soil, and climate drivers of drought sensitivity in forests and shrublands of the Pacific Northwest, USA. Scientific Reports 10 (1), 1–13.

Chang, X., Wang, S., Cui, S., Zhu, X., Luo, C., Zhang, Z., et al., 2014. Alpine grassland soil organic carbon stock and its uncertainty in the three rivers source region of the Tibetan Plateau. PLoS One 9 (5), e97140.

Ciccazzo, S., Esposito, A., Borruso, L., Brusetti, L., 2016. Microbial communities and primary succession in high altitude mountain environments. Annals of Microbiology 66 (1), 43–60.

Classen, A.T., Sundqvist, M.K., Henning, J.A., Newman, G.S., Moore, J.A., Cregger, M.A., et al., 2015. Direct and indirect effects of climate change on soil microbial and soil microbial-plant interactions: what lies ahead? Ecosphere 6 (8), 1–21.

Collins, M., Knutti, R., Arblaster, J., Dufresne, J.L., Fichefet, T., Friedlingstein, P., et al., 2013. Long-term climate change: projections, commitments and irreversibility. Climate Change 2013—The Physical Science Basis: Contribution of Working Group I to the Fifth Assessment Report of the Intergovernmental Panel on Climate Change. Cambridge University Press, pp. 1029–1136.

Crowther, T.W., Todd-Brown, K.E., Rowe, C.W., Wieder, W.R., Carey, J.C., Machmuller, M.B., et al., 2016. Quantifying global soil carbon losses in response to warming. Nature 540 (7631), 104–108.

Dawes, M.A., Schleppi, P., Hättenschwiler, S., Rixen, C., Hagedorn, F., 2017. Soil warming opens the nitrogen cycle at the alpine treeline. Global Change Biology 23 (1), 421–434.

Decaëns, T., 2010. Macroecological patterns in soil communities. Global Ecology and Biogeography 19 (3), 287–302.

Delgado-Baquerizo, M., Maestre, F.T., Reich, P.B., Trivedi, P., Osanai, Y., Liu, Y.R., et al., 2016. Carbon content and climate variability drive global soil bacterial diversity patterns. Ecological Monographs 86 (3), 373–390.

Donhauser, J., Frey, B., 2018. Alpine soil microbial ecology in a changing world. FEMS Microbiology Ecology 94 (9), fiy099.

Dunbar, J., Barns, S.M., Ticknor, L.O., Kuske, C.R., 2002. Empirical and theoretical bacterial diversity in four Arizona soils. Applied and Environmental Microbiology 68 (6), 3035–3045.

Egli, M., Poulenard, J., 2016. Soils of mountainous landscapes. International Encyclopedia of Geography: People, the Earth, Environment and Technology. Wiley, pp. 1–10.

Fierer, N., McCain, C.M., Meir, P., Zimmermann, M., Rapp, J.M., Silman, M.R., et al., 2011. Microbes do not follow the elevational diversity patterns of plants and animals. Ecology 92 (4), 797–804.

Frey, B., Rime, T., Phillips, M., Stierli, B., Hajdas, I., Widmer, F., et al., 2016. Microbial diversity in European alpine permafrost and active layers. FEMS Microbiology Ecology 92 (3), fiw018.

Gavazov, K., Ingrisch, J., Hasibeder, R., Mills, R.T., Buttler, A., Gleixner, G., et al., 2017. Winter ecology of a subalpine grassland: effects of snow removal on soil respiration, microbial structure and function. Science of the Total Environment 590, 316–324.

Giauque, H., Hawkes, C.V., 2013. Climate affects symbiotic fungal endophyte diversity and performance. American Journal of Botany 100 (7), 1435–1444.

Glassman, S.I., Wang, I.J., Bruns, T.D., 2017. Environmental filtering by pH and soil nutrients drives community assembly in fungi at fine spatial scales. Molecular Ecology 26 (24), 6960–6973.

Gruber, S., Haeberli, W., 2009. Mountain permafrost. Permafrost Soils. Springer, Berlin, Heidelberg, pp. 33–44.

Guelland, K., Hagedorn, F., Smittenberg, R.H., Göransson, H., Bernasconi, S.M., Hajdas, I., et al., 2013. Evolution of carbon fluxes during initial soil formation along the forefield of Damma glacier, Switzerland. Biogeochemistry 113 (1), 545–561.

Guo, G., Kong, W., Liu, J., Zhao, J., Du, H., Zhang, X., et al., 2015. Diversity and distribution of autotrophic microbial community along environmental gradients in grassland soils on the Tibetan Plateau. Applied Microbiology and Biotechnology 99 (20), 8765–8776.

Hagedorn, F., Martin, M., Rixen, C., Rusch, S., Bebi, P., Zürcher, A., et al., 2010. Short-term responses of ecosystem carbon fluxes to experimental soil warming at the Swiss alpine treeline. Biogeochemistry 97 (1), 7–19.

Han, D., Wang, N., Sun, X., Hu, Y., Feng, F., 2018. Biogeographical distribution of bacterial communities in Changbai Mountain, Northeast China. MicrobiologyOpen 7 (2), e00529.

Hartmann, D.L., Tank, A.M.K., Rusticucci, M., Alexander, L.V., Brönnimann, S., Charabi, Y.A.R., et al., 2013. Observations: atmosphere and surface. Climate Change 2013 the Physical Science Basis: Working Group I Contribution to the Fifth Assessment Report of the Intergovernmental Panel on Climate Change. Cambridge University Press, pp. 159–254.

Hättenschwiler, S., Jørgensen, H.B., 2010. Carbon quality rather than stoichiometry controls litter decomposition in a tropical rain forest. Journal of Ecology 98 (4), 754–763.

Hättenschwiler, S., Tiunov, A.V., Scheu, S., 2005. Biodiversity and litter decomposition in terrestrial ecosystems. Annual Review of Ecology Evolution and System 36, 191–218.

He, Y., Li, C.T., Yu, Y.C., He, H.P., Tao, X., 2021. Variation of subtropical forest soil microbial biomass and soil microbial community functional characteristics along an urban-rural gradient. Ying Yong Sheng tai xue bao = The Journal of Applied Ecology 32 (1), 93–102.

Hotaling, S., Hood, E., Hamilton, T.L., 2017. Microbial ecology of mountain glacier ecosystems: biodiversity, ecological connections and implications of a warming climate. Environmental Microbiology 19 (8), 2935–2948.

Khan, M.S.I., Oh, S.W., Kim, Y.J., 2020. Power of scanning electron microscopy and energy dispersive X-ray analysis in rapid microbial detection and identification at the single cell level. Scientific Reports 10 (1), 1–10.

King, A.J., Karki, D., Nagy, L., Racoviteanu, A., Schmidt, S.K., 2010. Microbial biomass and activity in high elevation (>5100 meters) soils from the Annapurna and Sagarmatha regions of the Nepalese Himalayas. Himalayan Journal of Sciences 6 (8), 11–18.

Kooch, Y., Noghre, N., 2020. The effect of shrubland and grassland vegetation types on soil fauna and flora activities in a mountainous semi-arid landscape of Iran. Science of the Total Environment 703, 135497.

Körner, C., 2007. The use of 'altitude' in ecological research. Trends in Ecology & Evolution 22 (11), 569–574.

Körner, C., Paulsen, J., Spehn, E.M., 2011. A definition of mountains and their bioclimatic belts for global comparisons of biodiversity data. Alpine Botany 121 (2), 73–78.

Kotas, P., Šantrůčková, H., Elster, J., Kaštovská, E., 2018. Soil microbial biomass, activity and community composition along altitudinal gradients in the High Arctic (Billefjorden, Svalbard). Biogeosciences 15 (6), 1879–1894.

Liu, L., Hart, M.M., Zhang, J., Cai, X., Gai, J., Christie, P., et al., 2015. Altitudinal distribution patterns of AM fungal assemblages in a Tibetan alpine grassland. FEMS Microbiology Ecology 91 (7), fiv078.

Liu, S., Zhang, F., Du, Y., Guo, X., Lin, L., Li, Y., et al., 2016. Ecosystem carbon storage in alpine grassland on the Qinghai Plateau. PLoS One 11 (8), e0160420.

Liu, M., Sui, X., Hu, Y., Feng, F., 2019. Microbial community structure and the relationship with soil carbon and nitrogen in an original Korean pine forest of Changbai Mountain, China. BMC Microbiology 19 (1), 1–14.

Looby, C.I., Martin, P.H., 2020. Diversity and function of soil microbes on montane gradients: the state of knowledge in a changing world. FEMS Microbiology Ecology 96 (9), fiaa122.

Ma, X., Chen, T., Zhang, G., Wang, R., 2004. Microbial community structure along an altitude gradient in three different localities. Folia Microbiologica 49 (2), 105–111.

Margesin, R., 2012. Psychrophilic microorganisms in alpine soils. Plants in Alpine Regions. Springer, Vienna, pp. 187–198.

Matson, A.L., Corre, M.D., Burneo, J.I., Veldkamp, E., 2015. Free-living nitrogen fixation responds to elevated nutrient inputs in tropical montane forest floor and canopy soils of southern Ecuador. Biogeochemistry 122 (2), 281–294.

Nagy, L., Grabherr, G., 2009. The Biology of Alpine Habitats. Oxford University Press on Demand.

Nie, Y.Y., Wang, H.H., Li, X.J., Ren, Y.B., Jin, C.S., Xui, Z.K., et al., 2018. Characteristics of soil organic carbon mineralization in low altitude and high altitude forests in Wuyi Mountains, southeastern China. Ying yong sheng tai xue bao = The Journal of Applied Ecology 29 (3), 748−756.

Nikrad, M.P., Kerkhof, L.J., Häggblom, M.M., 2016. The subzero microbiome: microbial activity in frozen and thawing soils. FEMS Microbiology Ecology 92 (6), fiw081.

Nottingham, A.T., Turner, B.L., Whitaker, J., Ostle, N., Bardgett, R.D., McNamara, N.P., et al., 2016. Temperature sensitivity of soil enzymes along an elevation gradient in the Peruvian Andes. Biogeochemistry 127 (2−3), 217−230.

Ohtsuka, T., Hirota, M., Zhang, X., Shimono, A., Senga, Y., Du, M., et al., 2008. Soil organic carbon pools in alpine to nival zones along an altitudinal gradient (4400−5300m) on the Tibetan Plateau. Polar Science 2 (4), 277−285.

Panikov, N.S., Flanagan, P.W., Oechel, W.C., Mastepanov, M.A., Christensen, T.R., 2006. Microbial activity in soils frozen to below − 39C. Soil Biology and Biochemistry 38 (4), 785−794.

Pauli, H., Gottfried, M., Lamprecht, A., Niessner, S., Rumpf, S., Winkler, M., et al., 2015. The GLORIA Field Manual−Standard Multi-Summit Approach, Supplementary Methods and Extra Approaches. GLORIA-Coordination, Austrian Academy of Sciences & University of Natural Resources and Life Sciences, Vienna. Globalo Bservation Research in Itiat.

Peay, K.G., von Sperber, C., Cardarelli, E., Toju, H., Francis, C.A., Chadwick, O.A., et al., 2017. Convergence and contrast in the community structure of Bacteria, Fungi and Archaea along a tropical elevation−climate gradient. FEMS Microbiology Ecology 93, 5.

Pepin, N., Bradley, R.S., Diaz, H.F., Baraër, M., Caceres, E.B., Forsythe, N., et al., 2015. Elevation-dependent warming in mountain regions of the world. Nature Climate Change 5 (5), 424−430.

Phillips, M., Schweizer, J., 2007. Effect of mountain permafrost on snowpack stability. Cold Regions Science and Technology 47 (1−2), 43−49.

Price, L.W., Harden, C.P., 2013. Mountain soils. In: Price, M.F., Byers, A.C., Friend, D.A., et al.,Mountain Geography. University of California Press, Berkeley, CA, pp. 167−182.

Rangwala, I., Miller, J.R., 2012. Climate change in mountains: a review of elevation-dependent warming and its possible causes. Climatic Change 114 (3), 527−547.

Rawat, N., Joshi, G.K., 2019. Bacterial community structure analysis of a hot spring soil by next generation sequencing of ribosomal RNA. Genomics 111 (5), 1053−1058.

Reyes, W.M., Epstein, H.E., Li, X., McGlynn, B.L., Riveros-Iregui, D.A., Emanuel, R.E., 2017. Complex terrain influences ecosystem carbon responses to temperature and precipitation. Global Biogeochemical Cycles 31 (8), 1306−1317.

Rime, T., Hartmann, M., Brunner, I., Widmer, F., Zeyer, J., Frey, B., 2015. Vertical distribution of the soil microbiota along a successional gradient in a glacier forefield. Molecular Ecology 24 (5), 1091−1108.

Romeo, R., Vita, A., Manuelli, S., Zanini, E., Freppaz, M., Stanchi, S., 2015. Understanding Mountain Soils: A Contribution from Mountain Areas to the International Year of Soils 2015. FAO, Rome.

Rousk, J., Bengtson, P., 2014. Microbial regulation of global biogeochemical cycles. Frontiers in Microbiology 5, 103.

Schimel, J., Schaeffer, S.M., 2012. Microbial control over carbon cycling in soil. Frontiers in Microbiology 3, 348.

Schneider, T., Keiblinger, K.M., Schmid, E., Sterflinger-Gleixner, K., Ellersdorfer, G., Roschitzki, B., et al., 2012. Who is who in litter decomposition? Metaproteomics reveals major microbial players and their biogeochemical functions. The ISME Journal 6 (9), 1749−1762.

Schöler, A., Jacquiod, S., Vestergaard, G., Schulz, S., Schloter, M., 2017. Analysis of soil microbial communities based on amplicon sequencing of marker genes. Biology and Fertility of Soils 53 (5), 485−489.

Schulz, S., Brankatschk, R., Dümig, A., Kögel-Knabner, I., Schloter, M., Zeyer, J., 2013. The role of microorganisms at different stages of ecosystem development for soil formation. Biogeosciences 10 (6), 3983−3996.

Shen, C., Liang, W., Shi, Y., Lin, X., Zhang, H., Wu, X., et al., 2014. Contrasting elevational diversity patterns between eukaryotic soil microbes and plants. Ecology 95 (11), 3190−3202.

Shen, C., Ni, Y., Liang, W., Wang, J., Chu, H., 2015. Distinct soil bacterial communities along a small-scale elevational gradient in alpine tundra. Frontiers in Microbiology 6, 582.

Shen, C., Gunina, A., Luo, Y., Wang, J., He, J.Z., Kuzyakov, Y., et al., 2020. Contrasting patterns and drivers of soil bacterial and fungal diversity across a mountain gradient. Environmental Microbiology 22 (8), 3287−3301.

Sherman, R.E., Fahey, T.J., Martin, P.H., Battles, J.J., 2012. Patterns of growth, recruitment, mortality and biomass across an altitudinal gradient in a neotropical montane forest, Dominican Republic. Journal of Tropical Ecology 28 (5), 483−495.

Siles, J.A., Margesin, R., 2016. Abundance and diversity of bacterial, archaeal, and fungal communities along an altitudinal gradient in alpine forest soils: what are the driving factors? Microbial Ecology 72 (1), 207−220.

Singh, B.K., Munro, S., Reid, E., Ord, B., Potts, J.M., Paterson, E., et al., 2006. Investigating microbial community structure in soils by physiological, biochemical and molecular fingerprinting methods. European Journal of Soil Science 57 (1), 72−82.

Singh, D., Takahashi, K., Adams, J.M., 2012. Elevational patterns in archaeal diversity on Mt. Fuji.

Smith, J.M.B., Cleef, A.M., 1988. Composition and origins of the world's tropicalpine floras. Journal of Biogeography 631−645.

Sousa Neto, E., Carmo, J.B., Keller, M., Martins, S.C., Alves, L.F., Vieira, S.A., et al., 2011. Soil-atmosphere exchange of nitrous oxide, methane and carbon dioxide in a gradient of elevation in the coastal Brazilian Atlantic forest. Biogeosciences 8 (3), 733−742.

Steinbauer, M.J., Field, R., Grytnes, J.A., Trigas, P., Ah-Peng, C., Attorre, F., et al., 2016. Topography-driven isolation, speciation and a global increase of endemism with elevation. Global Ecology and Biogeography 25 (9), 1097−1107.

Stocker, T.F., Qin, D., Plattner, G.K., Tignor, M.M., Allen, S.K., Boschung, J., et al., 2014. Climate Change 2013: The physical science basis. Contribution of working group I to the fifth assessment report of IPCC the intergovernmental panel on climate change.

Sundqvist, M.K., Sanders, N.J., Wardle, D.A., 2013. Community and ecosystem responses to elevational gradients: processes, mechanisms, and insights for global change. Annual Review of Ecology, Evolution, and Systematics 44, 261−280.

Tanner, E.V.J., Vitousek, P.A., Cuevas, E., 1998. Experimental investigation of nutrient limitation of forest growth on wet tropical mountains. Ecology 79 (1), 10−22.

Tedersoo, L., Bahram, M., Põlme, S., Kõljalg, U., Yorou, N.S., Wijesundera, R., et al., 2014. Global diversity and geography of soil fungi. Science (New York, N.Y.) 346, 6213.

Wang, J.T., Cao, P., Hu, H.W., Li, J., Han, L.L., Zhang, L.M., et al., 2015. Altitudinal distribution patterns of soil bacterial and archaeal communities along Mt. Shegyla on the Tibetan Plateau. Microbial Ecology 69 (1), 135−145.

Watteau, F., Villemin, G., 2018. Soil microstructures examined through transmission electron microscopy reveal soil-microorganisms interactions. Frontiers in Environmental Science 106.

Whitaker, J., Ostle, N., Nottingham, A.T., Ccahuana, A., Salinas, N., Bardgett, R.D., et al., 2014. Microbial community composition explains soil respiration responses to changing carbon inputs along an A ndes-to-A mazon elevation gradient. Journal of Ecology 102 (4), 1058−1071.

Xu, M., Li, X., Cai, X., Gai, J., Li, X., Christie, P., et al., 2014. Soil microbial community structure and activity along a montane elevational gradient on the Tibetan Plateau, European Journal of Soil Biology, 64. Academy of Sciences & University of Natural Resources and Life Sciences, 2015, pp. 6−14.

Yang, Y., Gao, Y., Wang, S., Xu, D., Yu, H., Wu, L., et al., 2014. The microbial gene diversity along an elevation gradient of the Tibetan grassland. The ISME Journal 8 (2), 430−440.

Yuan, Y., Si, G., Li, W., Zhang, G., 2015. Altitudinal distribution of ammonia-oxidizing archaea and bacteria in Alpine grassland soils along the south-facing slope of Nyqentangula Mountains, Central Tibetan Plateau. Geomicrobiology Journal 32 (1), 77−88.

Yue, H., Wang, M., Wang, S., Gilbert, J.A., Sun, X., Wu, L., et al., 2015. The microbe-mediated mechanisms affecting topsoil carbon stock in Tibetan grasslands. The ISME Journal 9 (9), 2012−2020.

Zhao, C., Long, J., Liao, H., Zheng, C., Li, J., Liu, L., et al., 2019. Dynamics of soil microbial communities following vegetation succession in a karst mountain ecosystem, Southwest China. Scientific Reports 9 (1), 1−10.

Zumsteg, A., Bååth, E., Stierli, B., Zeyer, J., Frey, B., 2013. Bacterial and fungal community responses to reciprocal soil transfer along a temperature and soil moisture gradient in a glacier forefield. Soil Biology and Biochemistry 61, 121−132.

Soil nutrient dynamics under selected tree species explains the soil fertility and restoration potential in a semi-arid forest of the Aravalli Mountain range

Shikha Prasad and Ratul Baishya

Ecology & Ecosystem Research Laboratory, Department of Botany, University of Delhi, New Delhi, India

O U T L I N E

Understanding Soils of Mountainous Landscapes
DOI: https://doi.org/10.1016/B978-0-323-95925-4.00018-2

6.1 Introduction

Soil and vegetation have a complex and very strong interrelationship because they interact over a long period of time. Vegetation has the greatest influence on soil properties (Zhao et al., 2017). Individual tree species can alter soil properties by explicitly absorbing nutrients and returning them to the soil (Prescott, 2002). Individual tree species, whether leguminous or non-leguminous, can sometimes play a critical role in restoring soil processes (Wang et al., 2010). It has been discovered that different tree species can help to restore degraded land by accelerating soil carbon sequestration and improving soil quality (Yan et al., 2020). The information on nutrient interaction and accumulation in soil would aid in understanding nutrient cycling and biogeochemical cycles (Pandit and Thampan, 1988). Among other things, the nutrient-richness of the soil is critical for proper plant growth. Thus, in the forest ecosystem, appropriate knowledge of different soil properties in close association with tree species is critical (Zhao et al., 2017).

Desert soils are distinguished by the fact that they have almost no water available for soil formation (pedogenesis) and mesophytic plant growth for extended periods of time (Abdelfattah, 2013). Desertification will spread due to anthropogenic activities and global climate change, as such soils cover one-third of the earth's surface (Costantini et al., 2016). Under semi-arid conditions, tree canopies affect the soil and make it more fertile (Singh and Lal, 1970; Aggarwal et al., 1976; Bernhard-Reversat, 1982). Several studies have been carried out to investigate the impact of tree canopy on soil fertility indices (organic matter, mineralizable nitrogen, extractable phosphorous, potassium, and calcium) and soil microbial biomass. They have been found to be higher in the canopy zone than in the outside canopy (Bernhard-Reversat, 1982; Hobbie et al., 2007; Vesterdal et al., 2012). The size of a tree's canopy expands as it grows, which may account for nutrient gradations within tree canopy zones.

Although it has not been investigated how different nutrients are enriched within the soil under various tree canopies, particularly in Delhi-ridge areas. Nonetheless, it is most likely due to nutrient inputs through litter produced by tree species, as trees transport nutrients from surrounding surface and subsurface soils to their canopy and transfer the nutrients in leaf and woody litter. To comprehend the forest community and its structure, one must first comprehend the relationship between soil nutrients and plant community composition, followed by the effect of tree canopies on the distribution and movement of soil nutrients. Therefore, the present study was conducted to answer the following questions (1) How do individual tree species affect the nutrient accumulation under its canopy? (2) What are the seasonal effects on soil nutrient accumulation?

6.2 Materials and methods

6.2.1 Study area

The study area is located in Sanjay-Van (28°31′42.0″ N and 77°10′21.0″ E), the South-Central ridge area of Delhi, India, and the northern extension of the Aravalli hills. The Aravali hills are known to be India's most ancient mountain range, extending from

Gujarat—Rajasthan to Haryana—Delhi. Delhi consists of patches of natural forest, which are also known as ridge areas. There are four ridge areas in Delhi viz., (1) Southern Ridge (6200 ha), (2) Central Ridge (864 ha), (3) South-Central Ridge (626 ha), and (4) Northern Ridge (87 ha). The present study site comes under a semi-arid area having sporadic rainfall events. Hence, the seasons are classified into four classes, which are (1) Pre-monsoon (March—May), (2) Monsoon (June—August), (3) Post-monsoon (September—November), and (4) Winter (December—February).

6.2.2 Basis of tree species selection

Tree species were chosen for this study based on visual field observations: (1) tree species abundant in the study site were chosen to observe the maximum effect of tree individuals on soil nutrients; (2) contribution to the forest floor through litter quantity and quality production; and (3) tree species with large canopies were chosen to observe the effect of tree canopies on soil nutrients. The current study included seven different tree species. They were further classified for their nativity and exotic nature to the Delhi ridge area as per their classification in the Flora of Delhi (Maheshwari, 1963). The tree species selected were *Ficus religiosa* (L.), *Vachellia leucophloea* (Roxb.) Maslin, Seigler and Ebinger, and *Millettia pinnata* (L.) among native species. In comparison, *Albizia lebbeck* (L.) Benth, *Prosopis juliflora* (Sw.) DC., *Azadirachta indica* (A.) Juss., and *Cassia fistula* (L.) are under the exotic category.

6.2.3 Soil sampling

Soil samples were collected seasonally during the peak period of each season to observe the season's maximum effect. For example, soil samples were collected in July to maximize the effect of rainfall on the samples. To investigate the direct effect of individual tree species on soil nutrients, soil samples were collected just beneath the tree canopy using a soil auger. During the peak of each season, soil temperature and moisture levels were measured. Samples were collected for three consecutive years (2014—17) from two different soil depths, i.e., 0—10 cm and 10—20 cm. Following collection, the samples were immediately brought back to the laboratory in zipper polythene bags for further analysis. For analyzing various soil nutrients, soil samples were air-dried, sieved using a 2 mm sieve, and all the larger granules were removed manually. Soil moisture, microbial biomass carbon, and soil pH were analyzed from the fresh soil samples within one week after the collection (Prasad and Baishya, 2021b).

6.2.4 Total carbon and soil nutrients

Total nitrogen and carbon were analyzed using a CHNS analyzer (Elementar Analysis Systems GmbH, Germany). The Total carbon was estimated (Prasad and Baishya, 2021b). Available phosphorus was estimated by the ammonium molybdate blue method (Allen et al., 1974), exchangeable potassium and sodium were analyzed using the flame photometer method (Allen et al., 1974), whereas calcium and magnesium were analyzed using EDTA (Ethylenediamine tetra-acetic acid) titration method (Allen et al., 1974).

6.2.5 Statistical analysis

The effects of seasons, tree species, and soil depth on various soil nutrients were investigated using a three-way ANOVA. To distinguish the differences at the 0.05 level, the Tukey post-hoc test was used. Pearson's correlation analysis was used to determine the relationship between soil nutrients and soil moisture, temperature, and organic carbon. SPSS version 21 was used for all statistical analyses.

6.3 Results

6.3.1 Soil nutrients under different tree species across various seasons

The soil nutrients were lowest during the monsoon and highest during the winter season. We observed statistically significant variation in soil nutrients across different seasons ($P < .01$, df = 3) and under different tree species ($P < .01$, df = 6). Across two depths also significant variation was observed but only in phosphorus and potassium.

The general trend observed in soil nutrients (except sodium) across seasons was winter > pre-monsoon > post-monsoon > monsoon.

1. Total nitrogen (TN): Among different tree species and across various seasons, total nitrogen varied significantly (F = 2.337, $P < .01$, df = 18). However, total nitrogen did not show any variation across different depths as the variation across depth was not significant statistically ($P = .07$). Total nitrogen was highest in the pre-monsoon (Table 6.1) and lowest in the monsoon (Table 6.2). It ranged from 0.39 Mg/ha (*F. religiosa*) to 3.54 Mg/ha (*A. lebbeck*) in 0−10 cm depth, whereas in 10−20 cm depth, it ranged between 0.14 Mg/ha (*F. religiosa*) and 3.07 Mg/ha (*A. lebbeck*) during the 3 years of study. The pattern followed by TN under tree species is as follows: *A. lebbeck* > *A. indica* > *C. fistula* > *P. juliflora* > *F. religiosa* > *M. pinnata* > *V. leucophloea*, whereas across seasons, the pattern is as follows: pre-monsoon > winter > post-monsoon > monsoon.
2. Available phosphorus (AP): Among different tree species and across various seasons, phosphorus varied significantly (F = 3.506, $P < .01$. df = 18). It also showed variation across different depths (F = 3.506, $P < .01$). Available Phosphorus was higher in winter (Table 6.3) and lower in the monsoon season (Table 6.2). It ranged from 25.24 Kg/ha (*F. religiosa*) to 12.26 Kg/ha (*A. indica*) in 0−10 cm depth, whereas in 10−20 cm depth, it ranged from 21.34 to 10.82 Kg/ha during the 3 years of study. The pattern followed by AP under tree species is as follows: *F. religiosa* > *V. leucophloea* > *M. pinnata* > *P. juliflora* > *A. lebbeck* > *C. fistula* > *A. indica* whereas across seasons the pattern is as follows: winter > pre-monsoon > post-monsoon > monsoon.
3. Exchangeable potassium (EK): Under different tree species and across various seasons, potassium showed statistically significant variations (F = 15.908, $P < .01$, df = 18). EK also showed variation across different depths (F = 15.908, $P < .01$). It was highest in

TABLE 6.1 Soil nutrients in *pre-monsoon season* under the canopy of different tree species *(SD represented in parenthesis)*.

Soil nutrients	Depth (cm)	Vachellia leucophloea	Ficus religiosa	Millettia pinnata	Albizia lebbeck	Prosopis juliflora	Azadirachta indica	Cassia fistula
Total nitrogen	0–10	1.19 (0.58)	2.31 (0.59)	1.30 (0.44)	3.54 (0.52)	3.01 (0.37)	3.26 (0.21)	2.94 (0.11)
	10–20	0.92 (0.44)	1.80 (0.66)	1.06 (0.37)	3.07 (0.37)	2.65 (0.21)	3.04 (0.16)	2.61 (0.16)
Available phosphorous	0–10	20.15 (1.92)	23.45 (0.58)	19.89 (1.48)	16.66 (1.15)	17.96 (0.06)	14.45 (0.66)	13.56 (0.62)
	10–20	18.70 (0.95)	20.65 (0.97)	18.56 (0.80)	13.45 (0.88)	12.45 (1.17)	11.23 (1.26)	10.45 (0.56)
Exchangeable potassium	0–10	184.56 (0.95)	195.67 (0.81)	170.67 (1.96)	164.89 (1.08)	157.78 (1.19)	145.34 (1.24)	150.58 (1.15)
	10–20	188.78 (0.71)	199.91 (1.53)	172.78 (0.82)	168.93 (0.89)	161.24 (0.89)	151.27 (1.07)	153.45 (0.93)
Exchangeable calcium	0–10	1.90 (0.48)	5.51 (2.45)	2.87 (2.08)	2.94 (0.87)	1.45 (0.20)	4.93 (0.45)	1.89 (1.11)
	10–20	2.57 (0.11)	6.56 (1.54)	3.70 (0.44)	1.96 (0.03)	1.87 (0.60)	4.68 (0.13)	2.42 (0.86)
Exchangeable magnesium	0–10	1.41 (0.26)	1.47 (0.12)	0.91 (0.08)	1.52 (0.14)	1.40 (0.15)	1.07 (0.02)	1.14 (0.37)
	10–20	1.58 (0.08)	4.06 (0.31)	1.02 (0.53)	1.81 (0.21)	1.30 (0.13)	0.99 (0.01)	1.32 (0.22)

winter (Table 6.3) and lowest in the monsoon (Table 6.2). It ranged from 105.78 Kg/ha (*A. indica*) to 202.34 Kg/ha (*F. religiosa*) in 0–10 cm depth, whereas in 10–20 cm depth, it ranged between 98.65 Kg/ha (*A. indica*) and 208.12 Kg/ha (*F. religiosa*) during the 3 years of study. The pattern followed by exchangeable potassium (EK) under tree species is as follows: *F. religiosa* > *V. leucophloea* > *M. pinnata* > *A. lebbeck* > *P. juliflora* > *C. fistula* > *A. indica* whereas across seasons the pattern is as follows: winter > pre-monsoon > post-monsoon > monsoon.

4. Exchangeable calcium (Ca): Among different tree species and across various seasons, calcium (Ca) varied significantly (F = 9.662, $P < .01$, df = 18). However, Ca did not show any variation across different depths ($P > .05$). Ca was highest in winter (Table 6.3) and lowest in the post-monsoon season (Table 6.4). It ranged from 0.52 Mg/ha (*C. fistula*) to 7.81 Mg/ha (*F. religiosa*) in 0–10 cm depth, whereas in 10–20 cm depth, it was lowest in the monsoon and highest in the pre-monsoon. It ranged between 0.96 Mg/ha (*C. fistula*) and 6.56 Mg/ha (*F. religiosa*) during the 3 years of study. The pattern followed by Ca under tree species was different in each season, whereas across seasons, the pattern is as follows: winter > pre-monsoon > monsoon > post-monsoon.

2. Soil microbial processes and nutrient dynamics

TABLE 6.2 Soil nutrients in *monsoon season* under the canopy of different tree species (*SD represented in parenthesis*).

Soil nutrients	Depth (cm)	Vachellia leucophloea	Ficus religiosa	Millettia pinnata	Albizia lebbeck	Prosopis juliflora	Azadirachta indica	Cassia fistula
Total nitrogen	0–10	18.90 (0.18)	17.52 (0.30)	19.32 (0.48)	16.56 (0.61)	15.78 (0.61)	12.62 (0.36)	14.56 (0.18)
	10–20	16.10 (0.11)	15.43 (0.06)	17.23 (0.40)	14.78 (0.58)	13.45 (0.55)	10.82 (0.34)	11.28 (0.06)
Available phosphorous	0–10	18.90 (0.79)	17.52 (0.68)	19.32 (0.59)	16.56 (0.56)	15.78 (0.91)	12.62 (0.66)	14.56 (1.09)
	10–20	16.10 (0.37)	15.43 (0.76)	17.23 (0.79)	14.78 (1.31)	13.45 (0.49)	10.82 (0.74)	11.28 (1.07)
Exchangeable potassium	0–10	131.89 (1.29)	143.18 (0.58)	132.78 (1.35)	129.90 (1.00)	124.23 (0.87)	105.78 (1.00)	112.78 (0.90)
	10–20	132.78 (1.00)	139.90 (0.56)	125.56 (1.10)	120.23 (0.75)	118.67 (1.06)	98.65 (0.95)	100.78 (2.11)
Exchangeable calcium	0–10	0.77 (0.16)	1.16 (0.18)	1.47 (0.55)	1.96 (0.69)	0.95 (0.31)	1.51 (0.42)	0.94 (0.02)
	10–20	0.88 (0.16)	1.34 (0.23)	1.58 (0.56)	2.40 (0.03)	1.30 (0.27)	1.97 (0.17)	0.96 (0.02)
Exchangeable magnesium	0–10	0.48 (0.02)	0.95 (0.32)	1.06 (0.40)	0.79 (0.03)	0.63 (0.38)	0.18 (0.02)	0.75 (0.06)
	10–20	0.53 (0.05)	1.52 (0.58)	2.05 (0.43)	0.87 (0.09)	1.63 (0.26)	0.24 (0.05)	0.84 (0.10)

5. Exchangeable magnesium (Mg): Among different tree species and across various seasons, magnesium (Mg) varied significantly (F = 11.978, $P < .01$, df = 18). However, Mg showed no variation across different depths ($P > .05$). Mg was highest in winter (Table 6.3) and lowest in the monsoon (Table 6.2). It ranged from 0.18 Mg/ha (*A. indica*) to 2.87 Mg/ha (*F. religiosa*) in 0–10 cm depth, whereas in 10–20 cm depth, it was lowest in monsoon and highest in pre-monsoon and ranged between 0.24 Mg/ha (*A. indica*) and 4.06 Mg/ha (*F. religiosa*) over the 3 years of study. The pattern followed by Mg across seasons is as follows: winter > pre-monsoon > post-monsoon > monsoon.

6. Exchangeable Sodium: Sodium did not show any variations across different seasons and tree species. It ranged between 0.0006% and 0.05% across different tree species.

6.3.2 Correlation analysis

In this study, a strong positive significant correlation was observed between soil nutrients and SOC ($P < .05$). Throughout the study period, as the SOC increased, so did all of the nutrients. All soil nutrients had a negative relationship with soil moisture and temperature

TABLE 6.3 Soil nutrients in *post- monsoon season* under the canopy of different tree species *(SD represented in parenthesis)*.

Soil nutrients	Depth (cm)	Vachellia leucophloea	Ficus religiosa	Millettia pinnata	Albizia lebbeck	Prosopis juliflora	Azadirachta indica	Cassia fistula
Total nitrogen	0–10	19.12 (0.11)	18.23 (0.28)	20.12 (0.56)	17.45 (0.98)	16.76 (0.53)	13.23 (0.53)	15.36 (0.58)
	10–20	15.34 (0.22)	14.23 (0.16)	18.12 (0.50)	16.56 (1.06)	15.45 (0.58)	11.36 (0.27)	12.32 (0.48)
Available phosphorous	0–10	19.12 (0.41)	18.23 (0.57)	20.12 (0.97)	17.45 (0.64)	16.76 (0.34)	13.23 (0.24)	15.36 (0.56)
	10–20	15.34 (0.59)	14.23 (0.48)	18.12 (0.40)	16.56 (0.97)	15.45 (0.42)	11.36 (0.56)	12.32 (0.52)
Exchangeable potassium	0–10	136.45 (1.47)	145.78 (1.13)	133.24 (1.82)	130.23 (0.88)	128.78 (0.66)	108.67 (1.97)	115.45 (1.14)
	10–20	138.45 (0.95)	148.78 (1.13)	135.45 (0.80)	138.56 (0.88)	129.23 (0.70)	112.34 (0.91)	118.45 (0.82)
Exchangeable calcium	0–10	1.54 (0.26)	4.23 (0.30)	2.12 (0.52)	2.52 (0.39)	1.28 (0.38)	6.48 (0.56)	0.52 (0.43)
	10–20	1.66 (0.25)	4.43 (0.33)	2.42 (0.55)	2.76 (0.40)	1.48 (0.20)	6.83 (0.58)	0.97 (0.61)
Exchangeable magnesium	0–10	1.75 (0.44)	0.74 (0.28)	0.44 (0.05)	2.09 (0.04)	1.43 (0.08)	0.39 (0.01)	2.68 (0.04)
	10–20	2.16 (0.58)	1.06 (0.49)	0.62 (0.16)	2.16 (0.06)	1.65 (0.24)	0.47 (0.04)	2.95 (0.39)

(Tables 6.5–6.11). Similar trends were observed in the entire 3 years of study. Values considered were the mean of seasons in 3 years of study.

6.4 Discussion

In the present study, marked seasonal variation was observed in soil nutrients $(P < .01)$ under all the seven tree species selected and a significant effect of tree species was observed $(P < .01)$. In semi-arid regions, the ecosystem processes are majorly controlled by water availability (Collins et al., 2008). Thus, TN and other nutrients analyzed in this study showed the highest value during winter (Table 6.3). The possible reason behind this could be the contribution of organic matter from senescing leaves and comparatively slow uptake of soil nutrients by plants (Weaver and Forcella, 1979) when the moisture content and temperature were low may have ceased the activity of microbes, resulting in the accumulation of more nutrients during dry periods as compared to wet periods (Singh et al., 2000). Microbe activity may increase during and immediately after the rainy season, resulting in fewer nutrients during the monsoon and post-monsoon. Plant species can influence

TABLE 6.4 Soil nutrients in *winter season* under the canopy of different tree species (*SD represented in parenthesis*)

Soil nutrients	Depth (cm)	*Vachellia leucophloea*	*Ficus religiosa*	*Millettia pinnata*	*Albizia lebbeck*	*Prosopis juliflora*	*Azadirachta indica*	*Cassia fistula*
Total nitrogen	0–10	22.32 (0.37)	25.24 (0.42)	20.04 (0.53)	16.70 (0.67)	18.98 (0.55)	12.90 (0.12)	14.08 (0.30)
	10–20	19.20 (0.66)	21.34 (0.43)	17.89 (0.70)	13.23 (0.58)	12.23 (0.48)	10.90 (0.96)	11.56 (0.21)
Available phosphorous	0–10	22.32 (0.56)	25.24 (0.80)	20.04 (1.00)	16.70 (0.28)	18.98 (1.03)	12.90 (1.06)	14.08 (0.18)
	10–20	19.20 (0.58)	21.34 (0.66)	17.89 (0.13)	13.23 (0.21)	12.23 (0.69)	10.90 (0.29)	11.56 (0.67)
Exchangeable potassium	0–10	188.67 (1.07)	202.34 (0.49)	175.48 (0.69)	170.45 (1.35)	162.56 (0.11)	152.98 (1.06)	155.23 (0.85)
	10–20	190.23 (0.89)	208.12 (0.23)	178.45 (0.61)	172.23 (0.47)	165.48 (0.61)	158.98 (0.57)	167.67 (0.58)
Exchangeable calcium	0–10	2.27 (0.39)	7.81 (2.80)	4.14 (0.34)	4.00 (0.57)	6.49 (1.73)	5.82 (1.28)	1.10 (0.22)
	10–20	2.51 (0.41)	6.21 (2.78)	4.55 (0.48)	4.37 (0.60)	5.51 (1.72)	6.42 (1.17)	1.12 (0.15)
Exchangeable magnesium	0–10	1.42 (0.09)	2.87 (0.05)	2.54 (0.06)	1.03 (0.02)	1.02 (0.22)	0.64 (0.12)	0.86 (0.13)
	10–20	1.53 (0.13)	3.03 (0.15)	2.75 (0.16)	1.22 (0.15)	1.45 (0.31)	0.88 (0.28)	0.84 (0.10)

TABLE 6.5 Pearson's correlation matrix between TN, AP, EK, Ca and Mg stocks with SOC, soil moisture and temperature under the canopy of *V. leucophloea* (in bold if significant, * = significant, ** = highly significant).

	TN	AP	EK	Ca	Mg	SOC	ST	SM
TN	1	0.547**	0.734**	0.789**	0.683**	0.645**	− 0.360*	−0.791**
AP		1	0.609**	0.347	0.009	0.700**	− 0.529**	−0.044
EK			1	0.837**	0.278	0.824**	− 0.427*	−0.568**
Ca				1	0.549**	0.583**	−0.221	− 0.737**
Mg					1	0.622**	−0.253	−0.770**
SOC						1	− 0.406*	−0.429

ecosystem biogeochemistry by varying the amount and chemistry of litter returned to the soil (Pastor and Naiman, 1992; Berendse, 1998; Prasad and Baishya, 2019; Prasad and Baishya, 2021b), and the related impacts on the identity, abundance, and activities of numerous soil heterotrophic organisms (Hassani et al., 2018; Prasad and Baishya, 2021b),

TABLE 6.6 Pearson's correlation matrix between TN, AP, EK, Ca and Mg stocks with SOC, soil moisture and temperature under the canopy of *F. religiosa* (*in bold if significant*).

	TN	AP	EK	Ca	Mg	SOC	ST	SM
TN	1	0.866**	0.649*	0.921**	0.496	0.778*	− 0.477*	− 0.089*
AP		1	0.479	0.756*	0.549	0.948**	− 0.508**	− 0.140**
EK			1	0.262	0.219	0.586*	− 0.407*	− 0.359*
Ca				1	0.685	0.661*	− 0.682	− 0.086*
Mg					1	0.644*	− 0.783	− 0.418*
SOC						1	− 0.003*	− 0.316*

TABLE 6.7 Pearson's correlation matrix between TN, AP, EK, Ca and Mg stocks with SOC, soil moisture and temperature under the canopy of *M. pinnata* (*in bold if significant*).

	TN	AP	EK	Ca	Mg	SOC	ST	SM
TN	1	0.606	0.675*	0.763	0.413	0.843**	− 0.749**	− 0.114**
AP		1	0.372	0.053	0.322	0.523*	− 0.214**	− 0.065*
EK			1	0.933**	0.597	0.577*	− 0.946*	− 0.227*
Ca				1	0.565	0.577*	− 0.661	− 0.441*
Mg					1	0.605*	− 0.895	− 0.203*
SOC						1	− 0.703*	− 0.176**

TABLE 6.8 Pearson's correlation matrix between TN, AP, EK, Ca and Mg stocks with SOC, soil moisture and temperature under the canopy of *A. lebbeck* (*in bold if significant*).

	TN	AP	EK	Ca	Mg	SOC	ST	SM
TN	1	0.031	0.843**	0.468	0.351	0.763*	− 0.646*	− 0.638**
AP		1	0.394	0.150	0.179	0.654**	− 0.627*	− 0.651*
EK			1	0.597	0.047	0.801*	− 0.692*	− 0.534*
Ca				1	0.170	0.559*	− 0.831	− 0.690*
Mg					1	0.720*	− 0.196	− 0.721*
SOC						1	− 0.220	− 0.671*

which led to a high amount of soil TN along with other nutrients. Carbon: nutrient ratios and secondary C compounds such as lignin influence litter-specific decay and nutrient uptake rates, which regularly minimize decomposition and mineralization rates (Melillo et al., 1982). Because N availability is such an important constraint on plant growth, plant traits that influence the N cycle, such as litter percentage of N and C chemistry, are

TABLE 6.9 Pearson's correlation matrix between TN, AP, EK, Ca and Mg stocks with SOC, soil moisture and temperature under the canopy of *P. juliflora (in bold if significant)*.

	TN	AP	EK	Ca	Mg	SOC	ST	SM
TN	1	0.371	**0.771***	0.450	0.043	**0.648***	**−0.680***	**−0.677***
AP		1	0.009	0.104	0.323	**0.619***	**−0.718****	**−0.778***
EK			1	0.690	0.059	**0.783***	**−0.552***	**−0.651***
Ca				1	0.126	**0.677***	**−0.977***	**−0.619***
Mg					1	**0.671***	**−0.718***	**−0.682***
SOC						1	−0.185	**−0.623***

TABLE 6.10 Pearson's correlation matrix between TN, AP, EK, Ca and Mg stocks with SOC, soil moisture and temperature under the canopy of *A. indica (in bold if significant)*.

	TN	AP	EK	Ca	Mg	SOC	ST	SM
TN	1	**0.620***	**0.791***	**0.800***	0.697	**0.666***	−0.615	**−0.638***
AP		1	0.079	0.083	0.177	**0.604***	−0.413	**−0.614***
EK			1	0.464	**0.881****	**0.666***	−0.529	**−0.625***
Ca				1	0.466	**0.713****	−0.485	**−0.673***
Mg					1	**0.631****	−0.478	**−0.690***
SOC						1	−0.528	**−0.619***

TABLE 6.11 Pearson's correlation matrix between TN, AP, EK, Ca and Mg stocks with SOC, soil moisture and temperature under the canopy of *C. fistula (in bold if significant)*.

	TN	AP	EK	Ca	Mg	SOC	ST	SM
TN	1	0.451	0.664	0.122	0.098	**0.766***	−0.582	**−0.746****
AP		1	−0.244	−0.552	0.205	**0.689***	−0.073	**−0.564***
EK			1	0.561	−0.335	**0.668***	−0.459	**−0.672***
Ca				1	−0.294	**0.687***	−0.304	**−0.783***
Mg					1	**0.608***	−0.039	**−0.651***
SOC						1	−0.519	**−0.648***

frequently the focus of research on species consequences of biogeochemical tactics (Wedin and Tilman, 1990; Pastor and Naiman, 1992; Berendse, 1998). Nitrogen is an essential component of all plant growth strategies, particularly in semi-arid regions. Nitrogen deficiency in plants can result in underdeveloped and stunted growth (Mitchell and Chandler, 1939). In most ecosystems, soil TN is considered the most limiting nutrient (Fenn et al., 1998).

As litter and herbage weights increase beneath each tree and foliar nutrient contents increase (Kellman, 1979), significant nutrient sequestration must occur in these compartments. However, the majority of N in soil is present as nitrates, which are soluble in moisture and mobile (Gupta and Sharma, 2009). The range of nitrogen in the current study is higher in the upper layer (01−0 cm) as compared to the lower layer (10−20 cm) (Tables 6.1−6.3), which could be mainly due to more water holding capacity (Prasad and Baishya, 2021b). A significant positive correlation of TN with SOC in the present study (Tables 6.5−6.11) supports that nitrogen availability depends mainly on the amount and properties of organic matter (Haan, 1977). Consequently, the high amount of organic matter under the surface layer (0−10 cm) (Prasad and Baishya, 2021b) led to the richness of N in the upper layer as compared to the lower layer (10−20 cm). As it has been observed that the major source of C and N is from soil organic matter, plant and animal debris, and C and N are closely linked to each other (Aber and Melillo, 1991). With the decomposition of the organic material, total N is also increased along with an increase in exchangeable base cations under tree species in both surface (0−10 cm) and subsurface layer (10−20 cm) (Kasongo et al., 2009).

Many studies have shown that along with SOC and increased litter, soil fertility is also increased under tree species (Campbell et al., 1994; Dunham, 1991; Isichei and Muoghalu, 1992; Kessler, 1992; Prasad and Baishya, 2019). Besides nutrient pools, microbial activity under tree canopies also influences the nutrient flux (Gairola et al., 2012). A negative correlation was observed in the current study between soil moisture, temperature with soil TN and other nutrients (Tables 6.5−6.11), which shows that as the soil moisture and temperature increases, microbial activity enhances, resulting in a higher rate of mineralization (Prasad and Baishya, 2021a) and decomposition.

Soil TN and other nutrients accumulate more in winter and pre-monsoon than in monsoon and post-monsoon (Tables 6.1−6.3). This could be due to organic matter accumulation under the canopy of tree species and reduced leaching, or it could be due to higher organic matter production by trees, which slowed the rate of mineralization due to a decrease in temperature during winter (Gairola et al., 2012). Nutrient sources under tree canopies may also include leachates from tree canopies, nutrient inputs from litter, and nutrient transport by tree roots from rooting zones to tree canopies (Gairola et al., 2012).

Available phosphorus is critical for vital plant growth processes, and P is present in almost all terrestrial structures in either organic or inorganic form. However, organic P forms are the most abundant phosphorus sources (Gairola et al., 2012). Little information exists on available P for Indian semi-arid soils, particularly underneath native vegetation (Austin et al., 2004). Available phosphorus was positively correlated with SOC in the current study (Tables 6.5−6.11), similar to Gupta and Sharma (2009). In many soils, the inorganic form of P is trapped in organic matter. Weathering rates also influence phosphorus cycling and availability (Bahuguna et al., 2012). The carbon-phosphorus and nitrogen-phosphorus ratios differ depending on the parent material and the level of weathering (Paul and Clark, 1996). Available phosphorus was high in winter and low in the monsoon in the current study (Tables 6.3 and 6.2). Microbes and plants compete with the soil for phosphorus during the monsoon, limiting its availability (Tuininga et al., 2002). The quantity of P shows the soil to enable specific plants to develop at a particular site, which is also useful for the vegetation type of the area. It has been reported that a massive percentage of P is saved in forms that are unavailable to plants (Murphy, 1958), for example, H_2PO_4, which becomes available at low pH values and suffers from fixation by hydrous oxides and silicate minerals (Soromessa et al., 2004). Plants can obtain phosphorus either through mineralization or through root association, and both of these processes are influenced by moisture availability (Shen et al., 2011).

The mineral potassium (exchangeable K) is found in all parts of plants in soluble form and is responsible for carbohydrate and protein formation. Potassium activates plant enzymes, which aid in plant metabolism, starch synthesis, nitrate reduction, and also plays a role in sugar degradation. Potassium regulates transpiration and respiration, influences enzyme action, synthesizes carbohydrates and proteins, and so on (Brady, 1996). The decrease in K is brought about via leaching and drainage, which results in the destruction of vegetation (Basumatary and Bordoloi, 1992).

Because no specific reason can be assigned to its differential quantities under different tree species and seasons, exchangeable K in soil is heavily reliant on the composition of parent rock material. Surface soil had a higher available K content because all of these characteristics are closely related to the presence and quantity of soil humus. This is supported by other authors' findings (Basumatary and Bordoloi, 1992; Boruah and Nath, 1992), who observed that a layer of organic matter significantly improves the retention of K in the soil.

Plant species can also influence C and nutrient cycling through litter and soils, influencing the distribution and concentrations of Ca, Mg, and K, which appear as cations in exchange reactions between soil solids and solution. Because Ca is a non-hydrolyzing cation that competes with H + cations for change sites on soil particle surfaces, primarily 2:1 layer-type clay and organic matter, its concentration in soil affects soil pH. Plant species, in general, have been found to influence mineral weathering of parent materials, as well as Ca mineralization and recycling (Quideau et al., 1996; Dijkstra et al., 2004). Litter and soil Ca content and pH can potentially influence C and N cycling because, in forest soils, greater pH is frequently associated with higher microbial biomass and, consequently, greater rates of litter decomposition, soil respiration, and N mineralization (Persson, 1989; Simmons et al., 1996; Andersson and Nilsson, 2001). High leaf litter Ca concentrations combined with a large volume of leaf litterfall may increase the amount of exchangeable Ca in sugar maple floor soils (Alban, 1982; Prasad and Baishya, 2019).

Sodium is more easily leached and shows no discernible trend over time (Kasongo et al., 2009). The current study found no clear trend across different tree species, seasons, or depth.

6.5 Conclusion

The seasonal variation in soil nutrients was primarily controlled by soil moisture, temperature, and soil microbes, resulting in lower nutrient values during the monsoon and higher nutrient values during the winter. The nutrient concentration under the tree species is largely determined by the quality and quantity of litter. Soil nutrient variation among tree species can be directly attributed to plant-mediated soil microclimate and soil biological properties. In general, under *F. religiosa*, all the nutrients showed higher value and lowest under *A. indica*, which can be directly related to the C/N ratio of their leaf litter and quantity of substrate. A higher concentration of soil nutrients under the upper layer (0−10 cm) of a few tree species like *F. religiosa* and *V. leucophloea* could be attributed to the high amount of organic matter compared to the lower layer (10−20 cm). Thus, changes in plant species composition caused by anthropogenic activities or global environmental change can alter the physical and chemical properties of the soil. Furthermore, suitable tree species for plantation can be chosen to improve soil quality and restore degraded forest areas.

Acknowledgments

We acknowledge the logistic support and research permission provided by the Forest Department, Govt. of Delhi, and Late Marshal Vinod, which helped conduct this study. The first author is grateful to the University of Delhi for providing a UGC-NON-NET fellowship. The corresponding author sincerely acknowledges the funds provided by the R&D grants and the Institution of Eminence (IoE), University of Delhi, for providing the Faculty Research Programme (FRP) grant (2020−2021).

References

Abdelfattah, M.A., 2013. Pedogenesis, land management and soil classification in hyper-arid environments: results and implications from a case study in the United Arab Emirates. Soil Use and Management 29 (2), 279−294.

Aber, J.D., Melillo, J., 1991. Terrestrial Ecosystems. Saunders College Publishing, Toronto.

Aggarwal, R.K., Gupta, J.P., Saxena, S.K., Muthana, K.D., 1976. Studies on soil physico-chemical and ecological changes under twelve years old five desert tree species of western Rajasthan. Indian Forester 102 (12), 863−872.

Alban, D.H., 1982. Effects of nutrient accumulation by Aspen, spruce, and pine on soil properties 1. Soil Science Society of America Journal 46 (4), 853−861.

Allen, S.E., Grimshaw, H.M., Parkinson, J.A., Quarmby, C., 1974. Chemical Analysis of Ecological Materials. Blackwell, Oxford, pp. 21−22.

Andersson, S., Nilsson, S.I., 2001. Influence of pH and temperature on microbial activity, substrate availability of soil-solution bacteria and leaching of dissolved organic carbon in a mor humus. Soil Biology and Biochemistry 33 (9), 1181−1191.

Austin, A.T., Yahdjian, L., Stark, J.M., Belnap, J., Porporato, A., Norton, U., et al., 2004. Water pulses and biogeochemical cycles in arid and semiarid ecosystems. Oecologia 141 (2), 221−235.

Bahuguna, Y.M., Gairola, S., Semwal, D.P., Uniyal, P.L., Bhatt, A.B., 2012. Soil physico-chemical characteristics of bryophytic vegetation residing Kedarnath Wildlife Sanctuary (KWLS), Garhwal Himalaya, Uttarakhand, India. Indian Journal of Science and Technology 5 (4), 2547−2553.

Basumatary, A., Bordoloi, P.K., 1992. Forms of potassium in some soils of Assam in relation to soil properties. Journal of the Indian Society of Soil Science 40 (3), 443−446.

Berendse, F., 1998. Effects of dominant plant species on soils during succession in nutrient-poor ecosystems. Biogeochemistry 42 (1−2), 73−88.

Bernhard-Reversat, F., 1982. Biogeochemical cycle of nitrogen in a semi-arid savanna. Oikos 321−332.

Boruah, H.C., Nath, A.K., 1992. Potassium status of three major soil orders of Assam. Journal of the Indian Society of Soil Science 40, 559−561.

Brady, N.C., 1996. The Nature and Properties of Soil, tenth ed. Prentice Hall, New Delhi.

Campbell, B.M., Frost, P., King, J.A., Mawanza, M., Mhlanga, L., 1994. The influence of trees on soil fertility on two contrasting semi-arid soil types at Matopos, Zimbabwe. Agroforestry Systems 28 (2), 159−172.

Collins, S.L., Sinsabaugh, R.L., Crenshaw, C., Green, L., Porras-Alfaro, A., Stursova, M., et al., 2008. Pulse dynamics and microbial processes in aridland ecosystems. Journal of Ecology 96 (3), 413−420.

Costantini, E.A., Branquinho, C., Nunes, A., Schwilch, G., Stavi, I., Valdecantos, A., et al., 2016. Soil indicators to assess the effectiveness of restoration strategies in dryland ecosystems. Solid Earth 7 (2), 397−414.

Dijkstra, F.A., Hobbie, S.E., Knops, J.M., Reich, P.B., 2004. Nitrogen deposition and plant species interact to influence soil carbon stabilization. Ecology Letters 7 (12), 1192–1198.

Dunham, K.M., 1991. Comparative effects of Acacia albida and Kigelia africana trees on soil characteristics in Zambezi riverine woodlands. Journal of Tropical Ecology 7 (2), 215–220.

Fenn, M.E., Poth, M.A., Aber, J.D., Baron, J.S., Bormann, B.T., Johnson, D.W., et al., 1998. Nitrogen excess in North American ecosystems: predisposing factors, ecosystem responses, and management strategies. Ecological Applications 8 (3), 706–733.

Gairola, S., Sharma, C.M., Ghildiyal, S.K., Suyal, S., 2012. Chemical properties of soils in relation to forest composition in moist temperate valley slopes of Garhwal Himalaya, India. The Environmentalist 32 (4), 512–523.

Gupta, M.K., Sharma, S.D., 2009. Effect of tree plantation on soil properties, profile morphology and productivity index-II. Poplar in Yamunanagar district of Haryana. Annals of Forestry 17 (1), 43–70.

Haan, S.D., 1977. Humus, its formation, its relation with the mineral part of the soil, and its significance for soil productivity. Soil Organic Matter Studies 1, 21–30.

Hassani, M.A., Durán, P., Hacquard, S., 2018. Microbial interactions within the plant holobiont. Microbiome 6 (1), 58.

Hobbie, S.E., Ogdahl, M., Chorover, J., Chadwick, O.A., Oleksyn, J., Zytkowiak, R., et al., 2007. Tree species effects on soil organic matter dynamics: the role of soil cation composition. Ecosystems 10 (6), 999–1018.

Isichei, A.O., Muoghalu, J.I., 1992. The effects of tree canopy cover on soil fertility in a Nigerian savanna. Journal of Tropical Ecology 8 (3), 329–338.

Kasongo, R.K., Ranst, Van, Verdoodt, E., Kanyankagote, A., Baert, G., P., 2009. Impact of Acacia auriculiformis on the chemical fertility of sandy soils on the Batéké plateau, DR Congo. Soil Use and Management 25 (1), 21–27.

Kellman, M., 1979. Soil enrichment by neotropical savanna trees. The Journal of Ecology 565–577.

Kessler, J.J., 1992. The influence of karité (Vitellaria paradoxa) and néré (Parkia biglobosa) trees on sorghum production in Burkina Faso. Agroforestry Systems 17 (2), 97–118.

Maheshwari, J.K., 1963. The Flora of Delhi. Council of Scientific Industrial Research.

Melillo, J.M., Aber, J.D., Muratore, J.F., 1982. Nitrogen and lignin control of hardwood leaf litter decomposition dynamics. Ecology 63 (3), 621–626.

Mitchell, H.L., Chandler, R.F., 1939. The nitrogen nutrition and growth of certain deciduous trees of Northeastern United States, with a discussion of the principles and practice of leaf analysis as applied to forest trees. Black Rock Forest Bulletin 11, 94.

Murphy, H.P., 1958. The Fertility Status of Some Soils in Ethiopia. College of Agriculture, Jimma.

Pandit, B.R., Thampan, S., 1988. Total concentration of P, Ca, Mg, N & C in the soils of reserve forest near Bhavnagar (Gujarat State). Indian Journal of Forestry.

Pastor, J., Naiman, R.J., 1992. Selective foraging and ecosystem processes in boreal forests. The American Naturalist 139 (4), 690–705.

Paul, E.A., Clark, F.E., 1996. Soil Microbiology and Biochemistry, second ed. Academic Press, San Diego, CA.

Persson, T., 1989. Role of soil animals in C and N mineralisation. Plant and Soil 115 (2), 241–245.

Prasad, S., Baishya, R., 2019. Interactive effects of soil moisture and temperature on soil respiration under native and non-native tree species in semi-arid forest of Delhi, India. Tropical Ecology 60 (2), 252–260.

Prasad, S., Baishya, R., 2021a. Effect of tree species and seasons on soil nitrogen transformation rates in the semi-arid forest of Delhi, India. Vegetos . Available from: https://doi.org/10.1007/s42535-021-00291-1.

Prasad, S., Baishya, R., 2021b. Seasonal dynamics and tree-species affects soil microbial biomass carbon in semi-arid forest of India (Accepted on 14 August 2021) International Journal of Ecology and Environmental Sciences.

Prescott, C.E., 2002. The influence of the forest canopy on nutrient cycling. Tree Physiology 22 (15–16), 1193–1200.

Quideau, S.A., Chadwick, O.A., Graham, R.C., Wood, H.B., 1996. Base cation biogeochemistry and weathering under oak and pine: a controlled long-term experiment. Biogeochemistry 35 (2), 377–398.

Shen, J., Yuan, L., Zhang, J., Li, H., Bai, Z., Chen, X., et al., 2011. Phosphorus dynamics: from soil to plant. Plant Physiology 156 (3), 997–1005.

Simmons, J.A., Fernandez, I.J., Briggs, R.D., Delaney, M.T., 1996. Forest floor carbon pools and fluxes along a regional climate gradient in Maine, USA. Forest Ecology and Management 84 (1–3), 81–95.

Singh, K.S., Lal, P., 1970. Effect of Khejari (Prosopis spicigera Linn) and Babool (Acacia arabica) trees on soil fertility and profile characteristics. Annals of Arid Zone.

Singh, G., Gupta, G.N., Kuppusamy, V., 2000. Seasonal variations in organic carbon and nutrient availability in arid zone agroforestry systems. Tropical Ecology 41 (1), 17–23.

Soromessa, T., Teketay, D., Demissew, S., 2004. Ecological study of the vegetation in Gamo Gofa zone, southern Ethiopia. Tropical Ecology 45 (2), 209–222.

Tuininga, A.R., Dighton, J., Gray, D.M., 2002. Burning, watering, litter quality and time effects on N, P, and K uptake by pitch pine (Pinusrigida) seedlings in a greenhouse study. Soil Biology and Biochemistry 34 (6), 865–873.

Vesterdal, L., Elberling, B., Christiansen, J.R., Callesen, I., Schmidt, I.K., 2012. Soil respiration and rates of soil carbon turnover differ among six common European tree species. Forest Ecology and Management 264, 185–196.

Wang, F., Li, Z., Xia, H., Zou, B., Li, N., Liu, J., et al., 2010. Effects of nitrogen-fixing and non-nitrogen-fixing tree species on soil properties and nitrogen transformation during forest restoration in southern China. Soil Science & Plant Nutrition 56 (2), 297–306.

Weaver, T., Forcella, F., 1979. Seasonal variation in soil nutrients under six Rocky Mountain vegetation types 1. Soil Science Society of America Journal 43 (3), 589–593.

Wedin, D.A., Tilman, D., 1990. Species effects on nitrogen cycling: a test with perennial grasses. Oecologia 84 (4), 433–441.

Yan, M., Fan, L., Wang, L., 2020. Restoration of soil carbon with different tree species in a post-mining land in eastern Loess Plateau, China. Ecological Engineering 158, 106025.

Zhao, D., Xu, M., Liu, G., Ma, L., Zhang, S., Xiao, T., et al., 2017. Effect of vegetation type on microstructure of soil aggregates on the Loess Plateau, China. Agriculture, Ecosystems & Environment 242, 1–8.

7

Soil nutrient dynamics under mountainous landscape: issues and challenges

S. Sivaranjani and Vijender Pal Panwar

Soil Science Discipline, Forest Ecology and Climate Change Division, Forest Research Institute, Dehradun, Uttarakhand, India

7.1 Introduction

Nutrient cycling is a major process in the forest ecosystem that aids in the production of organic matter. Primary production is typically influenced by nutrient availability, pattern, and rate under specific climatic conditions (Giweta, 2020). Water availability is also important for the survival of tree species and nutrient uptake. The trees absorb a large amount of nutrients, and much of what they absorb is returned to the soil through litterfall. After removal of litter, significant amount of nutrients are removed. Plantation nutrient removal increases as utilization intensifies (Lodhiyal and Lodhiyal, 1997).

Short rotation forestry species have a significant advantage over predictable forestry species. It consists of a higher yield per unit of land and labor production. Furthermore, the growth of dense populations aids in the production of dry matter and energy (Van Veen et al., 1981). To maximize the productivity, it is critical to estimate the concentration and level of nutrients. Macronutrients (N, P, K) are the most important limiting factor in agriculture and forest production (Razaq et al., 2017). The short-rotation plantations economy is highly influenced by its acceptance in forestry and agroforestry through its economic and energetic performance.

The plant species selected for commercial use are chosen primarily for their measurements to achieve higher development rates on various sites. Secondary considerations are given to biological versions and species acceptance to various stresses (Lodhiyal et al., 1995). Nutrient concentrations within the ecosystem are primarily determined by the practical balance of their nutrient cycling. These studies by species collected in a short-cycling period are critical for the sustainable management of forest plantations.

To understand high and sustained timber production, scientists in the forestry sector studied interactions between soils and forest tree species. It is about having a basic understanding of how nutrient availability affects tree growth and how soil parameters influence moisture and nutrient activities. The greater spatial and temporal variability in non-cultivated soils is a major constraint. It is also related to stand nutrient variation in requirements and vegetation pattern, which changes significantly in long rotation forests. The fastest takings of fine roots are nutrients that become available to the trees. Finally, nutrient integration in organic and inorganic forms with the help of mycorrhizal symbionts was discovered as a mode of nourishment (Schoonover and Crim, 2015). The site-specific information on nutrient interactions, nutrient availability driving mechanisms in forest ecosystem, and sustainable forest production emphasizes sustainable forest production.

Soils store more carbon than the atmosphere and terrestrial vegetation . It improves soil fertility and also acts as a sink or source for carbon emissions. Most tropical soils are deficient in mineral nutrients and rely on soil organic matter (SOM) to recover nutrients (Gougoulias et al., 2014). SOC contributes significantly to soil fertility by storing a wide range of nutrients such as NPK etc. It also increases the gaseous exchange and water holding capacity of soils. SOC loss indicates some degree of soil degradation. Carbon storage will help to reduce the amount of carbon in the atmosphere and thus help alleviating the problem of global warming and climate change.

Mountains are elevated landforms that rise above the surrounding landscape. It may combine above mean sea level (AMSL), slope steepness, and local elevation range, depending on the topographic criterion. UNEP-WCMC (2002) created the classification to represent the elevation gradient as a key component (Fig. 7.1).

Soils evolve slowly from mountains and are shallow due to low temperatures and limited biological activity. It is typically defined as skeletal, thin (shallow), acidic, poorly developed, and unproductive in comparison to other soil types. These soils become less fertile and developed as elevation increases (FAO, 2015). Altitude has a significant impact on soil properties and ecosystem processes. It causes biome differences due to varying temperatures and precipitation, which has an impact on vegetation. As altitude increases, temperature decreases, resulting in more sparse and diverse vegetation. However, in freezing elevation areas where recrystallization and weathering sequences reduce soil aggregation, the soil freezes at significantly higher altitudes. With an increase in the altitudinal gradient, such freezing conditions have an impact on the soil and plant nutrient cycles. The major nutrients, such as CNP and K, differed significantly from plains, but there was a consistent increase in soil C, total soil N, and microbial biomass. Nutrient cycling is heavily influenced by changes in the microbiome. Mountain soils support lush vegetation and biodiversity hotspots for plants and crops. It has several important functions, one of which is to control soil erosion. As previously stated, mountainous soils are less productive than low-lying soils, and farming activities require more labor and yield less productivity (FAO, 2015).

Polar, 10, 000 meters AMSL, Temperature does not exceed 0°C.	
Tundra, 3700-4500 meters above MSL, 0-18°C	
Taiga, 1500-3300 meters AMSL, 20-18°C	
Tropical forest, 900-1500 meters AMSL, 28-50°C	

FIGURE 7.1 Change of ecosystems along with altitude and temperature (Jeyakumar et al., 2020). Class 1: elevation ≥ 4500 m. Class 2: elevation 3500−4500 m Class 3: elevation 2500−3500 m. Class 4: elevation 1500−2500 m and slope ≥ 2° Class 5: elevation 1000−1500 m and slope ≥ 5 or local elevation range (LER) > 300 m Class 6: elevation 300−1000 m and LER > 300 m. Elevation classification (UNEP-WCMC, 2002).

Comprehensive knowledge of nutrient dynamics is required to understand nutrient cycling, organic carbon deposition, differences in microbiome, and soil microbial activity. It will also put pressure on the state of soils in forestry, which has been a neglected aspect (Miller, 1984). The present chapter deals with the following areas

- Nutrient availability
- Role of Nitrogen and Phosphorous
- Influence of standage on nutrient demands
- Nutrient cycling
- Retranslocation of nutrients within trees
- Nutrient cycling at higher altitude
- Litterfall
- Nutrient supply to the trees
- Leaching
- Additional factors

7.2 Nutrient availability

The number of nutrients that are readily absorbed and assimilated by growing plants is referred as available nutrients. Its availability is primarily influenced by two factors: soil and plant condition. Because plants actively absorb and assimilate nutrients, they are good indicators of available nutrients (Jacoby et al., 2017). Physical, chemical, and biological methods are used to estimate the potentially available nutrients for plant growth. It is more difficult to relate accessibilities to nutrient cycling processes under timberland produces where the plants grow for several years. Different methods have been developed to define nutrient accessibility and soil fertility in forest ecosystem. To define nutrient accessibility and soil fertility, various approaches, such as location classification, and soil and foliar examinations, are generally used. The site classes are determined by the mineralogy of the parent material, the conditions of the woodland floor, and the arrangement of the ground cover (Alban, 1982).

7.3 Soil analysis

The first step in characterizing soils at each site is to classify the forest soil into different study sites, followed by establishing relationships between expected tree growth and soil parameters. Physical and chemical edaphic characteristics are also used to identify growth-limiting factors in site-specific approaches. It is not applicable to all tree species at once, but it can be used to make management decisions for silvicultural operations (Rodrigues et al., 2018). Klinka et al. (1981) used various mathematical tools to identify nutrients that may be deficient to support healthy forest growth in specific locations. Because of the heterogeneity of soil limits in terms of individual nutrients, estimating available nutrients in the largest forest area is extremely difficult. The forest fertilization trials are not encouraging in terms of determining dosage rates in different forests. A significant variation was observed in growth in response to the fertilizer applied (Dong et al., 2012). Salonius and Mahendrappa

(1983) described the traditional methods for assessing growth response. There is no significant or reliable relationship between growth and fertilizer response . Many researchers worked on nitrogen availability, but only a few on phosphorus and potassium availability. The nitrogen availability is determined by the degree of mineralization of nitrogen under various conditions and the amount of nitrogenous compound present (Leye Samuel and Omotayo Ebenezer, 2014).

7.4 Foliar analysis

Following the soil analysis, the next step is to estimate the nutrient status of vegetation (foliage). Foliar analysis serves two primary purposes: determining deficiency and predicting tree response to fertilizer application conditions. In the absence of a reliable methodology for estimating N availability, nutrient concentration in foliage is widely used as a bioassay. Many researchers studied foliar N concentrations and needle units. It focuses on the interrelationships between parameters and growth, as well as their fertilizer response, in order to estimate site fertility (Nafiu et al., 2012). More emphasis was placed on standardizing sample preparation and the analytical technique used. The season in which the samples were collected, as well as the effect of climate, provide numerous suggestions for improvement. Powers (1983) gave the accurate appraising of forest nutrient status using response in nutrient concentration in foliage to a single fertilizer application.

7.5 Role of nitrogen and phosphorous

7.5.1 Nitrogen

Nitrogen is regarded as a critical cause of preventive tree development on temperate forest soils in the north. The fertilization trails contribute significantly to the prominence of nitrogen in the development of forests. It is not due to a scarcity of forest soils, but rather to a lack of availability. In podzolic soils, organic horizons are a major nitrogen reserve when compared to the amount of nitrogen present in standing crops. In temperate forests, however, the nitrogen content of organic horizons is high enough to support plant growth for more than two rotations. The nitrogen content of these soils is higher at mineral horizons.

The amount of nitrogen present in the mineral layer/horizons was greater than in the organic horizons. The amount of N collected under different forestry stands varied significantly within geographic regions due to differences in soil physicochemical properties (Santa Regina and Tarazona, 2001). The potential availability and accumulation of nitrogen is limited, by the amounts of nitrogen alteration and extractable mineral forms of N (Guignard et al., 2017).

7.6 Nitrogen availability indices

The quantities of water-soluble substances is correlated to N availability was estimated by using conventional methods. Whether the samples were incubated or not,

Mahendrappa (1980) methodology produced similar results. Most studies treat samples the same way, regardless of whether they are organic or mineral horizons. However, some analytical methods were deemed appropriate because they are suitable for agricultural systems and thus fit for forest soils. It was discovered in previous studies on nitrogen mineralization and tree nourishment. Therefore, the forest floor material (Litter + Fermented + Humus) layer needs evaluation.

In nitrogen mineralization studies on forest floors, general variability occurs due to sample and site variation, with seasonal, spatial, and analytical variability to be investigated further. The number of samples used for analysis should be significantly increased to obtain a representative estimate (Arp and Krause, 1984). However, there was no guaranteeof precise results. They include detailed sampling methodology and inevitable bias, and frequent sampling at the same site may be a problem. In inorganic soils, moistening and drying of the forest floor result in significant changes in the chemical and biological properties. These fluctuations have a significant impact on the N values during incubation (Salonius et al., 1982). The belongings of some variables have been practiced during the development of treated and non-treated samples in field conditions. Some researchers investigated it using exchange gums in nylon covers/bags that absorb cations and anions. When buried in soil, the resin bags are expected to function similar to plant roots, absorbing nutrient ions. In replication processes, alkali and heat treatment are used to convert composite organic compounds into simpler forms of N (Thiffault et al., 2000).

7.7 Phosphorus

The phosphorus requirement is least understood in temperate forest soils. There is also no scientific evidence that phosphorus treatment alone improves plant/tree growth. The combination of N and P treatment provides a superior response in terms of tree growth in inorganic soils of the boreal and other major regions (Zavisic et al., 2018). Phosphorous from natural resources is sufficient for plant growth, but N deficiency may be due to insufficient nitrogen sufficiency, where P was discovered to be an important variable in estimating location productivity (Radwan and Shumway, 1984).

7.8 Presence of extractable phosphorus in forest soils

Few studies recommend determining P take-up from various skylines within timberland soils, but some data can be obtained from tree establishing examples and soil scientific data. For example, extractable P increased in the soil beneath the B skyline to a level that was most likely related to the parent material's properties (Smeck, 1973). However, it is unclear how much P the trees could use from this part of the soil, and the entire fine root biomass was located beneath the skyline. Under these conditions, it was very likely correct that the trees determined a significant portion of the yearly phosphorous requirement from the natural surface skylines. Comparable examples of P transport have been accounted for in severely depleted soils in the Atlantic region (Krause, 1981) and Maine from north (Fernandez and Struchtemeyer, 1985). In the subsequent case, the main piece of P in the B skyline existed as bounded by iron (Fe) and aluminum (Al) phosphates.

The earth provides the fundamental commitments of the P necessities by trees. Overall, there isn't enough data to assess the significance of the carbon (C) skyline as a source of phosphorous (P) to tree nourishment. Setting up fundamental supplement spending plans, on the other hand, would suggest a requirement for phosphorous (P) contribution from the topographical substrate to make up phosphorous lost after accessible sources by plant take-up and the pedogenic interaction of impediment.

7.9 Indices of phosphorus availability

Many techniques for farming soils and harvests were developed; however, a portion of those were applied to backwood soils with varying success. The reaction of young pine ranches to P preparation was predicted by extracting samples with NH_4OAc (pH 7.8). In succeeding work, chemical comprising of a proportion of 0.05 N HCl and 0.025 NH_2SO_4 was utilized for a similar reason (Pritchett and Smith, 1972). Soil phosphorus supply is more precisely assessed with weaken mineral acids than with solid extractants, in the Bray tests (Zavisic et al., 2018). However, when extended periods of plantation development were considered, the above connection between extricated P and plant development was switched. In broad overviews of ranch development, Wilde (1964) related usefulness to soil phosphorous source as assessed by extracting with extremely weaken sulfuric acid (0.002 N, pH 3).

Dilute H_2SO_4 was also used as an extractant for accessible P to investigate dark tidy development, which revealed a critical link between backwoods efficiency and extractable P in woods floor materials. Several studies have revealed a close relationship between the P removed from the Ca and Al phosphate parts of soil. Because the Ca phosphate component is minimal to non-existent in corrosive woodland soils, the majority of the components used by trees in these conditions appear to originate from the Al phosphate component (Wuenscher et al., 2015). Maximum spatial changeability in forest soils complicates assessing plant-accessible P and accessible parts of different nutrients. Similarly, determining which portion of the sample should be tested is difficult. Fernandez and Struchtemeyer (1985) emphasized the need to test the whole soil profile. Notwithstanding, the horizontal and vertical variability, these examining prerequisites could not be functional for daily testing of woods soils.

7.10 Nutrient demands with respect to stand age

Regardless of the amounts of potentially accessible nutrients present in soils, the needs of woodland trees can change significantly during seedling foundation and development. Plant supplement prerequisites before crown closure are unmistakably different from those after crown closure. When data on supplement take-up and maintenance is used, this entails the assurance of the amount of plant-accessible nutrients.

7.11 Crown closure (before)

The early stages of tree development were described in small trees with minimal supplement requirements and rapid growth. The majority of the nutrients consumed are stored within the trees. Natural material aggregated from the previous pivot decays quickly at this stage due to the high temperature caused by high sun-based protection.

The nutrients/supplements transported to the decaying natural layer may be absorbed by non-crop plants or drained into lower horizons or groundwater. During these times, the plantation seedlings also retain supplements from the inorganic horizons. As a result, expecting a strong link between N accessibility and tree development during the early stages of development is ridiculous. Nonetheless, when supplement interest by trees may as low as 20 kg N/ha/yr, and plant-accessible N in whichever the natural or inorganic horizons or both are maximum.

7.12 Crown closure (after)

After canopy closure, leaf biomass and nitrogen demand in semi-mature stands will generally stabilize. Forest N cycle is based on litterfall, throughfall, fine root and natural material disintegration. During this stage of advancement, strong natural rivalry aimed at gradually recycling supplements (N) radiating from natural skylines. Consequently, the annual accumulation of biomass and nutrients by trees is much lower in mature and semi-mature stands with closed shelters than in open-crown new stands. Despite the fact that closed shelter stands are an important source of nitrogen from litter and humus, nitrogen uptake isn't always related to the absolute amount of nitrogen on the forest floor. It is not unusual due to the varying rates of N mineralization in various woodland floor types. As a result, when depicting site classes, the amount of plant-accessible supplements estimated utilizing conventional strategies should be deciphered with caution.

7.13 Nutrient cycling

The term nutrient cycling refers to a variety of physical, synthetic, biochemical, and natural cycles that occur in various components of forest ecosystems. These cycles can occur in solid, fluid, or gaseous stages and are so dynamic that determining the accessibility of a supplement based on soil tests collected at a specific time is difficult. Furthermore, some of the cycles, for example, mineral enduring, nitrogen obsession, and mycorrhizal supplement obtaining, are frequently overlooked as an important component of supplement cycles in the woodlands.

7.14 Nutrient addition in the forest

Nutrient take-up from wood crops is similar to that of plant or agricultural crops, but the amount collected from biomass accounts for approximately one-third of total take-up. Critical nitrogen content contrasts for the entire stand in various biological systems (Petrokofsky et al., 2012). Coniferous woodlands are generally less supplement demanding than hardwoods, but there is significant variation between the two groups. For example, the macronutrient content of a 40-year-old jack pine estate was not exactly 50% of that of a 40-year-old white spruce plantation on similar soil. How we interpret the supplement requirements of timberlands is based primarily on assessments of relatively unadulterated natural stands or ranches of single species. When compared to the pine (coniferous) environment, the maple (deciduous) biological system contained twice as much natural matter and multiple times as much N (Fig. 7.2). The majority of the nitrogen resources were found on the forest floor and in plant biomass in the various environments (McKay and Malcolm, 1988).

In comparison to local species, the amount of nutrients removed from the sites by reaping and filtering prior to the regeneration of forest vegetation. Assessing the overall effects of these misfortunes on destinations of varying usefulness is difficult due to our patchy data on the renewal of plant-accessible supplements from enduring and air inputs.

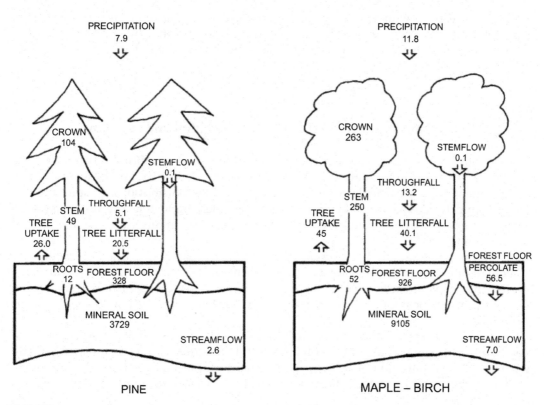

FIGURE 7.2   Nutrient flow in Pine and Maple Birch (Mahendrappa, 1980).

7.15 Nutrient input from atmospheric sources

A few experts have recorded the atmospheric contribution of nitrogen in the form of nitrate, ammonium, and other natural N compounds. A portion of the N input in soil regions occurs in the dry deposition structure. However, because of the filtering effects of trees, the amounts of N estimated as dry deposition in open regions may be only a small portion of the amounts of N added to backwoods. They rely on the canopy structure of vegetation (grass, bushes, or trees) to capture dry deposition (Fenn et al., 2003).

Limited amounts of N added from the climate as wet and dry deposition may appear to be insignificant when compared to the amounts of supplements in woods biomass and the amounts of N estimated to be potentially accessible in timberland soils. Nonetheless, during the normal revolution time of backwoods in the northern calm area of about 60–80 year old, even the most reduced paces of yearly nitrogen deposition documented can address a maximum extent of complete nitrogen in the forest ecosystem. Furthermore, in heavily polluted areas, the extra N may be sufficiently large to cause harmful effects due to a combination of increased deposition rates and the filtering effects of tree canopies. The deposition of significant levels of nitrates may oversaturate soils with N and cause a decrease in backwoods development. Furthermore, consistent nitrogen application in the summer can keep up with succulent foliage vulnerable to damage against an early frost. Therefore, nutrients added to woodland ecosystems from the atmosphere should be considered when determining potentially accessible supplements (N) to backwoods (Camargo and Alonso, 2006).

7.16 Nutrients returned through throughfall and stemflow

The amount of nutrients recycled through throughfall and stemflow is affected by the health of the trees (Schroth et al., 2001). They are regarded as critical pathways for nutrient movement to soil, particularly in terms of K and Mg contributions. Despite the fact that K is quickly filtered from foliage and is a versatile component in forest ecosystem, evidence suggests that K is conserved by forest vegetation. In any case, the amounts of nitrogen, phosphorous, and calcium moved to the soil are generally not the same as those returned to the soil by litterfall. The flaws in assessing nutrient recycling by throughfall and stemflow are related to the difficulties in distinguishing nutrients inferred inside plant local area (recycling) from those conserved on plant surface by moist and dry deposition, i.e., contribution to the biological system (Van Stan and Gordon, 2018).

The recycled nutrients may beincluded for the evaluations of nutrients accessible to trees. Some nutrients that are water-soluble natural structures can be assimilated directly by plants without ionization or disintegration. Traditional techniques for surveying nutrient accessibility exclude quickly absorbable natural mixtures containing varying amounts of nutrients (Miler, 1968).

7.17 Retranslocation of nutrients within trees

Throughout the development period, nutrients are constantly reallocated within trees and plants. Movement of supplements within plants is an essential component of the nutrient cycling procedure and a natural nutrient conservation mechanism. Before the foliage falls, a large amount of nutrients, primarily nitrogen and phosphorous, are removed from the foliage and redistributed within the tree. The method for determining the amount of nutrients retranslocated from falling foliage is complicated and unstandardized. Miller (1984) discovered retranslocation of foliar biomass and nutrient concenterations in foliage and litterfall. Along these lines, comparing the nutrient concentrations in the foliage and litterfall provides information on the overall amounts of nutrients retranslocated.

Hardwoods retranslocate more nitrogen before leaf fall than conifers, and re movement appears to be species-specific. The amount of supplements retranslocated within tree species is determined by their availability in soil. Return appears to be constrained by measures of sources and sinks. Trees in low supplement accessibility locations move a greater number of supplements from their leaves than species in rich destinations. Thus, the availability of supplements in the soil should not be regarded as an independent parameter in modelling nutrient dynamics in woodland ecosystem.

7.18 Nutrient recycling at higher elevation

The nutrient cycling at higher elevations varies entirely from lowlands. The generally unique environment of higher elevation imposes a more significant thrust on the nutrient cycle. The higher elevation's minimum temperature prevents decomposition. The soil surface had less microbial and enzymatic movement, which may have resulted due to unique ecosystem at higher elevations. Although the soil organic carbon content of soils increases with elevation, and soil nutrient availability decreases (Koerselman and Meuleman, 1996). The organic carbon content in plants remains constant, unaffected by climate change, but microbial biomass and soil C are declining. A decrease in organic matter degradation due to high carbon nitrogen ratios proportions of leaf litter, roots, and so on builds soil organic carbon (He et al., 2016). Decrease in temperature results in declining microbial and enzymatic activities. Soil texture, constant carbon inputs may be affected steadily by rising temperature (Bolstad et al., 2001). The rate of N_2 mineralization slows as elevation increases. A similar pattern is observed for soil phosphorus. The phosphorus content of soil correlates directly with elevation. At the highest elevations, the available phosphorus content decreases with soil depth (Johannessen, 1958). Furthermore, a large portion of the P in the soil remains in an unavailable form, which contradicts the fact that P is available in numerous organic forms and isfound in higher concentration under plantation cover.

The surface of the soil has a high exchangeable potassium content, which was later converted into soil solution. The alpine areas of the tundra ecosystem contain significant amounts of phosphorous, calcium, and magnesium, which increase with depth (Dangwal et al., 2012; Tian et al., 2017). There is usually a distinction between foliar nutrient content and soil nutrient content. The environment under montane forest, and the plants' access to available nutrients like nitrogen, phosphorous, potassium, and sulfur, decreased as elevation increased (Soethe et al., 2008). Total soil K varies with height as well. In the high- altitude environment, an even, tightly closed soil-plant-litter web exists. Patios and similar designs are used where the land is exceptionally steep to prevent disintegration and land corruption. Agrarian exercises are typically performed in halfway rise classes, whereas touching occurs at higher elevations. The polar ice covers are nearly always frozen. The supplement levels are low due to the lower temperature, which results in less vegetation and less microbial movement (Jonasson et al., 1996). The altitudinal vegetation belts (orobiomes) follow the environment zones, show plant types, propensities, and techniques to adjust to a specific climate (Table 7.1).

7.19 Litterfall

The total yearly litterfall, including ground vegetation, demonstrates comparable variations to those detailed for cool temperate and boreal zones. It is considerably linked to the basal area and age of the woodland. Deciduous and coniferous forests lose significant amounts of foliar biomass during the pre-winter season, but much less foliage and other plant parts in other months (Bray and Gorham, 1964). Because of the large contribution of herbs and bush litter comparative with ground cover, the yearly commitment of spices and bushes to nutrient content is significantly greater than their commitment to litterfall biomass (Landuyt et al., 2020). After canopy closure in regular woods, the number of living trees decreases all through the yield turn, but the contribution of tree mortality to litterfall interaction is mostly ignored. Similarly, the yearly expansion of organic matter and nutrients, as well as the demise of fine roots, are rarely remembered for nutrient cycling studies (Vogt et al., 1982).

The microbial decay of a litter part's sugars and nitrogenous constituents affects its ability to supply nitrogen. The carbon and nitrogen ratio proportion of leaf and needle litter varies between deciduous (maple) and coniferous (pine), and its underlying disintegration is described by N immobilization in microbial cells (Kriiska et al., 2021). As a result, the arrival of nitrogen and other nutrients from microbial matter is a critical stage in supplement recycling procedures. It includes microbial lysis, ecological effects (freezing-thawing, moistening-drying cycles), and soil fauna care. Soil full scale and microfaunal contact are particularly effective in assembling nutrients from infectious soil natural matter and tissue in litter (Anderson and Ineson, 1983).

When the carbon and nitrogen proportions fall below a certain threshold (25:1), inorganic nitrogen delivery to the soil increases. Over the C:N proportion, the underlying lignin grouping of litter and litter N concentration have been proposed as preferable indicators of deterioration rates. High lignin concentrations reduce the rate of leaf and needle litter decay, and in a few coniferous woods natural skylines, the accessibility of C can limit microbial movement (Kriiska et al., 2021).

TABLE 7.1   Major plant species present in different mountain ranges of the world (Jeyakumar et al., 2020).

Mountain ranges	Altitude	Tree species	References
Himalayan	1700–3800 m	Quercus semecarpifolia	Tiwari and Jha (2018)
	2000–2600 m	Pinus wallichiana	
	2700–4300 m	Betula utilis	
	2800–4400 m	Rhododendron campanulatum	
	3000–4000 m	Abies spectabilis	
	3700–4400 m	Juniperus sp.	
Chiricahua	1950 m	Pseudotsuga menziesii	Sawyer and Kinraide (1980)
	2100 m	Pinus ponderosa	
Andes	3560–4680 m	Polylepis sp.	Kessler et al. (2014)
	3300 m	Myrsine sp., Clusia spp.	Bader et al. (2007)
	3600 m	Weinmannia cochensis, Ilex colombiana	
Alborz	2400–2850 m	Quercus macranthera	Noroozi and Korner (2018)
	2000–3000 m	Juniperus excels	
Alps	2200 m	Alnus viridis	Carnelli et al. (2004)
	2100 m	Larix decidua, Pinus sp.	Loranger et al. (2016)
	2250–2540 m	Pinus cembra	Jochner et al. (2017)

The nature of the litter substance is critical in determining the rate of nutrient recycling due to decay. Temporal variations in the actual climate also have an impact on nutrient decomposer activities. Mineralization is hampered in many backwoods soils by cool temperatures and low humidity. Minimum temperatures in cool, mild, and boreal timberlands inhibit disintegration and collect supplements in litter and humic substances. Given the time required for timberlands to mature, the long-term supplement providing force of the soil is more fundamental than the easily quantifiable dissolvable nutrient supply (Powers et al., 2015). The tests would distinguish N-rlimited sites, and the board practices may be taken on that advance litter deterioration and arrival of nitrogen.

7.20 Nutrient supply to the trees

The nutrients taken up by plants go through two stages: transport to the root surface and retention by the tree. Plants attack a few components (Al, Cl, and Na) while retaining other nutrients such as nitrogen, phosphorous, potassium, calcium, and manganese. The exchange rate of particles from soil to root is a significant constraint in plant supplement take-up. As a result, root overflow is critical in tree nutrition. The appropriation, repetition, and profundity of establishing are critical components in supplement procurement. Fine root development is normally most critical in the surface soil horizons and generally decreases with depth. Binkley and Fisher (2019) investigated the root frameworks of jack pine, aspen, and dark tidy on moderately depleted medium-finished soils. When the physical and chemical properties of the soil are excellent, tree roots will investigate deeper soil layers. Regardless, the importance of earth nutrients to woodland sustenance has not been adequately evaluated. Organic horizon materials are more capable of mineralizing supplement components than inorganic layers. As a result, significantly more critical sums are stored in mineral soil on a unit region basis.

7.21 Leaching

The filtering of particles past the successful establishing zone reduces supplement take-up in the backwoods. The predominant anions governing cation filtering from the soil might be SO_4^{2-}, NO_3^-, HNO_2^- or natural acids. The anions of the environment in both wet and dry testimony mix with those produced by microbially interceded responses in the soil. Hydrology, topography, vegetation, soil type, and position relative to natural or anthropogenic outflows of synthetic substances can all have an impact on ionic organization in soil. In any case, in comparison to the rate of progress of nutrients through various nutrient cycling processes, the extents of supplement draining misfortunes are negligible under woods conditions (Uhlig and von Blanckenburg, 2019).

7.22 Additional factors

7.22.1 Mycorrhiza

Mycorrhizal roots, which become increasingly important under low ripeness conditions, promote supplement uptake by timberland trees. The predominant structure found in Pinaceae, Ectotrophic mycorrhizae, aids in the uptake of nutrients from soil adsorptive surfaces. Nutrient consumption may be reduced due to limited soil contact with low temperatures, roots, abundance dampness, or water pressure. Soil compaction during collection increases mechanical impedance to root stretching, thereby confining nutrient accessibility. Soils deficient in nutrients and available water, in woodland trees, should expend increased photosynthate and energy to deliver and maintain a functioning root system (Dell, 2002).

The increased surface area of dynamic roots increases nutrient take-up and decreases dry spell pressure due to their relationship with mycorrhizal growths. The standard P methodology revealed that immunizations with mycorrhizal societies increase P take-up. That, in general, is delegated P insufficient. Mycorrhizal parasites can use a wide range of basic natural nitrogen mixtures, confirming the possibility of direct supplement cycling in woods biological systems (Bowen and Smith, 1981).

Enduring yields base cations, for example, calcium, magnesium, potassium and sodium and a few acidic cations like iron, aluminum and manganese, and negligible measures of phosphorous. Contingent upon the idea of the parent material, the assessed commitments of supplements got from enduring reach under 10 to north of 700 kg/ha/yr (Parikh and James, 2012). This impressive variety may to some degree be credited to contrasts in the techniques utilized for assessing the commitments from enduring. Even the smallest assessments of supplement input from enduring can be considered to supply a lot of supplements over a revolution time of up to 100 year, rendering conventional strategies for assessing accessible supplements practically obsolete. While evaluating accessible supplements for pragmatic or displaying purposes, consider the commitment of supplements from enduring, particularly in soils that climate rapidly.

The mineral supplement holds in late glaciated soils are enormous, yet some of the materials enduring is somewhat sluggish. Enduring information sources are typically assessed (in situ) from either rock rot or catchment mass equilibrium estimations, despite the collection of supplements in woods biomass (Churchman and Lowe, 2012). In the primary gauge, the activity of roots and other biological agencies is ignored. Recognizing the nutrients released from inorganic and natural sources is a challenge for the second. The need to devise methods for determining particle discharge from essential minerals in the enduring climate of the establishing zone is widely regarded as a critical need for nutrient recycling.

7.23 Nitrogen fixation

In various parts of the wood's biological systems, basic nitrogen (N2) is converted into plant absorbable ammonia nitrogen. This interaction is carried out on tree boles and in the phyllosphere by free living and cooperative organic entities (Aasfar et al., 2021). The amounts of nitrogen entering the nutrient pattern of woodland environments through N obsession thus appear to rise and surpass the amounts cycled as throughfall and litterfall. Nitrogen fixed by microflora can meet the nitrogen requirements of some stands. Soils with a high N obsession typically have increased draining of base cations due to a high nitrification rate.

7.24 Conclusion

The nutrient requirements of trees differ significantly from those of crops. Extensive stretches are expected to arrive at the merchantable size and huge variations with nutrient demands when crown closure absolutely adds to the differences. The distinctions are also strongly linked to nutrient recycling in the woods (soil-plant-environment continuum). The rate of progress (transition) of certain nutrients through nutrient cycling processes outpaces the assessed supplement requests of trees or their soil . The amount of specific supplements returned to plantations each year by litterfall, stemflow, and throughfall is greater than the net amount of nutrients gathered from plantations each year. The transition of supplements from these cycles is related to tree development; it was independent of accessibility gauges. Plantations' ability to form relationships with mycorrhizal parasites and retain natural atoms containing supplements. Trees, in particular, have avoided the requirement for the mineralization cycle, which is the premise of accessible soil supplement appraisals. Trees benefit from a variety of data sources, including barometric evidence, N-obsession, and enduring. Mountain soils, particularly those at higher elevations, face a number of challenges, including low nutrient cycling, extreme climatic conditions, and increasing biological diversity. The physicochemical properties of soil vary greatly depending on the environment. Furthermore, with global climate change, it is critical to study nutrient cycling at higher elevations.

References

Aasfar, A., Bargaz, A., Yaakoubi, K., Hilali, A., Bennis, I., Zeroual, Y., et al., 2021. Nitrogen fixing Azotobacter species as potential soil biological enhancers for crop nutrition and yield stability. Frontiers in Microbiology 12, 354.

Alban, D.H., 1982. Effects of nutrient accumulation by aspen, spruce, and pine on soil properties. Soil Science Society of America Journal 46 (4), 853–861.

Anderson, J.M., Ineson, P., 1983. Interactions between soil arthropods and microorganisms in carbon, nitrogen and mineral element fluxes from decomposing leaf litter. In: Symposium of the British Ecological Society.

Arp, P.A., Krause, H.H., 1984. The forest floor: lateral variability as revealed by systematic sampling. Canadian Journal of Soil Science 64 (3), 423–437.

Bader, M.Y., Rietkerk, M.G., Bregt, A.K., 2007. Vegetation structure and temperature regimes of tropical alpine treelines. Arctic, Antarctic, and Alpine Research 39 (3), 353–364.

Binkley, D., Fisher, R.F., 2019. Ecology and Management of Forest Soils. John Wiley & Sons.

Bolstad, P.V., Vose, J.M., McNulty, S.G., 2001. Forest productivity, leaf area, and terrain in southern Appalachian deciduous forests. Forest Science 47 (3), 419–427.

Bowen, G.D., Smith, S.E., 1981. The effects of mycorrhizas on nitrogen uptake by plants. Ecological Bulletins 237–247.

Bray, J.R., Gorham, E., 1964. Litter production in forests of the world, Advances in Ecological Research, vol. 2. Academic Press, pp. 101–157.

Camargo, J.A., Alonso, Á., 2006. Ecological and toxicological effects of inorganic nitrogen pollution in aquatic ecosystems: a global assessment. Environment International 32 (6), 831–849.

Carnelli, A.L., Theurillat, J., Thinon, M., Vadi, G., Talon, B., 2004. Past uppermost tree limit in the Central European Alps (Switzerland) based on soil and soil charcoal. Holocene 14 (3), 393–405.

Churchman, G.J., Lowe, D.J., 2012. Alteration, Formation, and Occurrence of Minerals in Soils. CRC press, pp. 1–72.

Dangwal, L.R., Tajinder, S., Amandeep, S., Antima, S., 2012. Plant diversity assessment in relation to disturbances in subtropical Chir pine forest of the western Himalaya of District Rajouri, J&K, India. International Journal of Plant, Animal and Environmental Sciences 2 (2), 206–213.

Dell, B., 2002. Role of mycorrhizal fungi in ecosystems. CMU Journal 1 (1), 47–60.

Dong, W., Zhang, X., Wang, H., Dai, X., Sun, X., Qiu, W., et al., 2012. Effect of different fertilizer application on the soil fertility of paddy soils in red soil region of southern China.

FAO, 2015. Understanding Mountain Soils: A Contribution from Mountain Areas to the International Year of Soils 2015, by Romeo, R., Vita, A., Manuelli, S., Zanini, E., Freppaz, M., Stanchi, S. Rome, Italy.

Fenn, M.E., Haeuber, R., Tonnesen, G.S., Baron, J.S., Grossman-Clarke, S., Hope, D., et al., 2003. Nitrogen emissions, deposition, and monitoring in the western United States. Bioscience 53 (4), 391–403.

Fernandez, I.J., Struchtemeyer, R.A., 1985. Chemical characteristics of soils under spruce-fir forests in eastern Maine. Canadian Journal of Soil Science 65 (1), 61–69.

Giweta, M., 2020. Role of litter production and its decomposition, and factors affecting the processes in a tropical forest ecosystem: a review. Journal of Ecology and Environment 44 (1), 1–9.

Gougoulias, C., Clark, J.M., Shaw, L.J., 2014. The role of soil microbes in the global carbon cycle: tracking the below-ground microbial processing of plant-derived carbon for manipulating carbon dynamics in agricultural systems. Journal of the Science of Food and Agriculture 94 (12), 2362–2371.

Guignard, M.S., Leitch, A.R., Acquisti, C., Eizaguirre, C., Elser, J.J., Hessen, D.O., et al., 2017. Impacts of nitrogen and phosphorus: from genomes to natural ecosystems and agriculture. Frontiers in Ecology and Evolution 5, 70.

He, X., Hou, E., Liu, Y., Wen, D., 2016. Altitudinal patterns and controls of plant and soil nutrient concentrations and stoichiometry in subtropical China. Scientific Reports 6 (24261), 1—9.

Jacoby, R., Peukert, M., Succurro, A., Koprivova, A., Kopriva, S., 2017. The role of soil microorganisms in plant mineral nutrition-current knowledge and future directions. Frontiers in Plant Science 8, 1617.

Jeyakumar, S.P., Dash, B., Singh, A.K., Suyal, D.C., Soni, R., 2020. Nutrient cycling at higher altitudes. Microbiological Advancements for Higher Altitude Agro-Ecosystems & Sustainability. Springer, Singapore, pp. 293—305.

Jochner, M., Bugmann, H., Notzli, M., Bigler, C., 2017. Among-tree variability and feedback effects result in different growth responses to climate change at the upper treeline in the Swiss Alps. Ecology and Evolution 7 (19), 7937—7953.

Johannessen, C., 1958. Higher phosphate values in soils under trees than in soils under grass. Ecology 39, 373—374.

Jonasson, S., Michelsen, A., Schmidt, I.K., Nielsen, E.V., Callaghan, T.V., 1996. Microbial biomass C, N and P in two arctic soils and responses to addition of NPK fertilizer and sugar: implications for plant nutrient uptake. Oecologia 106 (4), 507—515.

Kessler, M., Toivonen, J.M., Sylvester, S.P., Kluge, J., Hertel, D., 2014. Elevational patterns of Polylepis tree height (Rosaceae) in the high Andes of Peru: role of human impact and climatic conditions. Front Plant Science 5 (194), 1—12.

Klinka, K., Green, R.N., Trowbridge, R.L., Lowe, L.E., 1981. Taxonomic Classification of Humus Forms in Ecosystems of British Columbia. First Approximation.

Koerselman, W., Meuleman, A.F., 1996. The vegetation N:P ratio: a new tool to detect the nature of nutrient limitation. Journal of Applied Ecology 33, 1441—1450.

Krause, H.H., 1981. Factorial Experiments with Nitrogen, Phosphorus and Potassium Fertilizers in Spruce and Fir Stands of New Brunswick: 10-Year Results. Maritime Forest Research Centre, Canadian Forestry Service, Department of the Environment.

Kriiska, K., Lõhmus, K., Frey, J., Asi, E., Kabral, N., Napa, Ü., et al., 2021. The dynamics of mass loss and nutrient release of decomposing fine roots, needle litter and standard substrates in hemiboreal coniferous forests. Frontiers in Forests and Global Change 91.

Landuyt, D., Ampoorter, E., Bastias, C.C., Benavides, R., Müller, S., Scherer-Lorenzen, M., et al., 2020. Importance of overstorey attributes for understorey litter production and nutrient cycling in European forests. Forest Ecosystems 7 (1), 1—11.

Leye Samuel, A., Omotayo Ebenezer, A., 2014. Mineralization rates of soil forms of nitrogen, phosphorus, and potassium as affected by organomineral fertilizer in sandy loam. Advances in Agriculture 2014.

Lodhiyal, L.S., Lodhiyal, N., 1997. Nutrient cycling and nutrient use efficiency in short rotation, high density central Himalayan Tarai poplar plantations. Annals of Botany 79 (5), 517—527.

Lodhiyal, L.S., Singh, R.P., Singh, S.P., 1995. Structure and function of an age series of poplar plantations in central Himalaya. II Nutrient dynamics. Annals of Botany 76 (2), 201—210.

Loranger, H., Zotz, G., Bader, M.Y., 2016. Early establishment of trees at the alpine treeline: idiosyn-cratic species responses to temperature-moisture interactions. AoB Plants 8, 1—14.

Mahendrappa, M.K., 1980. Relationships between different estimates of mineralizable N in the organic materials under black spruce stands. Canadian Journal of Forest Research 10 (4), 517—522.

McKay, H.M., Malcolm, D.C., 1988. A comparison of the fine root component of a pure and a mixed coniferous stand. Canadian Journal of Forest Research 18 (11), 1416—1426.

Miler Jr, R.J., 1968. Assimilation of Nitrogen Compounds by Tree Seedlings. Yale University.

Miller, H.G., 1984. Dynamics of nutrient cycling in plantation ecosystems. In: Bowen, G.D., Nambiar, E.K.S. (Ed.), Nutrition of Plantation Forests.

Nafiu, A.K., Abiodun, M.O., Okpara, I.M., Chude, V.O., 2012. Soil fertility evaluation: a potential tool for predicting fertilizer requirement for crops in Nigeria. African Journal of Agricultural Research 7 (47), 6204—6214.

Noroozi, J., Korner, C., 2018. A bioclimatic characterization of high elevation habitats in the Alborz mountains of Iran. Alpine Botany 128 (1), 1—11.

Parikh, S.J., James, B.R., 2012. Soil: the foundation of agriculture. Nature Education Knowledge 3 (10), 2.

Petrokofsky, G., Kanamaru, H., Achard, F., Goetz, S.J., Joosten, H., Holmgren, P., et al., 2012. Comparison of methods for measuring and assessing carbon stocks and carbon stock changes in terrestrial carbon pools. How do the accuracy and precision of current methods compare? A systematic review protocol. Environmental Evidence 1 (1), 1–21.

Powers, R.F., 1983. Estimating soil nitrogen availability through soil and foliar analysis. In: 6th North American Forest Soils Conference, June 1983. Department of Forestry, Knoxville, TX.

Powers, J.S., Becklund, K.K., Gei, M.G., Iyengar, S.B., Meyer, R., O'Connell, C.S., et al., 2015. Nutrient addition effects on tropical dry forests: a mini-review from microbial to ecosystem scales. Frontiers in Earth Science 3, 34.

Pritchett, W.L., Smith, W.H., 1972. Fertilizer responses in young pine plantations. Soil Science Society of America Journal 36 (4), 660–663.

Radwan, H.A., Shumway, J.S., 1984. Site index and selected soil properties in relation to response of douglas-fir and western hehl08k to nitrogen fertilizer.

Razaq, M., Zhang, P., Shen, H.L., 2017. Influence of nitrogen and phosphorous on the growth and root morphology of Acer mono. PLoS One 12 (2), e0171321.

Rodrigues, P.M.S., Schaefer, C.E.G.R., de Oliveira Silva, J., Ferreira Júnior, W.G., dos Santos, R.M., Neri, A.V., 2018. The influence of soil on vegetation structure and plant diversity in different tropical savannic and forest habitats. Journal of Plant Ecology 11 (2), 226–236.

Salonius, P.O., Mahendrappa, M.K., 1983. A comparison of growth promotion by controlled release ureas and conventional nitrogen fertilizers using two mensurational methods [Picea mariana in New Brunswick, Canada]. USDA Forest Service general technical report PNW United States, Pacific Northwest Forest and Range Experiment Station.

Salonius, P.O., Fisher, R.A., Mahendrappa, M.K., 1982. An alternative method of measuring fertilizer effects in forest stands. Canadian Journal of Forest Research 12 (2), 146–150.

Santa Regina, I., Tarazona, T., 2001. Nutrient cycling in a natural beech forest and adjacent planted pine in northern Spain. Forestry 74 (1), 11–28.

Sawyer, D., Kinraide, T., 1980. The forest vegetation at higher altitudes in the Chiricahua Mountains, Arizona. The American Midland Naturalist 104 (2), 224–241.

Schoonover, J.E., Crim, J.F., 2015. An introduction to soil concepts and the role of soils in watershed management. Journal of Contemporary Water Research & Education 154 (1), 21–47.

Schroth, G., Elias, M.E.A., Uguen, K., Seixas, R., Zech, W., 2001. Nutrient fluxes in rainfall, throughfall and stemflow in tree-based land use systems and spontaneous tree vegetation of central Amazonia. Agriculture, Ecosystems & Environment 87 (1), 37–49.

Smeck, N.E., 1973. Phosphorus: an indicator of pedogenetic weathering processes. Soil Science 115 (3), 199–206.

Soethe, N., Lehmann, J., Engels, C., 2008. Nutrient availability at different altitudes in a tropical montane forest in Ecuador. Journal of Tropical Ecology 24 (4), 397–406.

Thiffault, N., Jobidon, R., De Blois, C., Munson, A.D., 2000. Washing procedure for mixed-bed ion exchange resin decontamination for in situ nutrient adsorption. Communications in Soil Science and Plant Analysis 31 (3–4), 543–546.

Tian, L., Zhao, L., Wu, X., Fang, H., Zhao, Y., Yue, G., et al., 2017. Vertical patterns and controls of soil nutrients in alpine grassland: implications for nutrient uptake. Science and Total Environment 607, 855–864.

Tiwari, A., Jha, P., 2018. An overview of treeline response to environmental changes in Nepal Himalaya. Tropical Ecology 59 (2), 273–285.

Uhlig, D., von Blanckenburg, F., 2019. How slow rock weathering balances nutrient loss during fast forest floor turnover in montane, temperate forest ecosystems. Frontiers in Earth Science 7, 159.

UNEP-WCMC (United Nations Environment Programme-World Conservation Monitoring Centre), 2002. Mountain Watch: Environmental Change & Sustainable Development in Mountains. UNEP-WCMC, Cambridge, United Kingdom.

Van Stan, J.T., Gordon, D.A., 2018. Mini-review: stemflow as a resource limitation to near-stem soils. Frontiers in Plant Science 9, 248.

Van Veen, J.A., Breteler, H., Olie, J.J., Frissel, M.J., 1981. Nitrogen and energy balance of a short-rotation poplar forest system. Netherlands Journal of Agricultural Science 29 (3), 163–172.

2. Soil microbial processes and nutrient dynamics

Vogt, K.A., Grier, C.C., Meier, C.E., Edmonds, R.L., 1982. Mycorrhizal role in net primary priduction and nutrient cytcling in Abies amabilis ecosystems in Western Washington. Ecology 63 (2), 370−380.

Wilde, S.A., 1964. Changes in soil productivity induced by pine plantations. Soil Science 97 (4), 276−278.

Wuenscher, R., Unterfrauner, H., Peticzka, R., Zehetner, F., 2015. A comparison of 14 soil phosphorus extraction methods applied to 50 agricultural soils from Central Europe. Plant, Soil and Environment 61 (2), 86−96.

Zavisic, A., Yang, N., Marhan, S., Kandeler, E., Polle, A., 2018. Forest soil phosphorus resources and fertilization affect ectomycorrhizal community composition, beech P uptake efficiency, and photosynthesis. Frontiers in Plant Science 9, 463.

Microbial community structure and climate change: impact of agricultural management practices in mountainous landscapes

Anushree Baruah[1] *and Mitrajit Deb*[2]

[1]Department of Botany, The Assam Royal Global University, Guwahati, Assam, India
[2]Department of Zoology, The Assam Royal Global University, Guwahati, Assam, India

8.1 Introduction

Microorganisms play an important role in the biosphere by sustaining life and exhibiting a wide range of form and function. Through the various processes involved in the biogeochemical and nutrient cycles, these invisible (to the naked eye) but invincible organisms play an important role in regulating the climate. Microorganisms have existed on Earth for millions of years and will continue to exist in the future. However, the impact of climate change on microorganisms is not widely discussed (Cavicchioli et al., 2019). Human effects on microorganisms are less obvious and less well understood; however, one major concern is that changes in microbial biodiversity and activity will affect the resilience of all other organisms, and thus their ability to respond to climate change (Maloy et al., 2017).

In the agricultural ecosystem, one gram of topsoil contains thousands of microscopic cells from hundreds of different species. From performing numerous metabolic and physiological functions to producing and absorbing greenhouse gases (GHGs) such as methane and nitrous oxide, these microscopic organisms evolve into one of the most important and integral parts of soil health, crop fertility, pollutant remediation, carbon sequestration, and nutrient cycle regulation. Thus, soil microbial populations determine key soil functions, thereby directly influencing land value. Future climate scenarios may have an impact on soil microbial communities, resulting in soil carbon loss, changes in soil-borne greenhouse gas levels, and changes in critical plant-soil feedbacks that contribute to soil fertility (UK CEH, 2022).

Climate change has an impact on both aboveground and belowground ecosystems in terrestrial ecosystems, either directly or indirectly. Aboveground, the effects of global change are mostly direct: increased atmospheric carbon dioxide, as well as changes in temperature, precipitation, and nitrogen availability, result in changes in plant species abundance and altered land cover in unmanaged ecosystems (Tylianakis et al., 2008). Changes in seasonal climate and precipitation influence crop species choice, production, and stresses on those chosen species, which in turn influences management decisions for irrigation, fertilization, and pathogen dynamics in managed systems (Dixon, 2009). Land use, plant species composition, and plant productivity all indirectly influence plant communities through changes in the belowground microbial community (Balser et al., 2010).

Mountain soils are vulnerable to global changes, such as climate change, land-use change, deforestation, and overgrazing, which are affecting mountains in unprecedented ways. These changes, in turn, have an impact on the people who live in the mountains, as well as their livelihoods and food security. Floods, landslides, debris flows, and glacial lake outbursts are on the rise in most mountain regions, particularly those with rapidly growing populations and inadequate infrastructure. Climate change is exacerbating the impact of hazards by increasing the frequency of extreme events such as heavy rainfall, droughts, and glacier melt.

Individual soil microorganism species and consortia play critical roles in the many functional processes that support agricultural systems, and the soil microbiota becomes an integral part of the long-term sustainability of such systems. These functional processes include the acquisition and recycling of nutrients required for plant growth, soil structure maintenance, agrochemical/pollutant degradation, and biological pest control of plant and animal pests (Parkinson and Coleman, 1990; Lee and Pankhurst, 1992). Despite the importance of these functional processes in agricultural systems, little is known about the effects of agricultural practices on soil microbial community biodiversity. This is due in part to (1) the technical difficulties in sampling and quantifying the diversity of soil microorganisms, and (2) the fact that in many intensive agricultural systems, the use of agrochemicals and mechanical tillage overrides some of the functional processes carried out by soil microbial communities (Pankhurst et al., 1996). Agronomic measures of crop production over a several-month cropping season also integrate and obscure short-term events influenced by the specific activities of soil microbial populations (Anderson, 1994). It is critical to understand when and under what conditions the functional attributes of soil microbial communities change without jeopardizing agricultural system productivity.

Understanding mountain soils, their composition, dynamics, and value can assist in addressing the challenges of sustainable mountain development and ensuring better, more resilient livelihoods for mountain people (FAO, 2015). In contrast, agricultural management techniques to boost crop productivity and reduce GHG emissions from such techniques involve a variety of reversible as well as irreversible changes to the physical, chemical, and biological properties of soil, including a significant impact on soil productivity and sustainability. This chapter will discuss the effects of such management systems on the soil microbiota, their roles in reducing greenhouse gas emissions, and their effects overall.

8.2 Mountainous soil and climate change

Mountains cover approximately 22%−25% of the Earth's total land area (Grabherr and Messerli, 2011) and are an important ecosystem, providing refuge to many plants, animals, and mountain dwelling human communities. Mountainous soil has been shown in studies to play an important role in the proper functioning of bionetworks. The hydrology of highland belts is also supported by mountain soil. Because mountain soil serves so many functions, it is extremely vulnerable to climate change (Egli and Poulenard, 2016). Rapid climate change may cause unstable soil structure and reduce water holding capacity, resulting in less water available for mountain communities. Furthermore, there are concerns about the release of sequestered GHG stored in these mountainous regions' soils. This release of GHGs may exacerbate already deteriorating climatic conditions. Knowledge about the effects of climate change, as well as information about climate adaptation and vulnerability, has grown over time. More research is needed, particularly on the impact on land-water-ecosystem quality and its consequences for mountain communities (FAO, 2015).

Most mountain regions are warming twice as fast as the global average, accelerating microbial activity. As a result, the ratio of two key nutrients—nitrogen and phosphorus—is rapidly shifting, according to Nathan Sanders of the University of Vermont, who was part of an international team that studied soil chemistry in seven mountain regions around the world over the course of a growing season. This type of change will drastically alter or shift the existing range of organisms.

Mountain environments are extremely vulnerable to climate change (Beniston, 2003, 2005). They are among the most severely and rapidly impacted ecosystems, and are susceptible to changes in temperature and precipitation patterns at all scales (Zemp et al., 2009). Snow and ice are the primary control parameters of the hydrological cycle, particularly seasonal runoff, and have the potential to affect the entire geosystem (rocks, soils, vegetation, and river discharges). Water will most likely become less available as a result of climate change, with far-reaching' consequences beyond mountain regions (Lutz and Immerzeel, 2013). Similarly, climate change is likely to increase exposure to either natural or economic hazards, all the more so because poverty levels in many mountain areas are higher than in lowland areas, and food insecurity is more prevalent (Ives and Messerli, 1989; Kohler et al., 2014).

8.3 Climate change and soil microbial function

Soil microbes, as previously stated, play an important role in virtually all ecosystem processes, determining agricultural land sustainability, ecosystem resilience against nutrient mining, degradation of soil and water resources, and GHG emissions through their microbial abundance and activity (Wagg et al., 2014). Changes in the environment have a direct impact on their activity. Climate change is a relevant factor in this context, with the potential to affect the role of microbes in soil, which is critical to supporting agriculture worldwide. According to IPCC (2014), global surface temperatures are projected to increase in response to rising atmospheric concentrations of CO_2. There is considerable uncertainty in the direction and magnitude of the effects that climate change will bring upon the soil microbial communities (Castro et al., 2010; Xiao et al., 2018) which represents the driving force of a large number of ecosystem processes (Xiong et al., 2014) and also playing an important role in the response to climate change. The production of extracellular enzymes in soil microorganisms, which catalyzes the process of degradation, transformation and mineralization of organic molecules in soil organic matter affects carbon (C) and nutrient cycling (Xiao et al., 2018) and alter microbial function, soil quality (Marx et al., 2001) and ecosystem productivity (Sayer et al., 2013).

While we have a general understanding of the mechanisms by which climate change affects microbial communities, predicting microbial community responses to individual and interactive climate change drivers, let alone the consequences for ecosystem functioning, is difficult due to a significant knowledge gap. This gap can be attributed to activities under the control of soil microbial communities, which results in a balance between the storage and release of GHGs such as CO_2, CH_4, and N_2O, and thus can exacerbate or mitigate climate change (de Vries and Griffiths, 2018). However, the use of new technologies, such as isotope labeling and molecular approaches, has resulted in an exponential increase in studies assessing the mechanisms by which climate change affects soil functioning in the current decade.

Early research on soil system functioning and climate change focused on high CO_2 levels. After the 1990s, there was a greater emphasis on rising temperatures and changing precipitation patterns. While elevated CO_2 levels affect soil microbial communities indirectly by altering plant physiological processes, growth, and community composition, temperature changes not only influence soil microbial life directly but also indirectly through changes in evapotranspiration, plant physiology, root exudation, and vegetation (Dieleman et al., 2012; Lange et al., 2014). Ecosystem functions such as nitrogen fixation, nitrification (Isobe et al., 2011), denitrification and methanogenesis are regulated by specific microbial groups. Changes in the relative abundance of these organisms will directly the rate of that process. Temperature governs organic matter decomposition rates, denitrification, and nitrification processes of N_2O production at all levels. Soil temperature increases N_2O emissions by increasing the rate of microbial activity emission or by causing freezing/thawing events (Gregorich et al., 2006). A review by Smith et al. (2003) reported that higher temperatures increased the anaerobic fraction by increasing respiratory sink for O_2. There are many reports on effects of changing temperature on soil respiration rate (Kandeler et al., 1998; Bradford et al., 2008; Hartley et al., 2008), variation in total microbial biomass, including bacterial and fungal biomass (Bardgett et al., 1999;

Sheik et al., 2011; Wei et al., 2014; DeAngelis et al., 2015), but whether these patterns are the result of warming directly affecting soil microbial communities or indirectly through affecting plant growth is not clear. Many evidences, however, show that plant communities have made significant changes to ecosystem respiration in response to climate change. Microbial communities respond to warming and other perturbations by exhibiting resistance, which is enabled by microbial trait plasticity, or by reverting to their original composition after the stress has passed (Allison and Martiny, 2008).

Aside from temperature, other biotic and abiotic variables such as soil moisture, substrate availability, soil pH, soil texture, habitat, and geographical distance, among others, that comprise "the environment" and are affected by climate change, have a variety of effects on the structure and function of the microbial community. All of these climate change drivers can have an immediate or indirect impact on soil microbial communities by influencing plant physiological processes and plant community composition (Bardgett et al., 2013).

Soil moisture is an important driver of the composition and activity of soil communities, and the first studies to assess the effect of soil moisture fluctuations on soil communities did not do so from the perspective of global climate change. It is intuitively obvious that soil water content decreases with increasing temperature. Wet-dry cycles, drought, flooding, and smaller shifts in moisture can all occur. These various changes have varying structural and functional impacts on communities and are influenced by a community's native regime. While research on drought has historically lagged behind research on the effects of elevated CO_2 and warming on soil microbial communities, there is now mounting evidence that extreme weather events have a greater impact on ecosystem function than slow-moving changes in atmospheric CO_2 concentrations and temperature. There have been more research on drought as a climate change issue (Reichstein et al., 2013). Similar to drought, excessive moisture brought on by floods from prolonged periods of heavy rain has a negative impact on the makeup of the soil microbial population.

Diffusion and microbial interaction with the available substrate can control rates of microbial activity at higher temperatures (Zak et al., 1999). The slower-growing fungal population may take longer to react to moisture pulses than the bacterial colonies do (Cregger et al., 2014). Further, drought amplifies the differential temperature sensitivity of fungal and bacterial groups (Briones et al., 2014). Soil fungal communities may transition from one dominant member to another even with slight variations in soil moisture availability (30% loss in water holding capacity), but bacterial communities stay stable. These patterns suggest that during non-extreme wet-dry cycles, fungi are more plastic than bacteria (Kaisermann et al., 2015). If soil communities have been used to frequent wet-dry cycles or low water availability, there may be less of a compositional or functional shift in response to shifting water regimes (Evans et al., 2011). Climate change affects microbial composition and function because of interactions between microorganisms and background temperature and moisture regimes in every given place (Classen et al., 2015).

However, because a diversity of organisms drives some activities that take place at a finer scale, such as nitrogen mineralization, they are more closely connected with abiotic parameters like temperature and moisture than microbial community makeup (Hooper et al., 2005). Additionally, methanogenic archaea convert carbon into CH_4 in anaerobic, carbon-rich habitats including wetlands, ruminant cattle, and rice paddies. This process results in both naturally occurring and human-induced CH_4 emissions (Mesle et al., 2013). Methanotrophs play a significant role in controlling CH_4 fluxes in the atmosphere, but they are challenging to separate due to their sluggish growth rate and strong adhesion to soil particles.

8.4 Agricultural management practices: impacts on microbial community structure and function

The greatest threat to ecosystems, agricultural sustainability, and food security in the future decades is climate change. The FAO developed the idea of "Climate-smart agriculture" (CSA) to create agricultural strategies under climate change, with the main goals of (1) boosting agricultural production and revenue, (2) developing climate-resilient agriculture, and (3) lowering and/or eliminating GHG emissions. The effective management of the soil is crucial to the success of this program. In addition to providing essential ecosystem services including nutrient cycling, organic matter decomposition, and GHG emissions, soil also serves as a home for biodiversity. CSA is a strategy that will aid in minimizing these effects. CSA is an integrated method to developing agricultural solutions for adapting and strengthening the resilience of agricultural and food security systems, improving agricultural productivity sustainably, and lowering agricultural GHG emissions under climate change scenarios (Lipper et al., 2014; Paustian et al., 2016).

The requirement to produce more food is related to the growing human population. The number of people on earth is expected to reach over 10 billion by the middle of the century in 2050, up from around 2.1 billion in 1950. The continued innovation in crop production enhancements, soil management, and technology advancements all contribute to the growth in the human population. Each acre of land is anticipated to support 6.2 people by 2050 (on average worldwide). Since the Green Revolution, crop and soil management has become much more intensive. This is especially true for wealthy nations, since agronomic intensification began decades before it did in underdeveloped nations. The increase in agronomic management intensity has resulted in SOC losses of up to 50% globally (Benbi and Brar, 2009).

The main problem facing agricultural sciences is the development of technologies that, in addition to increasing crop productivity, also provide for dietary security and agricultural sustainability, especially in regions with limited resources (Gepstein and Glick, 2013; Patel et al., 2015; Hamilton et al., 2016). The current agricultural practices, which heavily rely on the extensive use of agrochemicals for high yield, also lead to environmental disturbances (Singh et al., 2011; Singh et al., 2021; Singh et al., 2019; Paul and Lade, 2014; Singh, 2015). Globally, agricultural practices generate 30% of GHGs emissions, particularly when land-use change is included in the estimate (FAO, 2012). Due to the difficulties for traditional agricultural methods brought on by declining resource quality and ongoing climate change scenarios, these methods need to be modified in a number of ways that strike a balance between traditional knowledge and contemporary cultivation methods (Singh et al., 2019).

A substantial source of GHG emissions is represented by agriculture and has been identified as a major driver of climate change (Vermeulen et al., 2012). Agriculture-related practices and activities, such as the use of synthetic and organic fertilizers, the growth of crops that fix nitrogen, and the application of animal manure to farmland and pasture, are the main human-related sources. These have the effect of raising the nitrogen content of soils and streams, which causes annual N_2O emissions of roughly 4.5 Tg N. Global food output has been impacted by the ongoing growth in human population, significant climatic change, reduction in agricultural areas, increasing urbanization, and widespread use of agrochemicals (Rashid et al., 2016).

A significant and growing source of N_2O in agriculture is the animal production system, which includes deposition of dung and urine from grazing animals on the pasture, storage of animal wastes in stables and buildings, application of waste to the ground, and burning of dung (Oenema et al., 2005). Globally, agricultural N_2O emissions have increased by nearly 17% from 1990 to 2005, and they account for about 60% of global anthropogenic N_2O emissions (IPCC, 2007) and is recognized as the largest anthropogenic source of the gas. This is a serious matter as more than half of the world's population is directly or indirectly dependent on agriculture for food security.

Microorganisms have a crucial role in enhancing the availability of nutrients to plants, which is connected to climate-smart agriculture practices (Pereg and McMillan, 2015; Hamilton et al., 2016). The management of soil nutrients, including the soil's macro- and micronutrient status, depends on organic additions (Barnawal et al., 2014). Microbes increase the effectiveness of manure and fertilizer applications as well as crop yields (Singh et al., 2011; Rashid et al., 2016). Organic amendments not only enhance the soil's physico-chemical state but also raise the probability of innovative bio-inoculants' viability and survival in order to revive strained agriculture and restore the environment (Rashid et al., 2016). The organic C compounds in the soil provide the microbial communities there with electron donors, energy for growth, and a starting point for the production of biological components. Drury et al. (1991) reported a strong correlation between the availability of organic C and denitrification. Palumbo et al. (2004) claimed that soil amendments (such as the application of organic materials) intended to improve carbon sequestration and water-holding capacity may actually increase N_2O emission. Indian farmers frequently use compost with artificial N fertilizer (urea) to increase soil fertility and increase production (Uprety et al., 2012).

Currently, adding biochar to agricultural soils is being investigated as a way to slow down climate change by storing carbon (C) while simultaneously enhancing the characteristics and capabilities of the soil (Zhang et al., 2012). However, the effects of biochar addition on N_2O emission are reported to be even more inconsistent, because the process of emission is very complicated, involving denitrification, autotrophic nitrification, and heterotrophic nitrification, among other processes (Wrage et al., 2001) which are also accompanied by the enhanced soil aeration (Rogovska et al., 2011), increased pH, and microbial immobilization of soil NO_3^- (Case et al., 2012). According to Berthrong et al. (2013), organic management changed the variety of the soil's microbial population, and these microbial variations produced novel dynamics between the C and N cycles. In response to short-term C inputs, the capacity for N mineralization was increased due to the changes in the microbiota, whereas organically maintained areas saw long-term buildup of N in the SOM. These responses show that modifications to long-term agricultural management techniques can take into account soil microbes and the biogeochemical processes that are related to them, and that these modifications can result in a microbial community that is better able to react to plant root inputs of labile C and supply endogenous mineral N. Growing dependence on linked C and N cycles may make agroecosystems more able to rely on endogenous N sources without as much reliance on fertilizer input.

In the natural world, plants respond to seasonal fluctuations and are sensitive to climatic changes, which affect their development and interactions with the soil. The primary variables that affect agricultural production by impairing rhizosphere function are environmental pressures and their unpredictability (Singh, 2015). Plant genotypes are responsible for the composition of root exudates, which influences microbial recruitment in the rhizosphere (Patel et al., 2015). A reliable biological mitigation technique is the selection and development of low-GHG emitting cultivars based on photosynthate allocation and yield characteristics, without sacrificing the health and richness of the soil. Baruah et al. (2010a,b) reported that rice varieties with lower grain productivity but profuse vegetative growth, recorded higher seasonal N_2O emission and lower yield while varieties with higher yield potential emitted low seasonal N_2O which can be recommended to plant breeders for varietal improvement programs as low GHG emitting rice varieties. Alternatively, crop spacing, tillage, or integrated inorganic fertilizer, residue, and SOM management may be used to modify denitrification through inputs into the plant rhizosphere, hence altering the composition of plant-derived carbon flux or nitrogen intake requirement (Thomson et al., 2012). Because legumes build atmospheric N_2 via biological N fixation, the addition of leguminous crops minimizes the demand for external fertilizer N inputs. Rochette and Janzen (2005) proposed that N_2O from leguminous crops was derived mainly from root exudates and crop residue decomposition; following this Carter and Ambus (2006) found, in a study with grass-clover mixtures grown in a 15 N_2^- enriched atmosphere, that only 2% of the total N_2O emitted from the sward came from recently fixed N. Including a fallow period in a crop rotation is advised to increase the soil microorganism populations with suitable moisture and temperature conditions and without any competition with crops. Including a legume in the cropping system and burying crop residues after harvesting maintains the initial soil organic matter levels (Moreno et al., 2012).

Land-use changes and the associated loss of beneficial microbial diversity are the major reasons for deterioration of soil fertility and agricultural productivity (Singh et al., 2010; Singh and Singh, 2012; Singh, 2014). Therefore, more research must be done in the areas connected to plant-microbe connections to tackle issues with stressed agriculture. In order to manage stressed agriculture, it is also necessary to investigate the variety of stress-tolerant microorganisms that host plant species, their habitat, and their geographic distribution (Vimal et al., 2017).

A emphasis on lowering area-scaled N_2O emissions without taking emissions per unit of crop production into account merits consideration as global demand for agricultural goods rises. The quantity of N_2O emitted per unit of crop output must be taken into account if agriculture is to reduce its total GHG effect while boosting crop productivity (Van Groenigen et al., 2010). There have been only a few studies that have directly reported yield-scaled emissions in corn or other cropping systems (Gagnon et al., 2011). Based on studies in cropland, it seems that improving nitrogen utilization efficiency (NUE) cannot consistently reduce N_2O emissions (Phillips et al., 2009), probably because the practices that improve NUE by reducing NH_3 and/or NO_3^- losses may make more N available in the soil for both N uptake in crops and soil N_2O production (Venterea et al., 2012).

8.5 Conclusion

When studying how plant and soil communities react to climate change or at a certain time point, interactions between them might be unexpected (Classen et al., 2015). An agricultural strategy utilizing constructed microbial communities could result from extending management approaches like Climate Smart Agriculture (CSA) to include the microbial components of crops and soils by combining breeding techniques and biotechnology with the use of co-adapted, mutualistic plant-microbial associations (endophytic and rhizosphere). As a consequence, it is possible to produce more crops that can adapt to climate change while also lowering production costs on both an economic and ecological level, such as by reducing the usage of agrochemicals and reducing greenhouse gas emissions. More powerful than the direct impact of changing temperature or precipitation on the microbial decomposition of soil carbon is the effect on soil carbon storage caused by paradigm alterations in the climate of a region and land management techniques impacting the above ground vegetation of that location. As a result of both warming and shifting tree lines or afforestation, the overall trends in carbon losses appear to be much stronger than the early rises. Furthermore, ecologically, how much redundancy will there be in these communities' biological compositions, and which elements must be preserved to preserve the functioning processes (Walker, 1992) and link community structure to system stability. This will be a fruitful area of research (Pankhurst et al., 1996).

References

Allison, S.D., Martiny, J.B.H., 2008. Resistance, resilience, and redundancy in microbial communities. Proceedings of the National Academy of Sciences 105 (Suppl. 1), 11512–11519. Available from: https://doi.org/10.1073/pnas.0801925105.

Anderson, J.M., 1994. In: Greenland, D.J., Szabolcs, I. (Eds.), Functional Attributes of Biodiversity in Land Use Systems in Soil Resilience and Sustainable Land Use. CAB International, pp. 267–290.

Balser, T.C., Gutknecht, J.L.M., Liang, C., 2010. How will climate change impact soil microbial communities? In: Dixon, G.R., Tilston, E.L. (Eds.), Soil Microbiology and Sustainable Crop Production. Springer Science + Business Media B.V., pp. 373–397. Available from: https://doi.org/10.1007/978-90-481-9479-7_10.

Bardgett, R.D., Kandeler, E., Tscherko, D., Hobbs, P.J., Bezemer, T.M., Jones, T.H., et al., 1999. Below-ground microbial community development in a high temperature world. Oikos 85 (2), 193–203.

Bardgett, R.D., Manning, P., Morrien, E., de Vries, F.T., 2013. Hierarchical responses of plant-soil interactions to climate change: consequences for the global carbon cycle. Journal of Ecology 101, 334–343.

Barnawal, D., Bharti, N., Maji, D., Chanotiya, C.S., Kalra, A., 2014. ACC deaminase-containing Arthrobacter protophormiae induces NaCl stress tolerance through reduced ACC oxidase activity and ethylene production resulting in improved nodulation and mycorrhization in Pisum sativum. Journal of Plant Physiology 171, 884–894.

Baruah, K.K., Gogoi, B., Gogoi, P., 2010a. Plant physiological and soil characteristics associated with methane and nitrous oxide emission from rice paddy. Physiology and Molecular Biology of Plants 16 (1), 79–91. Available from: https://doi.org/10.1007/s12298-010-0010-1.

Baruah, K.K., Gogoi, B., Gogoi, P., Gupta, P.K., 2010b. N_2O emission in relation to plant and soil properties and yield of rice varieties. Agronomy for Sustainable Development 30 (4), 733–742. Springer Verlag/EDP Sciences/INRA; https://doi.org/10.1051/agro/2010021. hal-00886534.

Benbi, D.K., Brar, J.S., 2009. A 25-year record of carbon sequestration and soil properties in intensive agriculture. Agronomy for Sustainable Development 29, 257–265. Available from: https://doi.org/10.1051/agro/2008070.

Beniston, M., 2003. Climatic change in mountain regions: a review of possible impacts. Climatic Change 59, 5–31.

Beniston, M., 2005. The risks associated with climatic change in mountain regions. In: Huber, U., Bugmann, H., Reasoner, M. (Eds.), Global Change and Mountain Regions: An Overview of Current Knowledge. Springer, Dordrecht, pp. 511–520.

Berthrong, S.T., Buckely, D.H., Drinkwater, L.E., 2013. Agricultural management and labile carbon additions affect soil microbial community structure and interact with carbon and nitrogen cycling. Soil Microbiology 158–170.

Bradford, M.A., Davies, C.A., Frey, S.D., Maddox, T.R., Melillo, J.M., Mohan, J.E., et al., 2008. Thermal adaptation of soil microbial respiration to elevated temperature. Ecology Letters 11, 1316–1327.

Briones, M.J.I., McNamara, N.P., Poskitt, J., Crow, S.E., Ostle, N.J., 2014. Interactive biotic and abiotic regulators of soil carbon cycling: evidence from controlled climate experiments on peatland and boreal soils. Global Change Biology 20, 2971–2982.

Carter, M., Ambus, P., 2006. Biologically fixed N_2 as a source for N_2O production in a grass-clover mixture, measured by 15N_2. Nutrient Cycling in Agroecosystems 74, 13–26.

Case, S.D.C., McNamara, N.P., Reay, D.S., Whitaker, J., 2012. The effect of biochar addition on N_2O and CO_2 emissions from a sandy loam soil—the role of soil aeration. Soil Biology and Biochemistry 51, 125–134.

Castro, H.F., Classen, A.T., Austin, E.E., Norby, R.J., Schadt, C.W., 2010. Soil microbial community responses to multiple experimental climate change drivers. Applied and Environmental Microbiology 76, 999–1007.

Cavicchioli, R., Ripple, W.J., Timmis, K.N., 2019. Scientists' warning to humanity: microorganisms and climate change. Nature Reviews. Microbiology 17, 569–586. Available from: https://doi.org/10.1038/s41579-019-0222-5.

Classen, A.T., Sundqvist, M.K., Henning, J.A., Newman, G.S., Moore, J.A.M., Cregger, M.A., et al., 2015. Direct and indirect effects of climate change on soil microbial and soil microbial-plant interactions: what lies ahead. Ecosphere 6 (8), 130. Available from: https://doi.org/10.1890/ES15-00217.1.

Cregger, M.A., Sanders, N.J., Dunn, R.R., Classen, A.T., 2014. Microbial communities respond to experimental warming, but site matters. PeerJ . Available from: https://doi.org/10.7717/peerj.358.

de Vries, F.T., Griffiths, R.I., 2018. Impacts of climate change on soil microbial communities and their functioning. Climate Change Impacts on Soil Processes and Ecosystem Properties. Elsevier. Available from: https://doi.org/10.1016/B978-0-444-63865-6.00005-3.

DeAngelis, K.M., Pold, G., Topcuoglu, B.D., van Diepen, L.T.A., Varney, R.M., Blanchard, J.L., et al., 2015. Long-term forest soil warming alters microbial communities in temperate forest soils. Frontiers in Microbiology.

Dieleman, W.I.J., Vicca, S., Dijkstra, F.A., Hagedorn, F., Hovenden, M.J., Larsen, K.S., et al., 2012. Simple additive effects are rare: a quantitative review of plant biomass and soil process responses to combined manipulations of CO_2 and temperature. Global Change Biology 18, 2681–2693.

Dixon, G.R., 2009. The impact of climate and global change on crop production. In: Letcher, T.M. (Ed.), Climate Change: Observed Impacts on Planet Earth. Elsevier, Oxford, UK/Amsterdam, Netherlands, pp. 307–324.

Drury, C.F., McKenney, D.J., Findlay, W.I., 1991. Relationships between denitrification, microbial biomass and indigenous soil properties. Soil Biology & Biochemistry 23, 751–755.

Egli, M., Poulenard, J., 2016. Soils of mountainous landscapes. International Encyclopedia of Geography: People, the Earth, Environment and Technology. John Wiley & Sons, Ltd, pp. 1–10. Available from: https://doi.org/10.1002/9781118786352.wbieg0197.

Evans, S.E., Byrne, K.M., Lauenroth, W.K., Burke, I.C., 2011. Defining the limit to resistance in a drought-tolerant grassland: long-term severe drought significantly reduces the dominant species and increases ruderals. Journal of Ecology 99, 1500–1507.

FAO, 2012. Live animal numbers, crop production, total nitrogen fertiliser consumption statistics for 2000–2008/2000–2010. Retrieved from https://www.fao.org/3/cb1329en/CB1329EN.pdf

FAO, 2015. In: Romeo, R., Vita, A., Manuelli, S., Zanini, E., Freppaz, M., Stanchi, S. (Eds.), Understanding Mountain Soils: A Contribution from Mountain Areas to the International Year of Soils. FAO, Retrieved from https://www.fao.org/news/story/en/item/294317/icode/.

Gagnon, B., Ziadi, N., Rochette, P., Chantigny, M.H., Angers, D.A., 2011. Fertilizer source influenced nitrous oxide emissions from clay soil under corn. Soil Science Society of America Journal 75, 595–604. Available from: https://doi.org/10.2136/sssaj2010.0212.

Gepstein, S., Glick, B.R., 2013. Strategies to ameliorate abiotic stress-induced plant senescence. Plant Molecular Biology 82, 623–633.

Grabherr, G., Messerli, B., 2011. An overview of the world's mountain environments. In: Austrian MAB Committee (Ed.), Mountain Regions of the World: Threats and Potentials for Conservation and Sustainable Use. Grasl Druck & Neue Medien, Bad Vöslau, Austria, pp. 8–14.

Gregorich, E.G., Rochette, P., Hopkins, D.W., McKim, U.F., St-Georges, P., 2006. Tillage-induced environmental conditions in soil and substrate limitation determine biogenic gas production. Soil Biology and Biochemistry 38, 2614–2628.

Hamilton, C.E., Bever, J.D., Labbe, J., Yang, X.H., Yin, H.F., 2016. Mitigating climate change through managing constructed microbial communities in agriculture. Agriculture, Ecosystems & Environment 216, 304–308.

Hartley, I.P., Hopkins, D.W., Garnett, M.H., Sommerkorn, M., Wookey, P.A., 2008. Soil microbial respiration in arctic soil does not acclimate to temperature. Ecology Letters 11, 1092–1100.

Hooper, D.U., Chapin III, F.S., Ewel, J.J., Hector, A., Inchausti, P., Lavorel, S., et al., 2005. Effects of biodiversity on ecosystem functioning: a consensus of current knowledge. Ecological Monographs 75 (1), 3–35, Wiley. Available from: https://doi.org/10.1890/04-0922.

IPCC, 2007. Causes of change. In: Pachauri, R.K., Reisinger, A. (Eds.), Climate Change 2007, Synthesis Report. Intergovernmental Panel on Climate Change, Geneva, Switzerland, pp. 35–42.

IPCC, 2014. In: Pachauri, R.K., Meyer, L.A. (Eds.), Climate Change Synthesis Report. Contribution of Working Groups I, II and III to the Fifth Assessment Report of the Intergovernmental Panel on Climate Change [Core Writing Team. IPCC, Geneva, Switzerland, 151 pp.

Isobe, K., Koba, K., Otsuka, S., Senoo, K., 2011. Nitrification and nitrifying microbial communities in forest soils. Journal of Forest Research 16, 351–362.

Ives, J.D., Messerli, B., 1989. The Himalayan Dilemma. Reconciling Development and Conservation. Routledge, London and New York, p. 324.

Kaisermann, A., Maron, P.A., Beaumelle, L., Lata, J.C., 2015. Fungal communities are more sensitive indicators to non-extreme soil moisture variations than bacterial communities. Applied Soil Ecology 86, 158–164.

Kandeler, E., Tscherko, D., Bardgett, R.D., Hobbs, P.J., Kampichler, C., Jones, T.H., 1998. The response of soil microorganisms and roots to elevated CO_2 and temperature in a terrestrial model ecosystem. Plant and Soil 202 (2), 251–262.

Mountains and climate change: a global concern. In: Kohler, T., Wehrli, A., Jurek, M. (Eds.), Sustainable Mountain Development Series. Centre for Development and Environment (CDE), Swiss Agency for Development and Cooperation (SDC) and Geographica Bernensia, Bern, Switzerland, 136 pp.

Lange, M., Habekost, M., Eisenhauer, N., Roscher, C., Bessler, H., Engels, C., et al., 2014. Biotic and abiotic properties mediating plant diversity effects on soil microbial communities in an experimental grassland. PLoS One 9, e96182.

Lee, K.E., Pankhurst, C.E., 1992. Soil organisms and sustainable productivity. Soil Research 30, 855–892. Available from: https://doi.org/10.1071/SR9920855.

Lipper, L., Thornton, P., Campbell, B.M., Baedeker, T., Braimoh, A., Bwalya, M., et al., 2014. Climate-smart agriculture for food security. Nature Climate Change 4 (12), 1068–1072, Springer Science and Business Media LLC. Available from: https://doi.org/10.1038/nclimate2437.

Lutz A.F., Immerzeel, W.W., 2013. Water availability analysis for the upper Indus, Ganges, Brahmaputra, Salween and Mekong River basins. Final Report to ICIMOD, September 2013. Future Water Report, No. 127. Wageningen, The Netherlands, Future Water.

Maloy, S., Moran, M.A., Mulholland, M.R., Sosik, H.M., Spear, J.R., 2017. Microbes and Climate Change: Report on an American Academy of Microbiology and American Geophysical Union Colloquium, Washington.

Marx, M.C., Wood, M., Jarvis, S., 2001. A microplate fluorimetric assay for the study of enzyme diversity in soils. Soil Biology and Biochemistry 33, 1633–1640.

Mesle, M., Dromart, G., Oger, P., 2013. Microbial methanogenesis in subsurface oil and coal. Research in Microbiology 164, 959–972.

Moreno, M.M., Moreno, C., Lacasta, C., Meco, R., 2012. Evolution of soil biochemical parameters in rainfed crops: effect of organic and mineral fertilization. Applied and Environmental Soil Science, Hindawi Publishing Corporation, Article ID 826236. Available from: https://doi.org/10.1155/2012/826236.

Oenema, O., Wrage, N., Velthof, G.L., van Groenigen, J.W., Dolfing, J., Kuikman, P.J., 2005. Trends in global nitrous oxide emissions from animal production systems. Nutrient Cycling in Agroecosystems 72 (1), 51−65, Springer Science and Business Media LLC. Available from: https://doi.org/10.1007/s10705-004-7354-2.

Palumbo, A.V., McCarthy, J.F., Amonette, J.E., Fisher, L.S., Wullschleger, S.D., Daniels, W.L., 2004. Prospects for enhancing carbon sequestration and reclamation of degraded lands with fossil-fuel combustion by-products. Advances in Environmental Research 8 (3−4), 425−438, Elsevier BV. Available from: https://doi.org/10.1016/s1093-0191(02)00124-7.

Pankhurst, C.E., Ophel-Keller, K., Doube, B.M., Gupta, V.V.S.R., 1996. Biodiversity of soil microbial communities in agricultural systems. Biodiversity and Conservation 5, 197−209.

Parkinson, D., Coleman, D.C., 1990. Microbial communities, activity and biomass. Agriculture, Ecosystems & Environment 34 (1−4), 3−33. Available from: https://doi.org/10.1016/0167-8809(91)90090-K. ISSN 0167-8809.

Patel, J.S., Singh, A., Singh, H.B., Sarma, B.K., 2015. Plant geno- type, microbial recruitment and nutritional security. Frontiers in Plant Science 6, 1−3.

Paul, D., Lade, H., 2014. Plant-growth-promoting rhizobacteria to improve crop growth in saline soils: a review. Agronomy for Sustainable Development 34, 737−752.

Paustian, K., Lehmann, J., Ogle, S., Reay, D., Robertson, G.P., Smith, P., 2016. Climate-smart soils. Nature 532, 49−57. Available from: https://doi.org/10.1038/nature17174.

Pereg, L., McMillan, M., 2015. Scoping the potential uses of beneficial microorganisms for increasing productivity in cotton cropping systems. Soil Biology & Biochemistry 80, 349−358.

Phillips, R.L., Tanaka, D.L., Archer, D.W., Hanson, J.D., 2009. Fertilizer application timing influences greenhouse gas fluxes over a growing season. Journal of Environmental Quality 38 (4), 1569−1579, Wiley. Available from: https://doi.org/10.2134/jeq2008.0483.

Rashid, M.A., Mujawar, L.H., Shahzad, T., Almeelbi, T., Ismail, I.M.I., Oves, M., 2016. Bacteria and fungi can contribute to nutrients bioavailability and aggregate formation in degraded soils. Microbiological Research 183, 26−41.

Reichstein, M., Bahn, M., Ciais, P., Frank, D., Mahecha, M.D., Seneviratne, S.I., et al., 2013. Climate extremes and the carbon cycle. Nature 500 (7462), 287−295, Springer Science and Business Media LLC. Available from: https://doi.org/10.1038/nature12350.

Rochette, P., Janzen, H.H., 2005. Towards a revised coefficient for estimating N_2O emissions from legumes. Nutrient Cycling in Agroecosystems 73, 171−179.

Rogovska, N., Laird, D., Cruse, R., Fleming, P., Parkin, T., Meek, D., 2011. Impact of biochar on manure carbon stabilization and greenhouse gas emissions. Soil Science Society of America Journal 75 (3), 871−879, Wiley. Available from: https://doi.org/10.2136/sssaj2010.0270.

Sayer, E.J., Wagner, M., Oliver, A.E., Pywell, R.F., James, P., Whiteley, A.S., et al., 2013. Grassland management influences spatial patterns of soil microbial communities. Soil Biology and Biochemistry 61, 61−68.

Sheik, C.S., Beasley, W.H., Elshahed, M.S., Zhou, X.H., Luo, Y.Q., Krumholz, L.R., 2011. Effect of warming and drought on grassland microbial communities. ISME Journal 5, 1692−1700.

Singh, J.S., 2014. Cyanobacteria: a vital bio-agent in eco-resto- ration of degraded lands and sustainable agriculture. Climate Change and Environmental Sustainability 2, 133−137.

Singh, J.S., 2015. Microbes: the chief ecological engineers in re- instating equilibrium in degraded ecosystems. Agriculture, Ecosystems & Environment 203, 80−82.

Singh, J.S., Singh, D.P., 2012. Reforestation: a potential approach to mitigate the excess CH_4 build-up. Ecological Management & Restoration 13, 245−248.

Singh, J.S., Kashyap, A.K., Singh, D.P., 2010. Microbial biomass C, N and P in disturbed dry tropical forest soils, India. Pedosphere 20, 780−788.

Singh, J.S., Pandey, V.C., Singh, D.P., 2011. Efficient soil microorganisms: a new dimension for sustainable agriculture and environmental development. Agriculture, Ecosystems & Environment 140, 339−353.

Singh, R., Kumari, T., Verma, P., Singh, B.P., Singh Raghubanshi, A., 2021. Compatible package-based agriculture systems: an urgent need for agro-ecological balance and climate change adaptation. Soil Ecology Letters. Available from: https://doi.org/10.1007/s42832-021-0087-1, Springer Science and Business Media LLC.

Singh, R., Singh, H., Raghubanshi, A.S., 2019. Challenges and opportunities for agricultural sustainability in changing climate scenarios: a perspective on Indian agriculture. Tropical Ecology 60 (2), 167−185, Springer Science and Business Media LLC. Available from: https://doi.org/10.1007/s42965-019-00029-w.

Smith, R.S., Shiel, R.S., Bardgett, R.D., Millward, D., Corkhill, P., Rolph, G., et al., 2003. Soil microbial community, fertility, vegetation and diversity as targets in the restoration management of a meadow grassland. Journal of Applied Ecology 40 (1), 51−64, Wiley. Available from: https://doi.org/10.1046/j.1365-2664.2003.00780.x.

Thomson, A.J., Giannopoulos, G., Pretty, J., Baggs, E.M., Richardson, D.J., 2012. Biological sources and sinks of nitrous oxide and strategies to mitigate emissions. Philosophical Transactions of the Royal Society B: Biological Sciences 367 (1593), 1157−1168, The Royal Society. Available from: https://doi.org/10.1098/rstb.2011.0415.

Tylianakis, J.M., Didham, R.K., Bascompte, J., Wardle, D.A., 2008. Global change and species interactions in terrestrial ecosystems. Ecology Letters 11, 1351−1363.

UK CEH, 2022. Why do soil microbes' matter? Retrieved from https://www.ceh.ac.uk.

Uprety, D.C., Dhar, S., Hongmin, D., Kimball, B.A., Garg, A., Upadhyay, J., 2012. Technologies for Climate Change Mitigation—Agriculture Sector. UNEP Risø Centre on Energy, Climate and Sustainable Development. Department of Management Engineering. Technical University of Denmark (DTU). TNA Guidebook Series.

Van Groenigen, J.W., Velthof, G.L., Oenema, O., Van Groenigen, K.J., Van Kessel, C., 2010. Towards an agronomic assessment of N_2O emissions: a case study for arable crops. European Journal of Soil Science 61 (6), 903−913, Wiley. Available from: https://doi.org/10.1111/j.1365-2389.2009.01217.x.

Venterea, R.T., Halvorson, A.D., Kitchen, N., Liebig, M.A., Cavigelli, M.A., Grosso, S.J.D., et al., 2012. Challenges and opportunities for mitigating nitrous oxide emissions from fertilized cropping systems. Frontiers in Ecology and the Environment 10 (10), 562−570. Available from: https://doi.org/10.1890/120062. Wiley.

Vermeulen, S.J., Campbell, B.M., Ingram, J.S.I., 2012. Climate change and food systems. Annual Review of Environment and Resources 37 (1), 195−222, Annual Reviews. Available from: https://doi.org/10.1146/annurev-environ-020411-130608.

Vimal, S.R., Singh, J.S., Arora, N.K., Singh, S., 2017. Soil-plant-microbe interactions in stressed agriculture management: a review. Pedosphere 27 (2), 177−192.

Wagg, C., Bender, S.F., Widmer, F., van der Heijden, M.G., 2014. Soil biodiversity and soil community composition determine ecosystem multifunctionality. Proceedings of the National Academy of Sciences of the United States of America 111, 5266−5270. Available from: http://doi.org/10.1073/pnas.1320054111.

Walker, B.H., 1992. Biodiversity and ecological redundancy. Conservation Biology 6 (1), 18−23, Wiley. Available from: https://doi.org/10.1046/j.1523-1739.1992.610018.x.

Wei, H., Guenet, B., Vicca, S., Nunan, N., AbdElgawad, H., Pouteau, V., et al., 2014. Thermal acclimation of organic matter decomposition in an artificial forest soil is related to shifts in microbial community structure. Soil Biology and Biochemistry 71, 1−12.

Wrage, N., Velthof, G.L., Van Beusichem, M.L., Oenema, O., 2001. Role of nitrifier denitrification in the production of nitrous oxide. Soil Biology & Biochemistry 33, 1723−1732.

Xiao, W., Chen, X., Jing, X., Zhu, B., 2018. A meta-analysis of soil extracellular enzyme activities in response to global change. Soil Biology and Biochemistry 123, 21−32.

Xiong, J., Sun, H., Peng, F., Zhang, H., Xue, X., Gibbons, S.M., et al., 2014. Characterizing changes in soil bacterial community structure in response to short-term warming. FEMS Microbiology Ecology 89, 281−292.

Zak, D.R., Holmes, W.E., MacDonald, N.W., Pregitzer, K.S., 1999. Soil temperature, matric potential, and the kinetics of microbial respiration and nitrogen mineralization. Soil Science Society of America Journal 63, 575−584. Available from: https://doi.org/10.2136/sssaj1999.03615995006300030021x.

Zemp, M., Hoelzle, M., Haeberli, W., 2009. Six decades of glacier mass balance observations − a review of the worldwide monitoring network. Annals of Glaciology 50, 101−111. Available from: https://doi.org/10.3189/172756409787769591.

Zhang, A., Bian, R., Pan, G., Cui, L., Hussain, Q., Li, L., et al., 2012. Effects of biochar amendment on soil quality, crop yield and greenhouse gas emission in a Chinese rice paddy: a field study of 2 consecutive rice growing cycles. Field Crops Research 127, 153−160, Elsevier BV. Available from: https://doi.org/10.1016/j.fcr.2011.11.020.

Biochar-mediated nutrients and microbial community dynamics in montane landscapes

Brahmacharimayum Preetiva, Abhishek Kumar Chaubey and Jonathan S. Singsit

School of Environmental Sciences, Jawaharlal Nehru University, New Delhi, Delhi, India

9.1 Introduction

Mountains have a variety of ecological and social significance. They are culturally significant to many ethnic groups and provide more than half of the world's total fresh water (Grumbine & Xu, 2021). Mountains cover one-fifth of the world's land area, housing one-tenth of the total human population, 33% of biodiversity, and half of biodiversity hotspots (Grumbine & Xu, 2021; Li, El Solh, & Siddique, 2019). People in the mountains rely heavily on agriculture for a living due to numerous constraints such as difficult topography,

inaccessibility, and poor infrastructure, making industrialization difficult (Li et al., 2019). Ironically,
300 million people in mountainous areas are food insecure, which means that one in every two people in remote mountain areas suffers from chronic hunger (Li et al., 2019). This is due to the infertile soil of these regions, which results in low productivity (40% less than the plains). Lower soil quality occurs as a result of (1) erosional losses of soil nutrients from hilly slopes during rainstorms, (2) decline in soil organic carbon due to clearing of forested areas (reducing aboveground biomass to 10% of its natural state), and (3) lower biomass input to soil (Dahal, Bajracharya, & Wagle, 2018). Mountain soils, in general, are shallow, acidic, poorly developed and have low water retention capacity (Romeo et al., 2015). As a result, increasing the fertility of these regions is critical in order to sustain the livelihoods of the people who live there. As a result, farmyard manure (FYM) and compost are commonly used soil amend-ments that are gradually being replaced by chemical fertilizers in order to intensify farming activities (Romeo et al., 2015; Dahal & Bajracharya, 2012; Bista, Ghimire, Chandra Shah, & RAJ Pande, 2010). These, however, are not sustainable options because they are labor intensive, require regular application, and emit large amounts of greenhouse gases. Biochar, a carbon-rich material derived from the pyrolysis of waste biomass, can serve as a more durable and sustainable alternative in this context (Gautam, Bajracharya, & Sitaula, 2017; Yan, Xue, Zhou, & Wu, 2021a, 2021b; Al-Wabel et al., 2018). The use of biochar on soil has been shown to increase the aboveground biomass and organic matter content of mountainous soil because it mineralizes slowly and thus serves as a nutrient source for an extended period of time (Vinh, Hien, Anh, Lehmann, & Joseph, 2014). Furthermore, because of its porous nature, it provides an ideal environment for microorganisms that promote the growth of bacterial and arbuscular mycorrhizal fungi (Yan et al., 2021a). Cold temperatures at high altitudes limit microbial activ-ity, which is directly related to soil fertility (Romeo et al., 2015). As a result, biochar applica-tion can significantly improve both nutrient status and microbial activity, thereby increasing mountain farming productivity. This chapter will go over the effects of using biochar in mountainous areas.

9.2 Traditional farming practices in mountainous system

Because soil fertility in mountain areas is generally low, amendments in the form of various organic manures such as farmyard manure and compost are commonly used. However, in recent years, farmers have shifted to inorganic fertilizers, which provide quick good output but degrade soil health in the long run, in order to increase productiv-ity (3—4 crops per year). This section will look at traditional manuring practices and how they can be replaced with more environmentally friendly and sustainable alternatives.

Mountain farming is also known as family farming or smallholder farming because cultiva-ble land areas are typically small (<10 ha) due to fragmented and dispersed topography with varying altitudes, and it employs less labor and is managed primarily by family members (Córdova, Hogarth, & Kanninen, 2019; Wymann Von Dach, Romeo, Vita, Wurzinger, & Kohler, 2013). These families are socially and economically marginalized, frequently food insecure, suf-fer from widespread poverty, and cannot afford expensive chemical fertilizers. As a result, they have primarily relied on traditionally prepared farmyard manure and compost for soil amendments on their farms (Negi & Bisht, 2017; Gosain, 2016; Wymann Von Dach et al., 2013). According to various studies, more than 80% of mountain farmers rely on FYM and compost for their fields' nutritional needs (Balla et al., 2014; Dahal & Bajracharya, 2012). This is due to the close proximity of farmlands to forested areas, which provides easy access to FYM raw materials (Balla et al., 2014). Nepal, for example, is a mountainous agricultural country with the highest livestock density per unit cultivated area. As a result, FYM is used as a soil amendment on more than 85% of farmland (Tripathi & Jones, 2010). FYM has several advantages for nutrient-depleted high-altitude soils. The release of hydroxyl ions during manure decomposition raises the acidic pH of mountainous soil. It also increases soil organic carbon, total nitrogen, and phosphorus and potassium availability (Abbasi & Tahir, 2012; Panwar, Ramesh, Singh, & Ramana, 2010; Nouraein, Skataric, Spalevic, Dudic, & Gregus, 2019). Every year, soil erosion and crop harvest deplete approximately 18 million tons of soil nutrients. Despite its high nutrient content, FYM only supplies about 20% of the nutrient requirements of farmlands for crop cultivation (Tiwari, Sitaula, Borresen, & Bajracharya, 2006; Shrestha, 2015). This is due to the loss of nutrients from FYM due to volatilization and leaching caused by poor manure management practices such as sun exposure, rainfall, and poor storage conditions (Thapa & Paudel, 2002). To top it all off, climate change and deforestation have resulted in a significant decline in forested areas of hilly reasons, reducing the production of raw materials for FYM production. Furthermore, a lack of knowledge and poor manure management, changing land-use patterns, and intensified cultivation systems have caused a continuous deterioration in the soil organic matter pool of mountain farmlands (Balla et al., 2014; Dahal & Bajracharya, 2012). Increased soil fertility degradation necessitates an increase in manure application, which increases the burden on an already overburdened household labor force (Thapa & Paudel, 2002). This has resulted in a shift toward chemical fertilizers, which are a major source of environmental and sustainability concerns in the long run. The race to produce more output for commercialization has reduced the use of FYM even further, and as a result, the cost ratio of chemical fertilizers to FYM has greatly increased (Sharma, Pathania, & Lal, 2006). In such an unsustainable scenario, biochar may be a better option because it can increase the productivity of water-stressed mountain soils by increasing water retention capacity and can also sequester carbon, assisting in climate change mitigation (Gautam et al., 2017).

9.3 Biochar—a boon for mountain farmers

Food security is one of the major global issues that has arisen as a result of the world's rapid population growth, particularly in developing countries (Palansooriya et al., 2019; Pradhan, Chan, Roul, Halbrendt, & Sipes, 2018). Furthermore, climate change has exacerbated the problem by reducing agricultural productivity and arable land, which can be a major source of concern for a large population by causing poverty and food insecurity (Pradhan et al., 2018). Biochar can be a great alternative in this situation because it can help increase soil productivity while also combating climate change by sequestering carbon. The beneficial impact of biochar as a soil amendment and plant growth inducer is gaining popularity (Hossain et al., 2020). A plethora of studies have documented the presence of readily available nutrients in biochar. The nutrient content is affected by a variety of factors, including the origin of the feedstock raw materials and the pyrolysis conditions (temperature, residence time, and gaseous environment) (El-Naggar et al., 2019; Hossain et al., 2020). The nutrient content of biochar corresponds to high nutrient content of the feedstock material, for example, biochars obtained from manure and sewage sludge feedstocks have high nutrient content (Hossain et al., 2020).

Mountain regions, as mentioned in previous sections, are difficult to cultivate due to a variety of constraints, including constantly deteriorating soil quality due to erosion of sloping mountainous top soil and landslides during heavy rains (Bishwakarma et al., 2015; Dahal et al., 2018). Furthermore, deforestation has reduced soil organic matter content, and increased use of nitrogenous fertilizers has degraded fertility by lowering soil moisture content. Because of the decrease in soil moisture, coarse hard clods form, reducing soil workability (Pilbeam, Mathema, Gregory, & Shakya, 2005). Despite soil cultivability issues, the majority of mountain people (particularly in Nepal and the Himalayan regions) rely on agricultural activities for a living (Dahal et al., 2018).

Biochar can help soil by acting as an organic additive, improving soil health, crop yield by improving low pH conditions, increasing water retention capacity, and ultimately reducing fertilizer/chemical inputs (Dahal et al., 2018; Glaser, Lehmann, & Zech, 2002; Lehmann & Rondon, 2006). At present, there are limited number of studies which report biochar's usage as inducer of soil health quality and plant growth in mountainous soils (Abujabhah, Bound, Doyle, & Bowman, 2016; Dahal et al., 2018; El-Naggar et al., 2018; Gao, Hoffman-Krull, Bidwell, & Deluca, 2016; Gautam et al., 2017; Kong, He, Chen, Yang, & Du, 2021; Sizmur, Wingate, Hutchings, & Hodson, 2011; Toma, Fujitani, Tsurumi, Tsurumi, & Ueno, 2019; Vinh et al., 2014; Yan et al., 2021a) (Fig. 9.1).

9.3.1 What is biochar?

Biochar is a carbonaceous substance derived from the thermo-chemical conversion of plant/animal waste residues by pyrolysis or carbonation (Lehmann & Joseph, 2015; Ahmad et al., 2014; Al-Wabel et al., 2018). International Biochar Initiative (IBI) recommends that biochar must have a H:C_{org} molar ratio of less than 0.7 (IBI, https://biochar-international.org/). The discovery of Amazonian Dark earths in the fertile soils of the Amazon basin containing high organic carbon known locally as "terra prata de Indio" sparked wide interest in biochar (Lehmann & Joseph, 2015). The presence of large proportion of fused aromatic C differentiates biochar from

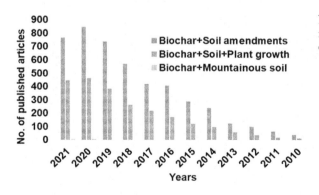

FIGURE 9.1 Graphical representation of number of published articles available on "web of science" database accessed on 06-10-2021.

other organic matter (Lehmann et al., 2011). Due to the presence of recalcitrant aromatic carbon groups, biochar is relatively stable (Cross & Sohi, 2013; Lehmann et al., 2011). Biochar possesses great soil amendment properties and carbon sequestering capability as well (Tarin et al., 2021). It is proven to be a reliable climate change mitigant (Lehmann & Joseph, 2015; Lehmann et al., 2011). The conversion of agricultural waste to biochar ensures lesser CO_2 emission, carbon capture, micronutrient retention and greater availability to plants, water retention, cation exchange capability (CEC) enhancement and supports the growth of beneficial microbes, etc. (Ding et al., 2016; Lehmann et al., 2011; Al-Wabel et al., 2018).

9.3.2 Beneficial properties of biochar as soil amendment (physical, chemical and biological)

A number of studies (Jien & Wang, 2013; Singh, Singh, & Purakayastha, 2019; Berihun, Tadele, & Kebede, 2017; Joseph et al., 2020; Blackwell, Krull, Butler, Herbert, & Solaiman, 2010) and a couple of reviews (Lehmann et al., 2011; Ding et al., 2016; Al-Wabel et al., 2018) assessing biochar role in soil has been carried out. The characteristics of biochar and its positive results on soil depends on its pyrolysis conditions, feedstock raw material used and soil properties (Ding et al., 2016; Gorovtsov et al., 2020; Lawrinenko & Laird, 2015). For instance, loss of nitrogen is reported with increasing temperature of pyrolysis (above 400°C), and it reaches half mark at 750°C. Aromatic characteristics, carbon, phosphorus and potassium content also shows direct proportionality with temperature (300°C—600°C) (Ding et al., 2016; Lawrinenko & Laird, 2015). Biochars which are intended for agricultural use are best prepared using slow pyrolysis techniques (Song & Guo, 2012). To maximize the utility of biochars in soil for agronomic performance, predominant limiting factors of soil, climate context of crop and the properties of biochar must be understood (Enders, Hanley, Whitman, Joseph, & Lehmann, 2012). Biochar is most widely documented as a positive soil ameliorating agent in acidic soil (Singh et al., 2019). Biochar produced from weeds like *Lantana* and *Parthenium* by virtue of their alkalinity, porosity and high C, K, P, N, Ca, and Mg content have been proven to reclaim acidic soils (Singh et al., 2019). The acid buffering capacity of biochar is also attributed to carbonate and organic anions content (Singh et al., 2019). As a result, a data library containing, but not limited to, pH, liming potential, buffer capacity, and elemental composition of biochars could be a useful repository for biochar soil application (Enders et al., 2012). A summary of biochar's influence on physicochemical parameters in soil is given in Table 9.1.

Biochar application is known to affect physical attributes of soil like the bulk density, aggregate stability, hydraulic conductivity, water-holding capacity, as it contains plenty of micro and macropores (Lehmann et al., 2011; Nelissen et al., 2015; Mukherjee & Lal, 2013). It also reduces the tensile strength of soil allowing easy germination while making movement of invertebrates easier by increasing soil aeration (Lehmann et al., 2011). It helps in the retention of moisture in the pores of soil particles, which improves the overall soil water retention capacity (Ding et al., 2016; Mukherjee & Lal, 2013). Berihun and colleagues studied the application of three different biochar produced from *Lantana camara* stem, corn cob, and *Eucalyptus globulus* to acidic soil (Berihun et al., 2017). Soil bulk density were reduced significantly while

TABLE 9.1 Biochar effect on physicochemical properties of soil.

Biochar [pyrolysis temp. (°C)]	Soil type	Application rate (t/ha)	Physical properties				Chemical properties				Reference
			Water holding capacity (%)		Bulk density (g/cm³)		pH		Organic matter		
			Untreated	Treated	Untreated	Treated	Untreated	Treated	Untreated	Treated	
Wood biochar (450–550)	Inceptisols	20	62	75	1.06		6.45	6.92	8.6		Gao and DeLuca (2018)
Leucaena leucocephala waste wood	Ultisols	100			1.42	1.15	3.95	4.65			Jien and Wang (2013)
		200			1.42	1.08		5.07			
Timberharvest residues	Sandy loam	20	38.50	47.94			5.88	6.88			Gao et al. (2016)
Maize straw (550)	Silt loam	20			1.34	1.29					Yan et al. (2019)
		40				1.26					
		60				1.22					
Barley straw (400)	Loam	10			1.4	1.3					Kang et al. (2018)
Poultry manure (450)	Alfisol	5			1.6	1.44					Are, Adelana, Fademi, and Aina (2017)
Wheat straw (500)		20					7	7.1	2.57	3.28	Wu et al. (2019)
		40						7.4		3.97	
Bamboo (450)		11.25					4.67	4.80	0.7	1.25	Tarin et al. (2019)
		45						4.95		1.90	
		180						5.30		3.55	
Hardwood (420)		11.25					4.67	4.70	0.7	1.13	Tarin et al. (2019)
		45						4.90		2.25	
		180						5.15		4.50	
Rice straw (500)		11.25					4.67	4.90	0.7	1.00	Tarin et al. (2019)
		45						4.95		1.90	
		180						5.45		2.55	

there was increments in total soil porosity (Berihun et al., 2017). Similarly, findings were reported for *Leucaena leucocephala* biochar which lowered the bulk density of soil with increasing application rates and increased porosity (Jien & Wang, 2013).

Biochar amendment has proven benefits on chemical properties of soil like pH, CEC, various nutrients, soil organic carbon, etc. (Enders et al., 2012; Jien & Wang, 2013; Yuan, Xu, & Zhang, 2011). This alteration depends on the pH of biochar used, which in turn, depends on the biochar's ash content and pyrolysis temperature (Enders et al., 2012). Biochar prepared from canola, peanut, corn and soybean straws pyrolyzed at 500°C are shown to reclaim acidic soils successfully (Yuan et al., 2011). The carbonates and organic anions contribute significantly to the alkalinity of biochar, helping in neutralizing soil acidity. Biochar's surface negative charge and high CEC ensures immobilization of minerals like Ca^{2+}, Mg^{2+}, NH_4^+ and K^+, inhibiting their loss (Yuan et al., 2011). Exchangeable K, Ca, and Mg content are also reported to increase with biochar addition (Jien & Wang, 2013). Similarly, *Leucaena leucocephala* wood biochar applied in highly weathered soils showed CEC increasing from 7.41 to 9.26 and 10.8 cmol (+)/kg following 2.5% and 5% biochar application. The total organic carbon and soil organic carbon also increased (Jien & Wang, 2013). The ability of biochar to buffer acidic soil is due to the presence of high basic cations in its ash content, which raises soil pH, soil organic carbon, total nitrogen, available phosphorus, and potassium (Berihun et al., 2017; Edenborn, Edenborn, Krynock, & Haug, 2015).

Biochar application indirectly improves microbial activity in soil by improving physical and chemical soil attributes, creating a favorable environment for microorganisms to thrive (Xu et al., 2014). The establishment of microbial communities with biochar addition is enabled by adhesion/sorption onto biochar and protection of bacteria and fungi from predators (Lehmann et al., 2011) The pores in biochar, pore size distribution, high labile carbon and nutrient content can provide suitable habitats for soil microbes to thrive (Singh et al., 2019; Gorovtsov et al., 2020). Enhanced soil aeration by biochar application is another factor which contributes towards the enhanced microbial activity (Singh et al., 2019). Microbial abundance increases with biochar, as measured by various tests and assays such as total genomic DNA, substrate induced respiration, phospholipid fatty acid (PLFA) extraction, culture and plate count and microbial biomass carbon (MBC) (Singh et al., 2019). Biochar has been shown to alter soil enzyme activities (cellulase, dehydrogenase, phosphatase), which are responsible for the soil's balanced nutrient cycling. It reduces the activity of some enzymes while increasing the activity of others. The effect of biochar on these enzymes varies greatly depending on the soil type and the nature of the enzyme-substrate interaction, which is influenced by the surface area and porosity of the biochar (Lammirato, Miltner, & Kaestner, 2011; Ameloot et al., 2014).

Studies which used sugarcane bagasse, rice husk, *Parthenium* and *Lantana camara* as precursor biomass for biochar preparation (applied at rates 2.25 and 4.50 g/kg in soil incubation study), all found to increase microbial biomass carbon (MBC), activities of cellulase (CA), dehydrogenase (DHA), acid phosphatase (AcPA). Lantana biochar amended soil showed the high enzyme activity (Singh et al., 2019). The presence of free radicals and the participation of biochar in electron transport has been linked with soil enzymatic activity (Gorovtsov et al., 2020). A short study of 21 days which measured microbial biomass carbon (MBC), found an increment from 835 to 1262 mg/kg when weathered soils were treated with 5% biochar amendment (Jien & Wang,

2013). The protective role of biochar on plants occurs due to sorption of enzymes emitted by phytopathogenic fungi (Jaiswal, Frenkel, Tsechansky, Elad, & Graber, 2018). However, this protective capability withers over time as the biochar ages.

9.3.3 Advantages over traditional compost and chemical fertilizers

Several studies reported biochar as a soil ameliorant which has the potential to serve as an alternative to chemical-based fertilizers (Ding et al., 2016; Blackwell et al., 2010). This could be due to the presence of humic and fluvic acid-like substances, as well as high nutrients content (nitrogen, phosphorus, and potassium). The beneficial effect of biochar on soil microbes, particularly mycorrhizal colonization, makes it an appealing choice as a soil amendment (Blackwell et al., 2010). Biochar application in soil is not limited to soil conditioning alone, they can also help in immobilization of heavy metals (Uchimiya, Klasson, Wartelle, & Lima, 2011), polycyclic aromatic hydrocarbons (Kołtowski et al., 2017) and other organic contaminants (Ahmad et al., 2014). The occurrence of O-donating ligands of carboxylic, carbonyl and phenolic groups found in biochar binds to contaminants in the soil (Uchimiya et al., 2011).

On a short-term or one-growth-cycle basis, chemical fertilizers and organic composts improve soil properties and plant growth. The majority of the nutrients are leached out, especially if the soil texture is sandy in a high-precipitation area. This will increase the need to add amendments more frequently, putting a significant financial and environmental burden on farmers (Schulz & Glaser, 2012). Biochar, on the other hand, has a more long-lasting and long-term effect. This is due to the higher concentration of black carbon, which is highly recalcitrant and thus less likely to undergo oxidative degradation.

The use of biochar in conjunction with fertilizers revealed that biochar not only increased crop productivity but also improved fertilizer efficiency. Biochar combined with chemical fertilizer increased grain yield by 17%. However, doubling the fertilizer rate resulted in no significant increase (Blackwell et al., 2010). A 3-year long-term study in an avocado orchard found that combining biochar with fertilizers and compost produced better results in terms of tree diameter, height, and fruit count than just fertilizer and compost application alone (Joseph et al., 2020). Treatments of biochar and compost together provides the best result for agronomic performance (Schulz & Glaser, 2012). Hence, biochar-compost combination can improve the soil organic matter fraction of soil deficit in it (Al-Wabel et al., 2018).

Biochar is mostly derived from waste (Lehmann & Joseph, 2015) making it a sustainable choice to use as a fertilizer alternative (Buss, Bogush, Ignatyev, & Mašek, 2020). For instance, Buss and colleagues produced biochar from doped sewage sludge with potassium, which showed enhanced properties like increase in water-extractable phosphorus content (Buss et al., 2020). Furthermore, biochar has a higher potential for climate change mitigation because it aids in the sequestration of significant amounts of carbon and provides stable soil storage. Rice husk and corn stover biochar can sequester 716% and 157% more carbon than straw biomass amendments, respectively (Mohan et al., 2018). Factors such as feedstock micromorphology, mineral composition and distribution, pyrolysis temperature, and environmental exposure all influence biochar's relative stability (Cross &

Sohi, 2013; Singh et al., 2019). Higher pyrolysis temperatures (550°C) produced more stable biochars as compared to lower temperatures (350°C). Increasing the residence time is inconsequential at higher temperatures whereas it improves stability at lower temperature (Cross & Sohi, 2013). Enders and colleagues specified some values to classify the stability of biochars by considering the volatile matter, O/C_{org} and H/C_{org} ratios. According to the study, when volatile matter content is greater than 80% it cannot sequester carbon, while volatile matter less than 80% with O/C_{org} ratio > 0.2 or $H/C_{org} > 0.4$, indicates a biochar with moderate sequestration potential. When O/C_{org} and H/C_{org} are lesser than these values, the biochar is considered to have high sequestration potential (Enders et al., 2012). With the loss of less heat resistant compounds at higher temperatures, an enrichment of aromatic lignin residues that are more stable occurs (Keiluweit, Nico, Johnson, & Kleber, 2010), attaining a lower $O:C_{org}$ value and hence with a higher carbon sequestration potential (Cross & Sohi, 2013). This introduces a new class of fertilizers made from waste material, which connects soil management, food production, waste management, and climate change. Biochar's beneficial physicochemical and biological effects on soil make it a viable addition.

9.3.4 Nutrients dynamics of biochar amendment on mountainous soil

Biochar application in soil is shown to alter soil organic matter, total carbon, total nitrogen, total phosphorus, total potassium, in a number of studies (Table. 9.2) (Clough, Condron, Kammann, & Müller, 2013; Gautam et al., 2017; Kong et al., 2021; Yan et al., 2021a; Zhou et al., 2019). A number of factors are responsible for these alterations, viz., concentration of basic cations in biochars used, their ability to consume protons and pH (Chintala, Mollinedo, Schumacher, Malo, & Julson, 2014). Since, most biochars are alkaline in nature, they can greatly help in the amelioration of acidic pH of mountain soils. During the process of pyrolysis, a good amount of carbonates of magnesium and calcium and $-COO^-$ $(-COOH)$ and $-O^-$ $(-OH)$ radicals are generated, which also helps in neutralizing soil acidity. Basically, reduction in pH is correlated with the calcium carbonate equivalent value of the biochar used (Berek & Hue, 2016; Clough et al., 2013; Kong et al., 2021). The surface of biochar contains negatively charged functional groups that consume protons during decarboxylation, causing pH to rise (Berek & Hue, 2016). It improves soil organic matter in degraded mountain soil by releasing its high organic carbon content after amendment (Kong et al., 2021). Total nitrogen and total phosphorus content increase as biochar addition reduces nitrogen leaching and act as a good source of phosphorus. Nitrogen leaching is reduced because biochar adsorbs organic nitrogen on its surface, ensuring nitrogen availability for a longer period of time (Clough et al., 2013). Some studies have found a decrease in plant available nitrogen forms - ammonium and nitrate - with biochar addition, but it has been linked to an increase in microbial activity, which can cause nitrogen immobilization (Wang et al., 2017). Reduction in phosphorus availability, also occurs as in some cases, biochar act as a phosphorus sink rather than source (Gautam et al., 2017). Another issue in mountainous acidic soils is aluminum toxicity. The negatively charged surface functional groups of biochar form complexes with the aluminum present in acidic soil, reducing the toxicity of aluminum (Berek & Hue, 2016).

TABLE 9.2 Application of biochar in different mountainous soils and their positive outcomes.

Location	Country	Biochar [pyrolysis temp. (°C)]	Application rate (t/ha)	Soil type	Changes in physicochemical properties	Significant outcomes	Reference
Puding County, Guizhou Province	China	Forest logging residues (500)		Sandy loam	Biochar treatment increased total nitrogen and phosphorous along with soil organic carbon	Increased soil nutrients and arbuscular mycorrhizal fungal diversity	Yan et al. (2021a)
Chandanpur, Panchkhal, and Talamarang	Nepal	Coffee waste materials (350–550)	5	Silt loam		Increased carbon stocks and also soil organic carbon	Dahal et al. (2018)
Sharswotikhel, Bhaktapur	Nepal	Grasses and *Eupatorium* sp. (300–500)	5		Increase in pH, soil organic matter and exchangeable potassium were reported	A significant positive change was observed for crop growth (height), soil chemical properties, and crop productivity. For immediate positive results, lower amount of biochar with farmyard manure can be useful	Gautam et al. (2017)
Parys Mountain Anglesey, Wales	U.K.		10% by volume	Contaminated soil (high concentration of Cu, Pb, and Zn)	The biochar addition increased the pH to 6.9 from initial soil pH of 2.7	Biochar addition with or without compost has been effective in controlling the availability and mobility of all three metals	Sizmur et al. (2011)
Central Sichuan (Hilly area)	China	Mixed crop i.e., wheat straw & peanut shell residues (500)	8	Coarse sand, fine sand, and silt/clay	Weathering/deterioration of biochar's pores edges and surfaces with relatively higher oxidized functional groups	Increase in soil organic carbon was reported after biochar application by 37% (fine sand), 42% (silt/clay), and 76% (coarse sand)	El-Naggar et al. (2018)
Toon city, Ehime prefecture	Japan	Bamboo biochar (800)	0.5, 0.15, and 0.3	Sandy clay loam	Reduction in soil pH compared to control	Approx. 31% brown rice yield was increased after using biochar at 600 kg 10a^{-1} application rate	Toma et al. (2019)

San Juan County, WA	USA	Mixture of douglas fir (80%), White fir (15%), and Western red cedar (5%)	20	Sandy loam		Soil total carbon was increased by 32%–33%, 45%–54% increase in soil available NH_4^+ was also obtained after biochar application	Gao et al. (2016)
Datan, Qilian Mountain, Gansu province	China	Maize crop residue (500)	30 g/kg soil	Inceptisol	Soil organic carbon and pH were increased after biochar addition	Improvement in soil quality and microbial biomass which further restores diversity and species richness	Kong et al. (2021)
Mountain River, Huon Valley, southern Tasmania	Australia	Acacia green waste (550)	47	Sandy loam	Biochar implementation increased soil moisture content by 13% and organic carbon level by 23%	Microbial abundance was improved after biochar addition and also 23% enhancement were reported in soil organic carbon	Abujabhah et al. (2016)
Thai Nguyen and Thanh Hoa	Vietnam	Rice straw, Bamboo, and tree branches	2.5–10	Plintic Acrisol and Haplic Acrisol		For rice, over time crop yields were improved using biochar but not in case of vegetables	Vinh et al. (2014)

Aluminum adsorption and coprecipitation as KAlSi3O8 on the surface of biochar are two other possible mechanisms for reducing aluminum toxicity (Qian & Chen, 2013).

9.3.5 Microbial dynamics of biochar amendment on mountainous soil

As mentioned in the preceding sections, mountain regions have lower temperatures than plain areas and are mostly acidic in nature. These are not ideal conditions for microbial communities. Biochar addition aids in overcoming these environmental conditions and promoting microbial activity. Nonetheless, biochar does not directly affect soil microbial health; rather, it alters soil chemical and physical properties, making the environment more conducive to microorganism survival (Xu et al., 2014). These properties include high surface area and porosity, high carbon content, ability to elevate pH, water holding capacity, CEC, total nitrogen and carbon/nitrogen ratio (Fig. 9.2) (Han, Lan, Chen, Yu, & Bie, 2017; Xu et al., 2014; Yan et al., 2021b; Zhou et al., 2019). Another reason biochar is an excellent inducer of microbial ability is its high-density surface negative charge, which attracts toxic chemicals from the soil to its surface, making the soil environment more hospitable to microbes (Kong et al., 2021). For example, a study in a karst mountainous area of Guizhou Province, China, found that adding biochar improved the bacterial community network and diversity, resulting in better carbon and other nutrient cycling. This improved nutrient cycling can be attributed to the addition of biochar, which increases actinobacteria (which maintains a high carbon level in the soil). The presence of more keystone taxa following biochar amendment can be attributed to the improvement in the community network. The stability of a microbial community is determined by keystone taxa.

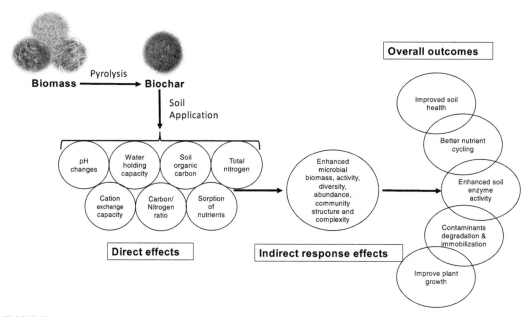

FIGURE 9.2   Biochar mediated microbial dynamics in soil.

The greater the number of these taxa, the greater the stability and thus the complexity of the community structure (Yan et al., 2021b; Zhou et al., 2019).

Another study found an increase in microbial biomass and diversity after adding biochar, which supports this finding. Biochar, due to its porous nature, provides a good habitat condition for microbes to grow and thrive by providing predator-free shelter. The pyrolysis process converts nutrients into usable forms, and its large surface area allows for better soil nutrient adsorption. Higher nutrient availability corresponds to higher microbial growth and, as a result, higher microbial biomass (Zhou et al., 2019). Kong et al. also report an increase in microbial biomass carbon as well as microbial biomass nitrogen with biochar amendment. Although the study also stated that biochar alone will not be able to sustain the increase, a combined amendment of biochar with nitrogenous fertilizers and manure is recommended in cases of severely degraded soil (Kong et al., 2021).

Despite increasing microbial diversity, biochar addition can reduce biomass. The increase in diversity has been linked to an increase in soil pH following biochar addition, among other things. Because bacteria are highly pH sensitive, the acidic mountainous soil becomes more favorable for microbial growth as the pH rises. The decrease in microbial biomass (observed at high biochar amendment rates) can be attributed to a decrease in soil organic carbon mineralization, as biochar is primarily composed of recalcitrant carbon, as well as the presence of polyaromatic hydrocarbons in biochar, which are toxic to microorganisms (Li et al., 2018). Biochar addition influences not only bacterial growth but also fungal population; however, bacterial population is more susceptible to changes in soil nutritional content than fungal population. A study found 74.73% bacterial variation compared to 35% with changing nutritional characteristics of soil (Kong et al., 2021; Yao et al., 2014). However, depending on application rates, biochar pyrolysis temperature, the type of precursor biomass used for biochar preparation, and other environmental factors in the soil, biochar amendment can have varying effects on soil microbial structure (Han et al., 2017; Li et al., 2018).

9.4 Conclusion

Mountains house a large portion of the world's population. Because of the harsh terrain and topography of mountainous areas, the potential for economic and industrial development is limited. As a result, the majority of mountain people rely on agricultural activities for a living. However, these regions have the most food insecure people. This is due to the low fertility and productivity of mountain soils. These changes, from organic manures as the primary form of amendment to a gradual shift toward chemical fertilizers, are brought about by commercialization and consumerism. It has raised the issue of sustainability. Biochar has been proposed as a sustainable alternative because of its ability to increase soil fertility without polluting the environment. However, there is a scarcity of knowledge in this field. There are currently few studies that describe biochar as a potential candidate for reducing fertility problems in mountainous areas. More long-term studies with different types of biochar and application rates are required before real-world applications.

Nonetheless, biochar has the potential to help poor mountain farmers achieve food security and sustainability.

References

Abbasi, M.K., Tahir, M.M., 2012. Economizing nitrogen fertilizer in wheat through combinations with organic manures in Kashmir, Pakistan. Agronomy Journal 104, 169–177.

Abujabhah, I.S., Bound, S.A., Doyle, R., Bowman, J.P., 2016. Effects of biochar and compost amendments on soil physico-chemical properties and the total community within a temperate agricultural soil. Applied Soil Ecology 98, 243–253.

Ahmad, M., Rajapaksha, A.U., Lim, J.E., Zhang, M., Bolan, N., Mohan, D., et al., 2014. Biochar as a sorbent for contaminant management in soil and water: a review. Chemosphere 99, 19–33.

Al-Wabel, M.I., Hussain, Q., Usman, A.R., Ahmad, M., Abduljabbar, A., Sallam, A.S., et al., 2018. Impact of biochar properties on soil conditions and agricultural sustainability: a review. Land Degradation & Development 29, 2124–2161.

Ameloot, N., Sleutel, S., Case, S.D., Alberti, G., Mcnamara, N.P., Zavalloni, C., et al., 2014. C mineralization and microbial activity in four biochar field experiments several years after incorporation. Soil Biology & Biochemistry 78, 195–203.

Are, K.S., Adelana, A.O., Fademi, I.O., Aina, O.A., 2017. Improving physical properties of degraded soil: potential of poultry manure and biochar. Agriculture and Natural Resources 51, 454–462.

Balla, M.K., Tiwari, K.R., Kafle, G., Gautam, S., Thapa, S., Basnet, B., 2014. Farmers dependency on forests for nutrients transfer to farmlands in mid-hills and high mountain regions in Nepal (case studies in Hemja, Kaski, Lete and Kunjo, Mustang district). International Journal of Biodiversity Conservation 6, 222–229.

Berek, A.K., Hue, N.V., 2016. Characterization of biochars and their use as an amendment to acid soils. Soil Science 181, 412–426.

Berihun, T., Tadele, M., Kebede, F., 2017. The application of biochar on soil acidity and other physico-chemical properties of soils in southern Ethiopia. Journal of Plant Nutrition and Soil Science 180, 381–388.

Bishwakarma, B.K., Dahal, N.R., Allen, R., Rajbhandari, N.P., Dhital, B.K., Gurung, D.B., et al., 2015. Effects of improved management and quality of farmyard manure on soil organic carbon contents in small-holder farming systems of the Middle Hills of Nepal. Climate and Development 7, 426–436.

Bista, P., Ghimire, R., Chandra Shah, S., RAJ Pande, K., 2010. Assessment of Soil Fertility Management Practices and Their Constraints in Different Geographic Locations of Nepal. Forum geografic.

Blackwell, P., Krull, E., Butler, G., Herbert, A., Solaiman, Z., 2010. Effect of banded biochar on dryland wheat production and fertiliser use in south-western Australia: an agronomic and economic perspective. Soil Research 48.

Buss, W., Bogush, A., Ignatyev, K., Mašek, O., 2020. Unlocking the fertilizer potential of waste-derived biochar. ACS Sustainable Chemistry & Engineering 8, 12295–12303.

Chintala, R., Mollinedo, J., Schumacher, T.E., Malo, D.D., Julson, J.L., 2014. Effect of biochar on chemical properties of acidic soil. Archives of Agronomy and Soil Science 60, 393–404.

Clough, T.J., Condron, L.M., Kammann, C., Müller, C., 2013. A review of biochar and soil nitrogen dynamics. Agronomy 3, 275–293.

Córdova, R., Hogarth, N.J., Kanninen, M., 2019. Mountain farming systems' exposure and sensitivity to climate change and variability: agroforestry and conventional agriculture systems compared in Ecuador's Indigenous Territory of Kayambi people. Sustainability 11, 2623.

Cross, A., Sohi, S.P., 2013. A method for screening the relative long-term stability of biochar. GCB Bioenergy 5, 215–220.

Dahal, N., Bajracharya, R.M., 2012. Effects of sustainable soil management practices on distribution of soil organic carbon in upland agricultural soils of mid-hills of Nepal. Nepal Journal of Science and Technology 13, 133–141.

Dahal, N., Bajracharya, R.M., Wagle, L.M., 2018. Biochar effects on carbon stocks in the coffee agroforestry systems of the Himalayas. Sustainable Agriculture Research 7, 103–114.

Ding, Y., Liu, Y., Liu, S., Li, Z., Tan, X., Huang, X., et al., 2016. Biochar to improve soil fertility. A review. Agronomy for Sustainable Development 36, 36.

Edenborn, S., Edenborn, H., Krynock, R., Haug, K.Z., 2015. Influence of biochar application methods on the phytostabilization of a hydrophobic soil contaminated with lead and acid tar. Journal of Environmental Management 150, 226–234.

El-Naggar, A., Awad, Y.M., Tang, X.Y., Liu, C., Niazi, N.K., Jien, S.H., et al., 2018. Biochar influences soil carbon pools and facilitates interactions with soil: a field investigation. Land Degradation & Development 29, 2162–2171.

El-Naggar, A., Lee, S.S., Rinklebe, J., Farooq, M., Song, H., Sarmah, A.K., et al., 2019. Biochar application to low fertility soils: a review of current status, and future prospects. Geoderma 337, 536–554.

Enders, A., Hanley, K., Whitman, T., Joseph, S., Lehmann, J., 2012. Characterization of biochars to evaluate recalcitrance and agronomic performance. Bioresource Technology 114, 644–653.

Gao, S., Deluca, T.H., 2018. Wood biochar impacts soil phosphorus dynamics and microbial communities in organically-managed croplands. Soil Biology and Biochemistry 126, 144–150.

Gao, S., Hoffman-Krull, K., Bidwell, A., Deluca, T., 2016. Locally produced wood biochar increases nutrient retention and availability in agricultural soils of the San Juan Islands, USA. Agriculture, Ecosystems & Environment 233, 43–54.

Gautam, D.K., Bajracharya, R.M., Sitaula, B.K., 2017. Effects of biochar and farm yard manure on soil properties and crop growth in an agroforestry system in the Himalaya. Sustainable Agriculture Research 6.

Glaser, B., Lehmann, J., Zech, W., 2002. Ameliorating physical and chemical properties of highly weathered soils in the tropics with charcoal—a review. Biology and Fertility of Soils 35, 219–230.

Gorovtsov, A.V., Minkina, T.M., Mandzhieva, S.S., Perelomov, L.V., Soja, G., Zamulina, I.V., et al., 2020. The mechanisms of biochar interactions with microorganisms in soil. Environmental Geochemistry and Health 42, 2495–2518.

Gosain, B.G., 2016. Some physico-chemical properties of Oak and Pine FYM based crop field soils: a study from mountain watershed, Kumaun Himalaya. International Journal of Applied Research 2, 713–718.

Grumbine, R.E., Xu, J., 2021. Mountain futures: pursuing innovative adaptations in coupled social—ecological systems. Frontiers in Ecology and the Environment.

Han, G., Lan, J., Chen, Q., Yu, C., Bie, S.J.S.R., 2017. Response of soil microbial community to application of biochar in cotton soils with different continuous cropping years. Scientific Reports 7, 1–11.

Hossain, M.Z., Bahar, M.M., Sarkar, B., Donne, S.W., Ok, Y.S., Palansooriya, K.N., et al., 2020. Biochar and its importance on nutrient dynamics in soil and plant. Biochar 2, 379–420.

Jaiswal, A.K., Frenkel, O., Tsechansky, L., Elad, Y., Graber, E.R., 2018. Immobilization and deactivation of pathogenic enzymes and toxic metabolites by biochar: a possible mechanism involved in soilborne disease suppression. Soil Biology and Biochemistry 121, 59–66.

Jien, S.-H., Wang, C.-S., 2013. Effects of biochar on soil properties and erosion potential in a highly weathered soil. Catena 110, 225–233.

Joseph, S., Pow, D., Dawson, K., Rust, J., Munroe, P., Taherymoosavi, S., et al., 2020. Biochar increases soil organic carbon, avocado yields and economic return over 4 years of cultivation. Science of the Total Environment 724, 138153.

Kang, S.-W., Kim, S.-H., Park, J.-H., Seo, D.-C., Ok, Y.S., Cho, J.-S., 2018. Effect of biochar derived from barley straw on soil physicochemical properties, crop growth, and nitrous oxide emission in an upland field in South Korea. Environmental Science and Pollution Research 25, 25813–25821.

Keiluweit, M., Nico, P.S., Johnson, M.G., Kleber, M., 2010. Dynamic molecular structure of plant biomass-derived black carbon (biochar). Environmental Science & Technology 44, 1247–1253.

Kołtowski, M., Hilber, I., Bucheli, T.D., Charmas, B., Skubiszewska-Zięba, J., Oleszczuk, P., 2017. Activated biochars reduce the exposure of polycyclic aromatic hydrocarbons in industrially contaminated soils. Chemical Engineering Journal 310, 33–40.

Kong, J., He, Z., Chen, L., Yang, R., Du, J., 2021. Efficiency of biochar, nitrogen addition, and microbial agent amendments in remediation of soil properties and microbial community in Qilian Mountains mine soils. Ecology and Evolution 1–14.

Lammirato, C., Miltner, A., Kaestner, M., 2011. Effects of wood char and activated carbon on the hydrolysis of cellobiose by β-glucosidase from *Aspergillus niger*. Soil Biology & Biochemistry 43, 1936–1942.

Lawrinenko, M., Laird, D.A., 2015. Anion exchange capacity of biochar. Green Chemistry 17, 4628–4636.

Lehmann, J., Rondon, M., 2006. Bio-char soil management on highly weathered soils in the humid tropics. Biological Approaches to Sustainable Soil Systems 113, 517–530.

Lehmann, J., Joseph, S., 2015. Biochar for environmental management: an introduction. In: Lehmann, J., Joseph, S. (Eds.), Biochar for Environmental Management Science, Technology and Implementation. Routledge.

Lehmann, J., Rillig, M.C., Thies, J., Masiello, C.A., Hockaday, W.C., Crowley, D., 2011. Biochar effects on soil biota — a review. Soil Biology and Biochemistry 43, 1812–1836.

Li, Q., Lei, Z.F., Song, X.Z., Zhang, Z.T., Ying, Y.Q., Peng, C.H., 2018. Biochar amendment decreases soil microbial biomass and increases bacterial diversity in Moso bamboo (*Phyllostachys edulis*) plantations under simulated nitrogen deposition. Environmental Research Letters 13, 044029.

Li, X., El Solh, M., Siddique, K., 2019. Mountain Agriculture: Opportunities for Harnessing Zero Hunger in Asia. Food and Agriculture Organization of the United Nations (FAO).

Mohan, D., Kumar, A., Sarswat, A., Patel, M., Singh, P., and Pittman Jr, C.U., 2018. Biochar production and applications in soil fertility and carbon sequestration – a sustainable solution to crop-residue burning in India. Royal Society of Chemistry 8, 508–520.

Mukherjee, A., Lal, R., 2013. Biochar impacts on soil physical properties and greenhouse gas emissions. Agronomy 3, 313–339.

Negi, G., Bisht, V., 2017. Promoting organic tea farming in mid-hills of North-West Himalaya, India. Tea 38 (2), 57–67.

Nelissen, V., Ruysschaert, G., Manka'abusi, D., D'hose, T., DE Beuf, K., Al-Barri, B., et al., 2015. Impact of a woody biochar on properties of a sandy loam soil and spring barley during a two-year field experiment. European Journal of Agronomy 62, 65–78.

Nouraein, M., Skataric, G., Spalevic, V., Dudic, B., Gregus, M., 2019. Short-term effects of tillage intensity and fertilization on sunflower yield, achene quality, and soil physicochemical properties under semi-arid conditions. Applied Sciences-Basel 9, 5482.

Palansooriya, K.N., Ok, Y.S., Awad, Y.M., Lee, S.S., Sung, J.-K., Koutsospyros, A., et al., 2019. Impacts of biochar application on upland agriculture: a review. Journal of Environmental Management 234, 52–64.

Panwar, N.R., Ramesh, P., Singh, A.B., Ramana, S., 2010. Influence of organic, chemical, and integrated management practices on soil organic carbon and soil nutrient status under semi-arid tropical conditions in Central India. Communications in Soil Science and Plant Analysis 41, 1073–1083.

Pilbeam, C.J., Mathema, S.B., Gregory, P.J., Shakya, P.B., 2005. Soil fertility management in the mid-hills of Nepal: practices and perceptions. Agriculture and Human Values 22, 243–258.

Pradhan, A., Chan, C., Roul, P.K., Halbrendt, J., Sipes, B., 2018. Potential of conservation agriculture (CA) for climate change adaptation and food security under rainfed uplands of India: a transdisciplinary approach. Agricultural Systems 163, 27–35.

Qian, L., Chen, B., 2013. Dual role of biochars as adsorbents for aluminum: the effects of oxygen-containing organic components and the scattering of silicate particles. Environmental Science & Technology 47, 8759–8768.

Romeo, R., Vita, A., Manuelli, S., Zanini, E., Freppaz, M., Stanchi, S.J.F., 2015. Understanding Mountain Soils: A Contribution from Mountain Areas to the International Year of Soils 2015, ROME.

Schulz, H., Glaser, B., 2012. Effects of biochar compared to organic and inorganic fertilizers on soil quality and plant growth in a greenhouse experiment. Journal of Plant Nutrition and Soil Science 175, 410–422.

Sharma, K., Pathania, M., Lal, H., 2006. Farming system approach for sustainable development of agriculture in mountain regions—a case of Himachal Pradesh. Agricultural Economics Research Review 19, 101–112.

Shrestha, S.K., 2015. Sustainable soil management practices: a key to combat soil desertification in the hills of Nepal. World Journal of Science, Technology and Sustainable Development 12, 13–24.

Singh, A., Singh, A.P., Purakayastha, T.J., 2019. Characterization of biochar and their influence on microbial activities and potassium availability in an acid soil. Archives of Agronomy and Soil Science 65, 1302–1315.

Sizmur, T., Wingate, J., Hutchings, T., Hodson, M.E., 2011. *Lumbricus terrestris* L. does not impact on the remediation efficiency of compost and biochar amendments. Pedobiologia-International Journal of Soil Biology 54, S211–S216.

Song, W., Guo, M., 2012. Quality variations of poultry litter biochar generated at different pyrolysis temperatures. Journal of Analytical and Applied Pyrolysis 94, 138–145.

Tarin, M.W.K., Khaliq, M.A., Fan, L., Xie, D., Tayyab, M., Chen, L., et al., 2021. Divergent consequences of different biochar amendments on carbon dioxide (CO_2) and nitrous oxide (N_2O) emissions from the red soil. Science of the Total Environment 754, 141935.

2. Soil microbial processes and nutrient dynamics

Tarin, M., Fan, L., Shen, L., Lai, J., Tayyab, M., Sarfraz, R., et al., 2019. Effects of different biochars ammendments on physiochemical properties of soil and root morphological attributes of Fokenia Hodginsii (Fujian cypress). Applied Ecology and Environmental Research 17, 11107–11120.

Thapa, G., Paudel, G., 2002. Farmland degradation in the mountains of Nepal: a study of watersheds 'with' and 'without' external intervention. Land Degradation & Development 13, 479–493.

Tiwari, K.R., Sitaula, B.K., Borresen, T., Bajracharya, R.M., 2006. An assessment of soil quality in Pokhare Khola watershed of the Middle Mountains in Nepal. Journal of Food, Agriculture & Environment 4, 276–283.

Toma, Y., Fujitani, W., Tsurumi, T., Tsurumi, K., Ueno, H., 2019. Effect of bamboo biochar application on rice growth and yield and a regional cycle model of bamboo materials in hilly and mountainous area in Japan. Wood Carbonization Research 16, 49–59.

Tripathi, B., Jones, J., 2010. Biophysical and socio-economic tools for assessing soil fertility: a case of western hills, Nepal. Agronomy Journal of Nepal 1, 1–9.

Uchimiya, M., Klasson, K.T., Wartelle, L.H., Lima, I.M., 2011. Influence of soil properties on heavy metal sequestration by biochar amendment: 1. Copper sorption isotherms and the release of cations. Chemosphere 82, 1431–1437.

Vinh, N., Hien, N., Anh, M., Lehmann, J., Joseph, S., 2014. Biochar treatment and its effects on rice and vegetable yields in mountainous areas of northern Vietnam. International Journal of Agricultural and Soil Science 2, 5–13.

Wang, Z.Y., Chen, L., Sun, F.L., Luo, X.X., Wang, H.F., Liu, G.C., et al., 2017. Effects of adding biochar on the properties and nitrogen bioavailability of an acidic soil. European Journal of Soil Science 68, 559–572.

Wu, L., Wei, C., Zhang, S., Wang, Y., Kuzyakov, Y., Ding, X., 2019. MgO-modified biochar increases phosphate retention and rice yields in saline-alkaline soil. Journal of Cleaner Production 235, 901–909.

Wymann Von Dach, S., Romeo, R., Vita, A., Wurzinger, M., Kohler, T., 2013. Mountain farming is family farming: a contribution from mountain areas to the International Year of Family Farming 2014, FAO.

Xu, H.-J., Wang, X.-H., Li, H., Yao, H.-Y., Su, J.-Q., Zhu, Y.-G., 2014. Biochar impacts soil microbial community composition and nitrogen cycling in an acidic soil planted with rape. Environmental Science & Technology 48, 9391–9399.

Yan, Q., Dong, F., Li, J., Duan, Z., Yang, F., Li, X., et al., 2019. Effects of maize straw-derived biochar application on soil temperature, water conditions and growth of winter wheat. European Journal of Soil Science 70, 1280–1289.

Yan, T., Xue, J., Zhou, Z., Wu, Y., 2021a. Impacts of biochar-based fertilization on soil arbuscular mycorrhizal fungal community structure in a karst mountainous area. Environmental Science and Pollution Research 1–15.

Yan, T.T., Xue, J.H., Zhou, Z.D., Wu, Y.B., 2021b. Effects of biochar-based fertilizer on soil bacterial network structure in a karst mountainous area. Catena 206, 105535.

Yao, M., Rui, J., Li, J., Dai, Y., Bai, Y., Heděnec, P., et al., 2014. Rate-specific responses of prokaryotic diversity and structure to nitrogen deposition in the *Leymus chinensis* steppe. Soil Biology & Biochemistry 79, 81–90.

Yuan, J.-H., Xu, R.-K., Zhang, H., 2011. The forms of alkalis in the biochar produced from crop residues at different temperatures. Bioresource Technology 102, 3488–3497.

Zhou, Z.D., Gao, T., Zhu, Q., Yan, T.T., Li, D.C., Xue, J.H., et al., 2019. Increases in bacterial community network complexity induced by biochar-based fertilizer amendments to karst calcareous soil. Geoderma 337, 691–700.

SECTION 3

Soil physicochemical parameters

Hair to canopy: role of organic debris in soil formation and succession of rock ecosystem

Solomon Kiruba[1], P. Maria Antony[2], Solomon Jeeva[3] and P.V. Annie Gladys[4]

[1]Department of Zoology, Madras Christian College, Tambaram, Chennai, Tamil Nadu, India
[2]ATREE's Agasthyamalai Community Conservation Centre (ACCC), Manimutharu, Tamil Nadu, India [3]Department of Botany, Scott Christian College, Nagercoil, Kanyakumari, Tamil Nadu, India [4]Department of English, Nesamony Memorial Christian College, Marthandam, Kanyakumari, Tamil Nadu, India

O U T L I N E

10.1 Introduction

The biotic and abiotic components are similar in their physicochemical nature. Human beings were created from ground dust, according to creation (The Holy Bible: Genesis 2:7), and to dust, he will return (Genesis 3:19) (New King James Version Bible, 1994). In terms of evolution, life began in water but in life, (along with its organic components) solids have a superior role over liquids. Remusat (2014) predicts that organic compounds and water are the potential precursors for the emergence of life. Water is depicted as a constituent of soil by Coleman et al. (2017). Soil can be thought of as a link between living and non-living things. However, in anthropocentric perspectives, the sacredness of soil is being underestimated, and soil is abused, and regarded as a filthy matter. Soil is seen as a source of material for construction, land filling, and rare elements in a materialistic worldview. As a result, the soil is uncovered for discovery, dislocated for construction allocation, and poisoned for production, affecting soil sanctity thus inducing soil erosion, a universal phenomenon that affects our biosphere at large (Bonell et al., 2010; Prasannakumar et al., 2011; Thomas et al., 2018; Miller et al., 2021). Humus-rich topsoil not only acts as a reservoir for water, but it also supports a diverse range of flora and fauna, which influences rainfall. Forests are mostly found in mountainous ecosystems in India, where they serve as a source of major perennial rivers. Even from an anthropocentric standpoint, the rock ecosystem should be prioritized for human survival and sustainability (Premchander et al., 2003; Ephraim and Murugesan, 2015).

Extensive changes in land use patterns and agricultural practices, particularly in mountainous habitats, have occurred, such as the introduction of exotic species, monoculture, and the continuous use of chemical fertilizers, pesticides, and weedicides (Brockerhoff et al., 2008; Mudappa and Raman, 2012). The latter is the most dangerous of all anthropogenic activities because it directly interferes with soil biota and affects soil pedology and edaphology (Edwards and Pimentel, 1989; Gómez et al., 2004; Sabatier et al., 2014; Liu et al., 2016). Poor land management practices harm native plants such as non-timber trees, lianas, rattans, and shade-loving understories of shrubs, woody herbs, and ferns. This reduces diversity and primary productivity, affecting the amount of humus, water percolation, and soil holding capacity, resulting in floods and landslides, which is an ongoing problem in hills (Kuriakose, 2010; Sajinkumar et al., 2011; Pradeep et al., 2015). Hence, conservation of these ecosystems with special reference to their soil is of prime importance.

Soil formation, colonization, succession, bottom up and top down effects, facilitation cascade, habitat cascade, and trophic cascade are all influenced by habitat type and species composition (Hunter and Price, 1992; Polis et al., 2000; Bruno et al., 2003; Ripple et al., 2016; Zhang and Silliman, 2019). Its operational time scale and complexities are higher in terrestrial ecosystems, particularly in rock ecosystems, than in wetlands (Polis et al., 2000). Hence studies are focusing on aquatic habitat rather than terrestrial habitat (Altieri et al., 2007; Carpenter et al., 2010; Angelini et al., 2015; Zhang and Silliman, 2019; Beheshti et al., 2021). Understanding the complex cascades of the rock ecosystem's pedogenesis and edaphogenesis is critical for its conservation, restoration, and long-term use. Rocks are always important in pedogenesis via abiotic and biotic processes, particularly in tropical rain forests (Hole, 1981; Bardgett, 2005; Jenny, 2012; Coleman et al., 2017). Scanning the possible mechanisms of soil formation and ecological succession in bare rock ecosystems, with a focus on the Southern Western Ghats (SWG), is therefore essential.

Experimental evidence is needed for developing hypotheses and theories. It is either difficult or impossible to imitate nature in experimental design, or the experimental techniques are unviable in terms of space, time, matter, energy, and economics. It is well known that theoretical approaches based on observations and available literature can, to some extent, fill these gaps. Loreau et al. (2001) contend that significant scientific advances occur when observational, experimental, and theoretical studies are conducted concurrently. The arts serve as an effective medium for communicating scientific research. To reflect complex natural processes such as soil formation and ecological succession, a multidisciplinary approach is required.

10.2 Rock, soil, and literature

The Renaissance represents the rebirth of learning, triggering a series of intellectual transformations that culminated in the revival of art and science. The renaissance spirit reached its pinnacle in the realm of art through observation and exploration of nature, aided by scientific insight. Furthermore, some of the most prominent minds in human history, such as Leonardo da Vinci, Wolfgang Goethe, Michelangelo, and Arthur Eddington, were known for their achievements in both art and science.

Leonardo da Vinci, the Renaissance icon, was an Italian painter who studied anatomy, botany, science, architecture, geology, mathematics, and other subjects. He could be credited with inventing scientific visualization (Pasipoularides, 2019). Leonardo, a tireless painter endowed with a keen sense of perception, used science and art to communicate his scholarly ideas. His formative statement on the confluence of art and science is worthwhile to refer: "To develop a complete mind: Study the science of art; Study the art of science. Learn how to see. Realize that everything connects to everything else." Kemp writes: "Leonardo mapped the human body. He charted its skeletal rocks, the course of its "rivers" and its fleshy soil, both within and without. He dissected the world, teasing out its bony rocks, its earthly flesh and its watery veins, both in its surface topography and deep within its core... Always, looking at his drawings of real sites, we can sense his urgent concern with the body of the Earth as a functioning system" (Bressan, 2014).

The soil is the source of our sustenance, sustaining both our past and our non-past. Soil covers nearly one-third of the Earth's total land area. It is a remarkably intricate and magnificently active crust. Life on Earth is influenced not only by the celestial environment, but also by the soil beneath our feet (Sabale et al., 2020). According to Hillel (1992), Adam, the first man in Semitic religion, was formed from clay and derived from adema, a Hebrew noun meaning earth or soil. Soil, "The great connector of lives, the source and destination of all," is the healer, restorer, and resurrector who transforms disease into health, old age into youth, and death into life. We cannot have a community if it is not properly cared for, and we cannot have life if it is not properly cared for (Berry, 2015). The natural and cultural history of soil exposes the use and abuse of it. American biologist and conservationist asserts: "We abuse land because we regard it as a commodity belonging to us. When we see land as a community to which we belong, we may begin to use it with love and respect" (Leopold, 1989). Hence, one has to toil and serve as soil stewards in order to maintain the soil's purity.

Earth, ground or soil lies like a mantle on the land. This mantle is a magnificent mixture of bits of rock, water, air, plants and animals and their residues. Soil is "the root domain of lively darkness and silence" on which land life depends. Many soil scientists have made attempts to publicize the subject of soil through composing songs to provide the common people a glimpse of "the earth beneath their feet" (Hole, 1994). Earth represents "the home of humanity's heritage" (Capra et al., 2017). Soil "resembles abstract art ... if you are used to thinking of soil as dirt, which is customary in our society, you are not keyed to find beauty in it." It is high time to realize and reclaim the companionship with the soil: "Because the earth beneath our feet is literally supporting and nourishing us every day of our lives, and stands ready to receive us back again when we die"(Hole, 1994). Moreover, human life commences and concludes with soil.

We are so much curious about the sky and star that we have deserted the earth. Little children are charmed by the verse: Twinkle, twinkle, little star... Like a diamond in the sky. A soil scientist and poet propose to have a similitude verse about the earth beneath our feet. He has exquisitely carved a poem on soil:

Darkle, darkle, little grain.

I wonder how you entertain

A thousand creatures microscopic.

Grains like you from pole to tropic

Support land life upon this planet.

I marvel at you, crumb of granite (Hole, 1994).

Rock is the ancestor of soil, from which myriad forms of life have evolved. Rocks, in the eyes of artists, represent power and dynamism, on which one can feel the pulse of determination and the flame of firmness. "On the Pulse of Morning" is a powerful poem written and performed by an acclaimed American poet and civil rights activist Maya Angelou (2015). She invites humanity to return to its roots and face the future with all-encompassing vitality. Moreover, she finds oneness with the nonhuman world of trees and rocks:

But today, the Rock cries out to us, clearly, forcefully,

Come, you may stand upon my

Back and face your distant destiny,

But seek no haven in my shadow,

I will give you no hiding place down here.

. . .

Clad in peace and I will sing the songs

The Creator gave to me when I and the

Tree and the rock were one.

3. Soil physicochemical parameters

Rocks are draped in moss, with crevices carrying loads of organic luggage; they support drooping tree branches. Water gushes over the rocks and cascades down the trees, causing the canopy to rise.

10.3 Hair to canopy: the complex cascade

The succession and soil formation of bare rocks occur in different stages (Figs. 10.1–10.3). The first and most important stage is the movement of organic debris uphill, which promotes the formation of organic soup, which then flows over the rocks during the rainy season, causing the formation of nano, micro, and mega organic coats depending on the slope and texture of the rocks. In the second stage, the organic debris settled in the bare rock promotes decomposition and attracts detritivores (Fig. 10.2) which pave the way to the primary succession, of lower plants including Algae, Fungi, and Bryophytes (Fig. 10.3) via facilitation cascade (Wilson, 1992; Bruno et al., 2003; Leibold et al., 2004; Altieri et al., 2007). Primary succession ultimately leads to secondary, tertiary, and quaternary succession resulting in climax canopy through complex cascades (Clements, 1936; Jones et al., 1997; Polis et al., 2000; Altieri et al., 2007; Thomsen et al., 2010; Ripple et al., 2016; Zhang and Silliman, 2019; Temmink et al., 2021).

10.3.1 Stage I: Uphill movement of organic debris

Because organic debris will hasten soil formation, it is critical to understand the various methods of organic debris uphill movement. Organic debris mobility varies depending on habitat, physicochemical nature of debris, and is lower in terrestrial habitats than in aquatic habitats such as oceans. Organic debris cannot move voluntarily; it must move involuntarily in accordance with Newton and Berkeley (1960) law of motion; it can be moved by applying sufficient external forces. Because it is moving against equatorial forces, altitudinal upward movement of organic debris requires more energy than latitudinal downward movement. As a result, tiny organic debris can only be moved altitudinally by biotic and abiotic forces.

10.3.1.1 Uphill movement of organic debris through abiotic forces

Over time, the presence of organic components (chemical nutrients) will facilitate the establishment of life on any substrate. The formation of organic soup eventually starts the momentum of ecological succession. Abiotic natural forces contribute minerals and transport lighter organic debris from lower to higher elevations, as well as from forested areas to bare rocks (Fig. 10.1). Wind action is paramount in shaping and fertilizing the rock through the process of erosion and deposition of minerals from and to the rocks. It also carries organic debris (Hindy et al., 2018), immobilizes the atmospheric gases along with organic volatile compounds (Karl et al., 2004; Yang et al., 2014), deposits salt (especially in the rock of coastal region) (Avis and Lubke, 1985; Moreno et al., 2006), and sows the planting materials such as spore and seeds of different flora (Gröger, 2000; Raju et al., 2009; Chang et al., 2014; Upadhyay et al., 2017). In desert ecosystem, winds carry tons of dust/day (Koren et al., 2006), can travel long distances including over oceans (Darwin, 1846; Favet et al., 2013) and displace

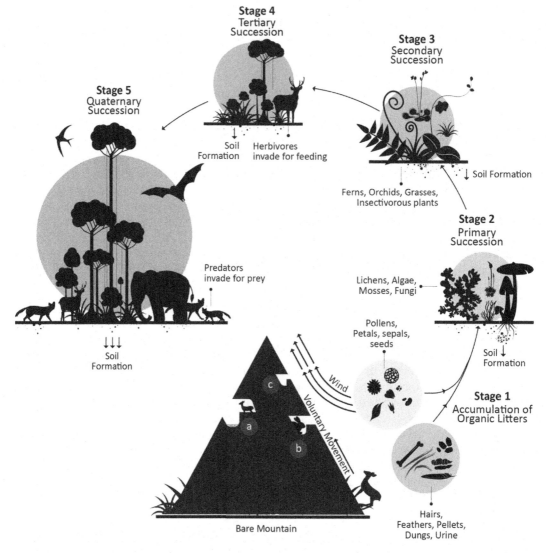

(A) Low Wind Zones (LWZ) , (B) Water Free Zones (WFZ), and (C) Sun Free Zone (SFZ)

FIGURE 10.1   The soil formation and succession of bare rock ecosystem mediated by uphill movement of organic debris.

eukaryotic microbes and spores of metazoans across the globe (Griffin et al., 2003; Kellogg et al., 2004; Lim et al., 2011; Favet et al., 2013; Incagnone et al., 2015; Ptatscheck et al., 2018).

Minute debris of Angiosperms like pollens, leaflets, and small leaves, petioles, petals and plant propagules (Nobel, 1981) can be carried by the wind (Valencia-Barrera et al., 2001; De Langre, 2008; Sobral-Leite et al., 2011; Muniz-Castro et al., 2012) from low to higher altitudes. The reception of such organic debris by rocks may be influenced by

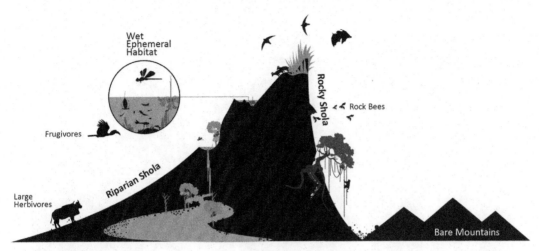

FIGURE 10.2 Fate of Succession in different terrain harboring varied biogeochemical: the succession of rock and riparian sholas.

physicochemical properties of the rock such as texture (cracks and crevices), slope, wind velocity, and moisture. In anthropocentric perspectives, these minute materials may not be given much weight. Pollen, in particular, is regarded as one of the most toxic components of the ecosystem (Knox and Suphioglu, 1996; Taylor et al., 2002; Glushakova et al., 2015; Visez et al., 2015; Aerts et al., 2021). However, in nature, each and every tiny material has its own direct and indirect positive effect on the ecosystem functioning.

Man understands nature through ubiquitous metaphors, a unique way of imagining human experience. For centuries, the metaphor of the leaf has influenced literary writing. From an artistic standpoint, green leaves represent growth, hope, life, and health, whereas fallen leaves represent not only loss and death, but also the sign of regeneration. For instance, in the visionary poem "Ode to the West wind" the poet portrays in the first stanza the momentum of fallen leaves by the blow of the wind. Similarly, he addresses the wind to "drive my thoughts over the universe/Like wither'd leaves to quicken new birth" and thereby to awaken the "unawaken'd earth." The scattered leaves merge with organic debris, and the decayed matters are sure to thrive again during the spring season, "If winter comes, can spring be far behind?" (Shelley and Gielgud, 1951). Even hair can become a canopy.

Various factors such as velocity, humidity, wind direction, and the shape and size of the debris will influence the uphill movement of organic debris through wind. As a result, the rock on one side that receives the most wind will wither, and the material carried away by the wind will be deposited elsewhere, preferably on the other side that receives less wind action—the low wind zone (LWZ). Except in a few tropical belt regions where there are two monsoons, the wind blows all year around the rock.

Physical components such as wind and light, as sources of energy and external pressure, play an important role in ecological succession, particularly in rock ecosystems (Bazzaz, 1979; Muniz-Castro et al., 2012; Hao et al., 2016). Wind's impact on rocks will vary in accordance with its physicochemical nature. Furthermore, the wind will facilitate ecological succession by bringing monsoon and minerals, but it will also cause

FIGURE 10.3 Observations from the rocks of Southern Western Ghats (except A) depict soil formation and succession. (A) Debris of a kid including hair flow from the roof of the house along with rainwater accelerated succession on a compound wall. (B) A Lithophytic *Ficus* sp. facilitate harvest of organic debris and water leads to soil formation and succession. (C) Primary succession on rock. (D) *Indirana* sp. tadpoles attached to the wet rock surface. (E) Bladderwort, a carnivorous plant of *Utricularia* genus flourishing in rock pools of forest streams. (F) Organic debris and soil formation.

desertification by eroding nutrients and moisture. Wind erosion is a well-known phenomenon and one of the major natural forces that contribute to desertification (Sterk and Stoorvogel, 2020). With respect to the SWG, it is predicted that the LWZ will be facilitating better ecological succession over the moderate and high wind zones (MWZ & HWZ). In the case of HWZ, the successions of pioneers like seasonal algae, lichens, and other microbes may occur. Drought-resistant plants such as scrubs and thorny bushes may be established in MWZ. LWZ and HLZ (High Light Zone) during winter, Water Free Zones (WFZ) and Moderate Light Zone (MLZ) during monsoon, and Low Light Zone (LLZ) and MWZ during summer will attract the uphill movement of mammals especially ungulates like deer, goats, and sheep. Pollard et al. (2002) reported that mountain goats in British Columbia's North Coast Forest prefer south-facing slopes during the winter season. This movement makes it easier to transport more organic debris uphill to areas where wind cannot transport such resources. Regardless of individual or group distribution, the giant honey bee, *Apis dorsata*, built nests on rock cliffs in south Karnataka, India, with their nests oriented east-west. This preference for direction is thought to be related to nests being exposed to moderate sunlight and low wind pressure, which facilitate normal colony activities, as well as the earth's magnetic field (Nagaraja and Yathisha, 2015).

10.3.1.2 *Animal dependent uphill movement of organic debris*

Organic debris moving uphill is critical in the ecological succession of bare rocks, and it is always an energy-intensive process (Wall et al., 2006; Venable et al., 2008). Mammals such as ungulates were forced to move upward due to biotic and abiotic forces like predator (Anderson et al., 2010; Wang et al., 2009; Schweiger et al., 2015) and ectoparasite pressure, wind, light, temperature, and moisture (Talbot, 1960; Hewson and Wilson, 1979; Rubenson et al., 2006; Boutros et al., 2007; Slagsvold et al., 2014; Baruzzi et al., 2017; Storz et al., 2020). Shepherd boy and King David says: "The high hills are for the wild goats; The cliffs are a refuge for the rock Badgers" (Ps. 104:18) (New King James Version Bible, 1994). Other than ungulates or herbivorous mammals, predators also move uphill in search of a hiding place to keep their young ones safe and also to get rid of ectoparasites prevailing in canopy-covered areas (Talbot, 1960; Hewson and Wilson, 1979; Boutros et al., 2007). The prophet states the significance of shelter and protection: "a hiding place from the wind, and a cover from the tempest,... rivers of waters in a dry place,... the shadow of a great rock in a weary land" (Isa. 32:2) (New King James Version Bible, 1994). Other than shadow, shelter, and hiding places available in the bare rocks or hilltops (Fig. 10.1), it also serve as isolated places which are un-inhabited and un-frequented by humans (Talbot, 1960) attract diverse fauna in varied seasons. At the uphill (bare rocks) the behavior of the mammals will be varying according to season and time of the day (Hewson and Hinge, 1990; Guan et al., 2013).

Along with excretion and secretion, organic debris carried on the external surfaces will be unloaded, either in the hiding space or at rest. When the mammal arrives at its resting place, the body surface is dried off if it was wet, shakes off its dust, and rubs along the rock to rid itself of ectoparasites. Goats prefer to rest on an elevated stone and scrape the ground with their forefeet before lying down (Ekesbo and Gunnarsson, 2018). Hairs become entangled in the rock or are shed down with the dust during this process. These mammals unintentionally transport various plant propagules (Janzen, 1984) in their various body parts (epizoochory) (Poinar and Columbus, 1992; Heinken et al., 2001; Heinken and Raudnitschka, 2002; Couvreur et al., 2004; Burgin and Renshaw, 2008; Couvreur et al., 2008; Kulbaba et al., 2009; Petersen and Bruun, 2019). It is well known that the hair on mammals' general body surfaces carries seeds of grasses and herbs, as well as spores of lower plants. It may also carry diatom, algae, and plankton spores that become attached in-between their hooves while consuming water or crossing wet and damp patches, puddles, pools, and streams along the way (Norton, 1992; Schulze et al., 2014; Petersen and Bruun, 2019). Sheep trampling, according to Eichberg and Donath (2018), strongly promotes seedling emergence in grazed open sand ecosystems by pressing seeds into the uppermost soil layer and improving microclimatic conditions.

While resting, the ruminant mammals used to ruminate (Anderson et al., 2010), spit out and regurgitate the seeds and unwanted materials (Corlett and Lucas, 1990; Corlett, 1998; Yumoto et al., 1998; Gill and Beardall, 2001). The proportion of roughages, hair, seeds, and spores in feces (such as pellets, poops, scats, and dung) varies according to season, habitat, and species habit. Cursorial mammals, which arch their backs and bend their hind legs while urinating or defecating, prefer rocky or plain substrates free of water and foliage (Coblentz, 1976). Frugivores like civets prefer to defecate in the hollows and depressions of rocks (Bearder and Randall, 1978; Ewer and Wemmer, 1974). Aside from cursorial mammals, aerial vertebrates and invertebrates aid in the movement of organic debris uphill. Nuptial flight of colonial insects, seasonal swarms, and insect migration attract predatory birds as well as non-flying predators such as arthropods, amphibians such as *Indirana* sp. (Fig. 10.3D), and reptiles such as rock lizards, geckos, and snakes (Downes and Shine, 1998; Encalada and Peckarsky, 2007; Carter et al., 2012). Rock bees used to build their combs in rock cliffs during the spring or pre-monsoon seasons, which attracted more predators, including insectivorous birds like swallows and swifts, to those habitats and contributed their excretions (Brown, 1988; Brown et al., 2008). Along with these insectivorous birds, large raptors like owls, eagles, and vultures too build their nest in the cliffs of rocks (Brown and Cooper, 1987; Beardsell et al., 2017). To feed their ever-growing young, they must bring more prey from lower-altitude locations. As a result, a load of organic debris such as excretions, slough, carcasses, nest materials, left portions, and regurgitations such as skeletons, elytra, wings, mandibles, scales, feathers, and hair will be moved upward.

In rock bee colony the curtain-forming bees are accustomed to excreting at regular intervals (Fig. 10.2). It is a common behavior of bees to excrete to reduce body weight while preparing for flight such as forage, swarm, and nuptial (Marshall, 1986). Although honey bee excretion is minimal, the large number of individuals compensates for this shortfall. Honey bee excretions are sticky and adhere strongly to rock surfaces. It also acts as a glue and promotes the accumulation of mineral and organic debris. Bats also play an important role in fertilizing rocks with their excretions and dispersing seeds where other animals cannot. In the spring, mother bats establish nursery colonies in caves, dead trees, and rock crevices (Tuttle and Moreno, 2005).

During the summer, the debris piles up, causing the cracks and crevices to fill up. Water runs through these crevices during the rainy season, soaking organic debris causing microbial decomposition. Detritivores, preferably pterygote insects, would be drawn to this. Organic soup rolling down the crevices will be dispersed to a different location. While dried pellets along with hair may choke the runoff, facilitating the formation of ephemeral habitat

(Fig. 10.2). This ephemeral habitat attracts ephemeral insects which initiate complex ecological processes like succession.

10.3.2 Stage II: Primary succession

The substrate's nutrient content and life-supporting surroundings encourage succession. Succession will vary depending on abiotic components such as soil, air, water, pH, temperature, and humidity. Carbon and nitrogen are essential components of plant and soil organic debris and play critical roles in soil formation, plant growth, and ecological succession (Sharrow and Ismail, 2004; Li et al., 2021). The ability to establish by overcoming all limiting factors and the ability to utilize the available limited resources are the appropriate factors that enable an autotroph to become a primary colonizer or successor. Pioneer plant species in any ecological succession should be able to survive in environments with limited abiotic resources and high physical fluctuation (Dagon and Schrag, 2019). The dominant primary successor of tropical climate will be in the following descending orders, Protista, Bryophytes, Pteridophytes, and Angiosperms.

The fate of organic soup and solids spilled by rain will be influenced by the terrain of rocks. The flow of these organic soups and solids may be comparatively slower in sloppy terrains through which ungulate mammals used to move. Cracks and crevices that receive more organic solids and low water holding capacity facilitate various types of ecological succession. Their fate will be different on the other side of the vertical slope, such as cliffs, where the cursorial animals cannot reach. Because the vertical cliffs are mostly inhabited by animals with wings, such as rock bees, birds, and bats, the components of their excretions will differ from those of other cursorial mammals.

10.3.3 Wet ephemeral habitat

The resistance provided by organic debris such as hair, leaves, petals, and pellets causes the unsteady and non-uniform flow of organic soup along the horizontal slope. As a result of this phenomenon, the rock softens, causing rock erosion, which promotes the formation of ephemeral habitat (Fig. 10.2). Zhang (2020) states that water sorption is one of the most significant causes of rock deformation. Because the organic debris contains animal excretions, different organisms' gut microbes will be inoculated into the ephemeral aquatic habitat alongside the organic soup. This causes decomposition, which starts the process of ecological succession all over again. Ephemeral pterygote insects will colonize these habitats. The diversity and abundance of aquatic invertebrates, including insects, are influenced by the complex habitat with high organic debris (Egglishaw, 1964; Rabeni and Minshall, 1977; Dudgeon and Wu, 1999; Subramanian and Sivaramakrishnan, 2005; Allan et al., 2020). Pterygote detritivorous insects are attracted to soaked debris lodged along the borders of these habitats. Carnivorous plants can be established in areas with less soil and nitrogen (Fig. 10.3E). Lack of soil nitrogen will be compensated by feeding a carnivorous diet (Friday, 1989; Friday and Quarmby, 1994; Rajasekar and Rajendran, 2018). These carnivorous plants, along with ephemeral Protista and Invertebrates, play an important role in the formation of streams and riparian forests, including Sholas. Waterless cracks and crevices received more organic solids, attracting a variety of detritivorous pterygote insects.

10.3.4 Organic debris and succession

The amount and quality of organic debris in the soil will affect the micronutrients and macronutrients. The diversity of the source may have a significant impact on the quality of the organic debris. If the source contains a diverse range of floral and faunal resources, organic debris may aid in the diversity of decomposers and the succession of various plant species. Organic debris can increase soil humidity and reduce temperature swings (Loydi et al., 2014). Physicochemical and mechanical properties of the organic debris affect the viability, germination, and growth of the seeds (Facelli and Pickett, 1991; Loydi et al., 2013). Variation in organic debris can have selective pressure on seed germination and the establishment of seedlings (Donath and Eckstein, 2008; Möhler et al., 2021). Thus, organic debris plays an unparalleled role in ecological succession.

10.3.5 Decomposition of organic debris and soil formation

Decomposition of organic debris may be affected by various biotic and abiotic factors such as altitude, steepness, temperature, moisture, solar radiation, litter quality, source of debris, diversity, and density of detritivores and decomposers (Austin and Vivanco, 2006; King et al., 2012; Lee et al., 2014; Bornman et al., 2015; Petraglia et al., 2019; Santonja et al., 2018). Hu et al. (2021) discovered that on the shaded aspect, the decomposition rate decreased with increasing slope steepness, whereas on the sunny aspect, the decomposition rate decreased with decreasing slope steepness. According to Pérez et al. (2021), litter decomposition is an ecological process critical to the functioning of forest headwater streams, with implications for the global carbon cycle. According to Hu et al. (2021), litter decomposition is critical for regulating the availability of soil nutrients and thus accelerates the productivity of terrestrial ecosystems. Culliney (2013) claims that, "the faeces of arthropods are the basis for the formation of soil aggregates and humus, which physically stabilize the soil and increase its capacity to store nutrients."

Microbes are instrumental in the mineralization of organic debris, a crucial step in soil formation (Manzoni et al., 2008; Culliney, 2013; Fanin and Bertrand, 2016; Marklein et al., 2016). Organic soup and solids may contain spores of various types. However, the weather conditions that persist in the bare rock environment may not support independent microbe succession. Detritivorous invertebrates, particularly pterygote insects, can either survive extreme weather conditions or exploit optimal weather conditions such as humidity, moisture, and temperature. The guts of detritivores provide a suitable environment for a diverse range of cultivable and non-cultivable microbes, and thus these detritivores are indirectly involved in mineralization. Invertebrate detritivores, particularly arthropods, can be thought of as mobile microbial incubators or bioreactors because of this dual function. If organic waste is not decomposed in a timely manner, it may undergo photo degradation, resulting in carbon emissions and global warming. According to Austin and Vivanco (2006), photo degradation is a process that uses solar radiation to break down organic debris and directly releases CO_2, causing the carbon to be discharged directly into the atmosphere rather than entering the pool of soil organic debris. These detritivorous organisms reduce the particle size of organic debris and inoculate more microbial spores, which induce mineralization and soil formation.

At this point, the million-dollar question is who can be considered the primary successor of bare rocks. Is it the animals that aid in the movement of organic debris uphill, the organisms involved in decomposition, or the plants (Protista, Bryophytes, or carnivorous angiosperms)? Pre-primary successors, pre-successors, predecessors, foundation species, or ecosystem engineers are organisms that contribute to the uphill movement of organic debris and decomposers, including detritivores and saprotrophs (Jones et al., 1997; Bruno and Bertness, 2001; Thomsen et al., 2010). These predecessors (here after facilitator) positively affect the primary colonizers via amelioration of the habitat and by providing limited resources (Altieri et al., 2007; Thomsen et al., 2010). Walker and Del Moral (2003) shows that, the primary succession starts when plants, animals and microbes colonize new surfaces.

10.3.6 Stage III, IV, and V: Secondary, tertiary and quaternary succession: the cascade effect

The colonizer is rarely found in climax communities, but it is usually replaced by further colonizers as a result of the primary colonizer's direct and indirect effects. The habitat cascade is caused by the indirect positive effect of the facilitator on the colonizer, which is mediated by the successive formation or modification of biogenic habitats. When ecological succession progresses, this simple cascade effect transforms into a multifocal and complex cascade (Hunter and Price, 1992; Pace et al., 1999; Altieri et al., 2007; Thomsen et al., 2010; Angelini and Silliman, 2014; Angelini et al., 2015; Thomsen et al., 2018; Zhang and Silliman, 2019; Temmink et al., 2021). Autotrophs' habitat-specific habits will determine the success of ecological succession. Primary succession increases organic debris through primary production, which promotes the uphill movement of consumer-like macroinvertebrates, primarily arthropods. Producer density and diversity influence productivity, which in turn influences consumer diversity. Production, consumption, and decomposition are the cyclical processes that allows for more soil formation and ecological succession. The larger mammals, which acted as a facilitator at first, now monitor the complex cascade effect by transporting propagules of various biotic components. It also directly contributes to ecological succession and soil formation by acting as a herbivore or predator.

10.3.7 Pollination, seed dispersal and succession

The success of sexually reproducing plants is inextricably linked to pollination, seed development, and dispersal. A suitable and successful plant avoids excessive specialization in pollination and seed dispersal. For example, the corolla must be open and suitable for anemophily or generalized entomophily. Rearing, supporting, and sustaining their own pollinators and seed dispersal agents helps plants that rely on specific zoochory and zoophily to thrive. It is unusual to find such a phenomenon in nature, where a single plant or group of plants of the same genus facilitates the survival of its pollinators and seed dispersing agents. Interestingly the figs of the genus *Ficus* have this dual adaptation, notably the keystone mutualism (pollination syndrome) between figs and their obligate wasp pollinator (Agaonidae, Chalcidoidea) (Herre et al., 2008; Gallery, 2014). Furthermore, fig plants have an association with ants, which helps to reduce abortion of developing figs, herbivory of figs, and parasitic wasp loads, resulting in increased pollination and seed production (Jander, 2015). Fig trees sustain and shape tropical frugivores by providing a consistent year-round nutritious food source when other food resources are scarce, earning the figs keystone status. Fig frugivores provide the most effective long-distance seed dispersal of any tropical tree species (Gallery, 2014). The seeds that are dispersed to a new and suitable habitat, with less competition, herbivory, disturbances and availability resources, can be colonized successfully (Terborgh et al., 2008; Wohllebben, 2016). Zoochorous seed dispersal is observed in the common trees of the typical tropical rain forest of SWG (Tables 10.1–10.6).

Ficus bengalensis, the Banyan tree is the most preferred roosting site for the Indian Flying Foxes *Pteropus medius*, especially in urban ecosystem. Figs also share resources with epiphytes, parasitic plants, and communities of frugivorous and folivorous vertebrates and invertebrates, resulting in the establishment of a complex food chain and food web. The hydraulic properties and physiological adaptations of figs' epiphytic, parasitic, lithophytic, and xerophytic adaptations show significant variation, including some of the highest photosynthetic rates ever recorded in nature (Patino et al., 1994; Zotz et al., 1995; Kalko et al., 1996; Shanahan et al., 2001; Harrison, 2005; Bercu and Popoviciu, 2014). These adaptations enable the fig plants as the most suitable pioneer as well as climax species of rock ecosystem.

Normally, frugivorous birds excrete before taking flight, with the majority of excretion concentrated in the roosting area. Because arboreal birds require a branch or twig to perch, they can only excrete in places where there is some canopy (Reid, 1989; Lu et al., 2008), making the birds useless for seed dispersal over bare rocks. However, because bats prefer to excrete while flying, they help the ecosystem by dispersing seeds where canopy covers are not available. Because the *P. medius* extracts only the juice from the fruits, it drops the seeds and thus contributes to ecological balance through seed dispersal. However, in the case of fig fruits, the parental investment in seeds is so low that the smaller seeds can be swallowed by these bats along with the juice. The smaller size of the seed, as well as the stickiness it develops when it passes through the gut of frugivores, allows the seeds to become attached even in sloppy rocks, and it can be washed down to suitable cracks and crevices during the rainy season, where it can colonize (Janzen, 1979; Milton et al., 1982; Slocum and Horvitz, 2000; Bashan et al., 2002; Sitaramam et al., 2009). Unfortunately, *P. medius* is migrating to urban ecosystems due to the appeal of a habitat with less competition, predator pressure, and the availability of high sugar fruits in home gardens all year. Instead of dispersing and fertilizing fig seeds on rocks, it disperses and fertilizes urban constructions in vain.

10.3.8 Palatability, herbivory, and succession

Herbivory is influenced by palatability, and palatability influences ecological succession. Ground-level palatable herbs, such as grasses, can withstand herbivory to a degree, but palatable seedlings of shrubs and trees attract heavy herbivory. "Light means sugar, which makes the young trees attractive to browsers" (Wohlleben, 2016). Unpalatable seedlings of woody herbs, shrubs, climbers, and trees outcompete palatable species for survival.

TABLE 10.1 Common flora of the Sholas of Southern Western Ghats (Sinha and Davidar, 1992; Ganesh and Davidar, 1999; Parthasarathy, 1999; Ayyappan and Parthasarathy, 1999; Sundarapandian and Swamy, 2000; Devy and Davidar, 2003; Ganesan and Davidar, 2003; Puyravaud et al., 2003; Raj, 2015). Upper story—species similar to evergreen forest (Height: 10−25 m).

Sl. No	Botanical name	Family	Ecological significance	Dispersal mode	Distribution	IUCN status
1	*Cinnamomum wightii* Meisn.	Lauraceae	Fruit	Zoochory	Southern Western Ghats	Endangered
2	*Elaeocarpus munronii* (Wl.) Masters	Elaeocarpaceae	Fruit	Zoochory	Southern Western Ghats	Lower risk/ near threatened
3	*Ficus amplocarpa* E. Govindarajalu & P. Masilamoney	Moraceae	Fruit	Zoochory	Western Ghats	Not evaluated
4	*Nothapodytes nimmoniana* (J. Grah.) D.J. Mabberley	Icacinaceae	Fruit	Zoochory	Indo-Malesia and China	Vulnerable
5	*Persea macrantha* (Nees) Kosterm.	Lauraceae	Fruit	Zoochory	Peninsular India and Sri Lanka	Vulnerable
6	*Rapanea thwaitesii* Mez	Primulaceae	Fruit	Zoochory	Western Ghats	Endangered
7	*Schefflera bourdillonii* Gamble	Araliaceae	Fruit	Zoochory	Southern Western Ghats	Endangered
8	*Symplocos pendula* Wight	Symplocaceae	Fruit	Zoochory	Indo-Sri Lanka to south east Asia	Least concern
9	*Syzygium densiflorum* Wall ex. Wight & Arn.	Myrtaceae	Fruit	Zoochory	Southern Western Ghats	Vulnerable
10	*Syzygium ramavarma* (Bourd.) Chitra	Myrtaceae	Fruit	Zoochory	Southern Western Ghats	Vulnerable

TABLE 10.2 Second story—typical species of sholas (Height: 5−10 m).

Sl. No	Botanical name	Family	Ecological significance	Dispersal mode	Distribution	IUCN status
1	*Ilex wightiana* Wall. ex Wight	Aquifoliaceae	Fruit	Zoochory	Peninsular India and Sri Lanka	Not evaluated
2	*Microtropis wallichiana* Wight ex Thw.	Celastraceae	Fruit	Zoochory	Indo-Malesia	Not evaluated
3	*Rapanea wightiana* (Wall. ex DC.) Mez.	Primulaceae	Fruit	Zoochory	Peninsular India and Sri Lanka	Not evaluated
4	*Symplocos cochinchinensis* (Lour.) S. Moore	Symplocaceae	Fruit	Zoochory	Indo-Malesia and China	Not evaluated
5	*Ternstroemia gymnanthera* (Wight & Arn.) Bedd.	Pentaphylacaceae	Fruit	Zoochory	Indo-Malesia and China	Not evaluated

3. Soil physicochemical parameters

TABLE 10.3   Third story—shrubby layers.

Sl. No	Botanical name	Family	Ecological significance	Dispersal mode	Distribution	IUCN status
1	*Anaphalis beddomei* Hook.f.	Asteraceae	Nectar	Autochory	Southern Western Ghats	Vulnerable
2	*Eurya* nitida Korth	Pentaphlacaceae	Fruit	Zoochory	Indo-Malesia and China	Not evaluated
3	*Gaultheria fragrantissima* Wall.	Ericaceae	Nectar	Autochory	Indo-Malesia	Not evaluated
4	*Ixora* lawsonii Gamble	Rubiaceae	Fruit	Zoochory	Peninsular India	Endangered
5	*Lasianthus* acuminatus Wight	Rubiaceae	Fruit	Zoochory	Southern Western Ghats	Not evaluated
6	*Pavetta breviflora DC.*	Rubiaceae	Fruit	Zoochory	Western Ghats	Not evaluated
7	*Polygala arillata* Buch.-Ham. ex D. Don	Polygalaceae	Fruit	Zoochory	Indo-Malesia and China	Not evaluated
8	*Psychotria truncata Wall*	Rubiaceae	Fruit	Zoochory	Western Ghats	Not evaluated
9	*Strobilanthes dupenii* Bedd. ex C.B. Clarke	Acanthaceae	Nectar	Autochory	Southern Western Ghats	Not evaluated
10	*Viburnum erubescens* Wall.	Adoxaceae	Fruit	Zoochory	India and Sri Lanka	Not evaluated

TABLE 10.4   Shola margins—ecotone species.

Sl. No	Botanical name	Family	Ecological significance	Dispersal mode	Distribution	IUCN status
1	*Berberis tinctoria Leschen.*	Berberidaceae	Fruit	Zoochory	India	Not evaluated
2	*Debregeasia longifolia* (Burm. f.) Wedd.	Urticaceae	Fruit	Zoochory	Indo-Malesia	Not evaluated
3	*Osbeckia reticulata* Bedd.	Melastomataceae	Nectar	Autochory	Southern Western Ghats	Not evaluated
4	*Rubus ellipticus Smith*	Rosaceae	Fruit	Zoochory	Indo-Malesia	Not evaluated
5	*Strobilanthes ciliatus Wall.* ex Nees	Acanthaceae	Nectar	Autochory	Southern Western Ghats	Not evaluated
6	*Strobilanthes luridus* Wight	Acanthaceae	Nectar	Autochory	Southern Western Ghats	Not evaluated
7	*Tarenna asiatica* (L.) Kuntze ex K. Schum.	Rubiaceae	Fruit	Zoochory	South Asia	Not evaluated

3. Soil physicochemical parameters

TABLE 10.5 Shola margins—ecotone species—invade grasslands.

Sl. No	Name	Family	Ecological significance	Dispersal mode	Distribution	IUCN status
1	*Ligustrum perrottetii* A.DC.	Oleaceae	Fruit	Zoochory	Western Ghats	Not evaluated
2	*Mahonia leschenaultii* (Wallich ex Wight & Arnott) Takeda	Berberidaceae	Fruit	Zoochory	India	Not evaluated
3	*Rhodomyrtus tomentosa* (Aiton) Hassk.	Myrtaceae	Fruit	Zoochory	Indo-Malesia and China	Not evaluated
4	*Symplocos anamallayana* Bedd	Symplocaceae	Fruit	Zoochory	Southern Western Ghats	Endangered
5	*Vaccinium leschenaultii* Wight	Ericaceae	Fruit	Zoochory	India and Sri Lanka	Not evaluated

TABLE 10.6 Common flora of Rock Shola.

Sl. No	Name	Family	Ecological significance	Dispersal mode	Distribution	IUCN status
1	*Ardisia rhomboidea* Wight	Primulaceae	Fruit	Zoochory	Southern Western Ghats	Not evaluated
2	*Beilschmiedia wightii* (Nees) Benth. ex Hook.	Lauraceae	Fruit	Zoochory	Southern Western Ghats	Endangered
3	*Daphniphyllum neilgherrense* (Wight) K. Rosenthal	Daphniphyllaceae	Fruit	Zoochory	Indo-Malesia	Not evaluated
4	*Ficus dalhousiae* (Miq.) Miq.	Moraceae	Fruit	Zoochory	India	Not evaluated
5	*Ficus tinctoria* G.Forst.	Moraceae	Fruit	Zoochory	Asia	Least concern
6	*Ficus tsjakela* Burm. f.	Moraceae	Fruit	Zoochory	Indo-China and Sri Lanka	Not evaluated
7	*Rhododendron arboreum* J.E. Smith *subsp. nilagirium* (Zenk.) Tagg.	Ericaceae	Nectar	Autochory	Southern Western Ghats	Vulnerable
8	*Schefflera capitata* (Wight & Arn.) Harms	Araliaceae	Fruit	Zoochory	Western Ghats	Not evaluated
9	*Schefflera racemosa* (Wight) Harms	Araliaceae	Fruit	Zoochory	Western Ghats	Not evaluated
10	*Schefflera rostrata* (Wt.) Harms	Araliaceae	Fruit	Zoochory	Western Ghats	Not evaluated

3. Soil physicochemical parameters

Common tree species found in SWG sholas, such as *Syzygium* and *Cinnamomum*, have unpalatable leaves (Tables 10.1). Plant antiherbivore defenses and plant palatability variability, according to Ripple et al. (2016) are sufficient to overcome herbivory in terrestrial ecosystems. Due to the inaccessibility of large mammalian cursorial herbivores, lithophytic and epiphytic adaptations help highly palatable fig species survive herbivory (Fig. 10.2). Despite the fact that the *Ficus* seed is dispersed throughout the vicinity of bare rock, it may not germinate successfully if disturbances continue. A less disturbed area is ideal for proper germination and growth. Despite the fact that it germinates on slow slopes, ungulates, including wild goats, prefer to feed on it due to its palatability. As a result, lithophytic figs can only be established on steep slopes where mammalian herbivory is limited.

A rock may react differently depending on the external forces. Rocks resist when a hammer is struck on their shoulder, but they respond positively when a tender baby root is struck softly on their heart, much like a father does to a young lad. Rocks are a well-balanced combination of placental and marsupial mother: it carries and nourishes before birth (germination), and it develops a placental relationship after birth, treats them as babies, and carries them throughout its life. The canopy of lithophytic figs of the genus *Ficus* dominated the SWG rock sholas (Moraceae) (Table 10.6). These figs can establish themselves in mild cracks and crevices during ecological succession. Through interactions that promote the accumulation of organic debris, including leaf litter, the roots of these figs can break and hold the rocks. More organic debris attracts more detritivores, resulting in more soil formation, which increases water holding capacity and promotes luxuriant growth, accelerating primary productivity. These figs' roots run downwards in search of potential water and soil resources. They may eventually discover water and soil sources. Arboreal mammals use these roots as bridges or tracks (Fig. 10.2). During the fruiting season, visits by arboreal mammals and birds attract more seeds of various species, including parasitic plants, lianas, and other fruit trees.

10.3.9 Prey, predation and succession

The uphill movement of pterygote insects, including detritivores, herbivores, pollinators, and swarms, mediates the movement of both macroinvertebrates and vertebrates. Amphibians, reptiles, and birds are among the vertebrate predators. The type of predator visit will vary depending on the habits and habitat of the prey community. The dry, rocky habitat inhabited by rock bees and other insects attracts reptiles such as rock agama lizards and birds such as swifts and swallows. During one of their field trips, the authors observed 10 Brahminy Kites *Haliastur indus* chasing rock agamas in a rocky habitat of Kanyakumari wildlife sanctuary. The density and diversity of these prey communities have an immediate impact on the density and diversity of predators. When the density and diversity of secondary consumers grows, it eventually invites higher consumers, including apex predators. This allows for more organic debris movement, which attracts more decomposers and speeds up soil formation and ecological succession. Predator habitat loss or the absence of large predators results in the dominance of large mammalian herbivores, which threatens the regeneration of highly palatable species. As a result, the presence of apex predators has a positive indirect effect on trophic cascades, whereas their absence has a negative indirect effect (Polis et al., 2000; Borer et al., 2005; Knight et al., 2006; Estes et al., 2010; Heath et al., 2014; Ripple et al., 2016; Beheshti et al., 2021). Different faunal species may make distinct contributions because they play distinct roles in ecosystems such as soil formation, seed dispersal, and ecological succession of various plant species, all of which have a direct impact on community composition and potentially forest carbon stocks.

The pollinator, herbivore, and predator discussed above do not have to be larger mammals; they can be small insects such as flies (Diptera), beetles (Coleoptera), and wasps (Hymenoptera). Ecologists and conservationists place a high value on apex predators such as cats and eagles, but smaller predators such as parasitoids are equally important. As a result of secondary, tertiary, and quaternary succession, bare rock that was once a resting and hiding place for mammals has become a feeding ground. The small ephemeral habitats were transformed into streams with riparian canopies. Waterless cracks and crevices can develop into rock sholas or typical wet evergreen forests.

10.4 Threats to soil and succession

The vital abiotic forces that influence soil formation and ecological succession are light, air, and water. It has a direct impact on the uphill movement of organic debris in two ways: through voluntary and involuntary natural actions. This organic debris is important in soil formation and ecological succession. The ever-growing list of extinction, critically endangered, and endangered biota leads us to believe in anthropogenic VI mass extinction (Leakey and Lewin, 1995; Barnosky et al., 2011; Payne et al., 2016; Ceballos et al., 2017; Ceballos et al., 2020; Palombo, 2021). Various anthropogenic activities pose serious threats to ecologically and economically important indigenous and endemic species. As a result, the cosmopolitan, invasive habitat generalist will supplant endemic and indigenous habitat specialists. It will have an impact on the mountain ecosystem's pedology and edaphology. Nilgiri Tahr *Nilgiritragus hylocrius*, the only mountain ungulate found in southern India (Predit et al., 2015), endemic to the SWG of Kerala and Tamil Nadu (Fox and Johnsingh, 1997) playing a crucial role in the uphill movement of organic debris is one of the critically endangered fauna of India. Anthropogenic activities including monoculture of exotic species, numerous hydroelectric projects affect the core tahr habitat, which comprises grasslands and sholas with rocky cliffs at elevations of around 300−2600 m above mean sea level (Schaller, 1977; Davidar, 1978). Fresh water ecosystems of mountains supporting high diversity and endemism are facing serious threats due to anthropogenic activities (Rogora et al., 2018; Farooq et al., 2021).

Soil, as a life-sustaining element, should be prioritized for conservation and sustainable use in nature-centered perspectives. Soil is an unavoidable component of ecological succession. Nature's deserted materials contribute to the formation of nutrient-rich soil. Diversity breeds diversity. Highly palatable trees native to SWG, such as jack fruit (*Artocarpus heterophyllus*), are becoming scarce. This demonstrates the dangers that the rock ecosystem faces. Because of its small and high quantity of seed production, *Artocarpus hirsutus*, which is endemic to SWG, has better seed dispersal ability. It is highly palatable during the seedling stage, especially in understories, and becomes unpalatable with very thick leaves when it reaches the canopy stage, but its contribution to primary productivity, humus production, and soil formation is irreplaceable.

Anthropogenic disturbances such as land-use changes have an impact on the diversity of faunal species, which has a direct impact on the success of habitat specialists. As a result, specialists will be replaced by generalists, affecting the diverse fauna that rely on specialists.

Because the biosphere is already facing habitat destruction and a decline in species composition, losing these organisms would be detrimental to both natural ecological succession (Angelini et al., 2011) and anthropogenic restoration efforts. Toxic constituents of particulate contaminants in the air can disrupt soil formation and ecological succession. The Arabian Sea, off India's west coast, is suffering from severe plastic pollution, including microplastics (Owens et al., 2022). Because the Western Ghats are so close to the West Coast, anthropogenic marine debris can have an impact on the rock habitat. Microplastics floating in the sea can be carried to higher altitude regions by tidal and wind action, affecting microbes as well as micro and macro flora and fauna. Native biodiversity conservation is critical for sustaining and promoting diversity (Wohlleben, 2016). We propose that the use of weedicides in the mountainous ecosystem be completely avoided. This should be monitored by the district green club/society in collaboration with state and central education and research institutes.

10.5 Conclusion

Diversity breeds diversity. The contribution of minute organic debris by various known and unknown organisms is unavoidable in soil pedology and edaphology. This observation can serve as a wake-up call to consider the importance of insignificant ecosystem components like hair in ecological succession and soil formation in bare rocks. All natural cascades are extremely complex to understand naturally and empirically, as it is an expensive, energy-intensive, and time-consuming process to mediate anthropogenically. To conserve the highly palatable habitat, the rock ecosystem, a high priority conservation strategy must be implemented. Because everything is interconnected, it is impossible to distinguish or prioritize specific groups of organisms such as pioneers, keystone species, and endemic species. However, it is critical to prioritize threatened species such as the Nilgiri Tahr, which promote soil formation and the ecological succession of rock ecosystems. Though studying this long-term multifocal spatiotemporal process is difficult, it is important to test these hypotheses experimentally for a better understanding of nature in nature-centered perspectives.

References

Aerts, R., Bruffaerts, N., Somers, B., Demoury, C., Plusquin, M., Nawrot, T.S., et al., 2021. Tree pollen allergy risks and changes across scenarios in urban green spaces in Brussels, Belgium. Landscape and Urban Planning 207, 104001.

Allan, J.D., Castillo, M.M., Capps, K.A., 2020. Stream Ecology: Structure and Function of Running Waters, third ed. Springer, The Netherlands.

Altieri, A.H., Silliman, B.R., Bertness, M.D., 2007. Hierarchical organization via a facilitation cascade in intertidal cord grass bed communities. The American Naturalist 169 (2), 195−206.

Anderson, T.M., Hopcraft, J.G.C., Eby, S., Ritchie, M., Grace, J.B., Olff, H., 2010. Landscape-scale analyses suggest both nutrient and antipredator advantages to Serengeti herbivore hotspots. Ecology 91 (5), 1519−1529.

Angelini, C., Altieri, A.H., Silliman, B.R., Bertness, M.D., 2011. Interactions among foundation species and their consequences for community organization, biodiversity, and conservation. Bioscience 61 (10), 782−789.

Angelini, C., Silliman, B.R., 2014. Secondary foundation species as drivers of trophic and functional diversity: evidence from a tree—epiphyte system. Ecology 95 (1), 185—196.

Angelini, C., van der Heide, T., Griffin, J.N., Morton, J.P., Derksen-Hooijberg, M., Lamers, L.P., et al., 2015. Foundation species' overlap enhances biodiversity and multifunctionality from the patch to landscape scale in southeastern United States salt marshes. Proceedings of the Royal Society B: Biological Sciences 282 (1811), 20150421.

Angelou, M., 2015. Maya Angelou: The Complete Poetry. Hachette UK.

Austin, A.T., Vivanco, L., 2006. Plant litter decomposition in a semi-arid ecosystem controlled by photodegradation. Nature 442 (7102), 555—558.

Avis, A.M., Lubke, R.A., 1985. The effect of wind-borne sand and salt spray on the growth of Scirpus nodosus in a mobile dune system. South African Journal of Botany 51 (2), 100—110.

Ayyappan, N., Parthasarathy, N., 1999. Biodiversity inventory of trees in a large-scale permanent plot of tropical evergreen forest at Varagalaiar, Anamalais, Western Ghats, India. Biodiversity & Conservation 8 (11), 1533—1554.

Bardgett, R., 2005. The Biology of Soil: A Community and Ecosystem Approach. Oxford University Press.

Barnosky, A.D., Matzke, N., Tomiya, S., Wogan, G.O., Swartz, B., Quental, T.B., et al., 2011. Has the Earth's sixth mass extinction already arrived? Nature 471 (7336), 51—57.

Baruzzi, C., Lovari, S., Fattorini, N., 2017. Catch me if you can: antipredatory behaviour of chamois to the wolf. Ethology Ecology & Evolution 29 (6), 589—598.

Bashan, Y., Li, C.Y., Lebsky, V.K., Moreno, M., De-Bashan, L.E., 2002. Primary colonization of volcanic rocks by plants in arid Baja California, Mexico. Plant Biology 4 (3), 392—402.

Bazzaz, F.A., 1979. The physiological ecology of plant succession. Annual Review of Ecology and Systematics 10 (1), 351—371.

Bearder, S., Randall, R., 1978. Comparative olfactory marking in the spotted hyaena and civets. Carnivore 1, 32—48.

Beardsell, A., Gauthier, G., Fortier, D., Therrien, J.F., Bêty, J., 2017. Vulnerability to geomorphological hazards of an Arctic cliff-nesting raptor, the rough-legged hawk. Arctic Science 3 (2), 203—219.

Beheshti, K.M., Wasson, K., Angelini, C., Silliman, B.R., Hughes, B.B., 2021. Long-term study reveals top-down effect of crabs on a California salt marsh. Ecosphere 12 (8), e03703.

Bercu, R., Popoviciu, D.R., 2014. Anatomical study of Ficus carica L. leaf. Annals of RSCB 19 (1), 33—36.

Berry, W., 2015. The Unsettling of America: Culture & Agriculture. Catapult.

Bonell, M., Purandara, B.K., Venkatesh, B., Krishnaswamy, J., Acharya, H.A.K., Singh, U.V., et al., 2010. The impact of forest use and reforestation on soil hydraulic conductivity in the Western Ghats of India: implications for surface and sub-surface hydrology. Journal of Hydrology 391 (1—2), 47—62.

Borer, E.T., Seabloom, E.W., Shurin, J.B., Anderson, K.E., Blanchette, C.A., Broitman, B., et al., 2005. What determines the strength of a trophic cascade? Ecology 86 (2), 528—537.

Bornman, J.F., Barnes, P.W., Robinson, S.A., Ballare, C.L., Flint, S.D., Caldwell, M.M., 2015. Solar ultraviolet radiation and ozone depletion-driven climate change: effects on terrestrial ecosystems. Photochemical & Photobiological Sciences 14 (1), 88—107.

Boutros, D., Breitenmoser-Würsten, C., Zimmermann, F., Ryser, A., Molinari-Jobin, A., Capt, S., et al., 2007. Characterisation of Eurasian lynx Lynx lynx den sites and kitten survival. Wildlife Biology 13 (4), 417—429.

Bressan, D., 2014. The Renaissance's contribution to geology: landscape painting. Scientific American Blog. https://blogs.scientificamerican.com/history-of-geology/the-renaissances-contribution-to-geology-landscape-painting/ (accessed 4 January 2022).

Brockerhoff, E.G., Jactel, H., Parrotta, J.A., Quine, C.P., Sayer, J., 2008. Plantation forests and biodiversity: oxymoron or opportunity? Biodiversity and Conservation 17 (5), 925—951.

Brown, C.R., 1988. Social foraging in cliff swallows: local enhancement, risk sensitivity, competition and the avoidance of predators. Animal Behaviour 36 (3), 780—792.

Brown, C.J., Cooper, T.G., 1987. The status of cliff-nesting raptors on the Waterberg, SWA/Namibia. Madoqua 1987 (3), 243—249.

Brown, C.R., Brown, M.B., Brazeal, K.R., 2008. Familiarity with breeding habitat improves daily survival in colonial cliff swallows. Animal Behaviour 76 (4), 1201—1210.

Bruno, J.F., Bertness, M.D., 2001. Habitat modification and facilitation in benthic marine communities. In: Bertness, M.D., Gaines, S.D., Hay, M.E. (Eds.), Marine Community Ecology. Sinauer Associates, Inc, Sunderland, MA, pp. 201—218.

Bruno, J.F., Stachowicz, J.J., Bertness, M.D., 2003. Inclusion of facilitation into ecological theory. Trends in Ecology & Evolution 18 (3), 119–125.

Burgin, S., Renshaw, A., 2008. Epizoochory, algae and the Australian eastern long-necked turtle Chelodina longicollis (Shaw). The American Midland Naturalist 160 (1), 61–68.

Capra, G.F., Ganga, A., Moore, A.F., 2017. Songs for our soils. How soil themes have been represented in popular song. Soil Science and Plant Nutrition 63 (5), 517–525.

Carpenter, S.R., Cole, J.J., Kitchell, J.F., Pace, M.L., 2010. Trophic cascades in lakes: lessons and prospects. Trophic Cascades: Predators, Prey and the Changing Dynamics of Nature. Island Press, pp. 55–69.

Carter, A., Goldizen, A., Heinsohn, R., 2012. Personality and plasticity: temporal behavioural reaction norms in a lizard, the Namibian rock agama. Animal Behaviour 84 (2), 471–477.

Ceballos, G., Ehrlich, P.R., Dirzo, R., 2017. Biological annihilation via the ongoing sixth mass extinction signaled by vertebrate population losses and declines. Proceedings of the National Academy of Sciences 114 (30), E6089–E6096.

Ceballos, G., Ehrlich, P.R., Raven, P.H., 2020. Vertebrates on the brink as indicators of biological annihilation and the sixth mass extinction. Proceedings of the National Academy of Sciences 117 (24), 13596–13602.

Chang, L., He, Y., Yang, T., Du, J., Niu, H., Pu, T., 2014. Analysis of herbaceous plant succession and dispersal mechanisms in deglaciated terrain on Mt. Yulong, China. The Scientific World Journal 2014.

Clements, F.E., 1936. Nature and structure of the climax. Journal of Ecology 24 (1), 252–284.

Coblentz, B.E., 1976. Functions of scent-urination in ungulates with special reference to feral goats (Capra hircus L.). The American Naturalist 110 (974), 549–557.

Coleman, D.C., Callaham, M., Crossley Jr, D.A., 2017. Fundamentals of Soil Ecology. Academic press.

Corlett, R.T., 1998. Frugivory and seed dispersal by vertebrates in the oriental (Indomalayan) region. Biological Reviews 73 (4), 413–448.

Corlett, R.T., Lucas, P.W., 1990. Alternative seed-handling strategies in primates: seed-spitting by long-tailed macaques (Macaca fascicularis). Oecologia 82 (2), 166–171.

Couvreur, M., Vandenberghe, B., Verheyen, K., Hermy, M., 2004. An experimental assessment of seed adhesivity on animal furs. Seed Science Research 14 (2), 147–159.

Couvreur, M., Verheyen, K., Vellend, M., Lamoot, I., Cosyns, E., Hoffmann, M., et al., 2008. Epizoochory by large herbivores: merging data with models. Basic and Applied Ecology 9 (3), 204–212.

Culliney, T.W., 2013. Role of arthropods in maintaining soil fertility. Agriculture 3 (4), 629–659.

Dagon, K., Schrag, D.P., 2019. Quantifying the effects of solar geoengineering on vegetation. Climatic Change 153 (1), 235–251.

Darwin, C., 1846. An account of the Fine Dust which often falls on Vessels in the Atlantic Ocean. Quarterly Journal of the Geological Society 2 (1–2), 26–30.

Davidar, E.R.C., 1978. Distribution and status of the Nilgiri Tahr (Hemitragus hylocrius) 1975–78. Journal of the Bombay Natural History Society 75, 815–844.

De Langre, E., 2008. Effects of wind on plants. Annual Review of Fluid Mechanics 40, 141–168.

Devy, M.S., Davidar, P., 2003. Pollination systems of trees in Kakachi, a mid-elevation wet evergreen forest in Western Ghats, India. American Journal of Botany 90 (4), 650–657.

Donath, T.W., Eckstein, R.L., 2008. Grass and oak litter exert different effects on seedling emergence of herbaceous perennials from grasslands and woodlands. Journal of Ecology 96 (2), 272–280.

Downes, S., Shine, R., 1998. Sedentary snakes and gullible geckos: predator–prey coevolution in nocturnal rock-dwelling reptiles. Animal Behaviour 55 (5), 1373–1385.

Dudgeon, D., Wu, K.K., 1999. Leaf litter in a tropical stream: food or substrate for macroinvertebrates? Archiv für Hydrobiologie 65–82.

Edwards, C.A., Pimentel, D., 1989. Impact of herbicides on soil ecosystems. Critical Reviews in Plant Sciences 8 (3), 221–257.

Egglishaw, H.J., 1964. The distributional relationship between the bottom fauna and plant detritus in streams. The Journal of Animal Ecology 463–476.

Eichberg, C., Donath, T.W., 2018. Sheep trampling on surface-lying seeds improves seedling recruitment in open sand ecosystems. Restoration Ecology 26, S211–S219.

Ekesbo, I., Gunnarsson, S., 2018. Farm Animal Behaviour: Characteristics for Assessment of Health and Welfare. CABI.

Encalada, A.C., Peckarsky, B.L., 2007. A comparative study of the costs of alternative mayfly oviposition behaviors. Behavioral Ecology and Sociobiology 61 (9), 1437–1448.

3. Soil physicochemical parameters

Ephraim, R., Murugesan, R., 2015. Opportunities for rural development in Musanze District, Africa: a rural livelihood analysis. International Journal of Business Management and Economic Research 6 (4), 231–248.

Estes, J.A., Peterson, C.H., Steneck, R.S., 2010. Some effects of apex predators in higher-latitude coastal oceans. Trophic Cascades: Predators, Prey, and the Changing Dynamics of Nature. Island Press, pp. 37–53.

Ewer, R.F., Wemmer, C., 1974. The behaviour in captivity of the African civet, Civettictus civetta (Schreber). Zeitschrift für Tierpsychologie.

Facelli, J.M., Pickett, S.T., 1991. Plant litter: its dynamics and effects on plant community structure. The Botanical Review 57 (1), 1–32.

Fanin, N., Bertrand, I., 2016. Aboveground litter quality is a better predictor than belowground microbial communities when estimating carbon mineralization along a land-use gradient. Soil Biology and Biochemistry 94, 48–60.

Farooq, M., Li, X., Tan, L., Fornacca, D., Li, Y., Cili, N., et al., 2021. Ephemeroptera (Mayflies) assemblages and environmental variation along three streams located in the dry-hot valleys of Baima Snow Mountain, Yunnan, Southwest China. Insects 12 (9), 775.

Favet, J., Lapanje, A., Giongo, A., Kennedy, S., Aung, Y.Y., Cattaneo, A., et al., 2013. Microbial hitchhikers on intercontinental dust: catching a lift in Chad. The ISME Journal 7 (4), 850–867.

Fox, J.L., Johnsingh, A.J.T., 1997. Status and Distribution of Caprinae by Region—India and the IUCN/SSC Caprinae Specialist Group. Wild Sheep and Goats and Their Relatives. Status Survey and Conservation Action Plan for Caprinae. IUCN, Gland, Switzerland and Cambridge, UK, p. 390.

Friday, L.E., 1989. Rapid turnover of traps in Utricularia vulgaris L. Oecologia 80, 272–277.

Friday, L.E., Quarmby, C., 1994. Uptake and translocation of prey derived N15 and P32 in Utricularia vulgaris L. New Phytology 126, 273–281.

Gallery, R.E., 2014. Ecology of tropical rain forests. Ecology and the Environment 1–22.

Ganesan, R., Davidar, P., 2003. Effect of logging on the structure and regeneration of important fruit bearing trees in a wet evergreen forest, southern Western Ghats, India. Journal of Tropical Forest Science 12–25.

Ganesh, T., Davidar, P., 1999. Fruit biomass and relative abundance of frugivores in a rain forest of southern Western Ghats, India. Journal of Tropical Ecology 15 (4), 399–413.

Gill, R.M.A., Beardall, V., 2001. The impact of deer on woodlands: the effects of browsing and seed dispersal on vegetation structure and composition. Forestry: An International Journal of Forest Research 74 (3), 209–218.

Glushakova, A.M., Kachalkin, A.V., Zheltikova, T.M., Chernov, I.Y., 2015. Yeasts associated with wind-pollinated plants—leading pollen allergens in Central Russia. Microbiology (Reading, England) 84 (5), 722–725.

Gómez, J.A., Romero, P., Giráldez, J.V., Fereres, E., 2004. Experimental assessment of runoff and soil erosion in an olive grove on a Vertic soil in southern Spain as affected by soil management. Soil Use and Management 20 (4), 426–431.

Griffin, D.W., Kellogg, C., Shinn, E., Gray, M., Garrison, G., 2003. Desert Storms and Their Ability to Move Microorganisms and Toxins Around the Globe. US Geological Survey, St. Petersburg, FL, pp. 25–28.

Gröger, A., 2000. Flora and vegetation of inselbergs of Venezuelan Guayana. Inselbergs. Springer, Berlin, Heidelberg, pp. 291–314.

Guan, T.P., Ge, B.M., McShea, W.J., Li, S., Song, Y.L., Stewart, C.M., 2013. Seasonal migration by a large forest ungulate: a study on takin (Budorcas taxicolor) in Sichuan Province, China. European Journal of Wildlife Research 59 (1), 81–91.

Hao, H.M., Lu, R., Liu, Y., Fang, N.F., Wu, G.L., Shi, Z.H., 2016. Effects of shrub patch size succession on plant diversity and soil water content in the water-wind erosion crisscross region on the Loess Plateau. Catena 144, 177–183.

Harrison, R.D., 2005. Figs and the diversity of tropical rainforests. Bioscience 55 (12), 1053–1064.

Heath, M.R., Speirs, D.C., Steele, J.H., 2014. Understanding patterns and processes in models of trophic cascades. Ecology Letters 17 (1), 101–114.

Heinken, T., Raudnitschka, D., 2002. Do wild ungulates contribute to the dispersal of vascular plants in central European forests by epizoochory? A case study in NE Germany. Forstwissenschaftliches Centralblatt Vereinigt Mit Tharandter Forstliches Jahrbuch 121 (4), 179–194.

Heinken, T., Lees, R., Raudnitschka, D., Runge, S., 2001. Epizoochorous dispersal of bryophyte stem fragments by roe deer (Capreolus capreolus) and wild boar (Sus scrofa). Journal of Bryology 23 (4), 293–300.

Herre, E.A., Jandér, K.C., Machado, C.A., 2008. Evolutionary ecology of figs and their associates: recent progress and outstanding puzzles. Annual Review of Ecology, Evolution, and Systematics 39, 439–458.

3. Soil physicochemical parameters

Hewson, R., Wilson, C.J., 1979. Home range and movements of Scottish Blackface sheep in Lochaber, north-west Scotland. Journal of Applied Ecology 743–751.

Hewson, R., Hinge, M.D.C., 1990. Characteristics of the home range of mountain hares Lepus timidus. Journal of Applied Ecology 651–666.

Hillel, D., 1992. Out of the Earth: Civilization and the Life of the Soil. University of California Press.

Hindy, K.T., Baghdady, A.R., Howari, F.M., Abdelmaksoud, A.S., 2018. A qualitative study of airborne minerals and associated organic compounds in Southeast of Cairo, Egypt. International Journal of Environmental Research and Public Health 15 (4), 568. Available from: https://doi.org/10.3390/ijerph15040568.

Hole, F.D., 1981. Effects of animals on soil. Geoderma 25 (1–2), 75–112.

Hole, F.D., 1994. The earth beneath our feet: explorations in community 11997 Soil Survey Horizons 38 (2), 40–53.

Hu, A., Duan, Y., Xu, L., Chang, S., Chen, X., Hou, F., 2021. Litter decomposes slowly on shaded steep slope and sunny gentle slope in a typical steppe ecoregion. Ecology and Evolution 11 (6), 2461–2470.

Hunter, M.D., Price, P.W., 1992. Playing chutes and ladders: heterogeneity and the relative roles of bottom-up and top-down forces in natural communities. Ecology 724–732.

Incagnone, G., Marrone, F., Barone, R., Robba, L., Naselli-Flores, L., 2015. How do freshwater organisms cross the "dry ocean"? A review on passive dispersal and colonization processes with a special focus on temporary ponds. Hydrobiologia 750 (1), 103–123.

Jander, K.C., 2015. Indirect mutualism: ants protect fig seeds and pollen dispersers from parasites. Ecological Entomology 40 (5), 500–510.

Janzen, D.H., 1979. How to be a fig. Annual Review of Ecology and Systematics 10 (1), 13–51.

Janzen, D.H., 1984. Dispersal of small seeds by big herbivores: foliage is the fruit. The American Naturalist 123 (3), 338–353.

Jenny, H., 2012. The Soil Resource: Origin and Behavior, vol. 37. Springer Science & Business Media.

Jones, C.G., Lawton, J.H., Shachak, M., 1997. Positive and negative effects of organisms as physical ecosystem engineers. Ecology 78 (7), 1946–1957.

Kalko, E.K., Herre, E.A., Handley Jr, C.O., 1996. Relation of fig fruit characteristics to fruit-eating bats in the New and Old World tropics. Journal of Biogeography 23 (4).

Karl, T., Potosnak, M., Guenther, A., Clark, D., Walker, J., Herrick, J.D., et al., 2004. Exchange processes of volatile organic compounds above a tropical rain forest: implications for modeling tropospheric chemistry above dense vegetation. Journal of Geophysical Research: Atmospheres 109 (D18).

Kellogg, C.A., Griffin, D.W., Garrison, V.H., Peak, K.K., Royall, N., Smith, R.R., et al., 2004. Characterization of aerosolized bacteria and fungi from desert dust events in Mali, West Africa. Aerobiologia 20 (2), 99–110.

King, J.Y., Brandt, L.A., Adair, E.C., 2012. Shedding light on plant litter decomposition: advances, implications and new directions in understanding the role of photodegradation. Biogeochemistry 111 (1), 57–81.

Knight, T.M., Chase, J.M., Hillebrand, H., Holt, R.D., 2006. Predation on mutualists can reduce the strength of trophic cascades. Ecology Letters 9 (11), 1173–1178.

Knox, B., Suphioglu, C., 1996. Environmental and molecular biology of pollen allergens. Trends in Plant Science 1 (5), 156–164.

Koren, I., Kaufman, Y.J., Washington, R., Todd, M.C., Rudich, Y., Martins, J.V., et al., 2006. The Bodélé depression: a single spot in the Sahara that provides most of the mineral dust to the Amazon forest. Environmental Research Letters 1 (1), 014005.

Kulbaba, M.W., Tardif, J.C., Staniforth, R.J., 2009. Morphological and ecological relationships between burrs and furs. The American Midland Naturalist 161 (2), 380–391.

Kuriakose, S.L., 2010. Physically Based Dynamic Modelling of the Effects of Land Use Changes on Shallow Landslide Initiation in the Western Ghats, Kerala, India. ITC.

Leakey, R.E., Lewin, R., 1995. The Sixth Extinction: Biodiversity and Its Survival. Phoenix Publisher, Phoenix, AZ, pp. 1–271.

Lee, H., Fitzgerald, J., Hewins, D.B., McCulley, R.L., Archer, S.R., Rahn, T., et al., 2014. Soil moisture and soil-litter mixing effects on surface litter decomposition: a controlled environment assessment. Soil Biology and Biochemistry 72, 123–132.

Leibold, M.A., Holyoak, M., Mouquet, N., Amarasekare, P., Chase, J.M., Hoopes, M.F., et al., 2004. The metacommunity concept: a framework for multi-scale community ecology. Ecology Letters 7 (7), 601–613.

Leopold, A., 1989. A Sand County Almanac, and Sketches Here and There. Oxford University Press, United States.

Li, J., Chen, Q., Li, Z., Peng, B., Zhang, J., Xing, X., et al., 2021. Distribution and altitudinal patterns of carbon and nitrogen storage in various forest ecosystems in the central Yunnan Plateau, China. Scientific Reports 11 (1), 1−11.

Lim, N., Munday, C.I., Allison, G.E., O'Loingsigh, T., De Deckker, P., Tapper, N.J., 2011. Microbiological and meteorological analysis of two Australian dust storms in April 2009. Science of the Total Environment 412, 223−231.

Liu, H., Blagodatsky, S., Giese, M., Liu, F., Xu, J., Cadisch, G., 2016. Impact of herbicide application on soil erosion and induced carbon loss in a rubber plantation of Southwest China. Catena 145, 180−192.

Loreau, M., Naeem, S., Inchausti, P., Bengtsson, J., Grime, J.P., Hector, A., et al., 2001. Biodiversity and ecosystem functioning: current knowledge and future challenges. Science (New York, N.Y.) 294 (5543), 804−808.

Loydi, A., Eckstein, R.L., Otte, A., Donath, T.W., 2013. Effects of litter on seedling establishment in natural and semi-natural grasslands: a meta-analysis. Journal of Ecology 101 (2), 454−464.

Loydi, A., Lohse, K., Otte, A., Donath, T.W., Eckstein, R.L., 2014. Distribution and effects of tree leaf litter on vegetation composition and biomass in a forest−grassland ecotone. Journal of Plant Ecology 7 (3), 264−275.

Lu, C., Zhu, Q., Deng, Q., 2008. Effect of frugivorous birds on the establishment of a naturally regenerating population of Chinese yew in ex situ conservation. Integrative Zoology 3 (3), 186−193.

Manzoni, S., Jackson, R.B., Trofymow, J.A., Porporato, A., 2008. The global stoichiometry of litter nitrogen mineralization. Science (New York, N.Y.) 321 (5889), 684−686.

Marklein, A.R., Winbourne, J.B., Enders, S.K., Gonzalez, D.J., van Huysen, T.L., Izquierdo, J.E., et al., 2016. Mineralization ratios of nitrogen and phosphorus from decomposing litter in temperate versus tropical forests. Global Ecology and Biogeography 25 (3), 335−346.

Marshall, E., 1986. Yellow rain evidence slowly whittled away: recently disclosed data from British and Canadian defense laboratories make the US case seem ever more doubtful. Science (New York, N.Y.) 233 (4759), 18−19.

Miller, V.S., Naeth, M.A., Wilkinson, S.R., 2021. Micro topography, organic amendments and an erosion control product for reclamation of waste materials at an arctic diamond mine. Ecological Engineering 172, 106399.

Milton, K., Windsor, D.M., Morrison, D.W., Estribi, M.A., 1982. Fruiting phenologies of two neotropical Ficus species. Ecology 63 (3), 752−762.

Möhler, H., Diekötter, T., Bauer, G.M., Donath, T.W., 2021. Conspecific and heterospecific grass litter effects on seedling emergence and growth in ragwort (Jacobaea vulgaris). PLoS One 16 (2), e0246459.

Moreno, T., Querol, X., Castillo, S., Alastuey, A., Cuevas, E., Herrmann, L., et al., 2006. Geochemical variations in aeolian mineral particles from the Sahara−Sahel Dust Corridor. Chemosphere 65 (2), 261−270.

Mudappa, D.I.V.Y.A., Raman, T.S., 2012. Beyond the borders: Wildlife conservation in landscapes fragmented by plantation crops in India. NCF Working Paper 1 (2012).

Muniz-Castro, M.A., Williams-Linera, G., Martínez-Ramos, M., 2012. Dispersal mode, shade tolerance, and phytogeographical affinity of tree species during secondary succession in tropical montane cloud forest. Plant Ecology 213 (2), 339−353.

Nagaraja, N., Yathisha, V., 2015. Nest orientation of Asian giant honeybee, Apis dorsata in plains of Karnataka, India. Journal of Entomological Research 39 (3), 197−201.

New King James Version Bible, 1994. Thomas Nelson Publishers, Nashville. (Original work published 1982) https://www.biblegateway.com/versions/New-King-James-Version-NKJV-Bible/.

Newton, I., 1960. The mathematical principles of natural philosophy. In: Berkeley, C.F. (Ed.), Sir Isaac Newton's Mathematical Principles. University of California Press.

Norton, T.A., 1992. Dispersal by macroalgae. British Phycological Journal 27 (3), 293−301.

Nobel, P.S., 1981. Wind as an ecological factor. Physiological Plant Ecology I. Springer, Berlin, Heidelberg, pp. 475−500.

Owens, K.A., Jaya, D.S., Conlon, K., Kiruba, S., Biju, A., Vijay, N., et al., 2022. Empowering local practitioners to collect and report on anthropogenic riverine and marine debris using inexpensive methods in India. Sustainability (in press).

Pace, M.L., Cole, J.J., Carpenter, S.R., Kitchell, J.F., 1999. Trophic cascades revealed in diverse ecosystems. Trends in Ecology & Evolution 14 (12), 483−488.

Palombo, M.R., 2021. Thinking about the biodiversity loss in this changing world. Geosciences 11 (9), 370.

Parthasarathy, N., 1999. Tree diversity and distribution in undisturbed and human-impacted sites of tropical wet evergreen forest in southern Western Ghats, India. Biodiversity & Conservation 8 (10), 1365−1381.

3. Soil physicochemical parameters

Pasipoularides, A., 2019. Emulating Leonardo da Vinci (1452–1519): the convergence of science and art in biomedical research and practice. Cardiovascular Research [online] 115 (14), 181–183.

Patino, S., Herre, E.A., Tyree, M.T., 1994. Physiological determinants of Ficus fruit temperature and implications for survival of pollinator wasp species: comparative physiology through an energy budget approach. Oecologia 100 (1), 13–20.

Payne, J.L., Bush, A.M., Heim, N.A., Knope, M.L., McCauley, D.J., 2016. Ecological selectivity of the emerging mass extinction in the oceans. Science (New York, N.Y.) 353 (6305), 1284–1286.

Pérez, J., Ferreira, V., Graça, M.A., Boyero, L., 2021. Litter quality is a stronger driver than temperature of early microbial decomposition in oligotrophic streams: a microcosm study. Microbial Ecology 82 (4), 897–908.

Petersen, T.K., Bruun, H.H., 2019. Can plant traits predict seed dispersal probability via red deer guts, fur, and hooves? Ecology and Evolution 9 (17), 9768–9781.

Petraglia, A., Cacciatori, C., Chelli, S., Fenu, G., Calderisi, G., Gargano, D., et al., 2019. Litter decomposition: effects of temperature driven by soil moisture and vegetation type. Plant and Soil 435 (1), 187–200.

Poinar, G.O., Columbus, J.T., 1992. Adhesive grass spikelet with mammalian hair in Dominican amber: first fossil evidence of epizoochory. Experientia 48 (9), 906–908.

Polis, G.A., Sears, A.L., Huxel, G.R., Strong, D.R., Maron, J., 2000. When is a trophic cascade a trophic cascade? Trends in Ecology & Evolution 15 (11), 473–475.

Pollard, B.T., Bio, R.P., Terrace, B.C., 2002. Mountain Goat Winter Range Mapping for the North Coast Forest District, pp. 1–26.

Pradeep, G.S., Krishnan, M.N., Vijith, H., 2015. Identification of critical soil erosion prone areas and annual average soil loss in an upland agricultural watershed of Western Ghats, using analytical hierarchy process (AHP) and RUSLE techniques. Arabian Journal of Geosciences 8 (6), 3697–3711.

Prasannakumar, V., Vijith, H., Geetha, N., Shiny, R., 2011. Regional scale erosion assessment of a sub-tropical highland segment in the Western Ghats of Kerala, South India. Water Resources Management 25 (14), 3715–3727.

Predit, P.P., Prasath, V., Raj, M., Desai, A., Zacharia, J., Johnsingh, A.J.T., Ghose, D., Ghose, P.S., Sharma, R.K., 2015. Status and Distribution of the Nilgiri Tahr Nilgiritragus hylocrius, in the Western Ghats. India. Technical Report. WWF-India.

Premchander, S., Jeyaseelan, L., Chidambaranathan, M., 2003. In search of water in Karnataka, India: degradation of natural resources and the livelihood crisis in koppal district. Mountain Research and Development 23 (1), 19–23.

Ptatscheck, C., Gansfort, B., Traunspurger, W., 2018. The extent of wind-mediated dispersal of small metazoans, focusing nematodes. Scientific Reports 8 (1), 1–10.

Puyravaud, J.P., Davidar, P., Pascal, J.P., Ramesh, B.R., 2003. Analysis of threatened endemic trees of the Western Ghats of India sheds new light on the Red Data Book of Indian Plants. Biodiversity & Conservation 12 (10), 2091–2106.

Rabeni, C.F., Minshall, G.W., 1977. Factors affecting microdistribution of stream benthic insects. Oikos 33–43.

Raj, A.D.S., 2015. The Shola of Kanyakumari. Baseline survey report submitted to the Forest Department of Kanyakumari District. p. 417.

Rajasekar, C., Rajendran, A., 2018. Prey composition of Utricularia striatula Sm.(Lentibulariaceae): lithophytic carnivore Southern Western Ghats, India. International Journal of Fisheries and Aquatic Studies 6, 382–388.

Raju, A.S., Ramana, K.V., Jonathan, K.H., 2009. Anemophily, anemochory, seed predation and seedling ecology of Shorea tumbuggaia Roxb.(Dipterocarpaceae), an endemic and globally endangered red-listed semi-evergreen tree species. Current Science 827–833.

Reid, N., 1989. Dispersal of misteltoes by honeyeaters and flowerpeckers: components of seed dispersal quality. Ecology 70 (1), 137–145.

Remusat, L., 2014. Organic material in meteorites and the link to the origin of life, BIO Web of Conferences, 2. EDP Sciences, p. 03001.

Ripple, W.J., Estes, J.A., Schmitz, O.J., Constant, V., Kaylor, M.J., Lenz, A., et al., 2016. What is a trophic cascade? Trends in Ecology & Evolution 31 (11), 842–849.

Rogora, M., Frate, L., Carranza, M.L., Freppaz, M., Stanisci, A., Bertani, I., et al., 2018. Assessment of climate change effects on mountain ecosystems through a cross-site analysis in the Alps and Apennines. Science of the Total Environment 624, 1429–1442.

3. Soil physicochemical parameters

Rubenson, J., Henry, H.T., Dimoulas, P.M., Marsh, R.L., 2006. The cost of running uphill: linking organismal and muscle energy use in guinea fowl (Numida meleagris). Journal of Experimental Biology 209 (13), 2395–2408.

Sabale, S.N., Suryawanshi, P., Krishnayaj, P.U., 2020. Soil metagenomics: concepts and applications. In: Hozzein, W.L. (Ed.), Metagenomics—Basic Methods and Applications. Intechopen, London, pp. 9–36.

Sabatier, P., Poulenard, J., Fanget, B., Reyss, J.L., Develle, A.L., Wilhelm, B., et al., 2014. Long-term relationships among pesticide applications, mobility, and soil erosion in a vineyard watershed. Proceedings of the National Academy of Sciences 111 (44), 15647–15652.

Sajinkumar, K.S., Anbazhagan, S., Pradeepkumar, A.P., Rani, V.R., 2011. Weathering and landslide occurrences in parts of Western Ghats, Kerala. Journal of the Geological Society of India 78 (3), 249–257.

Santonja, M., Pellan, L., Piscart, C., 2018. Macroinvertebrate identity mediates the effects of litter quality and microbial conditioning on leaf litter recycling in temperate streams. Ecology and Evolution 8 (5), 2542–2553.

Schaller, G.B., 1977. Mountain monarchs. Wild Sheep and Goats of the Himalaya. University of Chicago Press, Chicago.

Schulze, K.A., Buchwald, R., Heinken, T., 2014. Epizoochory via the hooves – the European bison (Bison bonasus L.) as a dispersal agent of seeds in an open-forest-mosaic. Tuexenia 34 (1), 131–144.

Schweiger, A.K., Schütz, M., Anderwald, P., Schaepman, M.E., Kneubühler, M., Haller, R., et al., 2015. Foraging ecology of three sympatric ungulate species—behavioural and resource maps indicate differences between chamois, ibex and red deer. Movement Ecology 3 (1), 1–12.

Shanahan, M., So, S., Compton, S.G., Corlett, R., 2001. Fig-eating by vertebrate frugivores: a global review. Biological Reviews 76 (4), 529–576.

Sharrow, S.H., Ismail, S., 2004. Carbon and nitrogen storage in agroforests, tree plantations, and pastures in western Oregon, USA. Agroforestry Systems 60 (2), 123–130.

Shelley, P.B., Gielgud, J., 1951. Ode to the West Wind. Columbia.

Sinha, A., Davidar, P., 1992. Seed dispersal ecology of a wind dispersed rain forest tree in the Western Ghats, India. Biotropica 519–526.

Sitaramam, V., Jog, S.R., Tetali, P., 2009. Ecology of Ficus religiosa accounts for its association with religion. Current Science 637–640.

Slagsvold, T., Hušek, J., Whittington, J.D., Wiebe, K.L., 2014. Antipredator behavior: escape flights on a landscape slope. Behavioral Ecology 25 (2), 378–385.

Slocum, M.G., Horvitz, C.C., 2000. Seed arrival under different genera of trees in a neotropical pasture. Plant Ecology 149 (1), 51–62.

Sobral-Leite, M., de Siqueira Filho, J.A., Erbar, C., Machado, I.C., 2011. Anthecology and reproductive system of Mourera fluviatilis (Podostemaceae): pollination by bees and xenogamy in a predominantly anemophilous and autogamous family? Aquatic Botany 95 (2), 77–87.

Sterk, G., Stoorvogel, J.J., 2020. Desertification—scientific versus political realities. Land 9 (5), 156.

Storz, J.F., Quiroga-Carmona, M., Opazo, J.C., Bowen, T., Farson, M., Steppan, S.J., et al., 2020. Discovery of the world's highest-dwelling mammal. Proceedings of the National Academy of Sciences 117 (31), 18169–18171.

Subramanian, K.A., Sivaramakrishnan, K.G., 2005. Habitat and microhabitat distribution of stream insect communities of the Western Ghats. Current Science 976–987.

Sundarapandian, S.M., Swamy, P.S., 2000. Forest ecosystem structure and composition along an altitudinal gradient in the Western Ghats, South India. Journal of Tropical Forest Science 104–123.

Talbot, L.M., 1960. A look at threatened species. Oryx 5 (4–5), 155–293.

Taylor, P.E., Flagan, R.C., Valenta, R., Glovsky, M.M., 2002. Release of allergens as respirable aerosols: a link between grass pollen and asthma. Journal of Allergy and Clinical Immunology 109 (1), 51–56.

Temmink, R.J., Angelini, C., Fivash, G.S., Swart, L., Nouta, R., Teunis, M., et al., 2021. Life cycle informed restoration: engineering settlement substrate material characteristics and structural complexity for reef formation. Journal of Applied Ecology 58 (10), 2158–2170.

Terborgh, J., Nuñez-Iturri, G., Pitman, N.C., Valverde, F.H.C., Alvarez, P., Swamy, V., et al., 2008. Tree recruitment in an empty forest. Ecology 89 (6), 1757–1768.

Thomas, J., Joseph, S., Thrivikramji, K.P., 2018. Assessment of soil erosion in a tropical mountain river basin of the southern Western Ghats, India using RUSLE and GIS. Geoscience Frontiers 9 (3), 893–906.

Thomsen, M.S., Wernberg, T., Altieri, A., Tuya, F., Gulbransen, D., McGlathery, K.J., et al., 2010. Habitat cascades: the conceptual context and global relevance of facilitation cascades via habitat formation and modification. Integrative and Comparative Biology 50 (2), 158–175.

3. Soil physicochemical parameters

Thomsen, M.S., Altieri, A.H., Angelini, C., Bishop, M.J., Gribben, P.E., Lear, G., et al., 2018. Secondary foundation species enhance biodiversity. Nature Ecology & Evolution 2 (4), 634–639.

Tuttle, M.D., Moreno, A., 2005. Cave-Dwelling Bats of Northern Mexico: Their Value and Conservation Needs. Bat Conservation International, pp. 1–75.

Upadhyay, S., Roy, A., Ramprakash, M., Idiculla, J., Kumar, A.S., Bhattacharya, S., 2017. A network theoretic study of ecological connectivity in Western Himalayas. Ecological Modelling 359, 246–257.

Valencia-Barrera, R.M., Comtois, P., Fernández-González, D., 2001. Biogeography and bioclimatology in pollen forecasting. Grana 40 (4–5), 223–229.

Venable, D.L., Flores-Martinez, A., Muller-Landau, H.C., Barron-Gafford, G., Becerra, J.X., 2008. Seed dispersal of desert annuals. Ecology 89 (8), 2218–2227.

Visez, N., Chassard, G., Azarkan, N., Naas, O., Sénéchal, H., Sutra, J.P., et al., 2015. Wind-induced mechanical rupture of birch pollen: potential implications for allergen dispersal. Journal of Aerosol Science 89, 77–84.

Walker, L.R., Del Moral, R., 2003. Primary Succession and Ecosystem Rehabilitation. Cambridge University Press.

Wall, J., Douglas-Hamilton, I., Vollrath, F., 2006. Elephants avoid costly mountaineering. Current Biology 16 (14), R527–R529.

Wang, G., Hobbs, N.T., Twombly, S., Boone, R.B., Illius, A.W., Gordon, I.J., et al., 2009. Density dependence in northern ungulates: interactions with predation and resources. Population Ecology 51 (1), 123–132.

Wilson, D.S., 1992. Complex interactions in metacommunities, with implications for biodiversity and higher levels of selection. Ecology 73 (6), 1984–2000.

Wohlleben, P., 2016. The Hidden Life of Trees: What They Feel, How They Communicate—Discoveries from a Secret World, vol. 1. Greystone Books.

Yang, M., Beale, R., Liss, P., Johnson, M., Blomquist, B., Nightingale, P., 2014. Air–sea fluxes of oxygenated volatile organic compounds across the Atlantic Ocean. Atmospheric Chemistry and Physics 14 (14), 7499–7517.

Yumoto, T., Noma, N., Maruhashi, T., 1998. Cheek-pouch dispersal of seeds by Japanese monkeys (Macaca fuscata yakui) on Yakushima Island, Japan. Primates; Journal of Primatology 39 (3), 325–338.

Zhang, N., 2020. Interaction between water and soft rocks. In: Kanji, M., He, M., Ribeiroe Sousa, L. (Eds.), Soft Rock Mechanics and Engineering. Springer, Cham. Available from: https://doi.org/10.1007/978-3-030-29477-9_9.

Zhang, Y.S., Silliman, B.R., 2019. A facilitation cascade enhances local biodiversity in seagrass beds. Diversity 11 (3), 30.

Zotz, G., Harris, G., Königer, M., Winter, K., 1995. High rates of photosynthesis in the tropical pioneer tree, Ficus insipida Willd. Flora 190 (3), 265–272.

3. Soil physicochemical parameters

11

Tourism and the properties of mountainous soil: a dynamic relationship

Kakul Smiti

School of Environmental Sciences, Jawaharlal Nehru University, New Delhi, Delhi, India

OUTLINE

Understanding Soils of Mountainous Landscapes
DOI: https://doi.org/10.1016/B978-0-323-95925-4.00011-X

11.1 Introduction

Tourism is defined as the voluntary tour of humans from their regular domestic surroundings to any other vicinity that takes less time over an extended distance than non-tourism types of comparable human mobility. Tourism has numerous crucial social and financial parameters like travelers, the tourism enterprise, humans, tourist destinations, and, regular domestic surroundings too.

The same old surroundings of a character, according to the United Nations (UN) and the United Nations World Tourism Organization (UNWTO, 2007), consist of the area of regular residence of the family to which he/she belongs, his/her very own workplace or study, and some other area that he/she visits often and often inside his/her current recurring of existence, even if this area is positioned a long way far from the place of regular residence. There are a variety of various forms of the tour which can be covered similarly to the extensively generic definition and they're as follows: traveling pals and members of the family, enterprise tour, fitness and clinical associated tour, training associated tour, pilgrimage, volunteer tour, etc. (Table. 11.1). Increased tourism-related activities could be attributed to an increase in the population of the affluent class in society, who can afford to take multiple tours and travel trips. With the improvement in people's average lifestyles worldwide, the concept of leisure activities is becoming more popular by the day. The tourism phenomenon is closely linked with the abiotic, biotic, human, financial, social, and cultural additives of the surroundings. Beaches, coastal regions, deserts, meadows, mountains, and plains are the most common physiographic visitor destinations, with mountains having a special significance in tourism (Beedie and Hudson, 2003). In 2000, simply seven hundred million worldwide traveler arrivals had been counted international UNWTO (2007). Ecotourism is the most appealing concept, which includes mountain trekking, camping, and other leisure activities. In the context of India, pilgrimage-related tourism is very common, and some states, such as Uttarakhand, see significant visitor flows each year; for example, 54% of pilgrims increased in Uttarakhand between 2000 and 2010

TABLE 11.1 Types of mobility and ideas behind tourism.

Non-tourism related mobility	Academic and technical categories of tourism	Popular ideas behind tourism
Diplomats	Visiting friends and relations	Holiday and vacation
Border workers	Travel to second home	Travel for leisure
Expatriate workers	Educational travel	—
Travel and military service	Business travel	—
Employment by an organization outside of the home	Health travel	—
Forced temporary migration	Shopping	—
Permanent migration	Pilgrimage	—

Source: *Modified from Price, M., Moss, L., Williams, P., 1997. Tourism and amenity migration. In: Messerli, B., Ives, D. (Eds.), Mountains of the World. A Global Priority. Parthenon, New York, pp. 249–280.*

(Sati and Gahalaut, 2013). The relaxation of regulations during the second wave of the COVID-19 pandemic in India resulted in a critical rush of travelers toward the mountains, such as Himachal Pradesh. Between 1997 and 2011, the number of visitors increased from 874,000 to 1,169,000, indicating a significant increase in tourism sites and visitors in Dehradun (Dey et al., 2018). Furthermore, a projected 2,089,000 tourists are expected to visit various destinations between now and 2025, demonstrating that tourism can have a significant impact on the world's environment. Despite the fact that an international interest of this magnitude is expected to harm the environment and the entire biosphere collectively, the effects have yet to be discovered. In this chapter, five predominant factors of the tourism-based effects on the environment are investigated: (1) Land cover and land-use changes (2) the usage of electricity and its related effects, (3) the extinction of wild species, (4) the dispersion of diseases, and (5) a physiological effect of the tour, the changes in the understanding of the environment because by travel. Roughly around 22% and 27% of the Earth's overall land vicinity is protected with the aid of using amazing mountains (Grabherr and Messerli, 2011).

Mountains are described as highlands and multiplied ice shields above 2500 m, observed with the aid of using hilly regions under this altitude and foothill vicinity (Grabherr and Messerli, 2011). Mountains are recognized for their first-rate scenery, scenic splendor, and specific amenity values, which is why mountains are regarded as one of the most popular tourist destinations. Mountains draw large crowds due to their pristine landscapes and biodiversity. The 1992 United Nations Conference on Environment and Development in Rio de Janeiro paved the way for the recognition of the importance of mountain ecosystems, despite the fact that mountain areas were not thoroughly covered in Agenda 21, and it became stored on par with deforestation, desertification, and weather change. The United Nations declared 2002 the International Year of Mountains, and the global community recognized their significance. Mountains, after coasts and islands, are the most important global tourism destinations, accounting for 15%–20% of the global tourism industry (Price et al., 1997).

11.2 Environmental implications of tourism: a general overview

The environmental impacts of various types of tourism, such as natural, adventure, and pilgrimage tourism, are significant. Environmental degradation in mountainous areas is accelerated by two major factors, namely, the increasing magnitude of tourist activities and the expansion of tourist infrastructure. Tourism is dependent on the environment because a natural resource base is required to attract visitors to tourist destinations. The common enhancement of visitor sites is also dependent on natural sources. The environment is both spatially and temporally variable and complex. In general, the impact of tourism in its most dynamic form, recreational tourism, is more intense and diverse than that of business or health tourism. The impact of mass individual tourism is less than that of planned mass tourism. The impact of non-institutionalized tourism is less than that of institutionalized tourism. Pathways are important cultural landscape components that have historically served as communication and connecting routes. Geotourism has grown in popularity in recent decades (Serrano and González-Trueba, 2005; Conway, 2010; Jorge et al., 2016; Rangel et al., 2019).

Trails have a couple of capabilities, which include getting access to natural environments, nature contemplation, pastime, and wearing sports (Hammitt and Cole, 1998; Leung and Marion, 1999; Lynn and Brown, 2003; Figueiredo et al., 2010).

Trampling is a major impact of huge tourist inflows in an area, and they can degrade trails (Lynn and Brown, 2003) and path mismanagement can exacerbate land degradation (Doran and Jones, 1996; Arshad and Martin, 2002). Many researchers have looked into trail soil degradation. Trampling reduces soil quality by affecting certain soil characteristics, such as soil compaction and erosion, chemical and organic depletion, with associated vitamin and soil natural count (SOM) losses, and reduced soil faunal interest (Wolf and Croft, 2014). The variety and depth of effects are determined by the character of the location. As a result, a few environments (e.g., simple regions and areas referred to as city settlements) can assist a large number of visitors because they have the adequate organizational structure to assist travelers. Tourism is linked to environmental degradation in a variety of ways, such as the transformation of the environment for the purpose of tourism-related development; the consumption of numerous resources for tourism-related developmental works; the strain exerted by congestion, overload, saturation, tension, conflict, pollution, devastation, and so on. If the consequences do not exceed the affordability threshold, they can be reversed and may return to near-original conditions. However, the enormous push for development has resulted in an increase in the production of motels, lodges, dhabas, and tea stalls alongside roads, particularly along watercourse valleys, which has unfavorable implications for sensitive landscapes. Climate change is the most important and significant cause of global change affecting mountain areas. The United Nations Intergovernmental Panel on Climate Change (IPCC, 1995) acknowledged the dangers of global climate change to mountain areas. If the consequences no longer exceed the affordability threshold, they are reversible, and the disturbed ecosystem can return to near-natural conditions. Tourism is extremely important financially to many mountain groups and is one of the world's fastest-growing financial sectors for mountain areas. Mountain ecosystems are also particularly vulnerable to global climate change. A wide variety of evaluations of the consequences of global climate change in mountain areas across the world (Price, 1999) and the technological know-how documenting ongoing environmental alternate in mountain areas (glacial retreat, melting permafrost, the elevation of the tree-line, modifications in species composition, non-local species introductions, expanded geomorphic techniques) is progressing steadily. Mountains are one of the most fragile ecosystems, which means that even minor disturbances can have a significant impact on the balance of mountainous ecosystems. Tourism is expanding globally, geographical boundaries are shrinking, and additional development is taking place in mountainous areas. Humans have walked and camped in the mountains since ancient times, but the intensity of tourism in such fragile ecosystems has increased dramatically in recent centuries. This has occurred as a result of people's improved lifestyles and increased financial security. With financial prosperity, the form of tourism in mountains and protected areas has also changed, and as a result, more modern forms of tourism such as horseback riding, cycling, sky diving, and trekking have emerged, which is one of the reasons why there is an increasing percentage of tourists in mountainous areas (Barros and Pickering, 2014). During the monsoon season, the entire Garhwal region of the Indian Himalayas receives widespread precipitation, and the rivers rise above their banks, resulting in large-scale erosion. The highland pilgrimages run alongside alpine meadows, which can be ecologically fragile. There have been numerous instances where major errors occurred, resulting in widespread damage to livelihood and property. On the 16th and 17th of June 2013, a downpour aided by debris-waft and flash floods washed away the Badrinath and Kedarnath pilgrimages, killing over 10,000 pilgrims as well as locals and destroying many homes (Sati and Gahalaut, 2013). This devastation on the route of those two highland shrines was caused primarily by large-scale construction along the course of the perennial rivers that flow through this region. Land transformation within valleys for the construction of settlements—hotels and motels—and for the production of accommodation for travelers right on the highways. Accommodation, tourism corridors, and facilities should not be developed in the area of steep slopes and fragile land on the side of the road. Furthermore, excessive erosion caused by massive deforestation and changes in land use have rendered the land less stable and vulnerable. Temperature fluctuations for tourists have been identified as the two most significant threats to the wilderness trek. Temperature fluctuations have a greater impact on high temperatures, and because long-distance trips are made at higher altitudes, the impact of temperature fluctuations is greater. The increased distance between travelers and tourists visiting the Garhwal Himalayas has also created difficulties for hikers (Sati and Gahalaut, 2013). Pilgrim pilgrimage to the Garhwal region has a long history. However, the advent of modern tourism has had serious consequences for the traditional economy, environment, and society; in fact, its impact has declined sharply in the system, and even sanctuaries and pilgrimages have experienced serious environmental effects (Sati and Gahalaut, 2013) Many amazing results of pilgrim travel include land reform in the valleys to produce cars and lodges and to provide travelers with accommodations near roadblocks, as well as landfills in the central highlands due to human migration. This is due to the valleys' low population density, as well as the low population density near the central and mountain ranges. As the number of tourists increases, so will the type of accommodation, leading to an increase in the production of illegal schemes along the river in mountainous areas such as Uttarakhand and Himachal Pradesh. New roads were built, and existing ones were widened, without regard for Himalayas' wearability (Kumar et al., 2015). In December 2000, national authorities restricted development within 200 m of the riverbanks, but the results were inconclusive. The mountains had been haphazardly reduced to make avenues, rendering the mountains volatile. The direct correlation has been established between the increase in tourism and the increase in a wide range of landslide incidents, which in the long run results in a decrease in visitors' enchantment closer to the vacation spot for their subsequent visits. After 2008, Uttarakhand experienced a slew of natural disasters. The frequency of disasters, for example, has increased. Cloudburst in Nachni near Munsiyari in Uttarakhand's Pithoragarh district in August 2009, massive cloudburst in Almora in September 2010 that rendered the entire village useless and drowned in September 2012, cloudburst in Kedarnath and Rambada vicinity in June 2013 (Kumar et al., 2015). Overpopulation has caused a slew of problems, including illegal production, unplanned growth, excessive garbage, encroachment on forest lands, sanitation and sewage issues, water scarcity, overcrowded roads, traffic jams, and vehicular pollution. The instability of the slope on which such heavy systems are built puts them at risk. Furthermore, a few unscrupulous contractors are breaking the rules by erecting condominium blocks in an attempt to make quick money. It is unusual to see resort marketers from the local community; these people are usually builders from outside the area. Mussoorie's motel industry is rapidly deteriorating. Even though there are no municipal sewer lines, building construction has been approved. As a result, the man or woman of the hill city has been misplaced, and it is difficult to distinguish it from the city conglomerate of the nearby Dehradun metropolis within the valley. Mussoorie and Kulri already resemble Dehradun Paltan Bazaar, a prime example of overcrowded, narrow, and unsanitary streets. The sanitary conditions are deplorable. Unchecked development within mountainous areas has resulted in the most commonplace environmental problem, known as soil erosion. Soil erosion has been extensively researched with the assistance of researchers. With the emphasis shifting to sustainable development, the importance of soil has become widely recognized. Unsustainable visitor practices have increased the likelihood and severity of soil erosion. Off-site deposits are formed by on- and off-site hydrology. This occurs as a result of soil loss, which can be detrimental (Pimentel et al., 1995). Buildings, transportation, communications, electricity, water, and waste management systems are just a few of the necessary physical facilities for mountain tourism to function. New tourism infrastructure, on the other hand, has the potential to have an impact on mountain communities and habitats both during and after construction (for example, earth movement) (e.g., an increase in population and corresponding resource needs, pollution, and so on). Ironically, new infrastructure that initially benefits tourism can cause enough negative cultural and environmental changes to make mountain areas unappealing to visitors (Mountain Forum/The Mountain Institute, 1998). Poor road planning can have serious ramifications for mountain ecology and water regimes. Soil erodes because it is subjected to a variety of environmental pressures, including fluvial, aeolian, organic, and human. Such forces reduce soil cohesiveness and thus exacerbate soil erosion. Several authors (Vashchenko and Biondi, 2013; Nepal and Nepal, 2004) identified environmental changes related to trails, including soil compaction, flower removal, alteration of current drainage systems through topsoil removal, and change of micro-topography. On the other hand, those changes have an impact on the microclimate. Furthermore, erosive techniques such as splash erosion, rill erosion, and even gullies can frequently be found on trails. These corrosive techniques can also degrade the essence of tourism and increase the risk of accidents. Fig. 11.1 represents the most common effects of tourism on tourist spots e.g., mountains.

11.3 Types of mountainous soil

Soil is an important component of the physical environment in tourist areas, as well as the primary target of tourist impacts. Soils are ecological traits, and their characteristics can accurately predict ecosystem characteristics. This susceptibility—or the rate and direction of change—is largely determined by the parent material properties, the gradient on which the soil forms, texture, and other parameters influenced by the five soil-forming processes (Helgath, 1975). The idea that soils are a function of independent state factors such as time, climate, topography, vegetation, and parent material governs soil formation. The following are the classifications for mountainous soils:

3. Soil physicochemical parameters

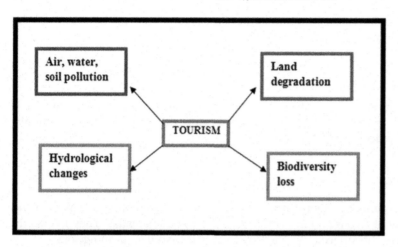

FIGURE 11.1 Possible negative impact of tourism on mountains.

11.3.1 Ultisols

The ultisol order of soils is mostly found in the Southeast Americas and contains a significant amount of clay. The Appalachian Mountain range, the Cascades of Washington, and the Cascades of Oregon are the main locations where we can find ultisols. Ultisols are more likely to be found in humid climates. Ultisols are typically acidic in nature and contain the majority of nutrients in the top few inches of soil. They are not good at retaining fertilizer.

11.3.2 Inceptisols

In hilly areas, inceptisols are among the most common soils. They can be found in the Appalachians, Cascades, Sierra Nevadas, and Rocky Mountains, though in lesser numbers. This group of soils is diverse and can be found in a range of climates, from humid to semi-arid. Because this type of soil has been very little weathered, the concentration of silicate clay in it is very low.

11.3.3 Spodosols

This soil order is not common in many mountain ranges, but it can be found in the Cascades and higher elevations of the Rocky Mountains. Spodosols are highly acidic, nutrient-deficient soils that are commonly found beneath coniferous forests. They are also low in organic matter.

11.3.4 Andisols

Andisols are fertile soils that effectively retain nutrients and water. They frequently contain volcanic debris and are typically found in areas with above-average rainfall. It's no surprise, then, that they're mostly found in the Cascades. They can be found in the northern Rocky Mountains as well.

11.3.5 Entisols

Entisols can be found in the Rocky Mountains and the southern Sierra Nevadas. They appear in areas prone to erosion, such as steep slopes and dunes. They are commonly composed of sand, metal oxides, and clay.

11.3.6 Alfisols

Alfisols are found in a number of mountainous areas, including the Appalachians and the Rocky Mountains. They are less common in the Cascades and Sierra Nevadas. They frequently form beneath trees and have subsurface clay. They are frequently excellent farming soils.

11.4 Soil organic matter

Due to heavy tourist traffic along the mountain trails, the most common thing to see is truncation, which is the increasing shallowness of the soil as the upper part of the soil profile is removed. In the mountains, visitor-caused physical erosion destroys the top horizons of soil, such as the O and A horizons. The destruction of soil aggregates as a result of the removal of the O and A horizons, and thus the discounting of soil natural count attention, results in a discounting of soil's ability to withstand the impact of compressing load (Soane, 1990). Trampling is the most commonplace place effect of tenting and other visitor sports, as are a few specific effects. Numerous studies have quantified the impact of trampling on campsite soil. Organic horizons on campsites are roughly one-third as thick as on controls, and mineral soil is typically exposed over a maximum of the campsites. Compaction has been linked to increased bulk density and penetration resistance in mineral soils. In the campsites, numerous parameters of soil like Organic horizon thickness (cm), Soil bulk density (g/cm^3), Soil penetration resistance (kPa) are different from the soil from control sites, wherein the Organic horizon thickness is lower than the control soils at the same time as the latter parameters have greater values than the control soil control sites (Reeves, 1997). Organic soil horizon abrasion and loss increase the intensity of soil compaction; organic soil horizons protect underlying mineral soil horizons from excessive compaction and erosion. Organic litter loss has a direct impact on plant and animal populations both above and below ground. If organic matter from the soil is lost, the water-holding capacity of the soil is significantly reduced, which harms soil microbial populations that rely on soil organic matter and root exudates from aboveground plants for energy. Microbial statistics contribute to ecosystem function in nutrient uptake, the transformation of soil organisms, production of phytohormones, and contributing to the soil nutrient web (Reeves, 1997) coarse soil, cohesiveness and erosion power. These visible soil changes affect soil chemistry and biota, but these changes are poorly understood. The organic matter content of the soil gradually increases as one moves away from the path. Since biodiversity is an important indicator of soil quality measurement (Reeves, 1997), they have reported a negative relationship between biodiversity and distance from key tourist destinations. Qin et al. (2006) discovered that trampling down the soil organic matter content is a plausible explanation for the decrease in organic matter content in the native areas surrounding the path. Camping sites promote the accumulation of soil litter, campgrounds maintain their physical properties best when established on medium-textured soils, and camping units should be designed to minimize user impacts (Lockaby and Dunn, 1984). Because of trampling and unintended human-mediated seed dispersal, plant invasions are especially vulnerable in high-elevation protected areas and mountains. Trampling may help non-native seeds establish and spread once they reach mountainous areas. Tourists and pack animals tramp on and off highland pathways, disrupting habitats that support the growth of non-native species (Barros and Pickering, 2014). As a result of anthropogenic disturbances, non-native plants are frequently introduced into mountainous environments, and where native vegetation is harmed, non-native plants often thrive (Barros and Pickering, 2014). The construction and maintenance of tourism infrastructures such as trails, roads, ski slopes, and campgrounds cause disruption and pave the way for the introduction of non-native plants (Barros et al., 2015). Normal tourism activities such as trampling on pedestrians and packing animals, mountain biking, and four-wheel drive contribute to reduced vegetation coverage, which may benefit non-native plants (Barros and Pickering, 2014). Adding nutrients, from the deliberate use of fertilizers, promote the growth of introduced species in gardens, ski slopes, and rehabilitation areas (Barros and Pickering, 2014). There is a general deforestation practice within each of the sample tourist destinations in those areas where tourist activity decreases significantly as the distance from each area increases, indicating that tourism harms forests. However, there has been a reversal in the trend from 2001 to 2013, indicating the importance of conservation. Interactions between land resources (tourism, climate), regions (national policies, major river management), and land (building and agriculture, energy and water resources to support the tourism industry) result in a distinct but complex pattern of land use and cover disruption (Boori et al., 2015). The removal of the upper part of the soil profile due to tourist activities results in the presence of a highly stony material in the uppermost part of the soil profile on the studied tourist trails in the Bieszczady mountains in southeast Poland (Drewnik et al., 2019). Soils impacted by tourists have degraded significantly over time, to varying degrees. The most common symptom is a truncated soil profile and soil compaction. 0.5 m is found to be the maximum depth of erosion. The following structures can be seen as a result of tourism-led truncation i.e., stones forming the "post-erosional pavement" play a crucial role in limiting deeper soil erosion. Drewnik et al. (2019) reported that soil regimes with high organic matter content, high cation exchange capacity, and formed in humid climates and high cation exchange capacity due to their loamy textures are capable of quickly restoring vegetation cover, protecting the soil from erosion.

11.4.1 SOIL pH

Compacted soils can also prevent seed germination and plant growth. The soil pH decreases as one moves away from the path. Soil pH at 0 m from the visitor path is substantially excessive than that at areas a long way from the direction. There had been no critical differences among soil hydrogen ion concentrations at ten and twenty m distances from the main trail path. Results of numerous types of research more endorse that the visitor population exaggerates soil pH, and this effect takes place within 10 m of the trail. The topsoil varies noticeably in a range of 2 m distant from the paths, reaching 20 cm below the surface of the soil, as compared to the control regions. In general, the influence on 0–10 cm soil is substantially bigger than on 10–20 cm soil. Tourist disruption has varying effects depending on where you go. There is a clear link between disturbance and effects in heavily populated areas, and the consequences are more severe. However, because of variations in landforms, flora, and other environmental conditions, the influence varies and is not always consistent with use. In an Illinois tenting ground, Young and Gilmore (1976) confirmed that inhabitation-fire ashes raised soil pH. Wang et al. (2003) found that tourism increased soil pH within the Yunmeng Mountain National Forest Park, whereas Jiang (1996) found that visitor disturbance decreased Mount Emei's soil pH. This discrepancy indicated that numerous disturbances e.g., tenting, driving, and appearance on trails, had absolutely extraordinary effects on soil hydrogen ion attention. Tourist disturbance primarily affects the soil close to the path, fixing soil wet and porosity by constantly changing the kind and amount of foreign matter (organic or inorganic matter carried by tourists' shoes and metric capacity unit thrown away) coming into the soil within the Tianchi scenic space. According to Fu et al. (2003) and Feng and Bao (1999) changes caused by visitor disturbance increased soil pH within the Tianchi scenic space, whereas areas far from the direction, with lower visitor interest effects, had low soil pH. The link between site visitors and hydrogen ion attention is vulnerable at the same time that soil density and pH are relatively related. According to some authors, soils in heavily visited areas are expected to be extra acidic due to the removal of organic matter and critical sound (Monti and Mackintosh, 1979). Indifference to the modern-day, Cole and Fichtler (1983) and Lockaby and Dunn (1984) argued that a high hydrogen ion concentration is encountered in heavily used areas (e.g., car park, main path, approaches to the cave) ought to be expected because of decrease in action originated by a decrease in infiltration caused by compaction. Furthermore, it would be due to the ashes from bonfires. The increase in visitor flows throughout the mountainous regions will increase the need to provide basic elements for sustenance such as meals and shelter, which will result in an increase in lodging systems such as motels, etc., in conjunction with the visitor trails. Kammer and Hegg (1990) mentioned that the soils right down to an intensity of 15 cm display better pH-values on snowed ski-runs as in comparison to control plots. The nutrient concentrations and soil pH of visitor spots are excessive because of the usage of synthetic snow (Kammer and Hegg, 1990).

11.5 Soil bulk density

The soil's bulk density near the path is significantly altered as a result of visitor disturbance, with values of this variable being higher near the path of the direction and decreasing grade by grade with increasing distance. This divergence was caused in particular by common trample by using travelers within the areas surrounding the get right of entry direction, resulting in soil compaction in those areas. According to Qiu et al. (2014) increasing soil bulk density modifies soil shape, resulting in the deterioration of soil microbial quality and, as a result, a decrease in soil microbial biomass. As a result, the increase in soil bulk density near the direction may be the cause of the deterioration of the soil microenvironment, hampering the growth of flowers and microbes. Disturbance in the soil microenvironment and reduction in nutrient availability affect plant root growth and development, with the main reason being a change in soil bulk density, which results in a reduction in soil organic matter inputs. The soil is exposed to erosive downpours because there are no plants or organic materials on the pathways. The track slope can also be used to measure severe track deterioration, and these findings are consistent with previous research that shows that erosion rates are highest on steeper grades and in areas where terrain and elevation combine to produce more precipitation (Marion and Leung, 2001).

11.6 Soil nitrogen content

The impact of tourist disturbance on soil total N content can be quantified because the disturbance decreased soil total N content, but the trend varies depending on the plot. At higher altitudes, the total N content of the soil increases significantly with increasing path distance, whereas it does not vary significantly at lower altitudes. The total N content of soil in Jinyun Mountain, China, decreased as the intensity of tourist activities increased. The lower altitude space has less trash and a lower organic matter content, indicating a low visitor intensity during this time. When compared to the control sites, soil water, soil porosity, soil clays, soil organic matters, total N, and obtainable N of all samples decrease by 22.79%, 42.03%, 40.31%, 39.76%, 37.99%, and 30.87%, respectively, while hydrogen ion concentration and bulk density increase by 6.59% and 56.64%. Soil total phosphorus, available phosphorus, total phosphorus, total compaction, and the reduction or removal of leaf litter and humus layers reduce soil ventilation, humidity, organic matter content, and microbial flora in frequently visited areas, all of which have a negative impact on plants. As a result, the vegetation in frequently visited areas is greatly disturbed (Qin et al., 2006).

11.7 Pollutants in the soil

According to soil samples collected along such areas, lower levels of aluminum, potassium, magnesium, and sodium in many of the "disturbed" samples can be attributed to severe leaching in dishwashing stations, as well as frequent discharge of wastewater from washing and cooking activities (Arocena et al., 2006). The soils surrounding the camping site wash stations have higher levels of phosphorus, copper, and zinc than the controls. Copper levels are higher in fire pit soils than in controls. Water disposal and campfires should be restricted to specific areas. Total K content has a low correlation with tourist disturbance intensity, while total P has the lowest correlation with tourist disturbance intensity. It is stated that the sensitivity of each index to tourist disruption varies (Arocena et al., 2006). In locations where tourism pressure is high, there is little contamination with persistent organic pollutants (POPs) and polyaromatic hydrocarbons (PAHs) in both soil and sediment samples (Guzzella et al., 2016). Chemicals used on ski runs melt the top layer of the snowfall, similar to salts used on roadways, affecting the snow quality

(Kobayashi et al., 2000). Salt is commonly used in mountainous areas to melt snow cover, but chemicals are now widely used, affecting both the underlying soil and the snow quality. As ski resorts, sports, and leisure activities become more intense, plant and soil qualities will change across a larger region (Freppaz et al., 2008).

11.8 Hydrological changes

The reduction of soil particles that could store water in the highlands causes a slew of issues. Soils in undisturbed or marginally disturbed pastures hold significantly more water as a result of slope disturbance, but soils in undisturbed or marginally disturbed pastures can hold significantly more water. The difference in soil water storage becomes even more substantial when the higher proportion of rock fragments (soil grains with a diameter greater than 2 mm) in many tourist areas, such as ski lines and resorts, is taken into account. According to a new research paper published after a study in Ruka, Finland, water bodies near a ski resort had greater nutrient retention and larger seasonal changes in nutrient concentrations than nearby control lakes (Finland).

11.9 Construction activities and mountainous soil

The development of tourist-related infrastructures generally causes modifications and topographic adjustments that lead to the degradation of soil and vegetation (Isselin-Nondedeu and Bédécarrats, 2007). The use of heavy machinery, such as earthmoving equipment, alters both ecological and geomorphologic processes, changing soil properties, biomass production, and plant species composition, increasing the risk of land degradation caused by erosion, landslides, and avalanches (Jamieson et al., 2002). Heavy machinery is used to remove obstacles like trees and boulders, as well as to smooth out rough or lumpy soil surfaces. During machine grading, upper soil layers and plants are destroyed or severely harmed (Haeberli, 1992). The soil surface of the dug hollows appeared to be well-drained and gritty due to the granite gravel. Massive rock removal has a significant impact on soil thermal properties, particularly in high alpine permafrost zones (Rixen et al., 2003). Machine grading has a negative impact on soil vegetation and water infiltration capacity, according to studies conducted in Europe (Wipf et al., 2005). Due to construction activity on mountains, the original soil thickness might be diminished, resulting in the loss of earlier soil horizonation and altered topsoil (Delgado et al., 2007). The anthropogenic effects of construction activities on mountains, according to Titus and Landau (2003), include the removal of the A horizon and a portion of the B horizon to establish suitable tourist spaces. The removal of topsoil and particles from such building regions exposes a whole surface of dolomitic limestone, lowering the water retention capacity. The presence of Lithic Leptosols was significantly higher in plots with the most disturbed soils. The same plots, on the other hand, had the highest proportion of Haplic Cambisols, indicating that geological substratum, relief, and plant management all played a role in the evolution of soils near tourist attractions. Haplic Cambisols were more common at the bottom of the hill, where disturbance and steepness were less of an issue (Barni et al., 2007). None of the physicochemical soil parameters studied (organic carbon, pH, aggregate breakdown) were found to be related to the age of infrastructural constructions, indicating that pedogenesis was extremely slow and site-specific in those conditions. Descroix and Mathys (2003) studied the impact of mountain management on alpine erosion and discovered that gully erosion and solifluctions (Freeze-thaw activity) is the collective name for progressive processes in which a mass slides down a slope (mass wasting) were frequently prompted by the expansion of ski resorts in the Northern Alps (Eglise and Ravoire Torrents, in the upper Isère valley, Savoie) (e.g., in Vars). Roadways and other construction sites frequently disturb the slopes and produce more sediments than less disturbed sites, which have low sediment formation. Construction activities are also associated with the loss of nutrient-rich topsoil, which is required for plant growth (Grismer and Hogan, 2005). Development of tourist infrastructure generally results in changes and topographic adjustments, which lead to soil and vegetation degradation (Isselin-Nondedeu and Bédécarrats, 2007). The use of heavy machinery, such as earth moving equipment, alters both ecological and geomorphological processes, changing soil properties, biomass production, and plant species composition, increasing the risk of land degradation caused by erosion, landslides, and avalanches (Jamieson et al., 2002; Arnaud et al., 2005). Trails via meadows contribute towards extra soil loss, greater uncovered soil and rock and much less vegetation than trails via steppe vegetation (Barros et al., 2013). Changes in surface soil parameters were studied for two vegetation units in the Boreal Forest Region of northwestern Ontario, containing *Pinus banksiana* and *Populus tremuloides*, along a gradient of increasing recreational activity. The compression and removal of surface leaf litter, followed by the formation of a thin, compacted mineral layer near the soil surface, were the two most notable results. According to porosity photograms, the content of macropores in the size range 300 to >3000 m is reduced by up to 60% in the top soil layer, while pores >2000 m are practically totally eradicated. Infiltration rates are lowered to 0.1 cm/h as a result (Monti and Mackintosh, 1979). New highway construction is the most common type of construction activity. Higher proportions of impermeable surfaces, such as roads, parking lots, and trails, accelerate rainfall-runoff processes, and faster surface runoff generates more energy for soil erosion (Harden, 2001).

11.10 Soil aggregation

Soil aggregation is a significant factor that controls percolation capacity, hydraulic conductivity, water-holding capacity, gas exchange, organic matter decomposition, and erosion resistance (Delgado et al., 2007). Barni et al. (2007) discovered a significant decrease in the aggregate stability of soils in popular tourist areas. In the same study, soil aggregate breakdown was found to be inversely related to organic matter concentration. The soil aggregates in the top strata are composed of fine material, organic substances, underground organs of some plant species, and soil organisms, and are known for their high water-holding capacity. Aggregates are destroyed, reducing soil water retention and, as a result, plant-available water, and increasing quick drainage. Soil aggregation factors control water-holding capacity and soil hydrological parameters (e.g., soil ability to hold water at a specific depth), so they must be taken into account for successful revegetation (Pintar et al., 2009). Construction activities in mountainous areas have a significant impact on the particle size distribution and organic carbon content, affecting areas. Because the presence of clasts and sand particles increases as a result of the mechanical crushing of stones during infrastructure construction, the soils of tourist destinations are frequently characterized by a high concentration of rock fragments (Gros et al., 2004). The clay and organic carbon content of Sierra Nevada natural soils decreased significantly as altitude increased. This pattern was not observed in ski-run soils, where management appeared to have a greater impact than environmental factors (Delgado et al., 2007). Soil deterioration is especially severe at high elevations, and geography, geology (e.g., tectonic activity), and climate all have an impact on trail conditions, correlating with previous findings (Bratton et al., 1978).

11.11 Artificial snow and soil erosion

Fertilization and artificial snow increase an area's total snow mass, which amplifies surface runoff due to the additional water, slowing soil erosion in the long run. Ski resort intensive fertilization causes nutrient leaching, which degrades the area's water and soil quality. The effect of artificial snow production on adding water, ions, microbes, and salts to the soil is unknown (Rixen et al., 2003). Constructional operations, according to Gros et al. (2004), degrade ecosystems, which require the restoration of linkages between soil physicochemical parameters, plant colonization, and soil microbial activity (Freppaz et al., 2008).

11.12 Conclusion and the way forward

Increased tourist inflows in popular mountain tourist destinations has accelerated the negative effects of anthropogenic activities in these areas, and mountains, as one of the most vulnerable ecosystems, cannot withstand such pressures for long. Camping, trampling, and other similar activities change soil bulk density, pH, nutrient content, and hydrology, making highland soils more vulnerable to even minor disturbances. To prevent further degradation of highland soils, we have made policy decisions and are working to put them into action on the ground. The isolation, limited access, ruggedness, height, climate, and other characteristics that impede the development and protection of mountains are also characteristics that attract visitors. Capitalizing on the area's inherent assets and advantages, as well as the people who live there, is one way to keep mountain tourism operations on a reasonable scale and effect level. This "asset-based approach" validates the concept of a "unique (tourist) selling point." Participation of stakeholders, particularly mountain residents, but also policymakers and administrators in government (in tourism and tourism-related issues), NGOs, the corporate sector, and, ideally, tourists (or mountain tourism market), is an important feature of sustainable mountain tourism planning. Governments should plan for mountain tourism by creating a 5- to 10-year long-term plan that includes national, regional, and local policies and initiatives. Communities involved in tourism development and management should develop local action plans with the assistance of non-governmental organizations (NGOs) or the government, as appropriate, and these plans should be coordinated with municipalities. Such plans should capitalize on individual areas' unique characteristics and capabilities, with the goal of spreading tourists throughout the area to share benefits and mitigate negative consequences, as well as collaborating to promote regional hotspots for repeat or longer-stay visits. It is impossible to overstate the importance of identifying and developing domestic and regional tourism markets. Tourism development planning should be integrated with other community development and conservation programs in alpine areas to promote a diverse range of livelihood options rather than an over-reliance on tourism. NGOs, trade associations, the commercial sector, community members, and organizations (at all levels, from national to local). To supplement the participatory approach, a strategy of decentralization of decision-making is required, delegating legal authority and responsibility for tourist management and plan execution to numerous stakeholders. Policies should encourage full participation of women in planning, decision-making, and benefit sharing, as well as broad representation of stakeholders from socioeconomic, ethnic, and cultural sectors. When evaluating the economic returns of development initiatives in mountainous areas, mountain tourist planning and decision-making should take into account the true worth and full economic and environmental costs and benefits of mountain resources. A portion of tourism revenue should be directed toward environmental and cultural education, as well as the conservation and preservation of natural and cultural mountain resources. Tourism will benefit communities. Mountain tourism policies and development rules should encourage transparency and equity in tourism development opportunities, leading to widespread benefit sharing (such as tourism taxes, conditions of development linked to local employment, or responsibilities for community development, for example). Government policy and legislation, financial and technical assistance, training and skill development, and other measures should be taken to assist small-scale, locally-owned tourism businesses. All mountain tourism development should be sustainable in terms of reducing impacts on biological resources and ecological diversity, as well as promoting mountain culture conservation and improving mountain people's well-being. Infrastructure (roads and other modes of transportation, power, water, and so on) should be phased in over time to meet expected needs. Demands for mountain tourism and other mountain development must be met while remaining within environmental and social constraints, as defined by comprehensive community development plans and environmental reviews of specific development applications. Infrastructure should be placed, sized, and constructed in such a way that it blends in with the natural and cultural environment while also protecting scenic views. To ensure that tourism issues are addressed, a regulatory structure that coordinates the assessment and approval of tourism development projects across key government ministries should be designed. Government tourism planners should collaborate with neighboring administrations whenever possible to improve transboundary tourism as a tourist attraction destination with development and benefit potential in distant areas.

References

Arnaud-Fassetta, G., Cossart, E., Fort, M., 2005. Hydro-geomorphic hazards and impact of man-made structures during the catastrophic flood of June 2000 in the Upper Guil catchment (Queyras, Southern French Alps). Geomorphology 66 (1–4), 41–67.

Arocena, J.M., Nepal, S.K., Rutherford, M., 2006. Visitor-induced changes in the chemical composition of soils in backcountry areas of Mt Robson Provincial Park, British Columbia, Canada. Journal of Environmental Management 79 (1), 10–19.

Arshad, M.A., Martin, S., 2002. Identifying critical limits for soil quality indicators in agro-ecosystems. Agriculture, Ecosystems & Environment 88 (2), 153–160.

Barni, E., Freppaz, M., Siniscalco, C., 2007. Interactions between vegetation, roots, and soil stability in restored high-altitude ski runs in the Alps. Arctic, Antarctic, and Alpine Research 39 (1), 25–33.

Barros, A., Pickering, C.M., 2014. Non-native plant invasion in relation to tourism use of Aconcagua Park, Argentina, the highest protected area in the Southern Hemisphere. Mountain Research and Development 34 (1), 13–26.

Barros, A., Gonnet, J., Pickering, C., 2013. Impacts of informal trails on vegetation and soils in the highest protected area in the Southern Hemisphere. Journal of Environmental Management 127, 50–60.

Barros, A., Monz, C., Pickering, C., 2015. Is tourism damaging ecosystems in the Andes? Current knowledge and an agenda for future research. Ambio 44 (2), 82–98.

Beedie, P., Hudson, S., 2003. Emergence of mountain-based adventure tourism. Annals of Tourism Research 30 (3), 625–643.

Boori, M.S., Voženílek, V., Choudhary, K., 2015. Land use/cover disturbance due to tourism in Jeseníky Mountain, Czech Republic: a remote sensing and GIS based approach. The Egyptian Journal of Remote Sensing and Space Sciences 18 (1), 17–26.

Bratton, S.P., Hickler, M.G., Graves, J.H., 1978. Visitor impact on backcountry campsites in the Great Smoky Mountains. Environmental Management 2 (5), 431–441.

Cole, D.N., Fichtler, R.K., 1983. Campsite impact on three western wilderness areas. Environmental Management 7 (3), 275–288.

Conway, J.S., 2010. A soil trail? A case study from Anglesey, Wales, UK. Geoheritage 2, 15–24.

Delgado, R., Sanchez-Maranon, M., Martin-Garcia, J.M., Aranda, V., Serrano-Bernardo, F., Rosua, J.L., 2007. Impact of ski pistes on soil properties, a case study from a mountainous area in the Mediterranean region. Soil Use and Management 23, 269–277.

Descroix, L., Mathys, N., 2003. Processes, spatio-temporal factors and measurements of current erosion in the French Southern Alps, a review. Earth Surface Processes and Landforms 28, 993–1011.

Dey, J., Sakhre, S., Gupta, V., Vijay, R., Pathak, S., Biniwale, R., et al., 2018. Geospatial assessment of tourism impact on land environment of Dehradun, Uttarakhand, India. Environmental Monitoring and Assessment 190 (4), 1–10.

Doran, J.W., Jones, A.J., 1996. Methods for Assessing Soil Quality. Soil Science Society of America Journal Special Publication, Madison, WI, p. 410.

Drewnik, M., Musielok, Ł., Prędki, R., Stolarczyk, M., Szymański, W., 2019. Degradation and renaturalization of soils affected by tourist activity in the Bieszczady Mountains (South East Poland). Land Degradation & Development 30 (6), 670–682.

Feng, X.G., Bao, H.S., 1999. Preliminary research on tourist activity influence upon the soil and cover plant of scenic spot. Journal of Natural Resources 14 (1), 75–78.

Figueiredo, M.A., Brito, I.A., Takeuchi, R.C., Almeida-Andrade, M., Rocha, C.T.V., 2010. Compactação do solo como indicador pedogeomorfológico para erosão em trilhas de unidades de conservação: Estudo de caso no Parque Nacional da Serra do Cipó, MG. Revista de Geografia 3, 236–247.

Freppaz, M., Filippa, G., Francione, C., Cucchi, M., Colla, A., Maggioni, M., et al., 2008. Artificial snow production with respect to the altitude, a case study in the Italian Western Alps. Geophysical Research Abstracts, 10, EGU 2008-A-9801, 2008. EGU General Assembly.

Fu, B., Wang, J., Chen, L., Qiu, Y., 2003. The effects of land use on soil moisture variation in the Danangou catchment of the Loess Plateau, China. Catena 54 (1–2), 197–213.

Grabherr, G., Messerli, B., 2011. An overview of the world's mountain environments. In: Austrian MAB Committee (Ed.), Mountain Regions of the World: Threats and Potentials for Conservation and Sustainable Use. Grasl Druck & Neue Medien, Bad Vöslau, Austria, pp. 8–14.

Grismer, M.E., Hogan, M.P., 2005. Simulated rainfall evaluation of revegetation/mulch erosion control in the lake Tahoe basin, 2. Bare soil assessment. Land Degradation & Development 16, 397–404. 64 The Impacts of Skiing on Mountain Environments.

3. Soil physicochemical parameters

Gros, R., Monrozier, L.J., Bartoli, F., Chotte, J.L., Faivre, P., 2004. Relationships between soil physico-chemical properties and microbial activity along a restoration chronosequence of alpine grasslands following ski run construction. Applied Soil Ecology 27, 7–22.

Guzzella, L., Salerno, F., Freppaz, M., Roscioli, C., Pisanello, F., Poma, G., 2016. POP and PAH contamination in the southern slopes of Mt. Everest (Himalaya, Nepal): long-range atmospheric transport, glacier shrinkage, or local impact of tourism? Science of the Total Environment 544, 382–390.

Haeberli, W., 1992. Construction, environmental problems and natural hazards in periglacial mountain belts. Permafrost and Periglacial Processes 3, 111–124.

Hammitt, W.E., Cole, D.N., 1998. Wildland Recreation: Ecology and Management, second ed. John Wiley & Sons, New York, p. 361.

Harden, C.P., 2001. Soil erosion and sustainable mountain development. Mountain Research and Development 21 (1), 77–83.

Helgath, S.F., 1975. Trail Deterioration in the Selway-Bitterroot Wilderness. Research Note INT-193. USDA Forest Service, Intermountain Research Station, Ogden, Utah.

Isselin-Nondedeu, F., Bédécarrats, A., 2007. Influence of alpine plants growing on steep slopes on sediments trapping and transport by runoff. Catena 71, 330–339.

Jamieson, B., Stethem, C., 2002. Snow avalanche hazards and management in Canada, challenges and progress. Natural Hazards 26, 35–53.

Jiang, W.J., 1996. Study on the effect of tourism on the ecological environment of Mountain Emei and protective strategies. Chinese Journal of Environmental Science 17, 48–51.

Jorge, M.C.O., Guerra, A.J.T., Fullen, M.A., 2016. Geotourism and footpath erosion—a case study from Ubatuba, Brazil. Geographical Review 29, 26–29.

Kammer, P., Hegg, O., 1990. Auswirkungen von Kunstschnee auf subalpine Rasenvegetation. Verhandlungen der Gesellschaft für Ökologie 19, 758–767.

Kobayashi, T., Endo, Y., Nohguchi, Y., 2000. Hardening of snow surface — 'natural/artificial'. In: Hjorth-Hansen, E., Holand, I., Löset, S., Norem, H. (Eds.), Proceedings of the Fourth International Conference on Snow Engineering, Snow Engineering. Recent Advances and Developments, Trondheim, Norway, pp. 99–103.

Kumar, D.S., Rana, G., Mairaj, H., 2015. Status and scenario of tourism industry in India—a case study of Uttarakhand. Sustainable Tourism Management 575–585.

Leung, Y., Marion, J.L., 1999. The influence of sampling interval on the accuracy of trail impact assessment. Landscape and Urban Planning 3, 167–179.

Lockaby, B.G., Dunn, B.A., 1984. Camping effects on selected soil and vegetative properties. Journal of Soil and Water Conservation 39 (3), 215–216.

Lynn, N.A., Brown, R.D., 2003. Effects of recreational use impacts on hiking experiences in natural areas. Landscape and Urban Planning 64 (1–2), 77–87.

Marion, J.L., Leung, Y.F., 2001. Trail resource impacts and an examination of alternative assessment techniques. Journal of Park and Recreation Administration 19 (3), 17–37.

Monti, P.W., Mackintosh, E.E., 1979. Effect of camping on surface soil properties in the boreal forest region of northwestern Ontario, Canada. Soil Science Society of America Journal 43 (5), 1024–1029.

Nepal, S.K., Nepal, S.A., 2004. Visitor impacts on trails in the Sagarmatha (Mt. Everest) national park, Nepal. AMBIO: A Journal of the Human Environment 33 (6), 334–340.

Pimentel, D., Harvey, C., Resosudarmo, P., Sinclair, K., Kurz, D., McNair, M., et al., 1995. Environmental and economic costs of soil erosion and conservation benefits. Science 267 (5201), 1117–1123.

Pintar, M., Mali, B., Kraigher, H., 2009. The impact of ski slopes management on Krvavec ski resort (Slovenia) on hydrological functions of soils. Biologia 64 (3), 639–642.

Price, M., 1999. Global Change in the Mountains. Parthenon, New York.

Price, M., Moss, L., Williams, P., 1997. Tourism and amenity migration. In: Messerli, B., Ives, D. (Eds.), Mountains of the World. A Global Priority. Parthenon, New York, pp. 249–280.

Qin, Y.H., Xie, D.T., Wei, C.F., Yang, J.H., Qu, M., 2006. Study on responses of soil ecological environment to impacts of tourist activities. Journal of Soil and Water Conservation 20 (3), 61–65.

Qiu, Y., Wang, Y., Xie, Z., 2014. Long-term gravel–sand mulch affects soil physicochemical properties, microbial biomass and enzyme activities in the semi-arid Loess Plateau of North-western China. Acta Agriculturae Scandinavica, Section B—Soil & Plant Science 64 (4), 294–303.

Rangel, L., Jorge, M.D.C., Guerra, A., Fullen, M., 2019. Soil erosion and land degradation on trail systems in mountainous areas: two case studies from south-East Brazil. Soil Systems 3 (3), 56.

Reeves, D.W., 1997. The role of soil organic matter in maintaining soil quality in continuous cropping systems. Soil and Tillage Research 43 (1–2), 131–167.

Rixen, C., Stoeckli, V., Ammann, W., 2003. Does artificial snow production affect soil and vegetation of ski pistes? A review. Perspectives in Plant Ecology, Evolution and Systematics 5 (4), 219–230.

Sati, S.P., Gahalaut, V.K., 2013. The fury of the floods in the north-west Himalayan region: the Kedarnath tragedy. Geomatics, Natural Hazards and Risk 4 (3), 193–201.

Serrano, E., González-Trueba, J.J., 2005. Assessment of geomorphosites in natural protected areas: the Picos de Europa National Park (Spain). Géomorphologie: Relief, Processus, Environnement 11 (3), 197–208.

Soane, B.D., 1990. The role of organic matter in soil compactibility: a review of some practical aspects. Soil and Tillage research 16 (1–2), 179–201.

Titus, J.H., Landau, F., 2003. Ski slope vegetation of Lee Canyon, Nevada, USA. The Southwestern Naturalist 48 (4), 491–504.

Vashchenko, Y., Biondi, D., 2013. Percepção da erosão pelos visitantes nas trilhas do Parque Estadual do Pico Marumbi, PR. Revista Brasileira de Ciências Agrárias 8 (1), 108–118.

Wang, Z.J., Cai, J., Zhang, Q.X., 2003. A Preliminary study on the effect of tourism disturbance on the soil of Yunmeng Mountain National Forest Park. The Journal of Hebei Forestry Science and Technology (5), 12–15.

Wipf, S., Rixen, C., Fischer, M., Schmid, B., Stoeckli, V., 2005. Effects of ski piste preparation on alpine vegetation. Journal of Applied Ecology 42 (2), 306–316.

Wolf, I.D., Croft, D.B., 2014. Impacts of tourism hotspots on vegetation communities show a higher potential for self-propagation along roads than hiking trails. Journal of Environmental Management 143, 173–185.

Young, R.A., Gilmore, A.R., 1976. Effects of various camping intensities on soil properties in Illinois campgrounds. Soil Science Society of America Journal 40 (6), 908–911.

12

Assessment of soil organic carbon stocks in Sahyadri mountain range Karnataka, India

Pavithra Acharya (G.M)[1], Anil K.S. Kumar[2], K.S. Karthika (Kavukattu)[2], M. Lalitha[2], Syam Viswanath[3], P.A. Lubina[3], Ravi Namasivaya[4], Sruthi Subbanna[4] and M.C. Sandhya[5]

[1]Department of Environmental Science, University of Mysore, Manasagangotri, Karnataka, India [2]ICAR-National Bureau of Soil Survey and Land Use Planning, Hebbal, Bengaluru, Karnataka, India [3]Kerala Forest Research Institute, Peechi, Thrissur, Kerala, India [4]Institute of Wood Science and Technology, Malleswaram, Bengaluru, Karnataka, India [5]Department of Plant Biotechnology and Cytogenetic, Institute of Forest Genetics and Tree Breeding, Coimbatore, Tamil Nadu, India

OUTLINE

12.1 Introduction

Global warming is defined as an increase in average global temperatures caused by natural events and human activities over a long period of time, primarily by increases in physical climate change drivers such as carbon dioxide (9%−26%), methane (4%−9%), water vapor and clouds (36%−72%), and ozone (3%−7%) (Piper, 2019). Its recent acceptance as a greenhouse gas mitigation strategy under the Kyoto Protocol has increased its visibility as a biological carbon(C) sequestration strategy (Saha et al., 2010). SOC's role in the global carbon cycle is gaining prominence as a potentially large and uncertain source of CO_2 emissions in response to projected global temperature increases, as well as a natural carbon sink that can reduce atmospheric CO_2 emissions (Churkina et al., 2010; Jain et al., 2012; Reinmann et al., 2016; Srinivasan et al., 2019). In this regard, mountains received much attention globally due to their rich and diverse biota, high-standing biomass, and greater productivity (Marín-Spiotta et al., 2014; Chen et al., 2019). This has been subjected to massive biotic disturbances and is undergoing fragmentation, habitat loss, and erosion of biodiversity (Schuman et al., 2002; Peres et al., 2010; Jin et al., 2013 Zang et al., 2020). There are several ways through which CO_2 reaches the atmosphere, which could be by combustion of fossil fuels, or from solid wastes, agricultural and forestry activities, deforestation, and other land-use changes (Lal et al., 1997; Tito et al., 2020), or as a result of certain chemical reactions. Soil reaction, texture, structure, and roots and pores constitute the indirect factors that contribute to soil carbon storage (Garcia-Pausas et al., 2008; Carter and Gregorich, 2010; Ajami et al., 2016; Mishra and Mapa, 2019; Yang et al., 2021).

SM spans 160,000 km^2 across the states of Karnataka, Goa, Maharashtra, Gujarat, Kerala, and Tamil Nadu, running parallel to the western coast. It is a UNESCO World Heritage Site as well as one of eight biodiversity hotspots rich in biodiversity listed in the IUCN threatened category (Subbiah and Asija, 1956; Billore and Hemadri, 1969; Bawa and Kadur, 2005; Kalaiselvi et al., 2021).

SOM dynamics in mountains vary with climate conditions, supporting biodiversity among forest species and the potential for carbon sequestration by forest vegetation and soil, implying that global C stocks should be high along the entire track (Manjaiah et al., 2000; Chiti et al., 2018; Simon et al., 2018). Many studies reported that land-use changes to a cultivated land decrease soil C storage (Houghton et al., 1999; Meena et al., 2018) influence of the release of more CO_2 to the atmosphere (Solomon et al., 2002). SM also has fewer disturbed landscape patterns as a result of agricultural practices. observed changes in land use (Liu, 2021; Guo and Gifford, 2002; Guo and Lin, 2018; McGuire et al., 2001; Amanuel et al., 2018). This study found that SOC stock values were higher in subsurface soils than in surface soils due to the absence of anthropogenic activities in forest reserves. Potential carbon sequestration in forest soils is permanently locked deep beneath the soil matrix (Gaikwad et al., 2014; Patwardhan et al., 2016; Kögel-Knabner and Amelung, 2021). More SOC is associated with SOM deposition at mountain ranges. PPP contributes to climate change, followed by SOC at adjacent agricultural land as crop cultivation expands, and proportionate land encroachment (Villarino et al., 2017; Ottoy et al., 2017). When compared to adjacent agricultural and degraded land-use types in western Karnataka, SOC stocks and carbon sequestration potential in PPPs in SM soils of WEF and MDF types are significantly higher. Increased human settlements, transportation infrastructure, dams, and power projects have resulted in significant landscape degradation of forest land (Leverett et al., 2021; Sitanggang et al., 2006). The extent to which these new agricultural lands replace forests, degraded forests, or grasslands has a significant impact on the environmental consequences of expansion (Gibbs et al., 2010; Gibbs and Salmon, 2015). PPPs are the possible solution to restore and rehabilitation in terms of global climate change (Tewari et al., 2014; Tewari and Gadow, 2012).

12.2 Material and methods

12.2.1 Study area

The current study is carried out from 2017 to 2020 at PPPs in wet evergreen and MDF types of SM landscape (Table. 12.1) and a comparison is made in adjoining agricultural and degraded land (Fig. 12.1). Sahyadri Mountain (SM) is well known as a cradle for many streams, Riparian Zones, Paddy Fields, Areca nut plantation, Sugarcane Fields, co-existence of human settlements, without undesirable changes by urbanization and modernization (Figs. 12.2 and 12.3). Tropical forest types in SM are primarily WEF and MDF, which form steep and gradual terrain landscapes at varying elevation levels. Located between 438 and 914 m above sea level. From 1939 to 1951, the Forest Survey of India (FSI) established linear tree increment plots (LTI) for the conservation of SM flora

TABLE 12.1 Description of the PPP in wet evergreen and moist deciduous forests in SM of Karnataka.

Forest type 5 PPP	Wet evergreen forest		Moist deciduous forest		
	Katlekan	Malemane	Karka	Bhagavati	Kulagi
Latitude	14°16′ 33.7″	14°16′17.91″	15°18′ 33.2″	15°07′92.3″	15°09′43.0″
Longitude	74°44′ 45.2.6″	74°42′00.39″	74°41′ 08.5″	74° 45′ 40.1″	74°40′25.2.5″
Elevation (msl)	914	903	552	505.9	438
Year of establishment	1939	1939	1950	1950	1951
Area	3.5	3.3	4	4	4
Annual rainfall (mm)	>4500	>4500	>2000	>1500	>2000
Terrain	Steep	Steep	Gradual	Gradual	Gradual
Water Logging	No	Yes	No	No	No
Roads	Paved 1956	Paved 1956	NA	NA	NA
Power lines	No	No	No	No	No
Fires	No	Occasional during summer			

3. Soil physicochemical parameters

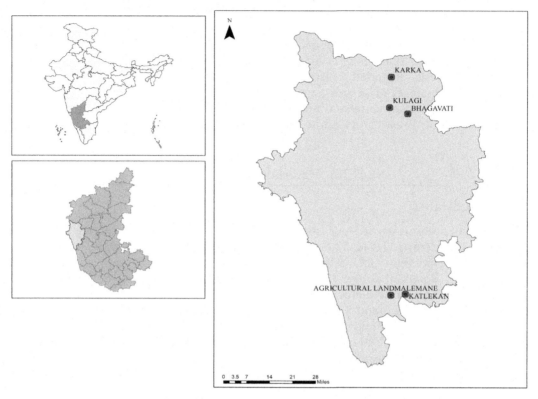

FIGURE 12.1   SM landscape, Central Western Ghats, Uttarakannada-Dharwad districts, Sirsi and Haliyal forest ranges.

FIGURE 12.2   Sampling sites of adjoining agricultural lands located at PPPs in WEF and MDF types (A) paddy fields, (B) areca nut plantation, and (C) sugarcane fields.

3. Soil physicochemical parameters

FIGURE 12.3 Sampling sites of adjoining landscapes (A) degraded lands, (B) human settlement, and (C) River-Riparian Zones.

and fauna. The annual rainfall minimum of 2000 mm varies up to 4500 mm. The identified site to soil sampling and recorded GPS location. A 1-m length, breadth, and depth pit was also dug to collect soil samples. In the protected areas, there are no forest fires, except in the summer due to anthropogenic activities and agricultural practices.

12.3 Laboratory analysis

Evaluation of SOC stocks 1 m depth soil profile dig and categorized into the surface and subsurface soil samples, 0–30 cm (surface soil samples), 30–100 cm (subsurface soil samples), 0–100 cm (soil profile), soil samples collected on-site data were recorded physical and soil morphological studies and brought to the lab for physio-chemical analysis.

Soil profiles were thoroughly examined for all physical and morphological characteristics in accordance with the USDA soil survey manual (Soil Survey Staff, 2010). Horizontal samples were taken for laboratory analysis. To identify dominant environmental factors that contributed to SM soil formation at this level on a regional basis, soils were grouped based on identification characteristics and classified using the USDA soil classification system. The bulk density of samples collected from each horizon was determined using a core sampler method. In 1:2.5 soil-water suspensions, the pH and electrical conductivity were measured (Jackson, 1973).

The soil samples collected horizon-wise were air-dried under the shade, powdered, and sieved using a 2 mm sieve. Particle size analysis was carried out by the international pipette method (Piper, 1966). Particle size distribution analysis made to understand Soil fractions percentage in each pedon USDA guidelines Total sand (2–0.05 mm), Very coarse sand (2–1 mm), Coarse sand (1–0.5 mm), Medium sand (0.5–0.25 mm), Fine sand (0.25–0.1 mm), Very fine sand (0.1–0.05 mm), Silt (0.05–0.002 mm), Clay (<0.002 mm).

Walkley and Black's (1934) Wet Digestion method was used to calculate SOC. Organic carbon in soil samples is oxidized by dichromate-sulfuric acid, and the amount of dichromate left after titration with a standard ferrous solution is determined. The Walkley–Black procedure provides the required temperature by using the heat of dilution of sulfuric acid. This, however, does not allow the dichromate to oxidize all the organically bound carbon. In the Walkley–Black method, an empirical recovery factor of 0.77 was proposed (Visconti et al., 2022; Ping et al., 2015).

In the oxidation of carbon-to-carbon dioxide, each atom (mole) of carbon reacts with two moles (four equivalents) of oxygen. Thus, the equivalent weight of carbon in this reaction is one-fourth of the atomic weight or 3.

$$C + 2(O) \rightarrow CO_2$$

$$\text{Organic carbon in soil } (\%) = \frac{10(V_1 - V_2) \times 0.003 \times 100 \times \text{mcf}}{V_1 \times W \times 0.77}$$

Where, Weight of sample $= W$ gm.

The volume of ferrous ammonium sulfate (FAS) used in blank $= V_1$ mL

The volume of FAS used in Titration $= V_2$ mL

Normality of FAS $= 10/V_1$, because 10 mL of 1.0 N dichromate solution was taken.

The factor 0.003 is for converting meq C into grams C and 077 is the recovery factor for the oxidation.

12.4 SOC stocks estimation

SOC stocks were calculated for each soil profile by adding the stocks of different layers in the proportion of their occurrence within the reference thickness (Hugelius et al., 2014; Meersmans et al., 2008; Grimm et al., 2008). The total organic carbon stock in kg/m^3 soil for each pedons was estimated using the general equation presented below (Grossman et al., 2001).

$$SOC = \frac{L_1 \times SOCP_1 \times \rho33_1 \frac{(1-^V>2_1)}{100} + L_2 \times SOCP_1 \times \rho33_2 \frac{(1-^V>2_2)}{100} + \ldots)}{10}$$

Where, SOC = Soil organic carbon in kg/m^3, SOC% = Soil organic carbon percent, L = Thickness of the layer in cm, ρ = Bulk density (oven dry), $V > 2$ = Volume percent of >2 mm (gravel).

12.5 Results and discussion

Due to the illuviation process that occurs during soil development, SM soil profile studies reveal that the subsurface horizons have higher clay content than the surface horizons (Sharma et al., 2004). The biomass production and fate of organic carbon favored by climate for carbon sequestration with enhanced rhizosphere and mycorrhizal biological activity is determined by high canopy and ground cover (Subbiah and Asija, 1956). SOC values were higher in pedons with a thicker soil horizon, lower bulk density values, and a lower volume of gravel content. These are the direct factors influencing SOC stocks.

Horizon A, B, AB, BA, Bt1, Bt2, Bt3, Bw1, Bw2 undulated structure of pedons strong, coarse, and prismatic parting, strong, small, sub-angular, and angular blocks formed by parent material Textural variations are primarily caused by the parent material, degree of weathering, topography, and time.

Particle size distribution analysis made to understand soil fractions percentage in each pedon U.S.D.A guidelines Total sand (2−0.05 mm), Very coarse sand (2−1 mm), Coarse sand (1−0.5 mm), Medium sand (0.5−0.25 mm), Fine sand (0.25−0.1 mm), Very fine sand (0.1−0.05 mm), Silt (0.05−0.002 mm), Clay (<0.002 mm). Surface and subsurface soils have been subjected to various environmental stress factors such as rainfall, soil erosion, biomass accumulation and rate of decomposition, and so on, as well as related microbial biome activities.

Clay content (in percent) estimated by Pressure Plate extraction (at 1/3 and 15 bar) Apparatus. Clay distribution based on SOM, particle structure or physical arrangement, surface and subsurface soils in PPPs, agricultural land, and degraded land.

Bulk Density is the ratio of mass of fine aggregates by the total volume they occupy, varies 0.86−1.66 mg/m^3. Table. 12.2 Shows coarse aggregates gradually increases, higher the weight of the soil, the pores decrease in turn bulk density increases at subsurface layers (30−100 cm) less in surface soil (0−30 cm).

In soil reaction studies, the clear supernatant of a 1:2.5 soil water suspension prepared for pH measurement can be used to estimate EC. The conductivity of the clear supernatant liquid varies from 4.49 to 5.68, indicating that mountain soils are highly acidic. The pH of all mountain soil samples was found to be slightly acidic (Yu et al., 2020).

Electrical conductivity in soil water system is a measure of the concentration of soluble salts and extent of salinity in the soil and is measured using a conductivity meter minimum 0.07 dS/m to a maximum of 0.116 dS/m (Ajami et al., 2016; Yang et al., 2021).

Organic Carbon varies from a minimum of 0.39% to a maximum of 4.87% found at mountain soil and adjoining land use types. These soils are categorized into four textural classes as defined in USDA: it follows the order of sandy-clay-loam (scl), loam (l), silt clay (sc), and clay (C).

The greater the values OC are found in the hilly zone as a result of high rainfall and high temperatures, which lead to increased biomass production provided by a thick surface horizon with a high percentage of organic carbon (Eusterhues et al., 2003). Forest with eucalyptus exhibited low SOC and SOC stocks due to allelopathic effect resulting in a reduction in biological activity, selective flora, and fauna, more decomposition and removal of C from solum coined by (Anil Kumar et al., 2009, 2015; Lorenz and Lal, 2018; Mandal et al., 2021).

The SOC Stocks in 1 m (0−100 cm profile) under natural forests of SM and, adjoining agricultural and degraded land use types (Table. 12.3).

Factors influencing carbon mineralization in mountain surface and subsurface soils. Carbon and nitrogen storage by deep-rooted tree species is also important in ecosystem surface and subsurface soil. Environmental factors influencing SOC storage in humid and sub-humid environments. Mineralogy, metal concentrations, surface charge, and oxidation state of suspended solids and sediments are all factors in SOM turnover. The application

TABLE 12.2　Study of soil profile under natural forests and adjoining agricultural and degraded land use types SM.

Depth (cm)	Horizon	Coarse fragments (%)	Clay (%)	Bulk Density (mg/m^3)	Soil reaction pH (1:2.5)	Electrical conductivity (dS/m)	Organic carbon (%)	Soil texture
SM PPPs in WEF type- soil profile 1								
0−17	AB	3.31	29.7	1.41	4.81	0.118	3.61	l
17−47	BA	3.31	29.7	1.25	4.79	0.119	3.57	l
47−69	Bt1	4.19	44.6	1.32	4.73	0.118	2.65	sc
69−101 +	Bt2	12.47	39.8	1.6	4.71	0.118	2.22	sc
SM PPPs in MDF type-soil profile 1								
0−23	AB	10.78	33.9	1.23	5.44	0.174	4.87	sc
23−49	BA	14.42	26.3	1.13	5.39	0.174	2.17	C
49−73	Bt1	7.57	23.5	1.33	5.32	0.174	2.49	C
73−101 +	Bt2	20.02	29.8	1.35	5.02	0.174	1.41	C
SM agricultural land-soil profile 1								
0−22	AB	10.89	36.1	0.86	5.55	0.17	4.76	l
22−47	BA	13.59	41.8	0.88	5.49	0.117	2.98	scl
47−74	Bt1	4.66	40.8	1.17	5.31	0.117	2.63	scl
74−101 +	Bt2	12.91	47.1	1.21	5.09	0.116	1.9	C
SM agricultural land soil profile 2								
0−19	A	4.78	53.2	1.34	5.68	0.071	4	C
19−41	Bt1	10.41	29.9	1.32	5.16	0.071	2.25	C
41−65	Bt2	3.29	29.6	1.18	5.22	0.07	2.17	C
65−101 +	Bt3	10.89	36.1	1.66	5.08	0.07	1.35	C
SM degraded land soil profile 1								
0−17	Ap	10.05	56.6	1.45	4.97	0.099	1.42	sc
17−47	AB	4.18	50.6	1.41	4.88	0.099	1.25	sc
47−69	Bt1	5.54	50	1.41	4.65	0.099	1.45	sc
69−101 +	Bt2	6.81	48.2	1.58	4.49	0.099	2.23	sc
SM degraded land soil profile 2								
0−21	A	23.6	18.2	1.5	4.97	0.115	0.62	l
21−43	BA	14.42	26.3	1.34	4.89	0.115	1.56	l
43−70	Bw1	7.57	23.5	1.47	4.81	0.115	0.39	scl
70−101 +	Bw2	4.78	53.2	1.39	4.74	0.116	0.8	C

3. Soil physicochemical parameters

TABLE 12.3 Depth-wise distribution of SOC stocks in SM under WEF and MDF PPP of central Western Ghats Karnataka, India.

Land use type selected for soil profile study WEF and MDF compared to adjoining agricultural and degraded land	Depth-wise distribution 0–30 (cm) 30–100 (cm) and 0–100 (cm) soil organic carbon (SOC) stocks in kg/m^3		
Soil profile depth (in cm)	0–30 (cm)	30–100 (cm)	0–100 (cm)
WEF PPP-1	12.07	28.01	40.09
WEF PPP-2	8.49	25.90	34.40
WEF PPP-3	7.99	26.18	34.17
WEF PPP-4	15.03	16.36	31.39
WEF agricultural land	10.62	14.67	25.30
WEF degraded land	7.60	14.71	22.31
MDF PPP	7.13	21.43	28.57
MDF agricultural land	10.87	13.86	24.73
MDF degraded land	4.91	11.31	16.22
MDF PPP	7.68	14.74	22.42
MDF agricultural land	10.31	9.18	19.48
MDF PPP	5.96	9.70	15.66
MDF agricultural land	3.35	9.86	13.22
MDF degraded land	3.78	7.62	11.40

of organic and synthetic fertilizer to increase productivity on agricultural land influences variation in soil physical and chemical parameters, including SOM status. SOC stocks range between 7.50 and 28.57 kg/m^3 and, in degraded land 4.08–25.90 kg/m^3. The SOC stock in layer 0–30 cm varied from 3.35 to 15.03 kg/m^3 with an average of 8.27 kg/m^3. Under different land-use types texture, structure, consistency, root distribution, effervescence, rock fragments, and reactivity compose the labile carbon pools in mountainous soil analysis. Depth-wise distribution of SOC stock in 30–100 cm: varied from 7.62 to 28.01 kg/m^3 with an average of 15.97 kg/m^3. The SOC stock in PPP soil profile 0–100 cm varied from a minimum of 11.40 kg/m^3 to a maximum of 40.09 kg/m^3. The average estimate of SOC stock in the forest region of WEF types with an average is 24.24 kg/m^3 in PPP due to high soil carbon dynamics and potential carbon sequestration by rangelands. The percentage of SOC gradually decreased as soil depth increased. However, expanding the scope of LTI Plots of forest regions and a proper strategic forest management system can sequester more carbon because a high SOC stock indicates improved soil and land quality and aids in sustainable development. Food and Agriculture Organization's tropical database of classified Landsat scenes was created to examine pathways of agricultural expansion across the major tropical forest regions in the 1980s and 1990s, and to use this

3. Soil physicochemical parameters

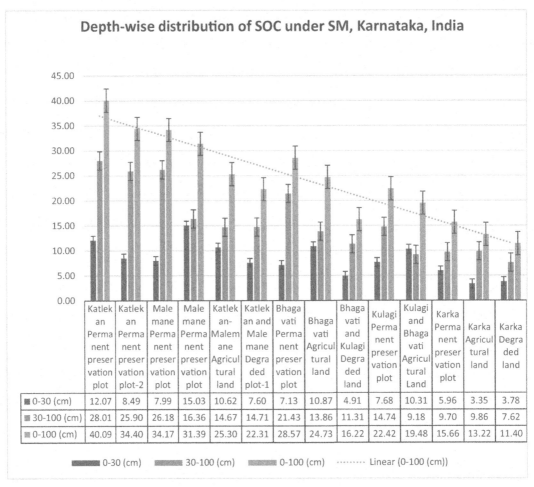

FIGURE 12.4   Graphical representation of SOC stocks at a 1-m depth soil profile, surface, and subsurface soil samples 0–100 cm (soil profile).

information to highlight future land conversions that will most likely be required to meet the mounting demand for agricultural products. Between 1980 and 2000, more than 55% of new agricultural land was created at the expense of intact forests, with the remaining 28% coming from disturbed forests (Fig. 12.4).

12.6 Conclusion

In the earth's atmosphere, biogeochemical cycles are an integral part of the mountain ecosystem. The SOC stocks (kg/m^3) values varied among different agro-climatic zones of southern Karnataka and different land-use systems. Organic carbon percentages are very high, and SOC stock values are also higher in mountainous soil types than in agricultural and degraded land.

This documentation calls for increased research focus on (1) expanding coverage of less studied ecosystems such as PPPs and related land-use types, and understanding the dynamics of land change, (2) Measuring the actual benefits of PPP ecosystem services in forest regions. The state forest department is protecting and establishing more such Linear Increment Plots (LTI) and observation plots for long-term conservation of mountain ecosystems, and (3) Evaluating effectiveness across land-use types and geographical regions.

References

Ajami, M., Heidari, A., Khormali, F., Gorji, M., Ayoubi, S., 2016. Environmental factors controlling soil organic carbon storage in loess soils of a subhumid region, northern Iran. Geoderma 281, 1–10.

Amanuel, W., Yimer, F., Karltun, E., 2018. Soil organic carbon variation in relation to land use changes: the case of Birr watershed, upper Blue Nile River Basin, Ethiopia. Journal of Ecology and Environment 42 (1), 1–11.

Anil Kumar, K.S., Shalima Devi, G.M., 2009. Soil organic carbon stocks as a land quality indicator under the coffee land use system in Karnataka. Journal of Coffee Research 37, 43–65.

Anil Kumar, K.S., Lalitha, M., Kalaiselvi, B., Siddaram, Patil, Nair, K.M., Sujatha, K., 2015. Soil organic carbon stocks: a potential land quality indicator for soils of Western Karnataka. Agropedology 25, 161–168.

Bawa, K., Kadur, S., 2005. Sahyadris: India's Western Ghats—A Vanishing Heritage. Ashoka Trust for Research in Ecology and the Environment (ATREE), Bangalore, India.

Billore, K.V., Hemadri, K., 1969. Observations on the Flora of Harishchandragarh, Sahyadri Range, Maharashtra. Nelumbo - The Bulletin of the Botanical Survey of India 11 (3–4), 335–346.

Carter, M.R., Gregorich, E.G., 2010. Carbon and nitrogen storage by deep-rooted tall fescue (*Lolium arundinaceum*) in the surface and subsurface soil of a fine sandy loam in eastern Canada. Agriculture, Ecosystems & Environment 136 (1–2), 125–132.

Chen, S., Arrouays, D., Angers, D.A., Chenu, C., Barré, P., Martin, M.P., et al., 2019. National estimation of soil organic carbon storage potential for arable soils: a data-driven approach coupled with carbon-landscape zones. The Science of the Total Environment 666, 355–367.

Chiti, T., Díaz-Pinés, E., Butterbach-Bahl, K., Marzaioli, F., Valentini, R., 2018. Soil organic carbon changes following degradation and conversion to cypress and tea plantations in a tropical mountain forest in Kenya. Plant and Soil 422 (1), 527–539.

Churkina, G., Brown, D.G., Keoleian, G., 2010. Carbon stored in human settlements: the conterminous United States. Global Change Biology 16 (1), 135–143.

Eusterhues, K., Rumpel, C., Kleber, M., Kögel-Knabner, I., 2003. Stabilisation of soil organic matter by interactions with minerals as revealed by mineral dissolution and oxidative degradation. Organic Geochemistry 34 (12), 1591–1600.

Gaikwad, S., Gore, R., Garad, K., Gaikwad, S., 2014. Endemic flowering plants of northern Western Ghats (Sahyadri Ranges) of India: a checklist. Check List 10 (3), 461–472.

Garcia-Pausas, J., Casals, P., Camarero, L., Huguet, C., Thompson, R., Sebastia, M.T., et al., 2008. Factors regulating carbon mineralization in the surface and subsurface soils of Pyrenean mountain grasslands. Soil Biology & Biochemistry 40 (11), 2803–2810.

Gibbs, H.K., Salmon, J.M., 2015. Mapping the world's degraded lands. Applied Geography 57, 12–21.

Gibbs, H.K., Ruesch, A.S., Achard, F., Clayton, M.K., Holmgren, P., Ramankutty, N., et al., 2010. Tropical forests were the primary sources of new agricultural land in the 1980s and 1990s. Proceedings of the National Academy of Sciences 107 (38), 16732–16737.

Grimm, R., Behrens, T., Märker, M., Elsenbeer, H., 2008. Soil organic carbon concentrations and stocks on Barro Colorado Island—digital soil mapping using Random forests analysis. Geoderma 146 (1–2), 102–113.

Grossman, R.B., Harms, D.S., Kinngsbury, D.F., Shaw, R.K., Jenkins, A.B., 2001. Assessment of soil organic carbon using the U.S Soil Survey. In: Lal, R., Kimble, J., Follet, R.F., Stewart, B.A. (Eds.), Assessment Methods for Soil Carbon. CRC Press, Boca Raton, pp. 87–102.

Guo, L., Lin, H., 2018. Addressing two bottlenecks to advance the understanding of preferential flow in soils, Advances in Agronomy, Vol. 147. Academic Press, pp. 61–117.

Guo, L.B., Gifford, R.M., 2002. Soil carbon stocks and land use change: a meta-analysis. Global Change Biology 8 (4), 345–360.

Houghton, R.A., Hacker, J.L., Lawrence, K.T., 1999. The US carbon budget: contributions from land use change. Science 285, 574−578. Available from: https://doi.org/10.1126/science. 285.5427.574.

Hugelius, G., Strauss, J., Zubrzycki, S., Harden, J.W., Schuur, E.A.G., Ping, C.L., et al., 2014. Estimated stocks of circumpolar permafrost carbon with quantified uncertainty ranges and identified data gaps. Biogeosciences 11 (23), 6573−6593.

Jackson, M.L. 1973. Soil Chemical Analysis. Prentice Hall of India (Pvt.) Ltd., New Delhi.

Jain, R., Urban, L., Balbach, H., Webb, M.D., 2012. Chapter Thirteen: contemporary issues in environmental assessment. Handbook of Environmental Engineering Assessment 361−447. Available from: https://doi.org/10.1016/B978-0-12-388444-2.00013-0.

Jin, V.L., Haney, R.L., Fay, P.A., Polley, H.W., 2013. Soil type and moisture regime control microbial C and N mineralization in grassland soils more than atmospheric CO2-induced changes in litter quality. Soil Biology & Biochemistry 58, 172−180.

Kalaiselvi, B., Hegde, R., Kumar, K.A., Vasundhara, R., Dharumarajan, S., Srinivasan, R., et al., 2021. Soil organic carbon stocks (SOCS) in different land uses of Western Ghats, Karnataka-A case study. Indian Society for Plantation Crops 49 (2), 146.

Kögel-Knabner, I., Amelung, W., 2021. Soil organic matter in major pedogenic soil groups. Geoderma 384, 114785.

Lal, R., Kimble, M., Follett, B., 1997. Land use and soil C pools in terrestrial ecosystems. Management of Carbon Sequestration in Soil. CRC Press, Boca Raton, pp. 1−9.

Leverett, R.T., Masino, S.A., Moomaw, W.R., 2021. Older eastern white pine trees and stands accumulate carbon for many decades and maximize cumulative carbon. Frontiers in Forests and Global Change 4, 40.

Liu, Z., 2021. Mountain soil characteristics and agrotourism management optimization based on distributed collaboration. Arabian Journal of Geosciences 14 (15), 1−19.

Lorenz, K., Lal, R., 2018. Soil carbon stock. Carbon Sequestration in Agricultural Ecosystems. Springer, Cham, pp. 39−136.

Mandal, U.K., Bhardwaj, A.K., Lama, T.D., Nayak, D.B., Samui, A., Burman, D., et al., 2021. Net ecosystem exchange of carbon, greenhouse gases, and energy budget in coastal lowland double cropped rice ecology. Soil and Tillage Research 212, 105076.

Manjaiah, K.M., Voroney, R.P., Sen, U., 2000. Soil organic carbon stocks, storage profile and microbial biomass under different crop management systems in a tropical agricultural ecosystem. Biology and Fertility of Soils 31, 273−278.

Marín-Spiotta, E., Gruley, K.E., Crawford, J., Atkinson, E.E., Miesel, J.R., Greene, S., et al., 2014. Paradigm shifts in soil organic matter research affect interpretations of aquatic carbon cycling: transcending disciplinary and ecosystem boundaries. Biogeochemistry 117 (2), 279−297.

McGuire, A.D., Sitch, S., Clein, J.S., Dargaville, R., Esser, G., Foley, J., et al., 2001. Carbon balance of the terrestrial biosphere in the twentieth century: analyses of CO2, climate and land use effects with four process-based ecosystem models. Global Biogeochemical Cycles 15 (1), 183−206.

Meena, V.S., Mondal, T., Pandey, B.M., Mukherjee, A., Yadav, R.P., Choudhary, M., et al., 2018. Land use changes: strategies to improve soil carbon and nitrogen storage pattern in the mid-Himalaya ecosystem, India. Geoderma 321, 69−78.

Meersmans, J., De Ridder, F., Canters, F., De Baets, S., Van Molle, M., 2008. A multiple regression approach to assess the spatial distribution of Soil Organic Carbon (SOC) at the regional scale (Flanders, Belgium). Geoderma 143 (1-2), 1−13.

Mishra, U., Mapa, R.B., 2019. National soil organic carbon estimates can improve global estimates. Geoderma 337, 55−64.

Ottoy, S., De Vos, B., Sindayihebura, A., Hermy, M., Van Orshoven, J., 2017. Assessing soil organic carbon stocks under current and potential forest cover using digital soil mapping and spatial generalisation. Ecological Indicators 77, 139−150.

Patwardhan, A., Pimputkar, M., Mhaskar, M., Agarwal, P., Barve, N., Gunaga, R., et al., 2016. Distribution and population status of threatened medicinal tree *Saraca asoca* (Roxb.) De Wilde from Sahyadri−Konkan ecological corridor. Current Science 1500−1506.

Peres, C.A., Gardner, T.A., Barlow, J., Zuanon, J., Michalski, F., Lees, A.C., et al., 2010. Biodiversity conservation in human-modified Amazonian forest landscapes. Biological Conservation 143 (10), 2314−2327.

Ping, C.L., Jastrow, J.D., Jorgenson, M.T., Michaelson, G.J., Shur, Y.L., 2015. Permafrost soils and carbon cycling. Soil 1 (1), 147−171.

3. Soil physicochemical parameters

Piper, C.S., 1966. Soil and Plant Analysis. Hans Publisher, Bombay.

Piper, C.S., 2019. Assessment of carbon sequestration potential of soils under different agro-climatic zones of southern Karnataka. M.Sc. Ag. Thesis, UAS, Bangalore, 163, In press.

Reinmann, A.B., Hutyra, L.R., Trlica, A., Olofsson, P., 2016. Assessing the global warming potential of human settlement expansion in a mesic temperate landscape from 2005 to 2050. The Science of the Total Environment 545, 512−524.

Saha, S., Nair, P.K.R., Nair, V.D., Kumar, B.M., 2010. Carbon storage in relation to soil size-fractions under tropical tree-based land-use systems. Plant and Soil 328, 433−446.

Schuman, G.E., Janzen, H.H., Herrick, J.E., 2002. Soil carbon dynamics and potential carbon sequestration by rangelands. Environmental pollution (Barking, Essex: 1987) 116 (3), 391−396.

Sharma, S.S., Totawat, K.L., Shyampura, R.L., 2004. Characterisation and classification of salt-affected soils of southern Rajasthan. Journal of the Indian Society of Soil Science 52, 209−213.

Simon, A., Dhendup, K., Rai, P.B., Gratzer, G., 2018. Soil carbon stocks along elevational gradients in Eastern Himalayan Mountain forests. Geoderma Regional 12, 28−38.

Sitanggang, M., Rao, Y.S., Nayan, Ahmed, Mahapatra, S.K., 2006. Bio-inoculants on growth and yield of betel vine 246 Characterization and classification of soils in watershed area of Shikohpur, Gurgaon district, Haryana. Journal of the Indian Society of Soil Science 54, 106−110.

Soil Survey Staff, 2010. Keys to Soil Taxonomy, Eleventh Edition. U.S.D.A.: Washington, DC.

Solomon, D., Fritzsche, F., Lehmann, J., Tekalign, M., Zech, W., 2002. Soil organic matter dynamics in the sub-humid agro-ecosystems of the Ethiopian Highlands: evidence from Natural 13C abundance and particle-size fractionation. Soil Science Society of America Journal 66 (3), 969−978. Available from: https://doi.org/10.2136/sssaj2002.0969.

Srinivasan, R., Natarajan, A., Anil Kumar, K.S., Lalitha, M., 2019. Carbon stocks in major cashew growing soils of coastal Karnataka, India. Journal of Plant Breeding and Crop Science 47, 55−61.

Subbiah, B., Asija, C.L., 1956. A rapid procedure for the estimation of available nitrogen in soils. Current Science 25, 270−280. In press.

Tewari, V.P., Gadow, K., 2012. Forest observational studies in India. Forest Observational Studies 20, 75.

Tewari, V.P., Sukumar, R., Kumar, R., Gadow, K., 2014. Forest observational studies in India: past developments and considerations for the future. Forest Ecology and Management 316, 32−46.

Tito, R., Vasconcelos, H.L., Feeley, K.J., 2020. Mountain ecosystems as natural laboratories for climate change experiments. Frontiers in Forests and Global Change 3, 38.

Villarino, S.H., Studdert, G.A., Baldassini, P., Cendoya, M.G., Ciuffoli, L., Mastrángelo, M., et al., 2017. Deforestation impacts on soil organic carbon stocks in the Semiarid Chaco Region, Argentina. The Science of the Total Environment 575, 1056−1065.

Visconti, F., Jiménez, M.G., de Paz, J.M., 2022. How do the chemical characteristics of organic matter explain differences among its determinations in calcareous soils? Geoderma 406, 115454.

Walkley, A., Black, C.A., 1934. Estimation of organic carbon by chromic acid titration method. Soil Science 37: 29−38. http://www.ucusa.org: The Planets' temperature is rising: Report published, 2016.

Yang, Y., Loecke, T., Knops, J., 2021. Surface Soil Organic Carbon Sequestration under Post Agricultural Grasslands Offset by Net Loss at Depth.

Yu, Z., Chen, H.Y., Searle, E.B., Sardans, J., Ciais, P., Penuelas, J., et al., 2020. Whole soil acidification and base cation reduction across subtropical China. Geoderma 361, 114107.

Zang, Z., Deng, S., Ren, G., Zhao, Z., Li, J., Xie, Z., et al., 2020. Climate-induced spatial mismatch may intensify giant panda habitat loss and fragmentation. Biological Conservation 241, 108392.

Soil development on alluvial fans in the mountainous arid regions: a case study of Spiti valley in North-western Himalaya, India

Amit Shoshta[1] and Sachin Kumar[2]

[1]Department of Geography, G B Pant Memorial Government College, Rampur Bushahr, Shimla, Himachal Pradesh, India [2]Department of Geography, Government Degree College, Shahpur, Kangra, Himachal Pradesh, India

O U T L I N E

13.1 Introduction

Soil is a very precious resource that is essential for the survival of life and human civilization on earth as well as cropland and livelihood. Soil formation occurs only at a few sites in the mountainous area, making it a scarce resource. Alluvial fans, along with river terraces, are one of those sites. Alluvial fans are prominent depositional landforms that form at the boundaries of a highland area and lowland plain area, and are a manifestation of a transitional environment between a degrading upland area and aggradational adjacent lowland area (Beaty, 1963; Harvey, 1997). They are suitable soil development sites due to their unique nature of trapping sediments transported by feeder channel/transporting agent from the catchment area, as well as their relatively gentle and sediment collecting surface. The purpose of this chapter is to make a modest attempt to study soils of such alluvial fans in the Spiti valley, which is located in the North-Western Himalayas. This research will most likely provide useful information about soil characteristics and soil formation processes in cold and arid mountainous regions in general. The chapter begins with a brief overview of pedogenesis on alluvial fans in general, followed by a description of the study area and methodology used. The following sections present and discuss the findings. Finally, the chapter summarizes the findings' implications for policy makers, development practitioners, and researchers.

13.2 Soil in arid mountainous region: a brief overview

Soil development is relatively a long process. Various factors such as climatic conditions (Parker, 1995; Zembo et al., 2012; Rashidi et al., 2018), organisms (vegetation cover and soil fauna) (Norton et al., 2003), topographic relief (Harden et al., 1991), parent material and time i.e., length of soil-forming process (Gile and Hawley, 1966; Machette, 1985; McFadden, 1988) influence soil formation and also the types of soil that result during soil formation process or pedogenesis (Retallack, 1988; Schaetzl and Anderson, 2005). Soil development does not occur uniformly in space and time because these factors vary spatially and temporally. Nonetheless, as several researchers have shown, soil development increases systematically with increasing surface age (Gile and Hawley, 1966; Machette, 1985; Reheis, 1987a,b; McFadden, 1988).

Erosion and leaching rates are low in arid regions. As a result, many arid region soils are characterized by the accumulation of materials such as salt and wind-blown dust, as these materials accumulate faster than they are eroded, particularly during dry and dusty periods. Local weathering products and organic matter also help with soil development. Due to the scarcity of vegetation cover, desert soils typically contain only trace amounts of organic material (Watson, 1992; Dunkerley, 2011). Because of the slow overall rate of landscape change, arid region soil-formation processes span long time periods.

Arid region soils generally exhibit the USDA classification system's aridisols, entisols, and inceptisols. In arid climates, the rate of rock alteration/soil development is relatively slow, taking tens of thousands of years. Soils with a very slight degree of soil development are one of the most common features of the arid landscape, either due to a lack of time or due to extremely unfavorable conditions. Entisols are the name given to such soils. The characteristics of the parent material are little altered by soil formation processes in these soils. As a result, these soils are as diverse as their parent materials, which include alluvium, till, and sand dunes, as well as a variety of rocks (Campbell and Claridge, 1992; Watson, 1992).

Inceptisols, or young soils, are another important arid region soil category that represents a stage in soil development beyond entisols but still falls short of the degree of development found in other soils. They could have accumulated clay in the subsurface horizon and moderately weathered it.

Aridisol is the third most important soil group in arid regions, with a higher degree of soil development than entisols and inceptisols. Aridisols have a light-colored, soft, and often vesicular surface horizon (McFadden et al., 1998). As precipitation in such regions is insufficient to leach/illuviate salts, thus these soils have shallow (less than 1 m deep) calcareous, gypsiferous or salty horizons. These types of cement accumulate in different forms such as powdery, nodules, or continuous layers. Although subsurface horizons that are not cemented by salts, carbonates, or sulfates may be similarly friable and silty, many aridisols have a clayey subsurface horizon. Clay and carbonate are thought to be derived from weathering and flushing down the profile of extremely fine-grained dust of feldspar and other easily weatherable minerals in these soils (Reheis et al., 1992; Alonso-Zarza et al., 1998).

Aside from these soil groups, desert pavement and rock varnish are common surficial features of arid region soils. Pavement forms on the surface as a result of fines being removed by wind and water, leaving the gravel as a lag deposit (Dregne, 1976); or by percolation of fine particles between surface stones with alternate freeze and thaw cycles, which in turn results in the consistent upward movement of large debris towards the surface (McFadden et al., 1998; Gustavson and Holiday, 1999; Ugolini et al., 2008). On the surface of a rock, rock varnish is a thin and darkly colored coating of amorphous silica and clay minerals, oxides of manganese and iron, and traces of some minor compounds such as copper and zinc oxides (Dorn and Oberlander, 1982; Dorn et al., 1987, 1992; Dorn, 1988, 1991; Nash, 2011; Aulinas et al., 2014). The amount of manganese oxide in the varnish determines its color. The higher the proportion of manganese oxide, the darker the varnish (Dorn and Oberlander, 1982; Dorn et al., 1987; Dorn, 1988).

Although all soil processes, including the production of complex compounds, mineral weathering, organic matter accumulation, and the liberation of various elements/compounds along with their accumulation in subsurface horizons, are active on mountains and in mountainous arid regions. However, pedogenesis in mountainous cold regions is less rigorous than in humid and sub-humid tropics (Chesworth, 1992). Mountainous cold regions are distinguished by low temperatures and, in some cases, permafrost. As a result, soil temperatures fluctuate significantly between day and night (Birse, 1980). Up to 165 freeze/thaw cycles per year were observed in the French Alps (at 2500 m) (Legros, 1992). This causes the bare rocks to fragment and screes to form. As a result, most glacial valleys in such areas are covered by thick layers of debris. Chemical weathering, which is the dominant weathering process in tropical and temperate regions, becomes less important as latitude increases, while physical processes become much more important in soil formation. As a result, the soils of cold climate regions differ greatly from those of tropical and temperate regions.

Campbell and Claridge (1975) have reported that weathering and soil formation in cold desert regions of the earth varies from almost nothing on young surfaces (<50,000 years) to very high on the oldest surfaces (Miocene or older). A surface stone pavement and several somewhat indistinct horizons, including a pale colored surface horizon, an oxidized horizon, and a salt horizon, can be found on well-weathered surfaces of typical cold desert soils. Finer material weakly consolidates coarse-textured, bouldery, or pebbly sands in soil. In these areas biological activity is at a very low level and has minimal impact on soil weathering processes (Campbell and Claridge, 1992) thus, organic matter content is extremely low. While many soils have detectable microorganism populations, many sites are devoid of plant or animal life (Campbell and Claridge, 1987).

Though mountainous cold region soils are similar to hot arid region soils to some extent, the former differ from the latter in terms of topography, surrounding environments, climatic conditions, precipitation type, and quantum of organic activities, among other things. In contrast to warm and humid regions, where thick soil covers a large portion of the landscape, the surfaces of mountainous arid regions/deserts are only partially covered with it. Because the landscape in the mountainous arid region is too steep for soil accumulation and development, it is mostly covered with detritus material and bed-rock outcrop. Only a few (relatively low and gently sloping) sites accumulate fluvial/aeolian sediments or finer particles. As previously stated, alluvial fans (along with river terraces) are examples of such sites. Soil development on alluvial fans, on the other hand, does not occur uniformly in these regions. Several factors influence the presence and development of soil on fans, which vary spatially and temporally. The presence of soil in the alluvial fan stratigraphic column represents a relatively stable period of fan exposure. This discontinuity in the depositional phase is caused by climate change, tectonics, or a change in base level (Bowman, 2019).

Various scientists have explored different aspects of soils in an alluvial fan environment. Such as chronosequence (McFadden, 1985; McFadden, 1988; Harden et al., 1991; Reheis, 1987a; Rashidi et al., 2018); pedogenesis (Talbot and Williams, 1979; Kesel and Spicer, 1985; Mills and Allison, 1995); desert pavement and varnish (McFadden et al., 1989; Dickerson et al., 2013; Regmi and Rasmussen, 2018); calcification (Lattman, 1973; Machette, 1985; Reheis et al., 1992); fertility (Bahrami and Ghahraman, 2019), etc. The effect of alluvial fan processes on the soil characteristics has also been evaluated by some authors (McCraw, 1968; Kesel and Spicer, 1985; Waters and Field, 1986; Field, 1992; Hill, 1993; Parker, 1995; Mills and Allison, 1995; White and Walden, 1997; Butterworth et al., 2000; Norton et al., 2003, 2007; Williams et al., 2013; Dickerson et al., 2015). Table 13.1 represents some studies dealing with soil development on alluvial fans in different parts of the world.

Mountains, as well as alluvial fans, cover a considerably large part of the world's arid regions (e.g., in southwest USA 38.1% and 31.4% of the total area of the arid landscape is occupied by mountains and alluvial fans, respectively) (Thomas, 2011b). Similarly, 24% of the total global area of arid regions is situated in cold deserts (Thomas, 2011b) that mainly spread from east (China) to west (Canada) in the mid-latitudes of the northern hemisphere (Thomas, 2011a). Furthermore, because the fans of Spiti valley (an intermountain cold desert) are least vegetated and altered by anthropogenic activities, they provide excellent conditions for understanding soil development on these features.

Furthermore, arid-region soils serve as a significant methane sink (which is a strong greenhouse gas). Likewise, carbonates that accumulate at depth in many desert soils play an important role in global carbon cycling and storage (Hirmas et al., 2010). As at lower temperatures, CO_2 dissolves easily (Clement and Vaudour, 1967), thus cold desert mountainous soils may act as an even better carbon sink.

3. Soil physicochemical parameters

TABLE 13.1 Studies dealing with soil development on alluvial fans in different parts of the world.

Area	Climatic conditions	Theme of the study	References
W. United States	Semi-arid	**Allogenic** control	Meek et al. (2020)
S Nevada	Semi-arid	Calcium carbonate cementation	Lattman (1973)
SE Spain	Semi-arid	Calcrete fossilization	Stokes et al. (2007)
Spain	Semi-arid	Calcrete development	Alonso-Zarza et al. (1998)
Nevada, US		Carbonate calcrete development	Reheis et al. (1992)
Iran	Arid	Chronology of soil development	Rashidi et al. (2018)
S. California	Arid	Chronosequence	McFadden (1985)
Southern Apennines, Italy	Humid Temperate	Climatic and tectonic control on pedogenesis	Zembo et al. (2012)
California	Arid	Climatic influence on soil development	Wells, McFadden and Dohrenwend (1987)
S. California	Arid	Climatic influence on soil development	McFadden (1988)
UAE and Oman	Arid	Desert pavement	Al-Farraj and Harvey (2000)
Mojave Desert, California	Arid	Desert pavements	McFadden et al. (1987)
Dead Sea region	Semi-Arid	Evolution of reg (gravelly) soil	Amit and Gerson (1986)
Southern New Mexico	Semi-arid	Geomorphic surfaces and surficial deposits	Ruhe (1967)
	Arid	Geomorphological effect on soil development	Parker (1995)
Tunisia	Semi-Arid	Iron oxide enrichment	White and Walden (1997)
		Modeling of infiltration on arid fans	Blainey and Pelletier (2008)
Hanaupah Fan, Death Valley, CA	Arid	Pedogenesis and geomorphology	Stadelman (1989)
Sado Basin, S-Portugal	Warm-Sub humid	Pedogenic processes	Pimentel (2002)
		Pedostratigraphic models for fan deposits	Wright and Zarza (1990)
Great Basin	Semi-Arid	Rate of soil development	Harden et al. (1991)
Sardinia, Italy	Warm Mediterranean	Relative age and climatic influence	Carboni et al. (2006)
Mojave Desert, California	Arid	Relative age estimation	McFadden et al. (1989)
Roan Mountain, North Carolina		Soil development	Mills and Allison (1995)

(Continued)

3. Soil physicochemical parameters

TABLE 13.1   (Continued)

Area	Climatic conditions	Theme of the study	References
Costa Rica	Tropical Humid region	Soil development on fans	Kesel and Spicer (1985)
Central Niger	Semi-arid (hot)	Soil development	Talbot and Williams (1979)
Iran	Arid	Soil fertility	Bahrami and Ghahraman (2019)
New Zealand	Humid	Soil development and pattern	McCraw (1968)
New Mexico	Arid	Soil-geomorphic relations	Gile (1975), Gile and Hawley (1966)
Costa Rica	Tropical Humid region	Soil-geomorphology	Camacho et al. (2020)
		Soil-geomorphology, pedology, and surficial processes	McFadden and Knuepfer (1990)
		Vesicular horizon	McFadden et al. (1998)

Despite these benefits, the soils of mountainous cold desert regions, particularly in alluvial fan environments, have received little attention. The current study in the Spiti Valley of the North-Western Himalayas is a small step in that direction.

13.3 Study area

Spiti Valley, the study area for this work, is located in the state of Himachal Pradesh in northwestern India. It is physically located in the trans-Himalaya region, between the Zanskar Mountains (in the north) and the Greater Himalaya range (in the south) . It runs north-west to south-east, parallel to the general trend (geological strike) of the western Himalayas. The valley extends between 31° 48′N to 32° 36′N latitudes and 77° 38′E to 78° 40′E longitudes. It is drained by the Spiti river, which is the largest tributary of the Satluj river in Himachal Pradesh. During its course, the Spiti river is joined by various tributaries *viz.* Pin, Lingti, Parechu, Ratang, Shila, Parilungbi, Talking, Gyundi, Mane etc. The length of the valley is about 181 km from the source of the Spiti river (at Kunzum La) to its confluence with the Satluj river (at Khab). The Spiti basin along with Kinnaur (in the East) and Lahaul (in the West), is the south-eastern extension of the Zanskar basin of Ladakh. These two segments combinely form the Spiti-Zanskar basin, which is the Largest Tethyan basin in the Indian Himalaya, and covers ∼35,000 km^2 area (Mazari and Bagati, 1991). Spiti valley is one of the most rugged and inhospitable areas in Himachal Pradesh. This valley has the distinction of having one of the highest inhabited villages (connected to the road) of the world i.e., Komic (4552 m AMSL). It also has Asia's highest post office at Hikkim (4342 m AMSL).

This region in the northwestern Greater Himalaya is most spectacular in terms of pale-ontology, tectonics, biostratigraphy, geomorphology, and climatology (Jhingran, 1981). The valley's current physiography and structure are closely related to the events that resulted in the formation of the Himalayan mountain system. The Spiti basin exposes a spectacular and nearly continuous Proterozoic-Palaeozoic-Mesozoic-Tertiary sequence of rocks mea-suring 11,000 m in thickness; as a result, the Spiti valley is also known as the "Museum of Indian Geology" (Fuchs, 1982; Bagati, 1990; Mazari and Bagati, 1991; Srikantia and Bhargava, 1998; Bhargava, 2008).

The older rocks are crystalline and highly metamorphic. The degree of metamorphism decreases as the age of the rocks decreases. Older rocks are generally found along the basin's margins. In general, the age of these rocks decreases towards the center. All of the depositional features along the Spiti river, which runs almost through the valley, are primarily Quaternary (Bhargava and Bassi, 1998).

In the study area, earlier mentioned rocks are generally folded into NW-SE trending anticlines and synclines. Though few crosses folds, i.e., striking NE-SW are also present (Ameta, 1979; Mazari and Bagati, 1991; Bagati and Suresh, 1991). Two sets of active faults are also present in the Spiti valley, which strike parallel to longitudinal (i.e., NW-SE) and transverse (i.e., NE-SW) fold axes (Ameta, 1979).

The study area is tectonically active and is undergoing upliftment consistently (Ameta, 1979; Bagati and Thakur, 1993; Anoop et al., 2012). Folds and faults, frequent earthquakes and seismites, subsidence and upliftment, occurrences of massive landslides, lacustrine deposits, lateral shifting of streams, strath terraces, fault scarps, thrusts, and other evidence of tectonic activity in the study area are plentiful (Ameta, 1979; Mazari and Bagati, 1991; Banerjee et al., 1997; Sah and Virdi, 1997; Bhargava and Bassi, 1998; Singh and Jain, 2007; Bhargava, 2008; Phartiyal et al., 2009; Anoop et al., 2012). Additionally, modern seismic data of the last five decades exhibit the region as seismically active (Khattri et al., 1978).

The overall drainage pattern of the Spiti river is semi-trellised (Ameta, 1979) and many of its important tributaries such as Ratang, Shilla, Pin, Lingti etc. flow through fault zones and join it at the right angle. During its course, the Spiti river itself flows through Hansa, Gyundi, and Spiti fault zones between Hansa and Mane villages. In the lower parts of the valley, Spiti river flows along Kaurik-Chango fault (Bhargava, 1990; Anoop et al., 2012). From its origin (at Kunzum, ~4950 m) to its mouth (at Khab, 2831 m), Spiti river experiences a notable fall of ~2119 m in a short distance of ~181 km. Thus, suggesting an average gradient of ~11.7 m/km.

Spiti valley experiences dry temperate to arctic weather conditions. The area's arid cli-matic conditions are primarily attributed to the very high altitudes of the consistently uplifting Pir Panjal and Greater Himalaya ranges. These ranges block the northward pene-tration of the southwest monsoon, resulting in a lack of moisture and vegetation in the area (Owen et al., 1998; Pant, 2003). Additionally, the valley's northern location (i.e. north of ~31.5°N latitude), and very high average elevation (~3900 m) also enhance arid and cold climatic conditions here (Marh and Rana, 2014). However, due to the W-E trending western disturbances, it receives fairly heavy precipitation in the form of snowfall during the winters. These disturbances move nearly parallel to the orientation of these mountain ranges, allowing them to penetrate deep into the intermountain valleys (Marh, 2000;

Burbank et al., 2003; Pant, 2003). The average elevation of the mountain ranges is over 5485 m AMSL (Ameta, 1979; Sah and Virdi, 1997). Higher parts of the valley remain snow-covered all year. Many glaciers and streams are fed by these snow-covered fields. As a result of the area's arid climatic conditions, vegetation is scarce and largely restricted to some gentler and fertile patches along river channels and canals. Slopes are steep, bare, and covered in weathered detritus material, and they are close to failure (Bookhagen et al., 2005).

13.4 Methodology

For general geomorphological analysis, a combination of topographic maps (by the Survey of India, at a scale of 1:50,000, as well as the U.S. Army Maps produced under series U 502, viz. NI 43–16, NH 43–4, NH 44–13 and NH 44-I), Digital Elevation Map of the Spiti valley (at 12.5 m resolution), and Google Earth imageries was used. On selected alluvial fans along the Spiti valley, detailed field work was carried out to examine vertical sections and surficial characteristics. Following that, reconnaissance and fieldwork were carried out, which included identification, qualitative observation, and quantitative measurements of the various features in the study area along the Spiti river.

13.4.1 Selection of alluvial fans

On the basis of accessibility, alluvial fans in the study area were divided into two groups. For this study, only accessible alluvial fans were taken into account. A list of all accessible alluvial fans, along with their elevations, was prepared to ensure that alluvial fans were represented from highest to lowest elevation. These fans were further classified into two groups based on the presence or absence of a perennial feeder stream. Both types of alluvial fans were chosen for the survey in order to better understand the development of soils/fans in both systems. Smaller alluvial fans were chosen so that they could be surveyed effectively in less time and with less labor input. The study focused on anthropogenically non/least disturbed alluvial fans. Furthermore, both multi-stage and single-stage alluvial fans were considered for selection. All of these criteria were obtained from Google Earth images prior to the field visit and were verified in the field during the reconnaissance survey. Five alluvial fans were chosen from Losar, Hansa, Kawang, Tabo, and Lari.

13.4.2 Fieldwork

To collect qualitative data, these alluvial fans were subjected to extensive fieldwork. Simple measuring tools (such as a clinometer and measuring tape), a digital camera, and GPS were employed. This was done in the following manner:

- To understand the general morphology of the alluvial fans, features such as the location of the intersection point, fan entrenchment, active depositional lobe, material deposited by different processes, abandoned channels, distributary streams, rills-gullies, terraces, scarps, and so on were observed in the study area.

- During the field survey, sedimentological evidences present on the surface of alluvial fans as well as preserved in vertical sections of deposits along roads, abandoned channels, and incised channels (from apex to distal part) were observed to help understand soil development.
- For distinguishing different depositional events, size of particles, sorting the shape of deposited material and individual particles, the angularity of particles, stratification, boundary contact (Hooke, 1987; Cooke et al., 1993), color and texture of deposits (Costa, 1984, 1988), presence of cut and fill structures and cross-bedding (Kochel, 1990) were observed in the field.
- Surficial characteristics such as the presence of desert pavement, desert varnish, vesicular horizon, and so on were also observed in the field.
- In the field, photographs of various features were taken for visual interpretation of qualitative data. Photographs of alluvial fans and their surrounding environments were taken from distant and higher locations to better understand them.
- About 150−200 GPS readings were taken at random on the surface of each alluvial fan, as well as along the margins of major geomorphic features. These readings were used to create general geomorphic maps of alluvial fans, demarcate features, sections, scarps, interpolate contours, and so on. The maps generated were also used to create profiles of fan surfaces in order to better understand the characteristics of these fans and the factors that control them.

13.5 Results and discussions

13.5.1 Soil development on alluvial fans

13.5.1.1 Soil development in vertical sections

Sedimentological evidences preserved in vertical sections were observed near the apices/proximal parts, middle and distal parts of all five fans to examine soil development at these locations (especially at sites where the walls of the channel were near vertical and thus the probability of deposition of fine particles which may obliterate the underlying reality was negligible). These vertical sections were examined along incised/abandoned channels, roads, and truncated parts/scarps on alluvial fan surfaces. Field observations show that large, angular, unstratified, and unsorted sediments dominate the entire vertical sections of the surveyed fans, not just the proximal or middle parts. On all surveyed fans, these sections (particularly at the proximal and middle parts) show negligible soil development (Fig. 13.1).

Many boulders with diameters greater than one meter were also found in these sections, indicating that these sediments were deposited by high energy depositional events. Furthermore, the relatively young age of these deposits/sections is indicated by their relatively fresh color, lower degree of weathering, and lower quantity of fine particles (particularly in the proximal and middle parts).

Vertical sections of the Tabo, Hansa, and Kawang fans show negligible soil development and are dominated by angular boulders and large sized sediments (Fig. 13.2). However, thick layers of fine particles exist in these sections of the Lari and Losar fans, particularly at the

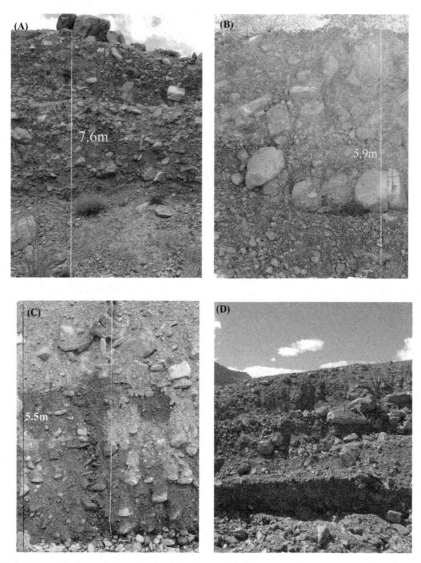

FIGURE 13.1   Vertical sections along the incised channel at the middle parts of Lari (A), Tabo (B), Hansa (C), and Losar (D), fans respectively. Observe the entire deposition in these sections by the high energy depositional events as suggested by large and highly angular boulders imbedded in these sections. Also observe the absence of fluvial strata and palaeo soils in these sections. Height of fan surface from stream bed in images "D" is 4.8 m.

truncated parts of these fans. Cross-bedding and impressions of current ripples on these sediments (Fig. 13.2D) indicate that these layers of fine particles are transported/deposited by a feeder channel or regional trunk stream, such as the Spiti river. Furthermore, the thickness of these layers decreases as one moves closer to the fan apex (as seen along the rill/gully). This fact once again confirms deposition by trunk stream, most likely during a flood. As a result, these fine particle layers can be described as transported soil.

3. Soil physicochemical parameters

FIGURE 13.2 Vertical sections at the distal parts of Tabo (A and C), Lari (B), and Losar (D) fans. Observe the dominance of large, angular, and unsorted boulders in "A" and "C" images. White colored deposits in image "C" (in ellipse) display calcretization. Boulders in image "B" are well rounded and matrix supported. The Spiti river has truncated the distal parts of all these fans and it's bed is also visible in image "D." Also observe the cross bedded deposits of fine particles capped by a relatively thin vesicular horizon in image "D."

All of these facts show that there was no significant soil development during the aggradational phase of these fans. This implies that aggradation occurred relatively quickly and continuously, and that the time between different depositional events was insufficient to cause soil development. This also suggests that the entire deposition was either caused by

3. Soil physicochemical parameters

a single large depositional event of very high energy, or by a series of depositional events of moderately high energy that occurred at irregular intervals.

The latter appears to be more likely. Previous research indicates that if the sedimentation rate is high, the fan surface will rapidly erode and either no soil formation will occur or the soils will be weakly developed. In contrast, at abandoned/inactive sites, soil development is relatively stronger (Blair and McPherson, 1994; Bowman, 2019).

Earlier studies that deal with and general geomorphology of the area (such as Ameta, 1979; Bhargava and Bassi, 1998; Srivastava et al., 2013; Marh and Rana, 2014), and different landforms along the Spiti river such as lacustrine deposits (Bookhagen et al., 2005; Phartiyal et al., 2009; Anoop et al., 2012), landslides (Bookhagen et al., 2005; Dortch et al., 2009; Bagati and Thakur, 1993) terraces (Mazari and Bagati, 1991), etc., have reported that these landforms are principally quaternary in origin. Bookhagen et al. (2005), Dortch et al. (2009), Phartiyal et al. (2009) and Srivastava et al. (2013) have suggested that high energy depositional events and mass wasting took place in the Spiti valley primarily during the phases of intensified monsoon (50−30 Ka) and (14−8 Ka) and caused valley filling. However, during 28−14 Ka, i.e., the last glacial phase of weak monsoon, sedimentation was caused by glacial processes. Increased sediment supply from glacial outwash caused valley aggradation during this phase (Srivastava et al., 2013). Short periods of low sediment to water ratio and higher stream power interrupted sedimentation, causing valley/fan incision.

These findings imply that the lack of soil development in vertical sections of surveyed alluvial fans is primarily due to the younger age of these landform features, as well as higher rates of sediment deposition by high energy events. In this arid landscape, the relatively short quiescent period between depositional events was insufficient to cause pedogenesis, resulting in insignificant soil development.

13.5.1.2 Calcrete encrustation

Though the vertical sections of all fans were dominated by angular clasts and devoid of organic matter, layers of calcretes were also present in the middle and distal parts of these sections. These horizons were observed at elevated sites (which remained inactive for an extended period of time) along the feeder channel on the Tabo and Hansa fans. A calcrete horizon was discovered in a vertical section of the abandoned channel on Lari fan. On the other two fans, Losar and Kawang, no calcrete horizon was discovered at exposed sites. Unlike the calcrete horizons of the arid and semi-arid region which usually exist at the depth of ∼1 m (Lattman, 1973; Campbell and Claridge, 1992), these horizons in the study area were found at a depth greater than ∼2 m from the surface (Fig. 13.2C).

Calcrete development is a typical feature of Pleistocene semi-arid fan surfaces (Alonso-Zarza et al., 1998) and occurs when the calcium-bearing minerals are leached downward from the upper soil horizon (Gile and Hawley, 1966). High temperatures and little precipitation characterize arid and semi-arid regions. Scant precipitation combined with a high rate of evaporation results in less intense and shallow leaching of arid region soils and, as a result, a lesser and shallow calcrete horizon (Dunkerley, 2011). Since calcrete horizon requires considerable time to develop, hence it is usually found at rather stable/abandoned sites.

In contrast, lower temperatures cause less evaporation of water from the surface in mountainous cold desert regions than in hot deserts. At low temperatures, CO_2 in the atmosphere and that produced by biological activity in the soil are easily dissolved, making rain or snowmelt waters more corrosive and thus accelerating the rate of solution action (Clement and Vaudour, 1967). Furthermore, precipitation in such areas is mostly in the form of snow. Furthermore, the valley's mountainous topography causes thick layers of snow to accumulate, particularly in relatively flat areas. The very low temperature of these regions causes longer residence time of snow on the surface and thus deeper penetration of water through the profile at such sites. These facts suggest that in mountainous cold deserts, sediment percolation is relatively enhanced, resulting in greater calcrete encrustation. Similarly, the greater depth of the calcrete horizon in the study area can be explained by the same phenomenon, namely greater and deeper water percolation through the profile. As a result, topography and climatic conditions may be blamed for increased and deeper calcrete encrustation in the study area.

13.5.2 Surficial characteristics/development

Desert pavements, a characteristic feature of arid region soils was observed on the surface of almost all surveyed fans in the Spiti valley (Fig. 13.3A). Desert pavements do not exist on the entire surface of fans, but only at abandoned sites that have not experienced any depositional activity/event for an extended period of time. Sections located on both sides of deeply incised channels, such as elevated terraces on the surface of fans, preserve well-developed desert pavements. On the multi-staged fans (e.g., Lari), relatively mature desert pavements were observed on the elevated and relatively older surfaces of alluvial fans (Fig. 13.3B), whereas on Kawang and Losar fan juvenile pavements were observed. The maturity of pavements in the study area generally increases with the increase in elevation of raised portions with respect to the local base level on the surface of each fan. Horizontally (i.e. along the valley), it generally decreases upstream as elevation from mean sea level increases. The best preserved pavements, however, were found not on the highest fan section, but on the second highest. This phenomenon may be attributed to the highest sections' susceptibility to rockfall events. The surface of these sections is obscured by fragmented and highly angular clasts deposited by rockfalls, according to field evidence. These deposits cover a significant area on these sections and obscure pavements as well as soil (if any) that was present on the surface prior to the depositional event.

Another important characteristic of mature surficial soil is A_v horizon which is thicker, clay-rich, and more vesicular (McFadden et al., 1998). When the reciprocal escape of air is hampered during the entry of water into soils during rain or run on events, vesicles form. Positive air pressure within the pore spaces of damp and deformable soil results in pores that are roughly spherical and up to a few millimeters in diameter. Even when the soil is saturated, these vesicles remain air-filled, reducing the volume of soil matrix available to conduct infiltrating water downwards through the soil (Dunkerley, 2011). Such vesicular soil horizons were also observed at elevated and abandoned sections on Lari (~ 17 cm thick), Hansa (~ 14 cm thick), Losar and Kawang fans (both ~ 8 cm thick) (Fig. 13.3C). The active surfaces were devoid of this horizon at these fans.

FIGURE 13.3   Pavements on the surface of Hansa fan in the Spiti Valley (A). Pavements (DP) on Lari multi-staged alluvial fan (B). A1 = Primary/highest fan section, A2 = secondary/2nd highest fan section A3 = tertiary/3rd stage fan section. All the desert pavement sites are located on elevated and abandoned terraces. Road in the image (B) indicates scale. Vesicular soil horizon and varnished desert pavement on the surface of Lari alluvial fan (C). Dark colored varnish on the surface of boulders on Tabo alluvial fan (D). Observe the absence of desert pavement caused by anthropogenic activities on this fan. Also, observe saplings planted on this part of the fan.

Aside from the presence of a vesicular horizon, the presence of dark-colored rock varnish on the exposed surface of pavement stones on the Lari alluvial fan attests to its age (Fig. 13.3). During the field survey, desert pavements were discovered to be absent on the surface of the Tabo alluvial fan, but very dark colored varnish on rocks was discovered on elevated terraces (~12 m high from the active channel). The lack of desert pavements on Tabo fan could be attributed to anthropogenic activities that destroyed these ancient land-form features in order to claim the land for agriculture and horticulture (Fig. 13.3D). Agriculture and horticulture are currently practiced on terraces on both sides of the Tabo *nullah*. Rock varnish, like pavements, was darkest and most mature on the highest sections of multi-staged fans and gradually decreased downfan. There was no or very little varnish on the surface of the tertiary fan segment, which is only 0.5—2 m above the active channel. Progressive darkening of varnish is a useful measure of the age or maturity of a surface (Al-Farraj and Harvey, 2000). The darkness of rock varnish on individual rocks also reflects the residence time of that rock on the surface. More is the residence time of rock, darker is the varnish (McFadden et al., 1987). This dark varnish on elevated sections of fans reflects their relatively older age and sheds light on the evolution of these fans' various sections.

As previously stated, the depositional features found along the Spiti River are primarily of Quaternary age. Furthermore, due to its higher elevation from sea level, the entire Himalayan region was covered by thick sheets of snow and glaciers during the late Pleistocene. As glaciers/ice sheets extended up to elevations as low as about 900 m in Kangra valley (Singh, 1971; Marh, 1986), so it is very unlikely that this area that is located above 3300 m AMSL (where there are plenty of glaciers even today) remained unaffected from advancing ice sheets. Glacier advancement was both destructive to older pavements and preventive (after advancement) to the development of newer pavements because the surface was now covered with thick sheets of snow. New pavements could only be built after the snow on the ground had melted, i.e. after the glacial period had ended. These findings also suggest that the current pavements of Spiti Valley (along with associated features like the vesicular horizon and rock varnish) appeared only after the late Pleistocene. Another indicator of this is the decreasing maturity of pavements as one moves upstream from the lower reaches of the Spiti valley. Pavements in the lower reaches are covered with well-developed rock varnish, which reflects the pavements' relatively older age as well as associated surface and landform features. Although juvenile pavements form on the fan surface in the upper reaches of the Spiti valley, there is no or very little varnish on the embedded clasts. Even on elevated abandoned fan surfaces, there is very little varnish. This could be explained in part by the fact that snow cover retreated relatively later from the upper and higher elevated areas of Spiti valley than from the lower areas. This delayed glacial retreat allowed for less time for primary processes to build and secondary processes to resculpt the alluvial fans and their surficial features in the upper valley.

Pavements, vesicular horizons, varnish, and calcrete horizons all suggest that only inactive and stable sites are suitable for soil development in an alluvial fan environment. Furthermore, the presence of these features and their relative maturity on abandoned sections suggest that incisement, segmentation, or shifting of the locus of deposition, or a combination of all of these processes, is largely responsible for the formation of these features. Incisement, fan surface segmentation, and locus of deposition shifting are primarily caused by tectonic activity or climatic conditions (Harvey, 1990; Blair and McPherson, 1994; Owen et al., 1997; Viseras et al., 2003; Pope and Wilkinson, 2005; Weissmann et al., 2005; Waters et al., 2010). Although, as previously stated, climatic conditions played a significant role in the construction and incision of fans in the study area, tectonic activities also played a significant role in the modification of fan morphology. Frequent earthquakes, seismites, uplift and subsidence, the occurrence of massive landslides, river lateral shifting, the presence of lacustrine deposits, fault scarps, thrusts, and other evidence of tectonic activity in the study area all add to the picture (Mazari and Bagati, 1991; Banerjee et al., 1997; Sah and Virdi, 1997; Bhargava and Bassi, 1998; Singh and Jain, 2007; Bhargava, 2008; Dortch et al., 2009; Phartiyal et al., 2009; Joshi et al., 2010; Anoop et al., 2012). These findings also suggest that the Spiti valley remained tectonically active during the previously mentioned aggradation and degradation/incisement phases. Further presence of strath terraces, narrow and deep gorges, complete incisement of fan surface, presence of almost continuous terraces along Spiti river for the major part of its course, gradual increase in the elevation of abandoned sections from apex to distal parts on entirely entrenched fans (e.g., Tabo and Hansa fans Figs. 13.1B and C and 13.2C), all combined suggest relative dominance of fall in base level in comparison to mountain front uplifting. All of these clearly demonstrate the relative dominance of incisement by the Spiti river. This incisement or active tectonic attitude is primarily responsible for the fall in local base level and, as a result, the segmentation or abandonment of fan surfaces in the study area, and thus exerts control over soil development here.

Thus, soil development on an alluvial fan is greatly influenced by tectonically induced incision by trunk stream and climatic conditions in the study area, and sheds light on the evolution of the Spiti valley's alluvial fan and landscape.

13.6 Recommendations

13.6.1 For policy makers

13.6.1.1 Implications for cultivation

The soils on the alluvial fans are shallow, as evidenced by the descriptions in the preceding sections, and thus unsuitable for plants with taproots and deep roots. Because these soils cannot retain moisture for long periods of time, they are also unsuitable for crops that require a large amount of water. While such plants may grow quickly in their early stages of development, they eventually dry out. Crops with fibrous and shorter roots should be preferred for these soils.

13.6.1.2 Implications for pastoral farming

One of the main activities in the Spiti valley is herding animals. Uncontrolled grazing is a potential threat to soil loss in this sensitive environment because it not only removes protective and sparse vegetation but also loosens particles from the surface, which are then displaced from their original location by different erosional agents (such as aeolian/fluvial). To preserve this valuable resource, controlled grazing should be encouraged.

13.6.1.3 Implications for road construction

Except for Kawang, State Highways 30 and 31 run through all of the surveyed fans. It should be noted that the active parts of fans are not suitable for road construction because they cause damage to the road during depositional events/peak discharge (either because of flash floods, snow melting, or debris flow). For example, a high energy flood event about ten years ago completely destroyed the iron bridge at Lari fan. Similarly, road construction is not permitted in the highest areas because it would destroy the valuable soils. Furthermore, the highest parts of fans are prone to rock falls. As a result, roads should be built on recently abandoned or old/inactive surfaces with minimal soil development.

13.6.2 For researchers

Only the development of soil on alluvial fans was considered in this study. This work can be expanded by investigating the chemical and biological properties of these soils in order to devote them to the best-suited purpose/operation. At the same time, the morphology of the alluvial fan influences not only soil development but also soil fertility.

There are extremely few studies such as Bahrami and Ghahraman (2019) that have explored this aspect. Morphological influence on soil fertility may be explored for the study area in order to better management of this resource.

13.7 Conclusions

- The study's findings show insignificant soil development in vertical sections of all surveyed fans, implying a relatively younger age of these features, rather rapid aggradation by high energy depositional events, and insufficient time between different depositional events to cause pedogenesis in the Spiti valley. Thus, no significant soil development occurred during the vertical aggradation phase of fans; however, soil development began following the cessation of vertical aggradation coupled with abandonment.
- The presence of a calcrete horizon at greater depth indicates that sediments are transferred more effectively and deeply through the soil profile in mountainous cold desert regions than in hot arid and semi-arid regions.
- At some abandoned sites, the presence of desert pavements, rock varnish, and vesicular horizons indicates the relative age of different sections of fans, as well as shifting locus of deposition and fan entrenchment/segmentation or abandonment. Abandoned surfaces remained abandoned for an extended period of time, resulting in the formation of pavements, varnish, and vesicular horizons. Nonetheless, all of these features are Quaternary in origin and appeared after the late Pleistocene.
- Furthermore, climatic conditions and active tectonic attitudes have had a significant impact on soil development in the Spiti valley.
- Furthermore, these soils are prone to degradation due to fluvial, aeolian, and anthropogenic activities, necessitating the use of appropriate agronomic techniques for optimal long-term utilization.

References

Al-Farraj, A., Harvey, A.M., 2000. Desert pavement characteristics on wadi terrace and alluvial fan surfaces, Wadi Al-Bih, UAE and Oman. Geomorphology 35, 279–297.

Alonso-Zarza, A.M., Silva, P.G., Goy, J.L., Zazo, C., 1998. Fan-surface dynamics and biogenic calcretes development: interactions during ultimate phases of fan evolution in the semi-arid SE Spain, Murcia. Geomorphology 24, 147–167.

Ameta, S.S., 1979. Some observations on the geomorphology of the Spiti valley, Lahaul and Spiti district, Himachal Pradesh. Himalayan Geology 9 (2), 646–656.

Amit, R., Gerson, R., 1986. The evolution of Holocene reg gravelly soils in the deserts – an example from the Dead Sea region. Catena 13, 59–79.

Anoop, A., Prasad, S., Basavaiah, N., Brauer, A., Shahzad, F., Deenadayalan, K., 2012. Tectonic versus climatic influence on landscape evolution: a case study from the upper Spiti Valley NW Himalaya. Geomorphology 145–146, 32–44.

Aulinas, M., Valles, M.G., Turiel, J.L.F., Gimeno, D., Saavedra, J., Gisbert, G., 2014. Insights into the formation of rock varnish in prevailing dusty regions. Earth Surfaces Processes and Landforms 40 (4), 447–458.

Bagati, T.N., 1990. Lithostratigraphy and facies variations in the Spiti Basin Tethys, Himachal Pradesh, India. Journal of Himalayan Geology 1 (1), 35–48.

Bagati, T.N., Suresh, N., 1991. Sedimentology of the lingti lacustrine sequence Spiti valley Himachal Pradesh. Journal of Himalayan Geology 2 (2), 125–132.

Bagati, T.N., Thakur, V.C., 1993. Quaternary basins of Ladakh and Lahul-Spiti in north-western Himalaya. Current Science 64, 898–903.

Bahrami, S., Ghahraman, K., 2019. Geomorphological controls on soil fertility of semi-arid alluvial fans: a case study of the Joghatay Mountains, Northeast Iran. Catena 176, 145–158.

Banerjee, D., Singhvi, A.K., Bagati, T.N., Mohindra, R., 1997. Luminescence chronology of seismites in Sumdo Spiti valley near Kaurik-Chango Fault, northwestern Himalaya. Current Science 73, 276–281.

Beaty, C.B., 1963. Origin of alluvial fans white mountains California and Nevada. Annals of Association of American Geographers 53, 516–535.

Bhargava, O.N., 1990. Holocene tectonics south of the Indus suture Lahaul-Ladakh Himalaya, India: a consequence of Indian plate motion, Tectonophysics, 174. pp. 315–320.

Bhargava, O.N., 2008. An updated introduction to Spiti geology. Journal of Palaeontological Society of India 53 (2), 113–129.

Bhargava, O.N., Bassi, U.K., 1998. Geology of Spiti-Kinnaur-Himachal Himalaya. Geological Survey of India Memoirs, Lucknow, p. 124.

Birse, E.L., 1980. Suggested amendments to the world soil classification to accommodate Scottish mountain and eolian soil. Journal of Soil Science 31, 117–124.

Blainey, J.B., Pelletier, J.D., 2008. Infiltration on alluvial fans in arid environments: influence of fan morphology. Journal of Geophysical Research 113, 1–18. Available from: https://doi.org/10.1029/2007JF000792.

Blair, T.C., McPherson, J.G., 1994. Alluvial fan processes and forms. In: Abrahams, A.D., Parson, A.J. (Eds.), Geomorphology of Desert Environments. Chapman and Hall, London, pp. 354–402.

Bookhagen, B., Thiede, R., Strecker, M.R., 2005. Abnormal monsoon years and their control on erosion and sediment flux in the high, arid northwestern Himalaya. Earth Planetary Science Letters 231, 131–146.

Bowman, D., 2019. Principles of Alluvial Fan Morphology. Springer Nature, Netherlands. Available from: http://doi.org/10.1007/978-94-024-1558-2.

Burbank, D.W., Blythe, A.E., Putkonen, J., Pratt-Sitaula, B., Gobet, E., Oskin, M., et al., 2003. Decoupling of erosion and precipitation in the Himalaya. Nature 426, 652–655.

Butterworth, R., Wilson, C.J., Herron, N.F., Greene, R.S.B., Cunningham, R.B., 2000. Geomorphic controls on the physical and hydrologic properties of soils in a valley floor. Earth Surface Processes and Landform 25 (11), 1161–1179.

Camacho, M.E., Quesada-Román, A., Mata, R., Alvarado, A., 2020. Soil-geomorphology relationships of alluvial fans in Costa Rica. Geoderma Regional 21, e00258. Available from: https://doi.org/10.1016/j.geodrs.2020.e00258.

Campbell, L.B., Claridge, G.G.C., 1975. Morphology and age relationships of Antarctic soils. In: Suggate, R.P., Cresswell, M.M. (Eds.), Quaternary Studies, 13. Royal Society, N Z, Bull, pp. 83–88.

Campbell, L.B., Claridge, G.G.C., 1987. Antarctica: Soils, Weathering Processes and Environment. Elsevier, Amsterdam.

Campbell, L.B., Claridge, G.G.C., 1992. Soils of cold climate regions. In: Martini, I.P., Chesworth, W. (Eds.), Developments in Earth Surface Processes 2 – Weathering, Soils & Paleosols. Elsevier Science Publishers, Amsterdam, pp. 183–202.

Carboni, S., Palomba, M., Vacca, A., Carboni, G., 2006. Paleosols provide sedimentation, relative age, and climatic information about the alluvial fan of the River Tirso Central-Western Sardinia, Italy. Quaternary International 156–157, 79–96.

Chesworth, W., 1992. Weathering systems. In: Martini, I.P., Chesworth, W. (Eds.), Developments in Earth Surface Processes 2 – Weathering, Soils & Paleosols. Elsevier Science Publishers, Amsterdam, pp. 19–40.

Clement, P., Vaudour, J., 1967. Observation on the pH of melting snow in the Southern French Alps. In: Wright, H.E., Osburn, W.H. (Eds.), Arctic and Alpine Environments. Indiana University Press, Bloomington, IN, pp. 205–231.

Cooke, R., Warren, A., Goudie, A., 1993. Alluvial fans and their environments. In: Goudie, A., Warren, A., Cooke, R. (Eds.), Desert Geomorphology. UCL Press Limited, London, pp. 168–186.

Costa, J.E., 1984. Physical geomorphology of debris flows. In: Costa, J.E., Fleisher, P.J. (Eds.), Developments and Applications of Geomorphology. Springer-Verlag, Berlin, Heidelberg, pp. 268–317.

Costa, J.E., 1988. Rheologic, geomorphic and sedimentologic differentiation of water floods, hydperconcentrated flows and debris flows. In: Baker, V.R., Kochel, R.C., Patton, P.C. (Eds.), Flood Geomorphology. Wiley, New York, pp. 113–122.

Dickerson, R.P., Forman, A., Liu, T., 2013. Co-development of alluvial fan surfaces and arid botanical communities, Stonewall Flat, Nevada, USA. Earth Surface Processes and Landforms 38, 1083–1101.

Dickerson, R.P., Bierman, P.R., Cocks, G., 2015. Alluvial fan surfaces and an age-related stability for cultural resource preservation: Nevada Test and Training Range, Nellis Air Force Base, Nevada, USA. Journal of Archaeological Science: Reports 2, 551–568.

Dorn, R.I., 1988. A rock varnish interpretation of alluvial-fan development in Death Valley, California. National Geographic Research 41, 56–73.

Dorn, R.I., 1991. Rock varnish. American Scientist 79 (6), 542–553.

Dorn, R.I., Oberlander, T.M., 1982. Rock varnish. Progress in Physical Geography 6, 317–367.

Dorn, R.I., DeNiro, M.J., Ajie, H.O., 1987. Isotopic evidence for climatic influence on alluvial–fan development in Death Valley, California. Geology 15, 108–110.

Dorn, R.I., Clarkson, P.B., Nobbs, M.F., Loendorf, L.L., Whitley, D.S., 1992. New approach to the radiocarbon dating of rock varnish, with examples from drylands. Annals of the Association of American Geographers 82 (1), 136–151.

Dortch, J.M., Owen, L.A., Haneberg, W.C., Caffee, M.W., Dietsch, C., Kamp, U., 2009. Nature and timing of large landslides in the Himalaya and Trans-Himalaya of northern India. Quaternary Science Reviews 28 (11–12), 1037–1054.

Dregne, H.E., 1976. Soils of Arid Regions. Elsevier, Amsterdam.

Dunkerley, D.L., 2011. Desert soils. In: Thomas, D.S.G. (Ed.), Arid Zone Geomorphology: Process, Form and Change in Dry-Lands, third ed. John Wiley & Sons Ltd., Chichester, United Kingdom, pp. 101–130.

Field, J.J., 1992. An evaluation of alluvial fan agriculture. In: Fish, S.K., Fish, P.R., Madsen, J.H. (Eds.), The Marana Community in the Hohokam World, 56. University of Arizona, Tucson, pp. 53–72. , Anthropological Papers.

Fuchs, G., 1982. The geology of the Pin Valley in the Spiti H P India. Geologisches Jahrbuch 124 (2), 325–362.

Gile, L.H., 1975. Holocene soils and soil-geomorphic relations in arid regions of Southern New Mexico. Quaternary Research 5, 321–360.

Gile, L.H., Hawley, J.W., 1966. Periodic sedimentation and soil formation on an alluvial-fan piedmont in southern New Mexico. Soil Science Society of America Journal 30, 261–268.

Gustavson, T.C., Holliday, V.T., 1999. Eolian sedimentation and soil development on a semiarid to subhumid grassland, Tertiary Ogallala and Quaternary Blackwater Draw formations, Texas and New Mexico high plains. Journal of Sedimentary Research 69, 622–634.

Harden, J.W., Taylor, E.M., Hill, C., Mark, R.K., McFadden, L.D., Reheis, M.C., et al., 1991. Rates of soil development from four soil chronosequences in the Southern Great Basin. Quaternary Research 35, 383–399.

Harvey, A.M., 1990. Factors influencing quaternary alluvial fan development in southeast Spain. In: Rachocki, A. H., Church, M. (Eds.), Alluvial Fans: A Field Approach. John Wiley & Sons Ltd., New York, pp. 247–269.

Harvey, A.M., 1997. The role of alluvial fans in arid zone fluvial systems. In: Thomas, D.S.G. (Ed.), Arid Zone Geomorphology: Process Form and Change in Drylands, second ed. Behaven Press and John Wiley & Sons, London, pp. 231–259.

Hill, R.B., 1993, Soil Landform Relationship on Bullock Creek Fan North Canterbury (Master of Applied Science Thesis). Lincoln University.

Hirmas, D.R., Amrhein, C., Graham, R.C., 2010. Spatial and process-based modeling of soil inorganic carbon storage in an arid piedmont. Geoderma 154, 486–494.

Hooke, R.L., 1987. Mass movement in semi-arid environment and the morphology of alluvial fans. In: Anderson, M.G., Richard, K.S. (Eds.), Slope Stability. Wiley & Sons, New York, pp. 505–529.

Jhingran, A.G., 1981. Geology of the Himalaya. In: Lall, J.S., Moddie, A.D. (Eds.), The Himalaya – Aspects of Change. India International Centre and Oxford University Press, New Delhi, pp. 77–97.

Joshi, M., Kothayari, G.C., Ahluvalia, A., Pant, P.D., 2010. Neo-tectonic evidences of rejuvenation in Kaurik-Chango fault zone Northern Himalaya. Geographical Information System 169–176.

Kesel, R.H., Spicer, B.E., 1985. Geomorphologic relationships and ages of soils on alluvial fans in the Rio General Valley, Costa Rica. Catena 12, 149–166. Available from: https://doi.org/10.1016/0341-8162(85)90007-4.

Khattri, K.N., Rai, K., Jain, A.K., Sinvhal, H., Gaur, V.K., Mithal, R.S., 1978. The Kinnaur earthquake, Himachal Pradesh, India, of 19 January 1975. Tectonophysics 49, 1–21.

Kochel, R.C., 1990. Humid fans of the Appalachian Mountains. In: Rachocki, A.H., Church, M. (Eds.), Alluvial Fans: A Field Approach. John Wiley & Sons Ltd, New York, pp. 109–129.

Lattman, L.H., 1973. Calcium carbonate cementation of alluvial fans in southern Nevada. Geological Society of America Bulletin 84, 3013—3028.

Legros, J.P., 1992. Soils of alpine mountains. In: Martini, I.P., Chesworth, W. (Eds.), Developments in Earth Surface Processes 2 — Weathering, Soils & Paleosols. Elsevier Science Publishers, Amsterdam, pp. 155—182.

Machette, M.N., 1985. Calcic soils of the southwestern United States. Geological Society of America 1—21. Special Paper 203.

Marh, B.S., 1986. Geomorphology of the Ravi River. Inter-India Publications, New Delhi.

Marh, B.S., 2000. Himachal Pradesh: physico-geographical set—up. In: Verma, L.R. (Ed.), Natural Resources and Development in Himalaya. Malhotra Publishing House, New Delhi, pp. 505—518.

Marh, B.S., Rana, V.S., 2014. Cold Desert Geomorphology in the Trans-Himalayan Region: A Preliminary Analysis of Landforms of the Spiti River Valley. Anamika Publishers and Distributers Pvt. Ltd., New Delhi.

Mazari, R.K., Bagati, T.N., 1991. Post-collision graben development in the Spiti Valley Himachal Pradesh. Journal of Himalayan Geology 22, 111—117.

McCraw, J.D., 1968. The soil pattern of some New Zealand alluvial fans. Soil Science 4, 631—640.

McFadden, L.D., 1985. Changes in the content and composition of pedogenic iron hydroxides in a chronosequence of soils in Southern California. Quaternary Research 23, 189—204.

McFadden, L.D., 1988. Climatic influences on rates and processes of soil development in quaternary deposits of southern California. Geological Society of America 216, 153—178. Special Paper.

McFadden, L.D., Wells, S.G., Jercinovich, M.J., 1987. Influences of eolian and pedogenic processes on the origin and evolution of desert pavements. Geology 15, 504—508.

McFadden, L.D., Ritter, J.B., Wells, S.G., 1989. Use of multi-parameter relative age methods for age estimation and correlation of alluvial fan surfaces on a desert piedmont, Eastern Mojave Desert. Quaternary Research 32, 276—290.

McFadden, L.D., Knuepfer, P.L.K., 1990. Soil geomorphology: the linkage of pedology and surficial processes. Geomorphology 3, 197—205.

McFadden, L.D., McDonald, E.V., Wells, S.G., Anderson, Q.J., Forman, S.L., 1998. The vesicular layer and carbonate collars of desert soils and pavements: formation, age and relation to climate change. Geomorphology 24, 101—145.

Meek, S.R., Carrapa, B., DeCelles, P.G., 2020. Recognizing allogenic controls on the stratigraphic architecture of ancient alluvial fans in the Western US. Frontiers in Earth Science 8. Available from: https://doi.org/10.3389/feart.2020.00215 article no. 215.

Mills, H.H., Allison, J.B., 1995. Weathering and soil development on fan surfaces as a function of height above modern drainage-ways, Roan Mountain, North Carolina. Geomorphology 14, 1—17.

Nash, D.J., 2011. Desert crusts and rock coatings. In: Thomas, D.S.G. (Ed.), Arid Zone Geomorphology: Process, Form and Change in Dry-Lands, third ed. John Wiley & Sons Ltd., Chichester, United Kingdom, pp. 131—180.

Norton, J.B., Sandor, J.A., White, C.S., 2003. Hillslope soils and organic matter dynamics within a Native American agro-ecosystem on the Colorado Plateau. Soil Science Society of America Journal 67, 225—234.

Norton, J.B., Sandor, J.A., White, S.C., Laahty, V., 2007. Organic matter transformations through arroyos and alluvial fan soils within a native American agroecosystem. Soil Science Society of America Journal 71 (3), 829—835.

Owen, L.A., Windley, B.F., Cunninghum, W.D., Badamgarav, J., Dorjnamjaa, D., 1997. Quaternary alluvial fans in the Gobi of southern Mongolia: evidences for neotectonics and climate change. Journal of Quaternary Science 123, 239—252.

Owen, L.A., Derbyshire, E., Fort, M., 1998. The quaternary glacial history of the Himalaya. In: Owen, L.A. (Ed.), Mountain Glaciation, Quaternary Proceeding No 6. John Wiley & Sons Ltd., Chichester, pp. 91—120.

Pant, G.B., 2003. Long-term climatic variability and change over monsoon Asia. Journal of Indian Geophysical Union 73, 125—134.

Parker, K.C., 1995. Effects of complex geomorphic history on soil and vegetation patterns on arid alluvial fans. Journal of Arid Environment 30, 19—39.

Phartiyal, B., Srivastava, P., Sharma, A., 2009. Tectono-climatic signatures during late Quaternary Period from Upper Spiti Valley NW Himalaya India. Himalayan Geology 30 (2), 167—174.

Pimentel, N.L.V., 2002. Pedogenic and early diagenetic processes in Palaeogene alluvial-fan and lacustrine deposits from the Sado Basin, S Portugal. Sedimentary Geology 148, 123—138.

Pope, R.J., Wilkinson, K.N., 2005. Reconciling the roles of climate and tectonics in Late Quaternary fan development on the Spartan piedmont, Greece. In: Harvey, A.M., Mather, A.E., Stokes, M. (Eds.), Alluvial Fans:

Geomorphology, Sedimentology, Dynamics. Geological Society, London, pp. 133–152. , Special Publications No, 251.

Rashidi, Z., Sohbati, R., Karimi, A., Farpoor, M.H., Khormali, F., Thompson, W., et al., 2018. Constraining the timing of palaeosol development in Iranian arid environments using OSL dating. Quaternary Geochronology . Available from: https://doi.org/10.1016/j.quageo.2018.04.006.

Regmi, N.R., Rasmussen, C., 2018. Predictive mapping of soil-landscape relationships in the arid Southwest United States. Catena 165, 473–486.

Reheis, M.C., 1987a. Gypsic soils of the Kane alluvial fans. US Geological Survey Bulletin 159-C.

Reheis, M.C., 1987b. Climate implications of alternating clay and carbonate formation in semi-arid soils of south-central Montana. Quaternary Research 29, 270–282.

Reheis, M.C., Sowers, J.M., Taylor, E.M., McFadden, L.D., Harden, J.W., 1992. Morphology and genesis of carbonate soils on the Kyle Canyon fan, Nevada, USA. Geoderma 523 (4), 303–342.

Retallack, G.J., 1988. Field recognition of paleosols. Geological Society of America Special Paper 216, 1–20.

Ruhe, R.B., 1967. Geomorphic surfaces and surficial deposits in southern New Mexico, N.M. Bureau of Mines and Mineral Resources Memoir no, 18.

Sah, M.P., Virdi, N.S., 1997. Geomorphic signatures of neotectonic activity along the Sumdo Fault, Spiti Valley, District Kinnaur, H, P. Himalayan Geology 18 (1–2), 81–92.

Schaetzl, R.J., Anderson, S., 2005. Soils: Genesis and Geomorphology. Cambridge University Press, Cambridge.

Singh, R.L. (Ed.), 1971. India: A Regional Geography. National Geographical Society of India, Varanasi.

Singh, S., Jain, A.K., 2007. Liquefaction and fluidization of lacustrine deposits from Lahaul-Spiti and Ladakh Himalaya: geological evidences of palaeoseismicity along active fault zone. Sedimentary Geology 19 (6), 47–57.

Srikantia, S.V., Bhargava, O.N., 1998. Geology of Himachal Pradesh. Geological Society of India, Bangalore.

Srivastava, P., Ray, Y., Phartiyal, B., Sharma, A., 2013. Late Pleistocene-Holocene morphosedimentary architecture, Spiti River, arid higher Himalaya. International Journal of Earth Sciences Geologische Rundschau 102, 1967–1984.

Stadelman, S.A., 1989. Pedogenesis and Geomorphology of Hanaupah Canyon Alluvial Fan, Death Valley, California (MSc thesis). Texas Tech University, Texas.

Stokes, M., Nash, D.J., Harvey, A.M., 2007. Calcrete 'fossilisation' of alluvial fans in SE Spain: the roles of groundwater, pedogenic processes and fan dynamics in calcrete development. Geomorphology 85, 63–84.

Talbot, M.R., Williams, M.A.J., 1979. Cyclic alluvial fan sedimentation on the flanks of fixed dunes, Janjari, Central Niger. Catena 6, 43–62.

Thomas, D.S.G., 2011a. Arid environments: their nature and extent. In: Thomas, D.S.G. (Ed.), Arid Zone Geomorphology: Process, Form and Change in Dry-lands, third ed. John Wiley & Sons Ltd., Chichester, UK, pp. 3–16.

Thomas, D.S.G., 2011b. Dryland system variability. In: Thomas, D.S.G. (Ed.), Arid Zone Geomorphology: Process, Form and Change in Dry-Lands, third ed. John Wiley & Sons Ltd., Chichester, UK, pp. 53–60.

Ugolini, F.C., Hillier, S., Certini, G., Wilson, M.J., 2008. The contribution of Aeolian material to an aridisol from southern Jordan as revealed by mineralogical analysis. Journal of Arid Environments 72, 1431–1447.

Viseras, C., Calvache, M.L., Soria, J.M., Fernandez, J., 2003. Differential features of alluvial fans controlled by tectonic or eustatic accommodation space, examples from the Betic Cordillera Spain. Geomorphology 50, 181–202.

Waters, M.R., Field, J.F., 1986. Geomorphic analysis of Hohokam settlement patterns on alluvial fans along the western flank of the Tortolita Mountains, Arizona. Geoarchaeology 1, 329–345.

Waters, J.V., Jones, S.J., Armstrong, H.A., 2010. Climatic controls on late Pleistocene alluvial fans, Cyprus. Geomorphology 115, 228–251.

Watson, A., 1992. Desert soils. In: Martini, I.P., Chesworth, W. (Eds.), Developments in Earth Surface Processes 2 – Weathering, Soils & Paleosols. Elsevier Science Publishers, Amsterdam, pp. 225–260.

Weissmann, G.S., Bennett, G.L., Lansdale, A.L., 2005. Factors controlling sequence development on Quaternary fluvial fans, San Joaquin Basin, California, USA. In: Harvey, A.M., Mather, A.E., Stokes, M. (Eds.), Alluvial Fans: Geomorphology, Sedimentology, Dynamics. Geological Society, London, pp. 169–186. Special Publications No, 251.

Wells, S.G., McFadden, L.D., Dohrenwend, J.C., 1987. Influence of late Quaternary climatic changes on geomorphic and pedogenic processes on a desert piedmont, Eastern Mojave Desert, California. Quaternary Research 27, 130–146.

White, K., Walden, J., 1997. The rate of iron oxide enrichment in arid zone alluvial fan soils, Tunisian southern atlas, measured by mineral magnetic techniques. Catena 30, 215–227.

Williams, A.J., Buck, B.J., Soukup, D.A., Merkler, D.J., 2013. Geomorphic controls on biological soil crust distribution: a conceptual model from the Mojave Desert USA. Geomorphology 195, 99–109.

Wright, V.P., Zarza, A.M.A., 1990. Pedostratigraphic models for alluvial fan deposits: a tool for interpreting ancient sequences. Journal of the Geological Society, London 147, 8–10.

Zembo, I., Trombino, L., Bersezio, R., Felletti, F., Dapiaggi, M., 2012. Climatic and tectonic controls on pedogenesis and landscape evolution in a Quaternary intramontane basin Val D'Agri Basin, Southern Apennines, Italy. Journal of Sedimentary Research 82, 283–309.

3. Soil physicochemical parameters

Soil organic carbon and soil properties for REDD implementation in Nepal: experience from different land use management in three districts of Nepal

Him Lal Shrestha[1], Roshan M. Bajracharya[1] and Bishal K. Sitaula[2]

[1]Department of Environmental Science and Engineering, School of Science, Kathmandu University, Dhulikhel, Nepal [2]Department of International Environment and Development Studies, Norwegian University of Life Sciences, Ås, Norway

14.1 Introduction

Climate change has become an increasingly pressing global concern in recent decades, requiring the global community's immediate attention. Its mitigation will undoubtedly include a reduction in emissions from forests, as land use, land use change, and forestry

(LULUCF) account for approximately 18% of global greenhouse gas (GHG) emissions (IPCC, 2006). Forests can act both as a sink and a source of carbon (C) emissions to the atmosphere (IPCC, 2006). If degraded forest land is restored and protected, it can act as a net sink for atmospheric C; however, if large areas of forest are converted to other land uses or degrade in quality, it can act as a net source of C to the atmosphere. Thus, the ongoing international discourse identifies the clear scope of emission reductions from the forestry sector by implementing "Reducing Emissions from Deforestation and Forest Degradation Plus" (REDD +) in developing countries to encourage their citizens to avoid deforestation and forest degradation (Dhital et al., 2015). According to this program, countries that actively reduce C emissions by avoiding deforestation or preventing forest degradation will receive financial compensation in a manner similar to Payment for Ecosystem Services (PES) (Dhital, 2009; Manandhar, 2013).

The current global issue of increasing CO_2 and other GHG concentrations in the atmosphere, as well as their negative effects on the environment and climate, has prompted more focused research on the quantities, types, distributions, and behavior of C in various systems (Eswaran and Berg, 1993; Johnson et al., 1991). Scientists have shifted their focus away from the atmosphere and toward the aquatic and terrestrial systems. Terrestrial carbon pools include biota, soils, and fossils. Biota can sequester approximately 620 billion tons of carbon, while soils can sequester approximately 2.5 trillion tons of carbon (Lal, 2009) as cited in (Dahal and Bajracharya, 2010), indicating that carbon stored in soil is nearly three times that of biota and twice that of the atmosphere (Eswaran and Berg, 1993).

Globally, 1576 pg. of carbon is stored in soil with nearly 506 pg. (32%) stored in tropical soil and 40% of carbon in the soil of the tropics is in forest soil (Eswaran and Berg, 1993). Because the carbon content of forest soil is high, changing land use systems results in the loss of sequestered carbon as well as the soil's ability to capture atmospheric carbon. The average loss of soil carbon after conversion of forest to cropland is 48%, 28% for grassland, and 35% for agricultural land (Buringh, 1984). The carbon stock changes not only during the conversion, but also as a result of how the land is managed afterward. Cole et al. (1997) estimated that the loss of carbon from soil due to cultivation was around 55 Pg (Milne et al., 2006). They stated that if half of this amount can be recovered, the global potential for C sequestration in cultivated soils through appropriate management could be between 20 and 30 Pg over the next 50–100 years. According to one study, grassland can contribute significantly to carbon sequestration if managed properly. Poor grassland management resulted in SOC losses ranging from 3% to 5% in temperate and tropical regions, respectively; however, changing management strategy could increase SOC content by 14% and 17% in temperate and tropical regions, respectively (Shrestha et al., 2012).

Because the tropical region has the highest population growth, conversion of natural forest to cultivable land is also high. As a result of the high rate of deforestation and degradation, Asia's forests are considered net CO_2 emitters (Dixon, 2004). This issue has emerged as a critical situation in developing countries such as India, Bangladesh, and Myanmar, resulting in a decrease in total standing C stock from 1880 to 1980 (Flint and Richards, 1991).

Stern (2006) states unequivocally that the increasing level of CO_2 concentration in the atmosphere is due to the burning of fossil fuels, deforestation, and forest degradation, as opposed to other sources of C emission. In response to rising CO_2 levels, scientists and environmentalists from around the world convened and concluded that it is economically feasible to control CO_2 emissions from forest degradation and deforestation, and thus proposed RED: Reducing Emissions from Deforestation. This was later changed to REDD: Reducing Emissions from Deforestation and Forest Degradation, which is a comprehensive approach mandated by the Bali Action Plan during COP 13, focusing primarily on GHG mitigation. (Manandhar, 2009). Under the UNFCCC, it is an emerging market-based approach (Dangi and Acharya, 2009). The REDD mechanism was then upgraded to REDD + to include forest enhancement, social and biodiversity co-benefits, as well as avoiding deforestation and forest degradation. REDD + could thus play a significant role not only in the mitigation of climate change issues, but also in adaptation by increasing ecosystem resilience (Dhital et al., 2015).

Increased international dialog on the implementation of the REDD + mechanism indicates that regular, consistent, and reliable monitoring of forests and other related parameters, such as governance, participation, and institutional development, is required for its success. Bio-physical and ecosystem services, such as biodiversity, forest fragmentation, C stocks, forest restoration or plantations, and other features, must be monitored on a regular and systematic basis (Shrestha et al., 2012). These concerns are related to REDD + implementation and are being raised from a variety of perspectives, including permanence, co-benefits, C content (from a carbon perspective), safeguards on social and environmental issues, Free Prior Informed Consent (FPIC), and sustainable forest management (non-carbon benefit perspective) (UNREDD, 2010, 2011).

In developing and underdeveloped countries such as Nepal, there is a lack of data as well as methodological inconsistencies in quantifying the various benefits of the forest (Molden and Sharma, 2013). Although forest cover and C stock have been monitored to some extent, with limited estimation accuracies, other benefits, such as ecosystem services, potential for climate change mitigation and adaptation, and so on, have not yet been adequately monitored in the region and are urgently needed in light of the impending climate change impacts. Full accounting of carbon stock is required as part of REDD + Monitoring, Reporting, and Verification (MRV) activities. The Net Ecosystem Carbon Balance needs to be quantified (NECB). Failure to account for NECB as a quantitative measure for evaluating REDD + activities will result in significant uncertainty in estimating total carbon stock. However, one of the most difficult issues for MRV activities is quantifying below-ground carbon dynamics, as factors such as SOC quantity distribution, quality, microbial activity, plant roots, and so on have a direct influence on C stocks in soil (Poudel et al., 2019). Studies reveal considerable changes in soil carbon after land use change (Don et al., 2011; Vargas et al., 2013). As a result, changes in different components of below-ground carbon dynamics must be accounted for in order to accurately calculate NECB and reduce uncertainty in C stock estimates. In order to develop effective methods for reducing carbon emissions and improving the framework for MRV under REDD +, detailed quantification of belowground carbon stocks is required (Vargas et al., 2013, Dhital et al., 2015).

Taking into account the influence of soil properties and land management on SOC status in terrestrial ecosystems, this study was carried out in three districts of central Nepal to assess the effects of forest, agroforestry, and conventional agricultural land use on SOC status and other soil properties.

3. Soil physicochemical parameters

14.2 Methods

14.2.1 Study area

Three watersheds from Gorkha, Chitwan, and Rasuwa districts were chosen as study sites. The sites were chosen from among 200 global eco-regions that differ in vegetation types, altitudinal variations, and physiographic zones. The study site, located within the Betrawati watershed of the Rasuwa district, represents the temperate forest at high altitudes in the Middle Mountain, with mixed deciduous and coniferous vegetation. The study site in the Gorkha district's Ludikhola watershed represents the hill Sal (*Shorea robusta*) forest with a moderate altitude level in the mid-hills, while the Chitwan district's Kayarkhola watershed site represents the low altitude forests of Nepal's foothills with *S. robusta* and associated species (Fig. 14.1).

14.2.2 Field sampling

Random sampling was carried out maintaining four samples in each of the land management types; Community forest (CF), Agroforestry (AGF) and Agriculture (AG). For the assessment of SOC, total 36 samples were laid over the 3 watersheds i.e. Ludikhola, Betrawati and Kayarkhola in three districts namely Gorkha, Rasuwa and Chitwan with four replicates in each Land management type.

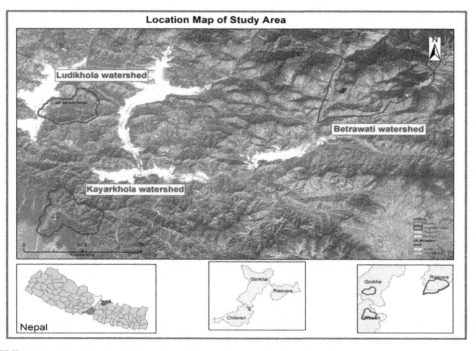

FIGURE 14.1   Map showing the location of the study watersheds in Nepal.

3. Soil physicochemical parameters

Total number of samples = 3 districts × 3 land management types × 4 replicates = 36.
Plots of 0.56 m radius plots were established at the site and sampling of soil was done in four replicated sites of each management types from the depth of 0−15 cm and 15−30 cm.

14.2.3 Soil organic carbon

Soil samples were collected at 0−15 cm and 15−30 cm depths. Samples of exactly 101.6 cm³ were taken and transferred to sampling bags. Samples were transported to the laboratory and oven dried (105°C) until constant weight to determine water content.

The carbon stock density of soil organic carbon is calculated as (Pearson et al., 2007):

$$SOC = \partial \times d \times \%C \qquad (14.1)$$

Where,

SOC = soil organic carbon stock per unit area (t/ha),
∂ = soil bulk density (g/cm³),
d = the total depth at which the sample was taken (cm), and
%C = carbon concentration (%)

14.2.4 Baseline soil properties

Other general baseline soil properties including pH, total N, available P and exchangeable K were determined using standard methods as given in the Methods of Soil Analysis (USDA Monograph No. 9). Soil pH was measured using a digital pH meter in a 1:1 soil: water mixture (McLean, 1982). Total N of the soil samples were determined using the Kjeldahl method (Bremner and Mulvaney, 1982), available P by a modified Olsen method (Olsen and Sommers, 1982) and exchangeable K by ammonium acetate extraction (Knudsen et al., 1982).

14.3 Results and discussion

Among the three land use types, the SOC contents in the study locations were generally highest in CF than in AGF and AG (Fig. 14.2). As cited by Bajracharya et al. (2015), other workers have also noted that forest soils tend to have high SOC, especially in the topsoil (Lal, 2005; Shrestha et al., 2012; Dahal and Bajracharya, 2010). Bationo et al. (2005), the total system carbon in different vegetation and land use types indicates that forests and woodlands have the highest total carbon contents, indicating potential for carbon sequestration. Chitwan, unlike Gorkha and Rasuwa, had higher SOC in AG than in AGF. It is worth noting that the differences in mean SOC content between land use types were not statistically significant (Table 14.1). Rasuwa, a comparatively higher and cooler eco-region, had the highest SOC contents in all three land use types among the study locations. The differences in mean SOC between different study sites or eco-regions were statistically significant as shown in Table 14.2. Similar result was derived in a study done by Liao et al. (2015) where SOC content of the cropland increased from 7.8 ± 1.6 to 11.0 ± 2.3 g/kg (41%). They discovered that incorporating crop residues significantly increased SOC, whereas increasing the mean annual temperature decreased SOC.

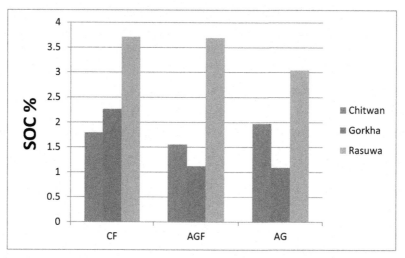

FIGURE 14.2 Mean values of SOC per cent for three land uses in the study districts. *CF*, community forest; *AGF*, agro-forest; *AG*, agricultural land.

TABLE 14.1 One-way analysis of variance of SOC according to land use types, $N = 72$.

		ANOVA			
	Sum of squares	*df*	**Mean square**	*F*	**Sig.**
Between groups	4.195	2	2.098	1.605	.208
Within groups	90.200	69	1.307		
Total	94.395	71			

TABLE 14.2 One-way analysis of variance of SOC according to eco-regions (districts), $N = 72$.

		ANOVA			
	Sum of squares	*df*	**Mean square**	*F*	**Sig.**
Between groups	55.767	2	27.884	49.808	.000
Within groups	38.628	69	0.560		
Total	94.395	71			

Except for agricultural land in Chitwan, which had a comparatively higher pH value, the pH in different land use types was found to be in the acidic group (Fig. 14.3). The pH of soil is affected by differences in land management practices and soil processes. The pH of soil in Chitwan and Gorkha is possibly increasing due to the application of agricultural

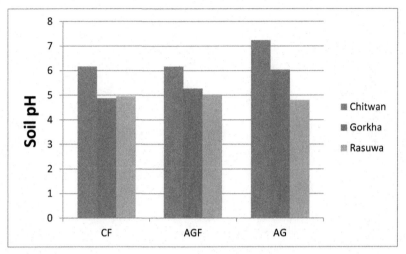

FIGURE 14.3 Mean values of soil pH in three land uses of the study districts. *CF,* community forest; *AGF,* agro-forest; *AG,* agricultural land.

TABLE 14.3 Pearson's correlation matrix for soil properties for all three eco-regions (districts) ($N = 72$).

	Correlations				
		SOC	pH	N	P	K
SOC	Pearson correlation	1	−0.392[a]	0.748[a]	0.025	−0.030
	Sig. (2-tailed)		0.001	0.000	0.836	0.803
pH	Pearson correlation	−0.392[a]	1	−0.144	0.515[a]	0.335[a]
	Sig. (2-tailed)	0.001		0.229	0.000	0.004
N	Pearson correlation	0.748[a]	−0.144	1	0.175	0.090
	Sig. (2-tailed)	0.000	0.229		0.140	0.454
P	Pearson correlation	0.025	0.515[a]	0.175	1	0.093
	Sig. (2-tailed)	0.836	0.000	0.140		0.439
K	Pearson correlation	−0.030	0.335[a]	0.090	0.093	1
	Sig. (2-tailed)	0.803	0.004	0.454	0.439	

[a]*Correlation is significant at the 0.01 level (2-tailed).*

lime ($CaCO_3$), which is used to improve or avoid further increases in soil acidity (Tripathi, 1999). Correlation analysis revealed that increase in soil pH significantly decreases the SOC potential of the soil (Table 14.3).

The total N contents were found to be higher in CF and AG land use types as shown in Fig. 14.4. This could be due to soil management practices such as mulching and the use of

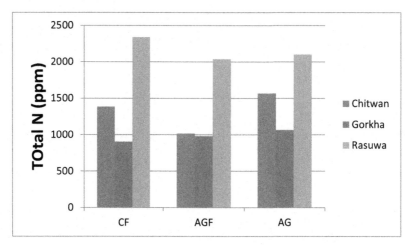

FIGURE 14.4   Mean values of total N in three land uses of the study districts. *CF*, community forest; *AGF*, agro-forest; *AG*, agricultural land.

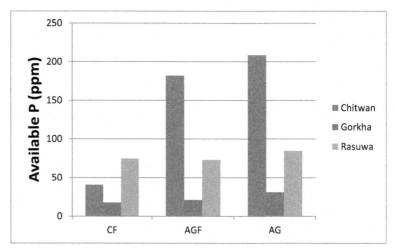

FIGURE 14.5   Mean values of available P in three land uses of the study districts. *CF*, community forest; *AGF*, agro-forest; *AG*, agricultural land.

nitrogen-fixing plants, as well as the addition of nitrogen-containing fertilizers, as reported in studies by Gami et al. (2009) and Geissen et al. (2009), where fertilizer application was cited as the reason for increased total nitrogen and SOC. According to Regmi et al. (2000), continuous application of FYM can increase total N in farmland, which in their study increased from 0.09% to 0.17% in long-term application. According to the correlation analysis, total N content is significantly correlated with SOC content (Table 14.3).

The available P was found to be similar in all the land use types except in Chitwan where it is higher in AGF and AG (Fig. 14.5). This is due to the use of agricultural

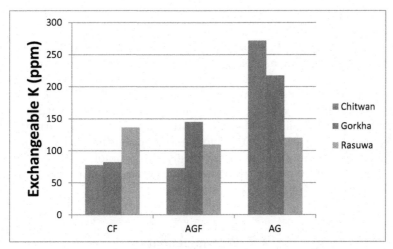

FIGURE 14.6 Mean values of exchangeable K in three land uses of the study districts. *CF*, community forest; *AGF*, agro-forest; *AG*, agricultural land.

fertilizers, as well as the fact that very little P is lost through leaching and crop removal, allowing it to accumulate in the soil for a longer period of time (Tisdale et al., 1985). The correlation analysis revealed that available P is positively correlated with SOC content, but the correlation is not statistically significant.

The exchangeable K content of AG in Chitwan and Gorkha was found to be the highest, implying that farmers applied significant amounts of fertilizer or manure to their lands. Only Rasuwa showed the opposite trend, with exchangeable K being lower in AG than in CF. This could be due to the presence of different types of rock or parent material that geology of CF contains as fertilizers or geology has been the source of exchangeable K in soil. The correlation analysis revealed that exchangeable K is inversely related to SOC, but the relationship was not statistically significant (Fig. 14.6).

14.4 Conclusions

According to the above findings, SOC contents were higher in CF land use compared to AGF and AG, and they were significantly positively correlated with total N. SOC, on the other hand, was significantly negatively correlated with soil pH but not with available P or exchangeable K. SOC content was generally higher in Rasuwa, which is higher in elevation than Gorkha and Chitwan, based on eco-regions. The high SOC in the Rasuwa district could be attributed to the district's cool, moist climate and slow organic matter decomposition rates.

The study found that forests and agroforestry land use have higher SOC and carbon sequestration than agriculture land use, despite being statistically insignificant. The SOC of agroforestry land is as high as the forests of other lower eco-regions in high altitude ecological regions with cool moist climate, and the difference in SOC between eco-regions is statistically significant (Bajracharya et al., 2015).

Forests are a natural blessing that provide numerous benefits to both humans and the environment, including carbon sequestration. Carbon sequestration is an important approach for reducing global C emissions, and it may eventually help to mitigate the effects of climate change (Shrestha et al., 2017). The ecosystem services provided by forests are part of the REDD implementation. As a result, with increasing anthropogenic intervention in forest lands for agricultural purposes, agroforestry systems may prove to be a better alternative. Agroforestry creates a win-win situation for both nature and humans because we believe that both should be able to survive and grow in harmony (Bajracharya et al., 2015). Given the limited capacity of forest and agricultural areas to mitigate climate change and meet people's food needs, respectively, the promotion of agroforestry systems may alleviate the problem of climate mitigation through C sequestration on farms and pastures while also providing livelihood enhancement.

14.5 Recommendations

Longer-term studies with repeated measurement of the framed indicators in a broader range of forest/land management types are required to validate the findings. The methods used in this study can be replicated in future studies to incorporate multiple inputs for forest and soil C assessment, such as field measurement, remote sensing and GIS data, and GPS tracking. Thus, the study establishes the foundation for the future development of a National Forest Monitoring System (NFMS) for the implementation of REDD or another similar mechanism.

Acknowledgement

Authors extend their gratitude to Mr Udeepta Raj Bhandari for his contribution during framing the structure of the paper. We would also like to acknowledge NORAGRIC for the financial support during the study.

References

Bajracharya, R.M., Shrestha, H.L., Shakya, R., Sitaula, B.K., 2015. Agro-forestry systems as a means to achieve carbon co-benefits in Nepal. Journal of Forest and Livelihood 13 (I), 58–70.

Bationo, A., Kihara, J., Vanlauwe, B., Waswa, B., Kimetu, J., 2005. Soil organic carbon dynamics, functions and management in West African Agro-systems. Agricultural Systems, Sicence Direct .

Bremner, J.M., Mulvaney, C.S., 1982. Nitrogen total. In: second ed. Page, A.L., Miller, R.M., Keeney, D.R. (Eds.), Methods of Soil Analysis Part 2, 9. American Society Agronomy Monograph, Madison, WI, pp. 595–610. Chemical and Microbiological Properties.

Buringh, P., 1984. Organic carbon in soils of the world. The Role of Terrestrial Vegetation in the Global Carbon Cycle. John Wiley & Sons Ltd, pp. 91–109.

Cole, C.V., Duxbury, J., Freney, J., Heinemeyer, O., Minami, K., Mosier, A., et al., 1997. Global estimates of potential mitigation of greenhouse gas emissions by agriculture. Nutrient Cycling in Agroecosystems 49 (1), 221–228.

Dahal, N., Bajracharya, R., 2010. Prospects of soil organic carbon sequestration: implications for Nepal's mountain agriculture. Journal of Forest and Livelihood 45–56.

Dangi, R.B., Acharya, K.P., 2009. A quick review of potential benefits and costs of REDD in Nepal. Ready for REDD? Taking Stock of Experience, Opportunities and Challanges in Nepal. Nepal Foresters' Associaltion, Kathmandu, pp. 9–20.

Dhital, N., Shrestha, H.L., Shrestha, B.M., Gautam, S., Rijal, B.R., 2015. REDD + as a development tool to improve rural livelihoods in Nepal. In: Adhikari, A.P., Dahal, G.P. (Eds.), Sustainable Livelihood Systems in Nepal: Principles, Practices and Prospects. IUCN and CFFN, Kathmandu, Nepal, 2015.

Dhital, N., 2009. Reducing emissions from deforestation and forest degradation (REDD) in Nepal: exploring the possibilities. Journal of Forest and Livelihood 8 (1), 56–62.

Dixon, R.K., 2004. Carbon pools and flux of global forest ecosystems. Science 263 (5144), 185–190.

Don, A., Schumacher, J., Freibauer, A., 2011. Impact of tropical land-use change on soil organic carbon stocks—a meta-analysis. Global Change Biology 17 (4), 1658–1670.

Eswaran, H., Berg, E.V., 1993. Organic carbon in soils of the world. Soil Science. USDA Soil Conservation Service, Washington, pp. 192–194.

Flint, E.P., Richards, J.F., 1991. Historical-analysis of changes in land-use and carbon stock of vegetation in south and southeast Asia. Canadian Journal of Forest Research-Revue.anadienne de Recherche Forestier 21 (1), 91–110.

Gami, S.K., Lauren, J.G., Duxbury, J.M., 2009. Soil organic carbon and nitrogen stocks in Nepal long-term soil fertility experiments. Soil and Tillage Research 106 (1), 95–103.

Geissen, V., Sanchez-Hernandez, R., Kampichler, C., Ramos-Reyes, Sepuveda-Lozada, A., Ochoa-Goana, S., et al., 2009. Effects of land-use change on some properties of tropical soils-an example from Southeast Mexico. Geoderma 151 (3–4), 87–97.

IPCC, 2006. 2006 IPCC Guidelines for National Greenhouse Gas Inventories, Prepared by the National Greenhouse Gas Inventories Programme. In: H.S. Eggleston, L. Buendia, K. Miwa, T. Ngara, K. Tanabe (Eds.), IGES, Japan. http://www.ipcc-nggip.iges.or.jp, V 4, p. 595.

Johnson, M.G., Kern, J.S., Lammers, D.A., Lee, J.J., Liegel, L.H., 1991. Sequestering carbon in soils: A workshop to explore the potential for mitigating global climate change. Held in Corvallis, Oregon on February 26–28, 1990 (No. PB-91-216390/XAB). ManTech Environmental Technology, Inc, Corvallis, OR (United States).

Knudsen, D., Peterson, G.A., Pratt, P.F., 1982. Potassium. In: Pages, A.L., Miller, R.H., Keeney, D.R. (Eds.), Methods of Soil Analysis Part 2, second ed. Madison, USA, pp. 225–246. , Chemical and Microbiological Properties.

Lal, R., 2005. Climate change, soil carbon dynamics, and global food security. In: Climate Change and Global Food Security, CRC Press, Boca Raton (FL), pp. 113–143.

Lal, R., 2009. Soil Carbon Sequestration for Climate Change Mitigation and Food Security.

Liao, Y., Wu, W.L., Meng, F.Q., Smith, P., Lal, R., 2015. Increase in soil organic carbon by agricultural intensification in Northern China. Biogeosciences 12, 1403–1413.

Manandhar, U., 2009. REDD, REDD + and co-benefits. Ready for REDD? Taking Stock of Experience, Opportunities and Challanges in Nepal. Nepal Foresters' Association, Kathmandu, pp. 1–8.

Manandhar, U., 2013. Forest monitoring, measurement, reporting and verification: from principle to practice. Journal of Forest and Livelihood 11 (2), 46–54.

McLean, E.O., 1982. Soil pH and lime requirement. methods of soil analysis. Part 2, second ed. Chemical and Microbiological Properties, 9. American Society of Agronomy, Madison, USA.

Milne, E., Easter, M., Cerri, C.E., Paustian, K., Williams, S., 2006. Assessment of Soil Organic Carbon Stocks and Change at National Scale. Gunther Fischer and Francesco Tubiello, Laxenburg.

Molden, D., Sharma, E., 2013. ICIMOD's strategy for delivering high-quality research and achieving impact for sustainable mountain development. Mountain Research and Development 33, 179. e183.

Olsen, S.R., Sommers, L.E., 1982. Phosphorous. In: A. L. Page, R. M. Miller and D. R. Keeney (Eds.), Methods of Soil Analysis. Part 2 (second ed.) Chemical and Microbiological Properties (pp. 403–416). American Soc. of Agron. Monograph No. 9, ASA-SSSA, Inc., Madison, WI.

Pearson, T.R.H., Brown, S.L., Birdsey, R.A., 2007. Measurement Guidelines for the Sequestration of Forest Carbon, USDA Forest Service, 47 p.

Poudel, A., Shrestha, H.L., Bajracharya, R.M., 2019. Quantification of carbon stock under different land use regimes of Chitwan district, Nepal. Banko Janakari 29 (2), 13–19.

Regmi, A.P., Pandey, S.P., Joshy, D., 2000. Effects of long-term application of fertilizers and manure on soil fertility and crop yields in rice-rice-wheat cropping system in Nepal. Long-term Soil Fertility Experiments Paper Series 6, 120–138.

Shrestha, H.L., Bajracharya, R.M., Sitaula, B.K., 2012. Forest and soil carbon stocks, pools and dynamics and potential climate change mitigation in Nepal. Journal of Environmental Science and Engineering B, David publishing, 1 (6), 800–811. doi:10.17265/2162-5263/2012.06.012 http://www.davidpublisher.org/Public/uploads/Contribute/551e48f21b67a.pdf.

3. Soil physicochemical parameters

Shrestha, H.L., Bhandari, T.S., Karky, B.S., Kotru, R., 2017. Linking soil properties to climate change mitigation and food security in Nepal. Environments 4 (2), 29. Available from: http://www.mdpi.com/2076-3298/4/2/29.

Stern, N., 2006. Stern Review: The Economics of Climate Change.

Tisdale, S.L., Nelson, W.L., Beaton, J.D., 1985. Soil Fertility and Fertilizers. MacMillan Publishers Co, New York.

Tripathi, B.P., 1999. Review on Acid Soil and its Management in Nepal, Lumle Seminar Paper, vol. 99/1.

UNREDD, 2010. Methods for assessing and monitoring change in the ecosystem derived benefits of afforestation, reforestation and forest restoration, UNREDD Programme, Multiple Benefits Series 6.

UNREDD, 2011. Identifying and mapping the biodiversity and ecosystem-based multiple benefits of REDD[+]: a manual for the Exploring Multiple Benefits tool.

Vargas, R., Paz, F., Jong, Bd, 2013. Quantification of forest degradation and below ground carbon dynamics: ongoing challanges for monitoring, reporting and verification activitied for REDD + . In Carbon Management 579–582.

Land use and land cover change

Urbanization and resilience in mountain soil ecosystem: case of outwash fan area of Leh, Ladakh, India

Sunny Bansal[1], Nazish Abid[2], Shivangi Singh Parmar[3,4] and Joy Sen[5]

[1]School of Architecture, Anant National University, Ahmedabad, Gujarat, India [2]Department of Architecture and Interior Design, College of Engineering, University of Bahrain, Bahrain [3]RCG School of Infrastructure Design and Management, IIT Kharagpur, Kharagpur, West Bengal, India [4]School of Architecture, Vellore Institute of Technology (VIT), Vellore, Tamil Nadu, India [5]Department of Architecture and Regional Planning, IIT Kharagpur, Kharagpur, West Bengal, India

15.1 Introduction

Mountain regions have a fragile ecosystem and are sensitive to environmental changes (Emmer et al., 2021; Yu et al., 2021). The exploitation of the mountain region's ecosystem services causes soil erosion, biodiversity loss, and frequent natural disasters. It is critical to understand the trade-off relationship between the mountain soil ecosystem and anthropogenic activities (Briner et al., 2013). Because of extreme topography, climate, altitude, and multiple hazards such as flash floods, landslides, debris flow, and earthquakes, the mountain ecosystem is extremely vulnerable to climate change (Hart and Hearn, 2018). According to the Census 2011, India is having about 31% population living in urban areas with 3.1% of the decadal growth (Ahluwalia, 2016; Bhagat, 2011). As a result, the rapidly growing urban population leads to unplanned urban development. This unplanned growth depletes ecosystem services and harms the environment. The negative effects of unplanned urbanization are not uniform across the Indian subcontinent. Mountain regions are particularly vulnerable to anthropogenic changes. As a result, understanding the negative consequences of unplanned urbanization in mountain areas such as the Indian subcontinent's Himalayan region is critical for long-term and resilient development (Anhorn et al., 2015; Dame et al., 2019a,b; Hewitt and Mehta, 2012). This chapter examines urbanization and its effects on the mountain soil ecosystem of Leh, Ladakh, in the Trans-Himalayas. Because of political-economic shifts and an increase in tourism, it is emerging as a rapidly growing urban center. The main concerns about changes in mountain soil ecosystems are their effects on environmental, sociocultural, and economic changes. The resulting anthropogenic pressure on Leh's mountain ecosystem is a source of concern for its resilient development.

Leh, the capital of Ladakh, is located in the outwash fan in the region's center and east. Floods caused by cloudbursts, glacial lake outbursts, river flooding, and extreme snow blizzards all threaten Leh (Mueller et al., 2019). The increased frequency of cloudbursts and rainfall in the region as a result of anthropogenic pressure and climate change has emerged as a threat to Leh's soil and settlements. Leh today is dealing with transitioning from a remote historical capital to a political capital, as well as increased natural disasters (Suri, 2018).

15.1.1 Study area

Leh Town is the capital of the newly formed Union Territory of Ladakh, India, in 2019. It is the largest town and historic capital of the Ladakh region. Geographically, it is situated in Trans-Himalayan Region, having undulated topography and high altitude passes (Fig. 15.1), 32–36°N latitude and 75–80°E longitude, at an altitude of 2300–5000 m above mean sea level (Giri et al., 2019). Leh lies in the cold-desert biome and is characterized by arctic and desert climates (WHC-UNESCO, 2015). It lies in the valley, the rain shadow of Ladakh Range of Trans Himalaya in the north, Zanskar range on the south, and the Tibetan plateau in the east, which serves as the orographic barrier (Sant et al., 2011).

According to the 2011 Census, the population of Leh Town is 30,870 people, with 51% urban residents and 29.97% decadal population growth. Leh Town is divided into 13

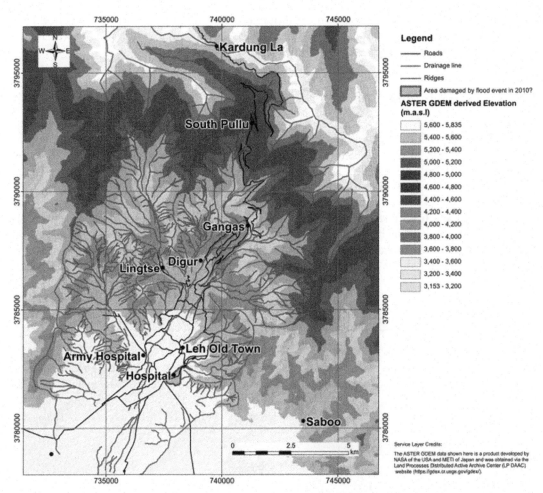

FIGURE 15.1   Mountain watersheds of Leh region (Hart and Hearn, 2018).

wards and has a total area of 9 square kilometers under Municipal Committee Leh. It was isolated from the rest of the world due to its geographical location and rugged terrain before the Ladakh region was opened to mass tourism in the mid-1970s (Guru et al., 2017). In the summers, Leh Town is well connected with the motorable road Leh-Manali highway to the mainland, so the floating population of Leh Town triples due to the arrival of tourists (Alexander, 2005). It has limited access to the mainland because the Zozila, Rotang, Baralacha, and Changla passes are closed for more than 7 months during the winter season, forcing it to rely on less frequent air services.

Another whirlwind of change occurred with the transfer of political power to Leh as the capital of the newly formed union territory. With a workforce participation rate of 56%, Leh's traditional economic base is agriculture and animal husbandry, while new economic sectors include information technology, tourism, and small scale and cottage industries. The region's physiographic uniqueness has resulted in the emergence of a distinct culture and indigenous knowledge systems of building, construction, water harvesting, and agriculture (WHC-UNESCO, 2015).

15.1.2 Geography and regional settings

The Ladakh region's landscape is made up of glacial landforms, debris-flow dominated alluvial fans, fluvial landforms, and dunes (Fig. 15.2). According to Ladakh's geological history, the region was formed by the collision of the Indian tectonic plate and the Eurasian plate (Shah et al., 2018). It also caused the formation of an intermontane basin, with alluvium filling a wide valley between mountain ranges (Shah et al., 2018). The Leh Town is located in the outwash fan of the intermontane basin of the Indus, a Trans-Himalayan River, originated in the Gangdise Range of Tibet. Several mountain streams confluence at the outwash fan terrace, spread over 5 km upstream to Leh (Hart and Hearn, 2018). The significant slip rate of the Karakoram fault and debris flows in the intermontane basin's outwash fan increase the risk of earthquakes and floods.

15.2 Urbanization: history, current trends, and evidence

More than half of the world population live in the cities and towns (UN News, 2018). This ongoing trend can be seen not only in larger metropolises or urban agglomerations, but also in small and medium-sized cities and towns (Loibl et al., 2018). The pace of

FIGURE 15.2 Landscape of Ladakh (Mueller et al., 2019).

urbanization is not only limited to plains but is also evident in the foothills of the mountains and mountain towns (Singh et al., 2020). The challenges of haphazard urbanization in sensitive mountain ecosystems are numerous. Rapid urbanization in mountain towns has been hampered by a lack of suitable land for construction activities. It results in the exploitation of environmentally sensitive land prone to flooding, landslides, soil erosion, and so on (Tse-ring et al., 2010). Due to their specific topographical setting mountain towns are quite susceptible to severe pollution of air, water, and land, especially soil.

Humans have lost touch with the mountain soil ecosystem and the life it supports as a result of this rapid urbanization. Increased urban activity has a significant impact on the physical characteristics and pollution levels of mountain soil. These mountain town urban processes and activities have varying effects on the biochemical properties of the soil, which in turn affects its life-supporting services. As a result, understanding the urbanization process and scale in mountain towns is critical for further research into the impacts on mountain soil ecosystems. Understanding the urbanization process entails recognizing the spatial-temporal characteristics of urbanization, its form and pace, as well as the sustainability challenges it faces by researching historical urban expansion patterns, current trends and evidence, and future challenges.

As previously discussed, Leh Town in Ladakh, India, will be used as a case study to understand and interpret urbanization in mountain towns. Most Himalayan towns, including Leh, had historical ties to religious and/or political sites and functions. Historically, such towns served as vital trans-Himalayan transit hubs connecting Central and South Asia. The current geopolitical situation along the Himalayan ranges is a major driver of urban development and infrastructure enhancement in these urban settlements. Over the last few decades, Leh Town has become very appealing to migrants and tourists from both within and outside the Ladakh region (Müller and Dame, 2019). Rapid and haphazard urbanization, scarcity of water, and an increase in pollutants have all had an impact on the mountain soil ecosystem, increasing the risk of natural disasters (Dame et al., 2019a,b).

15.2.1 Urbanization history

The history and growth of the Ladakh region with its then political status and other details are presented in Table 15.1.

15.2.2 Trends of urbanization

Seasonal changes affect the population and urban activities in Leh Town. The arrival of tourists during the summer months causes a significant increase in the number of migrant workers working in agricultural, tourism, and infrastructure sectors. These rural seasonal migrants and army personnel are not counted in the population (Dame, 2018). The town's built-up area has more than fivefold increased in the last five decades, from 36 to 196 ha, with an increase of 18,000 new buildings (Dame et al., 2019a,b). There are three trends characterized in the urbanization process of Leh Town:

TABLE 15.1   History and growth of Ladakh (Dame et al., 2019a,b; Goodall, 2004; Wester et al., 2019).

Year	Region	Details
950–1842	Capital of Kingdom of Ladakh	• Trade center for South and Central Asia (Trans-Himalayan and Trans-Karakoram region)
1842–1947	Province of Kingship of Jammu and Kashmir	• Invaded by Dogra army in 1842 and came under indirect British rule • About 2500 inhabitants (rising to 4000 during summer) and 400 residential buildings by 1888
1947–2019	Province of the Indian State of Jammu and Kashmir	• The population rose to 30 thousand by the year 2011 • Sensitive border situation with China and Pakistan resulted in urban development due to the stationing of military troops in the form of massive infrastructural developments, especially roads • Ladakh Autonomous Hill Development Council (LAHDC) was created in 1995 to gain more autonomy
2019 onwards	Capital and largest town of the Indian Union Territory of Ladakh	• Transfer of control directly to of the Government of India, the town is receiving huge infrastructural investments especially in the education, tourism, and medical sector

1. *Expansion of Leh town towards barren areas*: The huge barren lands have been converted into a concretized environment in the eastern, south-eastern, and western areas of Leh Town. Most of these new settlements are for military, permanently settled migrants, nomadic groups, and administrative infrastructure (Dame et al., 2019a,b).
2. *Expansion of Leh town towards agricultural areas*: The second major expansion of urban areas is into arable lands. The agricultural land loss percentage increased from 1% to 8% in the last five decades. The majority of this expansion includes roads, restaurants, and hotels (Goodall, 2004).
3. *Densification of old Leh town*: Modernization and densification of the settled areas is the third urbanization characteristic in the form of multistory concrete buildings (Dame and Nüsser, 2008).

15.2.3 Drivers of urban change

For every town and city, certain triggers drive the urban change and catalyze the urbanization process. Four dominant drivers of urban change in Leh Town are discussed below:

• *Administration, infrastructure, and services*: With the formation of LAHDC, many job opportunities arose, as did the establishment of several administrative and NGO complexes. The availability of infrastructure facilities such as a central bus station, airport, healthcare, education, and a market make Leh Town appealing to both temporary and permanent migrants (Tiwari and Tiwari, 2018).
• *Tourism growth*: Due to its scenic landscape and cultural heritage, Ladakh is becoming a popular tourist destination, with tourist arrivals increasing from 28,400 to 323,590

between 2003 and 2018. Leh Town is the Ladakh tourism arrival point due to its tourist facilities and infrastructure, which includes hotels and guesthouses (Kapoor, 2021).

- *Diffusion of urban lifestyle*: The shift away from agriculture and toward international tourism has resulted in a shift in urban sociocultural lifestyles. Urban-rural migration, an increase in nuclear families, and visitor influence from other Indian urban agglomerations all contribute to Leh Town's urbanization (Smith, 2017).
- *Geopolitical significance*: Ladakh is located in a geopolitical sensitive zone of India between the neighboring countries of Pakistan and China. Leh Town serves as the logistical base for Ladakh and Siachen Glacier for their military transport and road infrastructure (Dame and Nüsser, 2008).

15.2.4 Challenges of urban governance

Leh Town's urban development has four governance bodies—Leh Municipal Corporation, Ladakh Autonomous Hill Development Council, Indian Army, and NGOs (Müller and Dame, 2019). Their roles have been listed in Fig. 15.3.

The biggest challenge for Leh Town today is the growing vulnerability to natural hazards. (Masson, 2015). Illegal construction on alluvial fans and soils near riverbeds has exacerbated this. These vulnerable areas are vulnerable to debris flows and flash floods. Inadequate planning, particularly inefficient risk management, and a poor choice of construction location are major challenges for Leh Town (Keilmann-Gondhalekar et al., 2015). The city desperately needs technological solutions like rainwater retention basins, river regulation, and soil protection and remediation.

15.3 Anthropogenic dimension of ecosystem change

There are several anthropogenic dimensions of ecosystem change, but in the context of Leh, three are discussed here: environmental, sociocultural, and economic.

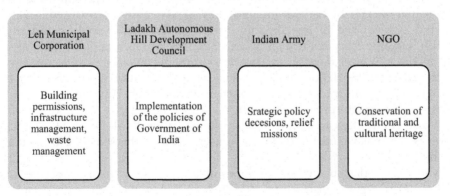

FIGURE 15.3   Role of different bodies in urban management in Leh (Müller and Dame, 2019).

15.3.1 Dimensions of environmental change

Despite being the historical capital for centuries, Leh Town has emerged as the capital of the newly formed Union Territory of Ladakh since 2019. This shift has created a plethora of development opportunities as well as future challenges to the region's delicate mountain ecosystem. Understanding the environmental change in Leh and its dimensions is critical for resilient development. The section that follows explains the dimensions of environmental changes in terms of land use and land cover, soil exploitation and pollution, mountain hazards, and climate change.

15.3.1.1 Land cover and land-use

Changes in land cover are a good indicator of the degree of anthropogenic interference in the ecosystem. Leh Town's land cover classification includes a river system, snow-clad mountains, grasslands, forests, and barren soil. Leh town's land use is divided into agriculture areas, residential areas, open areas for recreation, forest areas, and so on. Because Leh Town is the capital of the Union Territory of Ladakh, it is emerging as the region's political and economic center. Ladakh is a major tourist destination of the Indian subcontinent (Müller and Dame, 2019). As a result, tourism, along with political shifts and economic growth, is a driving factor in Leh Town's land-use change. The degree of urbanization stresses the ecosystem and its balance (Doichinova et al., 2006).

Because the mountain soil ecosystem is extremely sensitive to change, changes in land use and land cover have a significant impact on it. The severity of risk is determined by natural hazards and their interaction with vulnerable populations, and the nature of risk is location-based. Due to a lack of background information on the hazards and their interactions with changes in land use and land cover, the mountain ecosystem deteriorated further. Planning of sustainable tourism and urbanization will lead to the controlled land use and land cover change (Xu et al., 2012); and will help in maintaining the sensitive mountain soil ecosystem of Leh Town.

15.3.1.2 Mountain soil issues and urbanization

Ladakh has sparse vegetation due to its arid and cold climate, but Leh Town's location in the outwash fan of the Indus intermontane basin provides the geographical advantage of fertile soil. The soils of Ladakh are generally sandy, coarsely textured, and prone to severe wind erosion due to low organic carbon and moisture-holding capacity (Singh et al., 2019). Organic moss content in unconsolidated alluvium and the moisture-holding capacity of the alluvium deposit are two important factors responsible for agricultural patches in Leh Town (Spalzin et al., 2020). The region's traditional economy was based on pastors, agriculture, and trades. As an economic hotspot, Leh Town has seen a significant decline in pastor-agriculture practices. Increased connectivity of the region from the mainland also contributed to a shift in agricultural patterns (Müller et al., 2020). Natural vegetation is the major contributor in the conservation of the sensitive mountain soil of Leh Town.

In Leh town, the presence of sharp slopes of deep gully results in unstable roadsides (Singh et al., 2019). Road construction and other construction activities on sloppy erodible steep mountains can result in landslides. To check the ground's instability and avoid further disasters, appropriate methods such as contour wattling, geotextile, crib structure, and terracing are used. The increasing population growth in Leh Town has pushed people to settle in vulnerable areas such as along streams and in the highlands. As a result, uncontrolled urbanization in extremely vulnerable areas with unstable ground and flash flood prone-lower Indus outwash fan may increase the multihazard risk. Solid waste management in the mountain soil ecosystem is one of the major challenges of increasing anthropogenic pressures such as rapid urbanization, tourism, and MSMEs in Leh Town (Thakur et al., 2021). The poor solid waste management in Leh Town has been a threatening issue of the region (Alexander, 2005).

15.3.1.3 *Mountain ecosystem and climate change*

Climate change is the matter of concern of this century (Prosekov and Ivanova, 2018; Wilts et al., 2021). The effect of climate change is not equal in all geographical areas. Disturbance in the mountain ecosystem is considered the indicator of climate change (Chevuturi et al., 2018). Mountain regions have a higher impact due to climate change and are extremely vulnerable to resultant temperature and precipitation pattern change (Jhajharia and Singh, 2011). The climate of Leh Town situated in the Trans-Himalayas is affected by climate change impact at various levels (Páez-Curtidor et al., 2021).

Despite Leh Town's arid climate, there has been an overall increase in temperature and precipitation days in the region since the 1990s, which will amplify the occurrence of extreme unexpected events such as the multiple cloudbursts of 2010 (Páez-Curtidor et al., 2021; Thayyen et al., 2013). The 2010 flash floods due to cloudburst resulted in hundreds of losses of lives and infrastructure damages. Leh Town faced multiple flash floods, debris flow, and thick deposits of mud. The location of Leh Town (Fig. 15.4) on an outwash fan terrace situated in the valley of Ladakh Range, a south-eastern part of Karakoram; and the confluence of multiple mountain streams amplify the vulnerability to increased precipitation of the region (Ziegler et al., 2016). Leh outwash fan area was extremely affected by the 2010 flash floods.

15.3.2 Dimensions of sociocultural changes

15.3.2.1 *Tourism*

Leh City is at an elevation of 3500 m above sea level in Ladakh's Trans-Himalayan area, along an Indus River tributary to the north. Ladakh's critical political situation has resulted in the closure of trade relations across its international borders, leaving the region heavily reliant on commodities imported and subsidized by the government of India (Government of India). This shift from reliance on existing agriculture and international trade relations to reliance on external economies, primarily tourism, has exposed Ladakh's local economy to fluctuations in regional and global markets (Goodall, 2004). This is due to the political and geological grip that the region holds.

To promote regional economic integration and prevent a large-scale exodus of Ladakhis to the plains, Ladakh opened its border to tourists in 1974, after which it was completely restricted due to its strategic location in relation to Pakistan and China; though Ladakh has never been isolated in the past, it was geographically and politically barred to the rest

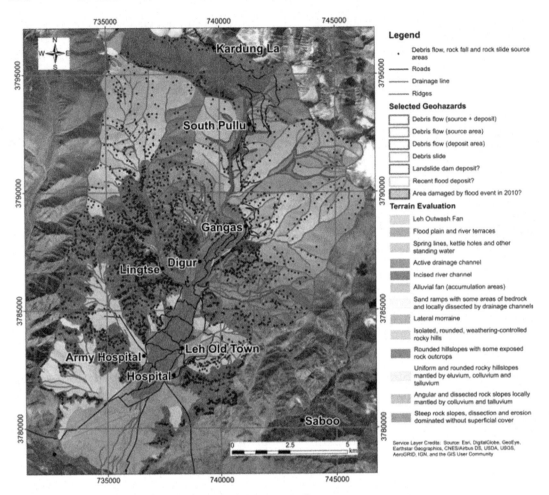

FIGURE 15.4   Geomorphology of Leh (Hart and Hearn, 2018).

of the world from 1948 to 1974. The second most important factor driving urbanization is the expansion of the tourism industry. The rapid expansion of leisure travel in the Ladakh region has contributed significantly to urbanization and anthropogenic pressure; tourism has become the backbone of the local Leh economies.

Since tourists have been allowed to enter this region, it has become a popular tourist destination due to its cultural heritage and picturesque landscape, and tourism has become a major revenue generator not only for the inhabitants but also for the surrounding region of Jammu and Kashmir. The recent tourist boom began in 2003, with 28,300 visitors, increasing to 79,100 in 2010, and to 323,950 in 2018 (through November), owing primarily to an increase in domestic visitors (Dame et al., 2019a,b) as shown in Fig. 15.5.

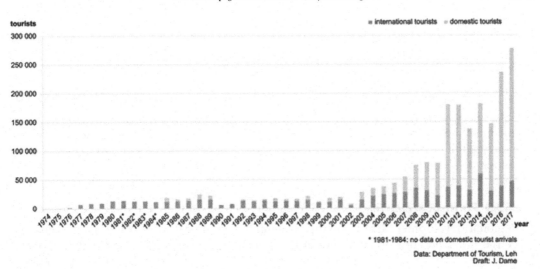

FIGURE 15.5   Tourist arrivals in Ladakh between 1974 and 2016 (Dame et al., 2019a,b).

15.3.2.2 *Demography and physical setting*

Leh Town, the most developed region of the Leh district, has seen significant rural-urban migration over the years due to better employment opportunities, education, facilities, healthcare, and infrastructure development. The tourism industry in Ladakh is not evenly distributed, and the majority of the population still lives outside of the commercial hub. As a result, the divide between the core and periphery regions is growing, and development efforts are concentrated primarily in areas with a noticeable influx of tourists. During the previous few decades, the urban agglomeration has attracted migrants from both within and outside of Ladakh, as well as visitors (Sherratt, 2014).

As per the 2011 Indian census, Leh district's urban population is increased from 2895 in 1911 to 30,870 in 2011 (Müller et al., 2020). The actual population of Leh Town is expected to be significantly higher because the census excluded large army settlements and rural seasonal migrants who have a second home in Leh Town. Furthermore, because census data did not account for the large number of army personnel and rural seasonal migrants who have a second home in Leh Town, the total population of the urban region is expected to be significantly higher. The population of Leh Town grows dramatically during the summer months due to an increase in tourism, which attracts a large number of migrants looking for work in tourism and agriculture.

The rapid spatial development of housing settlements on both former agricultural and barren lands, as well as the densification of built-up zones, demonstrates the region's rapid demographic growth. The pace of construction in Leh Town more than doubled between 2003 and 2017, with 9400 new buildings built in 14 years. From 1969 to 2004, nearly the same number of structures were built in the 35 years preceding this era. The number of guesthouses and hotels surged from around 35 in 1975 (Schettler and Schettler, 1977) to around 190 in 2005 (Thoma and Passang, 2005), and further to about 649 in 2015 (Department of Tourism, Leh).

Between 1967 and 2017, the built-up region of Leh Town quintupled, from 36 to 196 ha; between 1969 and 2017, 18,660 new buildings were added to the town's urban fabric (Dame et al., 2019a,b); and agricultural land loss due to the construction activities increased from 1.2% in 1969 to 8.4% in 2017, highlighting the speedy demographic expansion in construction activities and infrastructure development.

Fig. 15.6 indicates terraced agriculture plots dominating Leh Valley until 1969. The built-up density remained highest in the old town and bazaar area, and it became denser as one moved towards scattered settlements like Sankar, Ganglas, and Gonpa, which were only accessible by a few roads. Large military settlements and the Leh airport are located in the region's western and southern regions (Dame et al., 2019a,b).

15.3.2.3 Political scenario

Regional development has been supplemented by political and socioeconomic changes since the independence of India and consequent partition in 1947 when Ladakh became part of the Indian state of Jammu and Kashmir (Dame and Nüsser, 2008). Attempts to acquire more political autonomy were partially realized with the establishment of the Ladakh Autonomous Hill Development Council (LAHDC) in 1995 (Beek, 1999). Due to the sheer region's geopolitical significance and the sensitive border situation with China and Pakistan (Aggarwal and Bhan, 2009; Baghel and Nüsser, 2015), a considerable number of armed forces positioned nearby have significantly aided Leh Town's urban development and converted the area into "a mountain fortress for the Indian nation-state" (Gagné, 2017). Massive investments in public infrastructure have been made in the region, particularly in road construction to improve access from the Indian plains to the Himalayan range. The Srinagar-Leh highway, which opened in 1962, and the Manali-Leh highway, which opened in the 1970s, were built for military use but were later opened to civilians (Demenge, 2012).

15.3.3 Dimensions of economic change

15.3.3.1 Economic growth

The Ladakh region has seen significant economic development over the last few decades. An increase in trade and tourism activities, resource extraction, and labor migration has created new opportunities for local development and income generation, as well as new challenges (Pal, 2018). Even so, the region's common people remain economically vulnerable, as mountain poverty persists. Agriculture and animal husbandry are the most common occupations. However, the growing population is putting a strain on the mountain's limited resources. Leisure and religious tourism jobs are also on the rise in the region, and as significant drivers of environmental and socioeconomic change, they pose a risk of over-exploitation. Trade is another important source of income, but it performs below its potential due to a complex sociopolitical environment. Rapid globalization and constrictive mountainous characteristics such as inaccessibility and fragility necessitate the development of strong mountain-specific policies (Jodha, 2000).

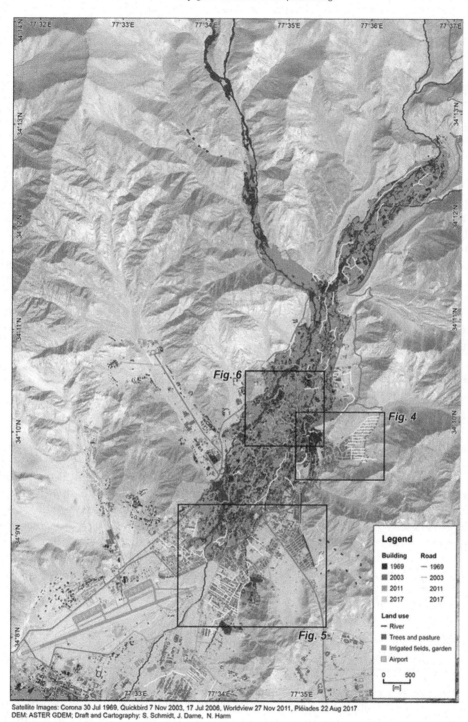

Satellite Images: Corona 30 Jul 1969, Quickbird 7 Nov 2003, 17 Jul 2006, Worldview 27 Nov 2011, Pléiades 22 Aug 2017
DEM: ASTER GDEM; Draft and Cartography: S. Schmidt, J. Dame, N. Harm

FIGURE 15.6 Urbanization of Leh (Dame et al., 2019a,b).

4. Land use and land cover change

15.3.3.2 *Infrastructure development*

Infrastructure and development have a positive relationship. If the provision of infrastructural services responds efficiently to demand, it can improve a region's overall economic scenario, alleviate poverty, and aid in the improvement of environmental conditions (Singh and Chudasama, 2020). Infrastructure development to improve accessibility in the Ladakh region has been a major driver of change. It has aided in reshaping the region's sociopolitical and economic situation. This increased access has also resulted in resource exploitation, increased reliance on the lowlands, environmental degradation, migration, and sociocultural conflicts (Dame and Nüsser, 2008).

Physical infrastructure in the Ladakh region has long been underdeveloped due to unfavorable geography, high construction costs, and proximity to sensitive international borders. Rather than serving civilians, early road construction and expansion were aimed at strategic military interests. However, these roads have improved overall accessibility for mountain people. However, due to the adverse conditions, road development in the region is slower. In addition, the region is prone to landslides, making road maintenance an expensive endeavor. The region's accessibility and connectivity are dependent not only on the mountain roads, but also on a network of trails, ropeways, suspension bridges, porters, and animal transport. Furthermore, air transport is essential for such remote locations. The region also has enormous potential for renewable energy sources. If properly utilized, this has the potential to cause a socioeconomic transformation in people's lives (Wang et al., 2019).

15.3.3.3 *Urban and peri-urban growth*

Urbanization is an independent driver of change in the Ladakh region with cross-cutting implications for mountain sustainability. Urbanization is the result of global and regional socioeconomic processes. In this context, urbanization is a socioeconomic phenomenon that includes demographic changes, migration, and market economy diversity. It cannot be restricted to urban sprawl, expansion, or the establishment of new towns and cities. This is because general urbanization overlooks aspects such as multilocality or temporary outmigration, second homes for leisure, or internet-based occupation in ostensibly rural areas. Change is also driven by urbanization. The city of Leh is important for its surroundings. The city benefits from public infrastructure and technology. It attracts people seeking a better life through higher incomes and livelihoods, greater freedom, and larger social networks. It also makes it easier to create economies of scale, which creates jobs, attracts reinvestment, and encourages in-migration from rural hinterlands. Cities eventually become congested, resulting in higher concrete buildings rather than vernacular ones (Barrett and Bosak, 2018).

15.4 Risks and vulnerability

The extreme topography, climate change, and anthropogenic activities in Leh Town have resulted in frequent geohazards (Hart and Hearn, 2018). The pressure of urbanization in Leh is exacerbating the situation. Rapid urbanization as a result of in-migration from other parts of Ladakh in search of a better quality of life is driving settlement growth in risky terrains. According to The Sendai Framework for Disaster Risk Reduction 2015–2030 (United Nations, 2015), understanding and assessment of disaster risk and vulnerability; management and governance of disaster risk; and preparedness are crucial for resilient development.

15.4.1 Risks to mountain ecosystem

As a result of rapid urbanization and infrastructure development, the town of Leh has been subjected to a number of environmental and urban concerns. Urbanization in highland regions is characterized not only by local population growth and migration from rural villages, but also by a scarcity of suitable construction land, which generally leads to construction activities in landslide or flood-prone zones (Ziegler et al., 2016). Furthermore, due to their unique topographical situation, mountain towns are prone to air-pollution, soil degradation, and water-related issues (Borsdorf et al., 2015). As a result of the rural-urban exodus that has resulted in a demographic paradigm shift, the town is now subject to water scarcity and soil erosion in the region. Water consumption in Leh Town has increased exponentially as the number of tourists has increased, resulting in a lack of water for agriculture during the summer months. Following torrential rains, tourism-related construction near riverbeds and on alluvial fans was vulnerable to flash floods and debris flows, exposing the town to catastrophic natural risks (Dame et al., 2019a,b).

The region's opening to transnational tourism and declining dependence on agriculture has resulted in the spread of urban lifestyle trends (Dame et al., 2019a,b). With the influx of tourists from major Indian cities or Ladakhi students who choose to pursue their education in metropolitan clusters outside the mountain region, a societal pattern of a new urban elite is emerging (Smith, 2017). Over the years, the fragmentation of multigenerational families into small nuclear households has resulted in the rapid construction of new residences. Because of the seasonal nature of the available employment in the region, the majority of migrants have established second homes in Leh.

The uncontrolled proliferation of temporary accommodations in Ladakh has harmed the fragile physical environment. The expansion of the tourism industry has resulted in massive construction, as well as alluvial fans and riverbeds, making the region vulnerable to floods (which occurred in 2006, 2008, 2010, 2015, and 2018) and debris flow following heavy rain, resulting in massive loss of human life and property. Water contamination has emerged as a major concern for the region. In the winter, most neighborhoods rely on water tankers, and in the spring, public and private boreholes. Increased tourism and urbanization have resulted in increased water demand, particularly during the summer months when there is a severe water shortage. The discharge of septic tank effluent into canals and groundwater has contaminated the water and soil, as well as created problems with clean water supply (Dame et al., 2019a,b).

While the tourism industry boosted the region's economy, it also harmed Leh Town's architecture. Traditional mud and timber structures were demolished and replaced with two to three-story brick and concrete structures that housed cafes, restaurants, emporiums, souvenir shops, and other tourist attractions. Simultaneously, guest lodges and hotels began to spring up on the outskirts of town, many of which were built in agricultural fields. Parts of the old town's heart remained deserted, with mud dwellings deteriorating into ruins as a result of neglect. Many of these homes were demolished to make way for godowns and tenements that were rented to migrant laborers who came to work primarily in the construction industry every summer.

Debris flows from nearby hills and streams wreaked havoc on built-up communities on infertile land, where a large number of Tibetan refugees and labor migrants live. Rapid population growth and uncontrolled urban development, combined with insufficient regulation, have resulted in (re)constructions in hazard-prone areas severely impacted by the 2010 flood, posing critical governance and urban planning challenges. Aside from increased tourism, new urban sociocultural lifestyles and spending patterns are also contributing to the region's increasing vulnerability (Dame et al., 2019a,b).

Because of Leh's sensitive political situation, the expansion of military bases has resulted in the construction of road infrastructure and other amenities to support these large communities, which are frequently spreading into agricultural and barren land that is not suitable for heavy construction. As a result, agricultural heritage has been lost, and the region has become overly reliant on imported food due to a scarcity of fertile soil (Dame et al., 2019a,b). There is no land development plan; nevertheless, property-owners need to get their plans signed by a municipal authority, but approval is granted without delay or complications (Dame et al., 2019a,b).

15.4.2 Vulnerability of soil ecosystem

The functioning of natural mountain ecosystems and the ecological balance they maintain are impacted by urbanization. It has a significant impact on the mountain soil ecosystem. The specific microclimate of the urban area influences soil changes. As a result, the soil varies from being completely natural in some places to being modified as a result of human activity in others. Human activity has a wide range of effects on the top layer of soil. It can contaminate the soil with dust, imported soils, and other materials from buildings with low clay differentiation. It may also alter the soil's properties (physical and chemical) by altering the water retention capacity and increasing compaction. It may also cause the soil's upper layers to become more alkaline. Furthermore, organic wastes, compounds, residues, and/or potentially toxic elements may be added to the soil. Urbanization also increases the soil's potential for erosion, which can lead to desertification. If sensitive mountain soil is disturbed, its strength may be altered, increasing the likelihood of landslides (Doichinova et al., 2006).

Understanding and assessing the effects of urbanization in Leh Town is critical for streamlining development to be resilient and sustainable in the face of climate change, unplanned urbanization, and increased tourism pressure. Economic growth, conservation of sensitive mountain soil ecosystems, and climate risk should all be balanced. One potential solution is to change current construction techniques and return to indigenous construction traditions. The combination of indigenous knowledge systems and modern techniques such as remote sensing may aid in the understanding of the nature and behavior of the terrain and mountain soil ecosystem in order to define the future course of resilient development based on risk assessment and preparedness.

15.5 Indigenous knowledge systems for construction—a solution

Ladakhi houses are simple yet elegant structures that have been polished over time. They are the best examples of building with local materials and to accommodate the local climate. Posts, beams, and window frames are made of the limited wood available, while thick masonry walls provide structural support and insulation. Mud is a common binding element found in unburned bricks and rammed earth, and it is used to finish both vertical and horizontal surfaces. The location has an impact on the construction of the house, but not only because of the resources available. Since imported labor was more difficult to find, every hamlet had to rely heavily, if not entirely, on local masons and carpenters. Masons were frequently required to work with wood when carpenters were unavailable.

The settlement's old town, Leh, is now in ruins. Because almost none of the original owners still live in town, the current residents are less sensitive to the systems at work in the buildings. Houses have recently become more of a product than a personal expression of care and trained expertise. This is what has changed dramatically over the last century, especially in house construction. Today's construction activity is more intense and advanced than ever before, a phenomenon of this magnitude never seen in Ladakh before.

This has put the settlement at risk of extinction because the traditional buildings that remain are mostly in disrepair, and there has been talk of demolishing them to make way for modern construction. More than 30% of the town's buildings are in poor condition and may be demolished soon. People prefer to live near the market due to the economic shift and large tourist influx, and the settlement pattern is more nucleated now than the traditional settlement planning. The authorities cater to Newtown as well, and the old town is regarded as a slum. To accommodate the growing urban population, the town's extensions are multistory structures. These tall (mid-rise) buildings ruin the view of the town. In contrast to the old town architecture, these structures make extensive use of concrete. Furthermore, hotels constructed by the tourism industry use extensive concrete blocks and brickwork, making the structure heavy and unsuitable for the mountain soil ecosystem (Khan, 2013).

The region has some of the most severe climatic conditions for human survival. Despite this, it has been continuously inhabited by humans for thousands of years. This is due to indigenous building knowledge, which results in highly efficient, environmentally friendly, and climate-responsive vernacular architecture (Fig. 15.7). Certain key points must be considered for modern construction in the region that is consistent with traditional indigenous construction knowledge systems and has negligible negative impacts on the mountain soil ecosystem (Khan, 2013):

- *Construction materials*: Local materials such as sun-dried mud bricks, timber, mud plaster, and stones should be used because they not only complement the landscape but also keep the interiors warm during the winter and at night.
- *Solar gain*: Every year, the region receives more than 320 sun days. To maximize the benefits of the warm sun, strategies such as passive solar gain by constructing the Trombe wall will be used.
- *Topography*: The construction must adhere to the region's natural topography and thus not interfere with natural drainage. Construction on slopes is more advantageous when facing north-south, as maximum solar heat can be gained from the south façade and the north mountainside can protect from harsh winds.

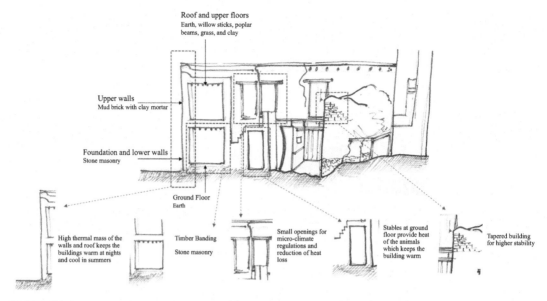

FIGURE 15.7 Indigenous knowledge systems of construction highlighting material and techniques, climate-responsive design, and safety measures in construction. (Architecture Sans Frontiers—UK, 2011).

- *Earthquake and landslides*: Ladakh is prone to earthquakes and landslides due to its natural setting and increased human activity. To withstand the seismic load, earthquake-resistant building materials and techniques such as timber frames must be used.
- *Rains and floods*: Despite being in the Himalayan rain shadow, Ladakh does not receive much rain. However, due to climate change, it has recently experienced high precipitation, resulting in floods. Traditional mud houses become unsuitable in heavy rains, and new constructions should use stabilized compressed earth blocks (SCEB).
- *Winds and extreme temperatures*: Aside from solar gain and building orientation, doors and windows are kept relatively small and few in number to prevent lesser winds from entering the interiors.
- *Snowfall*: Flat roof with a gradual slope covered with hay for additional insulation is a distinctive architectural feature of Ladakh designed for dry climate and snow in winters.
- *Spatial layout*: Individual houses are no more than three stories tall, with relatively small rooms on the upper floors for dining, living, prayer, and so on, and the ground floor for cattle and human waste collection. The bottom walls are thick to increase stability on the slopes. Spaces that are used more frequently are kept on the south side, while other spaces are kept on the north.

15.6 Conclusion

The current chapter advances understanding of the negative effects of urbanization, related anthropogenic activities, and mountain ecosystem soils. Urbanization is a broad phenomenon that encompasses nearly all human activities around the world. It has had various effects on the environment all over, but has had a greater impact on eco-sensitive zones such as mountains. The complexity and vulnerability of the mountain ecosystem have been established through a case study of Leh Town in India's Union Territory of Ladakh. A thorough investigation into the history, current trends, and drivers of urban change, as well as urban governance and challenges, has been presented. In addition, three anthropogenic dimensions of mountain ecosystem change are investigated. Land cover and land use, mountain soil issues, hazards, and climate change are all discussed in Dimensions of Environmental Change. Dimensions of sociocultural change examine the tourism industry, the demographic and physical environment, and the political scenario. Dimensions of Economic Change investigates economic growth, infrastructure development, and urban and peri-urban growth. A thorough understanding of the urbanization process and its consequences aids in assessing the risks to the mountain ecosystem and soil vulnerability. The study revealed that a more planned and efficient approach to urbanization is urgently needed for the sensitive mountain ecosystem. One such solution is the increased use of indigenous knowledge systems of construction in modern times. This will not only enhance the landscape but will also have little negative impact on the mountain ecosystem. Finally, recommendations for modern construction in the mountain ecosystem are presented.

References

Aggarwal, R., Bhan, M., 2009. "Disarming violence": development, democracy, and security on the borders of India. The Journal of Asian Studies 68 (2), 519–542.

Alexander, A., 2005. Leh Old Town, Ladakh—A Participatory Approach to Urban Conservation, Community-based Upgrading and Capacity-building. [Online] Available at: <http://www.tibetheritagefund.org/media/download/leh05_sml.pdf> (accessed 7.10.21.).

Ahluwalia, I.J., 2016. Challenges of urbanisation in India. In: Besley, T. (Ed.), Contemporary Issues in Development Economics. International Economic Association Series. Palgrave Macmillan, London, pp. 163–177.

Anhorn, J., Lennartz, T., Nüsser, M., 2015. rapid urban growth and earthquake risk in Musikot, mid-western hills, Nepal. Erdkunde 69 (4), 307–325. Available from: http://www.jstor.org/stable/24585779.

Architecture Sans Frontiers—UK, 2011. Learning in Leh. Architecture Sans Frontiers—UK, London.

Baghel, R., Nüsser, M., 2015. The vertical dimension of the Siachen conflict between India and Pakistan in the Eastern Karakoram. Political Geography 48, 24–36.

Barrett, K., Bosak, K., 2018. The role of place in adapting to climate change: a case study from Ladakh, Western Himalayas. Sustainability 10 (4).

Beek, M.v., 1999. Hill councils, development, and democracy: assumptions and experiences from Ladakh. Alternatives 24 (4), 435–460.

Bhagat, R., 2011. Emerging pattern of urbanisation in India. Economic and Political Weekly 46, 10–12.

Borsdorf, A., et al., 2015. Impacts and risks of global change. In: Grover, V.I., et al., (Eds.), Impact of Global Changes on Mountains: Responses and Adaptation. Taylor & Francis Group, Boca Raton, London, New York, pp. 33–76.

Briner, S., et al., 2013. Trade-offs between ecosystem services in a mountain region. Ecology and Society 18 (3).

Chevuturi, A., Dimri, A.P., Thayyen, R.J., 2018. Climate change over Leh (Ladakh), India. Theoretical and Applied Climatology 131, 531–545.

Dame, J., 2018. Food security and translocal livelihoods in high mountains: evidence from Ladakh, India. Mountain Research and Development 38 (4), 310–322.

Dame, J., Nüsser, M., 2008. Development perspectives in Ladakh, India. Geographische Rundschau International Edition 4, 20–27.

Dame, J., Schmidt, S., Müller, J., Nüsser, M., 2019a. Is it all downhill from here for Leh?. [Online] Available at: <https://www.indiawaterportal.org/articles/mountain-urbanization-case-leh> (accessed 14.10.21.).

Dame, J., Schmidt, S., Müller, J., Nüsser, M., 2019b. Urbanisation and socio-ecological challenges in high mountain towns: insights from Leh (Ladakh), India. Landscape and Urban Planning 189, 189–199.

Demenge, J., 2012. The Political Ecology of Road Construction in Ladakh. Institute of Development Studies, Brighton and Hove.

Doichinova, V., Zhiyanski, M., Hursthouse, A., 2006. Impact of urbanisation on soil characteristics. Environmental Chemistry Letters 3, 160–163.

Emmer, A., Cook, S.J., Frey, H., Shugar, D.H., 2021. Editorial: geohazards and risks in high mountain regions. Frontiers in Earth Science 9, 1–3.

Gagné, K., 2017. Building a mountain fortress for India: sympathy, imagination and the reconfiguration of Ladakh into a border area. South Asia: Journal of South Asian Studies 40 (2), 222–238.

Giri, A., et al., 2019. Utility of multivariate statistical analysis to identify factors contributing river water quality in two different seasons in cold-arid high-altitude region of Leh-Ladakh, India. Applied Water Science 9 (2), 26–40.

Goodall, S.K., 2004. Rural-to-urban migration and urbanization in Leh, Ladakh: a case study of three nomadic pastoral communities. Mountain Research and Development 24 (3), 220–227.

Guru, B., Seshan, K., Bera, S., 2017. Frequency ratio model for groundwater potential mapping and its sustainable management in cold desert, India. Journal of King Saud University—Science 29 (3), 333–347.

Hart, A.B., Hearn, G.J., 2018. Mapping geohazards in the watersheds above Leh, Ladakh: the use of publicly-available remote sensing to assist risk management. International Journal of Disaster Risk Reduction 31, 789–798.

Hewitt, K., Mehta, M., 2012. Rethinking risk and disasters in mountain areas. Journal of Alpine research | Revue de Géographie Alpine 100.

Jhajharia, D., Singh, V.P., 2011. Trends in temperature, diurnal temperature range and sunshine duration in Northeast India. International Journal of Climatology 31 (9), 1353–1367.

Jodha, N.S., 2000. Globalization and fragile mountain environments: policy challenges and choices. Mountain Research and Development 20 (4), 296–299.

Kapoor, S., 2021. Sustainable Tourism in the Indian Himalayan Region—A Case Study of the Union Territory of Ladakh. Indian Institute of Public Administration, New Delhi.

Keilmann-Gondhalekar, D., Nussbaum, S., Akhtar, A., Kebschull, J., 2015. Planning under uncertainty: climate change, water scarcity and health issues in Leh Town, Ladakh, India. In: Filho, W., Sumer, V. (Eds.), Green Energy, Technology: Sustainable Water Use and Management. Springer, s.l.

Khan, N., 2013. Vernacular architecture and climatic control in the extreme conditions of Ladakh. Bathinda, Advancements in Sustainable Practices and Innovations in Renewable Energy.

Loibl, W., et al., 2018. Characteristics of urban agglomerations in different continents: history, patterns, dynamics, drivers and trends. In: Ergen, M. (Ed.), Urban Agglomeration. IntechOpen, London, pp. 29–63.

Masson, V.L., 2015. Considering vulnerability in disaster risk reduction plans: from policy to practice in Ladakh, India. Mountain Research and Development 35 (2), 104–114.

Mueller, S., et al., 2019. Disaster scenario simulation of the 2010 cloudburst in Leh, Ladakh, India. International Journal of Disaster Risk Reduction 33, 485–494.

Müller, J., Dame, J., 2019. Small town, great expectations: urbanization and beautification in Leh. South Asia Multidisciplinary Academic 14.

Müller, J., Dame, J., Nüsser, M., 2020. Urban mountain waterscapes: the transformation of hydro-social relations in the Trans-Himalayan Town Leh, Ladakh, India. Water 12 (6).

Páez-Curtidor, N., Keilmann-Gondhalekar, D., Drewes, J.E., 2021. Application of the water—energy—food nexus approach to the climate-resilient water safety plan of Leh Town, India. Sustainability 13 (19).

Pal, B.K., 2018. Seasonal labour migration: a case study of Leh-Town, Ladakh. Economic Affairs 63 (2), 481–487.

Prosekov, A.Y., Ivanova, S.A., 2018. Food security: the challenge of the present. Geoforum 91, 73–77.

Sant, D.A., et al., 2011. Morphostratigraphy and palaeoclimate appraisal of the Leh valley, Ladakh Himalayas, India. Journal of the Geological Society of India 77, 499–510.

Schettler, M., Schettler, R., 1977. Kaschmir + Ladakh — Globetrotter-Ziele beiderseits des Himalayas (Kashmir and Ladakh. Destinations for globetrotters on both sides of the Himalayas). s.l.:s.n.

Shah, A.A., et al., 2018. Living with earthquake and flood hazards in Jammu and Kashmir, NW Himalaya. Frontiers in Earth Science 6.

Sherratt, K., 2014. Social and economic characteristics of Ladakh, India. s.l.: Geology for Global Development.

Singh, P.K., Chudasama, H., 2020. Evaluating poverty alleviation strategies in a developing country. PLoS One 15 (1).

Singh, R., et al., 2019. Soil erosion-status, causes and preventive measures in Trans-Himalayan region of Leh-Ladakh in J&K. Journal of Soil and Water Conservation 18 (3).

Singh, S., Hassan, S.M.T., Hassan, M., Bharti, N., 2020. Urbanisation and water insecurity in the Hindu Kush Himalaya: insights from Bangladesh, India, Nepal and Pakistan. Water Policy 22 (S1), 9—32.

Smith, S., 2017. Politics, pleasure, and difference in the intimate city: Himalayan students remake the future. Cultural Geographies 24 (4), 573—588.

Spalzin, S., Dhyani, A., Shantanu, K., Uniyal, P.L., 2020. Diversity and distribution pattern of mosses in cold desert of Leh, Ladakh. Journal of Himalayan Ecology and Sustainable Development 15.

Suri, K., 2018. Understanding historical, cultural and religious frameworks of mountain communities and disasters in Nubra valley of Ladakh. International Journal of Disaster Risk Reduction 31, 504—513.

Thakur, A., et al., 2021. Solid waste management in Indian Himalayan Region: current scenario, resource recovery, and way forward for sustainable development. Frontiers in Energy Research 9.

Thayyen, R.J., Dimri, A.P., Kumar, P., Agnihotri, G., 2013. Study of cloudburst and flash floods around Leh, India, during August 4—6, 2010. Natural Hazards 65, 2175—2204.

Thoma, H., Passang, S., 2005. Leh valley map with hotel/guesthouse register. Leh: PH 2005 Publication.

Tiwari, P.C., Tiwari, A., 2018. Urban growth in Himalaya: understanding the process and options for sustainable development. Journal of Urban and Regional Studies on Contemporary India 4 (2), 15—27.

Tse-ring, K., Sharma, E., Chettri, N., Shrestha, A., 2010. Climate Change Vulnerability of Mountain Ecosystems in the Eastern Himalayas. International Centre for Integrated Mountain Development, Kathmandu.

UN News, 2018. Around 2.5 billion more people will be living in cities by 2050, projects new UN report. [Online] Available at: https://www.un.org/development/desa/en/news/population/2018-world-urbanization-prospects.html.

United Nations, 2015. Sendai Framework for Disaster Risk Reduction 2015—2030. United Nations Office for Disaster Risk Reduction, Sendai.

Wang, Y., et al., 2019. Drivers of change to mountain sustainability in the Hindu Kush Himalaya. In: Wester, P., Mishra, A., Mukherji, A., Shrestha, A.B. (Eds.), The Hindu Kush Himalaya Assessment: Mountains, Climate Change, Sustainability, and People. Springer, Cham, pp. 17—56.

Wester, P., Mishra, A., Mukherji, A., Shrestha, A.B., 2019. The Hindu Kush Himalaya: Mountains, Climate Change, Sustainability and People, first ed. Springer Nature, Cham.

WHC-UNESCO, 2015. Cold Desert Cultural Landscape of India. [Online] Available at: <https://whc.unesco.org/en/tentativelists/6055/> (accessed 26.11.21.).

Wilts, R., Latka, C., Britz, W., 2021. Who is most vulnerable to climate change induced yield changes? A dynamic long run household analysis in lower income countries. Climate Risk Management 33.

Xu, Y., Qi, S., Wang, G., Shang, G., 2012. Urbanisation induced landscape change of urban hills in Jinan City, Karst Geological Region of North China. Landscape Research 37 (6).

Yu, Y., et al., 2021. Response of multiple mountain ecosystem services on environmental gradients: how to respond, and where should be priority conservation. Journal of Cleaner Production 278.

Ziegler, A.D., et al., 2016. A clear and present danger: Ladakh's increasing vulnerability to flash floods and debris flows. Hydrological Processes 30, 4214—4223.

Understanding the urbanization induced issues in mountainous ecosystems of India: a comparative study between Nilgiris (Tamil Nadu), and Lower Himalayas (Uttarakhand), India

Vidhu Bansal[1,3], Sharmila Jagadisan[2] and Joy Sen[1]

[1]Department of Architecture and Regional Planning, IIT Kharagpur, Kharagpur, West Bengal, India [2]School of Architecture, Vellore Institute of Technology, Vellore, Tamil Nadu, India [3]Anant National University, Ahmedabad, Gujarat, India

Understanding Soils of Mountainous Landscapes
DOI: https://doi.org/10.1016/B978-0-323-95925-4.00016-9

16.1 Introduction

Many natural resources are stored in mountainous regions around the world. Minerals, flora, fauna, rare herbs, and a variety of other natural treasures can be found in these remote terrains. The gift of mountains is immense, and human civilization has been fortunate to have access to it all at this point. However, with the onset of rapid urbanization, this scenario is rapidly changing. The process of urbanization has engulfed not only the urban and rural areas of countries worldwide, but also their remote terrains. The hunger for urbanization is almost insatiable, and it will burn the very earth that feeds it. The need of the hour is to assess the current conditions that are slowly driving these fragile ecosystems to their demise. The current study considers the aforementioned scenario and attempts to determine the extent of urbanization-induced stresses in mountainous regions. The chapter investigates it through two case studies from India, one from the country's northernmost and one from its southernmost regions. The two case studies present two geographical perspectives set in different sociocultural contexts, with the goal of extrapolating a broader view of the situation. The comparative analysis technique will highlight the issues that are arising in these regions. The study would conclude with the authors' recommendations for the current situation and a future course of in-depth research.

Mountains are considered hotspots for biological and cultural diversity because they provide a variety of ecosystem services all over the world. According to the UNEP report, by 2021, mountain regions will cover roughly one-quarter of the Earth's land surface and house 15% of the world's population. According to Korner and Ohsawa, the mountain ecosystem covers the majority of the world's continental land surface, and it is estimated that 83% of the world's 237 countries include mountains (Korner and Ohsawa, 2005). Their impact extends far beyond their immediate surroundings. They enable numerous ecological (with a diverse range of habitats), hydrological, climatic, social, and economic activities to have an impact on a variety of lives and livelihoods (UNEP, 2021). Many communities rely on healthy mountainous areas to maintain a healthy lifestyle. Mountainous areas are an important spatial carrier for a wide range of people's activities and have a significant impact on sustainable economic growth (Ding and Peng, 2018, p. 2). Most mountain ranges have astonishing and perplexing biodiversity in terms of taxonomic groups and endemic species, accounting for one-quarter of the world's terrestrial biodiversity (Conservation International, 2021). Though the mountain environment varies greatly across the globe, they share many of the same problems as the most fragile ecosystems on the planet.

16.2 Importance of mountainous regions and mountain ecosystems

In the current situation, mountain regions face a double bind. On the one hand, there are opportunities for employment, a variety of socioeconomic services, and infrastructure expansion that will aid in regional development. At the same time, the negative effects of urbanization are causing a shift in the delicate ecosystems' balance (Tiwari et al., 2018). A slew of issues have engulfed these areas, the most notable of which are haphazard urbanization with serious consequences for watershed regions and other hydrological processes, climate change, availability of potable water and water for other purposes, deforestation, soil erosion, loss of natural habitat, increased risk of natural disasters, and so on (Tiwari et al., 2018; Jamwal, 2020). These issues need to be tackled sooner than later if the fragile balance of these ecosystems needs to be restored (Fig. 16.1).

Migration pressure across regions and unrelenting commercialization have put enormous strain on the carrying capacity of the hills, resulting in soil erosion, increased pollution, natural habitat loss, and indigenous people marginalization. According to IUCN data, 52% of the 6109 Key Biodiversity Areas in mountains worldwide are less than 30% protected, and 40.4% are completely unprotected (IUCN, 2020). This finding shows that mountains are considered a fragile environment, and they were unfortunately induced by anthropogenic perturbations that are threatening our planet with rampant pollution, collapsing biodiversity, and climate change. One of the most pressing global issues is urbanization, which necessitates the creation of a framework with the lofty goal of conserving biodiversity. The government must envision a nature-positive future for our urban century by preserving remnant natural habitats and restoring the modified environment to improve the long-term health of the mountain ecological system.

The Himalayas (with distinct subdivisions), Satpuras and Vindhyas, Aravallis, Eastern Ghats, and Western Ghats are the most prominent mountainous zones in India. The Himalayas are the world's youngest mountain ranges, but they contain the majority of the world's highest peaks. They contribute to the formation of a barrier that protects Northern India from the harsh cold weather found further north. The Satpuras and Vindhyas are located in central India and are home to numerous scenic spots and temples. They are made up of numerous small mountains. The Aravallis are India's oldest mountain ranges, located on the western side of the country. They act as a barrier to monsoon clouds, diverting them eastward to regulate the rivers of the Sub-Himalayan region. The Western Ghats, also known as the Sahyadri range, are well-known in India and have been designated as one of the eight unique biodiversity hotspots by UNESCO. The Eastern Ghats run parallel to the Bay of Bengal and contain some well-known pilgrimage sites (India Today, 2017). The authors chose the Western Ghats and the Himalayas as sites for comparative analysis for this study because they provide two distinct geographical accounts. It is being investigated here how the geography and related sociocultural scenario have any common problems or are typical to the area.

16.2.1 The Western Ghats or Sahyadri Range—an outstanding universal value

India is recognized as one of the world's 12 mega-diverse countries, with the Western Ghats (much older than Himalayan mountains) and the Eastern Himalayan region as the

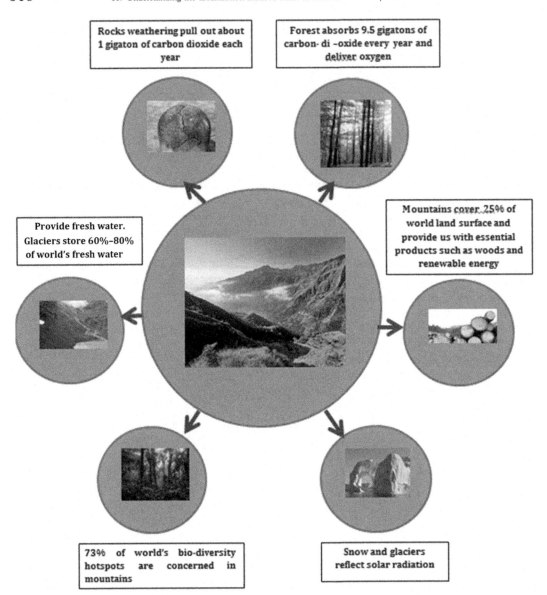

FIGURE 16.1   Redrawn by the author from Mountain Matter (Kilian Jornet Foundation, 2021).

most prized natural assets, serving as a repository for several exceptional levels of economically important plant and animal diversity (Myers, 1988 (1990)). The Western Ghats, also known as the Sahyadri range, are a series of mountain ranges that run parallel to the Indian peninsula's western ridge. According to phytogeographers, this region, known as the "Malabar Botanical Province," is internationally recognized as a region of exceptional universal value for the conservation of biological diversity, as well as containing areas of unique cultural, geological, and esthetic values (Rao, 2021). This Ghat runs through six different states: Kerala, Tamil Nadu, Karnataka, Maharashtra, Goa, and Gujarat. These mountains cover approximately 140,000 km^2 over a 1600 km stretch, and the entire range is a contiguous landform except for a 30 km wide interruption in Kerala known as the "Palaghat gap."

The proposed Western Ghats property has received heritage status and is made up of 30 parts divided into seven subclusters. Agasthyamalai, Periyar, Anaimalai, and the Nilgiris in Kerala and Tamil Nadu, Talacauvery and Kudremukh in Karnataka, and Sahyadri in Maharashtra are among the seven nominated spots or subclusters, which reflect the property's Outstanding Universal Value in terms of endemism and species richness. Menon and Bawa estimated the gross deforestation rate in the Western Ghats to be 0.57% per year from 1920 to 1990, while Prasad et al. estimated a 0.90% annual decline in Kerala's natural forest cover from 1961 to 1988 (Menon & Bawa, 1998).

16.2.1.1 Threats to the Western Ghats

1. Mining:
   a. Because of India's increasing demand for low-quality iron ores, iron-ore mining has become a contentious issue in Goa, often in violation of all laws, resulting in a profound impact on the environment and social disruption.
   b. Illegal sand mining has emerged as a serious threat in Kerala, which is heavily dependent on unsustainable sand mining. As a result, Kerala has become one of the major landslide-prone areas, causing significant damage to agriculture and water sources, negatively impacting people's lives and livelihoods in those areas.
2. **Livestock Grazing**: Large areas of habitat bordering the protected area are highly degraded. Livestock grazing (cattle and goats) is a major contributor to habitat loss and fragmentation across the Western Ghats.
3. **Human-wildlife conflict:** Given that the Western Ghats habitats are interspersed with the matrix of an intensely human-dominated mountain landscape, humans and wildlife are currently at odds, posing a serious threat. For example, villagers living near the Bhadra Wildlife Sanctuary in the state of Karnataka lose approximately 11% of their crop depredation each year due to elephant raids (Gopal, 2018).
4. **Extraction of forest produces:** Communities living in or near protected areas in the Western Ghats are frequently reliant on the extraction of NTFP (Nontimber forest products) for subsistence. The sustainability of NTFP is a critical issue, given rising population and increased demand for domestic consumption patterns.
5. **Plantations**: Plantations owned by private individuals and the corporate sector continue to grow in the Western Ghats and constitute an important source of fragmentation of natural habitat (CEPF, 2021).
6. **Infringement by human settlements:** A significant threat has arise from human settlements occurring both within and outside protected areas all across the Western Ghats.
7. **Pollution**: The unrestricted use of agrochemicals in the vicinity of forests, particularly in tea and coffee estates, causes serious damage to aquatic and forest ecosystems.

8. **Hydropower projects and large dams**: Large dam projects in the Western Ghats have resulted in environmental and social disruption despite cost-benefit analyses and environmental impact assessments being done by the government and companies.
9. **Deforestation:** Conversion of forest land into agricultural land or for commercial purposes like tourism, illegal logging for timber has had significant negative effects on biodiversity.
10. **Climate change:** The changes in land use and deforestation have led to big variations in the duration and intensity of rainfalls. Climate change has been considered a cause of floods in many regions in the recent past.

16.2.2 The Himalayas: hotspot of diversity

Rapid urbanization has engulfed the entire globe. This phenomenon is now encroaching on even the most remote parts of the globe. India is not an exception. The Himalayan region's haphazard urban growth is now making an impression. The Himalayan region is suffering as a result of natural resource depletion and the resulting disruption of ecosystem services (Tiwari et al., 2018).

16.2.2.1 Threats to Himalayan region

1. **Depleting natural resources:** There are numerous factors that have contributed to the depletion of the Himalayan flora and fauna. Scientists have cited a variety of reasons, ranging from "any kind of disruption in the natural process" to "the encroachment of modern civilizations" to something like "Anthropogenic discharge leading to climate change." The fundamental rule of an exorbitant increase in demand and an unchanging supply of resources has resulted in the current situation (Singh et al., 2021; Williams, 1989).
2. **Disruption of ecosystem services:** The impacted forest cover is the primary cause of the disruption of ecosystem services in the Himalayas. When the forest cover begins to thin, it exacerbates the associated problems in a negative way. Soil erosion, landslides, loss of soil fertility, decrease in fodder and fuel, siltation of rivers, lack of groundwater recharge, and other problems multiply (Das et al., 2016; Singh et al., 2021).
3. **Increasing the socioeconomics and environmental inequality:** Increased population leads to increased cropland and livestock, both of which are dependent on forests. This increase in fodder and fuel demand has resulted in a reduction in forest cover. Many researchers have linked the thinning of forest cover on the outskirts of settlements in mountainous areas to an increase in invasive species of flora (McDougall et al., 2011).
4. **Rapid and unplanned urbanization and increased fragility of slopes:** The Himalayan slopes have been repeatedly attacked for their abundant forest resources. In the first phase, these assaults in the name of urbanization and development resulted in widespread deforestation, increasing the fragility of slopes and resulting in widespread landslides. It also results in the deprivation of people (forest tribes) who rely on forest resources for their daily activities (Narain, 2013). The forest conservation act of the 1980s as a result of this phenomenon.
5. **Rapidly changing climatic conditions:** The Himalayas are also known as the "Roof of the World." Climate change is threatening this "Roof." Melting glaciers and rapid

icecap loss are exposing this permafrost at an exponential rate. Many studies have identified these as significant losses (ICIMOD, 2009). Climate change will eventually cause unstable slopes and disrupt the hydrological cycles of urban watershed areas. Urbanization-induced stress would put additional strain on these vulnerable areas, causing them to collapse completely.

6. **Climate change-induced hydrological hazards (in hinterlands):** The importance of hydroelectricity in the development of the country cannot be overstated. However, the issue arises in the selection of the location and the construction of such projects, which require extreme caution. These types of projects are being completed at a rapid pace in Himalayan regions. These projects must be thoroughly examined from an ecological standpoint. Many floods in Uttarakhand have been closely associated with dam construction, and experts believe that careful construction of such projects can reduce the severity of hydrological hazards (Narain, 2013).

7. **Invasive species:** Invasive alien plant species are species that are not indigenous to the area and can wreak havoc on the natural surroundings of a location, either economically, environmentally, or ecologically, where they have been introduced either accidentally or deliberately (Lamsal et al., 2018). The invasion of alien species is also becoming more visible in the Himalayas. According to studies, increased tourism and the construction of roads that wind through these alpine forests have introduced these species to previously unreachable ranges. The trend will worsen as global temperatures rise and human interference increases (Pathak et al., 2019). The irreplaceable loss in the Himalayan context would be the biodiversity of the forest that makes these mountain ranges so unique.

8. **Pollution:** The Himalayan mountains, like any other place on the planet, are under serious threat from various types of pollution. Many people rely on glaciers and snow-fed rivers for their freshwater needs. The timing of which has serious economic and social consequences. "Black carbon" is the reason for this. Soot, also known as black carbon, is a byproduct of incomplete combustion of fuel, which can be biofuel, fossil fuel, or biomass. These carbon particles are observed to be carried by the winds and deposited on the surfaces of Himalayan glaciers. As a result, by darkening the surface, it increases the absorption of sunlight. This process eventually raises the temperature, hastening ice melting in high-altitude areas (Mani, 2021).

9. **Tourism:** In terms of tourism, the Himalayas have a comparative advantage. It is a rapidly expanding field. The issue in the Himalayan context is one of activity regulation. It is plagued by ineffective regulatory mechanisms and institutions, which could lead to serious environmental issues in the future. If such tourism, whether adventure tourism, leisure tourism, eco-tourism, or any other type, is planned scientifically, it can lead to the region's long-term development (Khawas, 2009).

16.3 Case studies

The proposed chapter will concentrate on the ecosystems of the Nilgiri and Uttarakhand Himalayan Mountain ranges. The two regions provide an in-depth look at a few issues that are causing problems in these areas. The research would be a significant contribution to addressing issues such as global climate change, urbanization, and biodiversity conservation, as well as providing a new perspective on these ecosystems and their natural disturbance regimes.

16.3.1 Nilgiris: an overview

The Nilgiris hills are home to the magnificent Blue Mountain, also known as "Nilgiri Mountain" or Ooty, which is part of the Western Ghats, as well as the Nilgiris Bio-Sphere Reserve, which spans the states of Tamil Nadu and Kerala (Murugesan, 2013). It is located at a high altitude of 1800–2500 m, which covers nearly 65% of its geographical area. Because of its altitude, the climate in this district remains between 21°C and 25°C in the summer and between 10°C and 12°C in the winter. Nilgiris was once home to tribes such as the Todas, Kurumbas, Irulas, Paniyas, and Katunayake (Rural Transformation Project, 2020). Doddabetta, the highest peak, is approximately 2623 m high. Nilgiris is located at the confluence of the two Ghat ranges of the Sahayadri Hills and has a breathtaking view of Kerala on the west, Mysore State on the north, and Coimbatore district on the east and south. The district contains six subdistrict or talukas namely: Udhagamandalam, Panthalur, Coonoor, Kundah, Kotagiri (Murugesan, 2013).

Area Details: The ninefold classification of land use pattern in the Nilgiris district reveals that the largest forest cover area of 56% comes under the Nilgiris Biosphere Reserve (Rural Transformation Project, 2020).

According to the graph above, the land area distribution in Nilgiri district is as follows: 56% is forest, 30% is net cropped. The main crops grown here are tea, coffee, spices, and vegetables (Fig. 16.2).

16.3.1.1 Soil characteristics

The Nilgiris district is divided into five agro-climatic zones: Coonoor, Kotagiri, Kullakamby, Kundah, and Ooty. Lateritic red loam, Black Soil (best as rich loamy soil), Sandy Coastal Alluvium, Brown Soil (clayey loam which is the second best), Yellow Soil (with higher clay content not suitable for agriculture), Red Loam are the most common soil types in Nilgiris (fragile soil structure compared to yellow soils) (Subburaj, 2008). The current study shows that soil in this district has a low pH content due to the abundance of acidic metamorphic rocks such as Charnockite and Genesis. The soil also has a high phosphorus fixation capacity, indicating extensive leaching due to the region's climate and topography.

Higher altitude soil type: Lateritic

Classification of soil region: Red and Laterite soil region II

Soil description: Nilgiris soil has a sand content of 45.80% and an organic carbon content of 2.50% and it lacks in Lime Content (Madhu et al., 2017). There is total 17 soil series whose percentages are indicated in the figure below with their predominant soil series (Fig. 16.3).

16.3.1.2 Issues in Nilgiris

Over the last few decades, Nilgiris has struggled with massive environmental issues such as pollution, deforestation, forest encroachment, and extreme weather conditions that contribute to massive flooding and landslides. Furthermore, exotic invasive species such

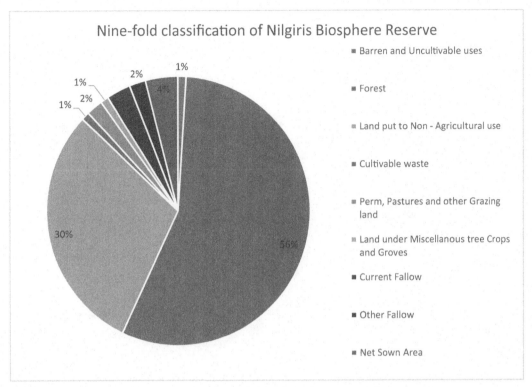

FIGURE 16.2 Redrawn by the author from Ninefold classification of Nilgiris Biosphere Reserve (Rural Transformation Project, 2020).

as wattle and eucalyptus have been identified as a threat to native biodiversity in the Western Ghats (65%–75%). As a result, native dry moist deciduous and thorn forest, including grass used as fodder by elephants and gaur, have been threatened, forcing these wild animals to venture into villages and towns (Nina, 2021).

16.3.1.2.1 Landslides

Landslides are the most common problem worldwide, and in recent years, the Nilgiris region has been subjected to high to severe landslide hazards, which are disastrous and have been increasing exponentially. The division of terrain into zones and their ranking according to degrees of actual/potential hazard caused by mass movement is referred to as landslide hazard mapping or zonation (Varnes, 1984). Previous surveys in the Nilgiris district show that this region has a long history of several catastrophic landslides that have caused significant destructions and acute economic loss related to personal property and infrastructure; however, the loss of life is very low when compared to landslide impacts in other parts of India. From the review of literatures it is very evident that most of the landslides in the Nilgiris are triggered by the heavy intense rainfall and anthropogenic activities (Thennavan and Ganapathy, 2020).

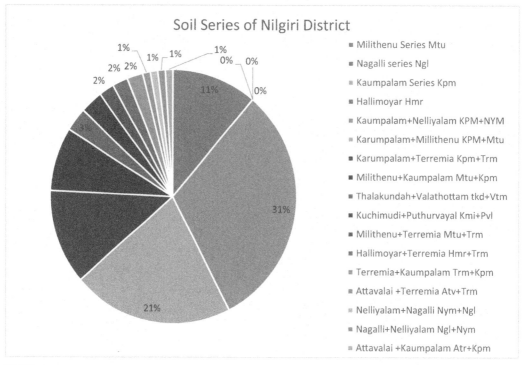

FIGURE 16.3   Redrawn by the author from Soil series in Nilgiris district (SPA Bhopal, 2018).

The Nilgiris district is the first biosphere in the Western Ghats region, with endangered flora and fauna and frequent landslides. Landslides have had disastrous consequences, destroying infrastructure and causing economic loss due to material and natural resource damage. The magnitude and intensity of destruction are determined by the existing land use pattern and the level of urbanization. The problem with landslide hazards in places like Nilgiris is that man-made slopes on hillsides are close to existing buildings (Thirumalai et al., 2015). Priority to settle on slopes by inhabitants is due to hike inland cost, hype in the property price, change in land use pattern, the proliferation of land grabs, urban pressures, workplace proximity, livelihood, and social and economic compulsions.

16.3.1.2.2 Excavation of slope at the toe

As previously discussed, previous road widening work on the Coonoor-Mettupalayam highway and excavation work on natural slopes at the toe may deform the slope face, resulting in a reduction in shear strength and causing slope failure and landslides. It has been discovered that excavation of the slope at the toe reduces shearing resistance and increases shearing stress, which, when combined with rainfall infiltration, can lead to slope failure. Toe is a critical component of slope for stability or buttressing effect (Chandrasekaran et al., 2016).

16.3.1.2.3 Unplanned vertical cut

A haphazard vertical cut along the slope creates instability and reduces lateral support. The removal of lateral resistance causes a rapid decrease in shear strength and an increase in shear stress, resulting in slope failure. In such high-risk landslide-prone areas, mitigation strategies include avoiding further development, limiting existing-use rights to rebuild, and limiting the use of built space.

16.3.1.2.4 Removal of vegetation and dumping of loose soil

Plant roots, in general, form a dense network of interlocking roots that hold unconsolidated materials together and prevent flow and soil erosion (Coppin and Richards, 1990). Furthermore, the plant's roots have a significant impact on soil strength by removing excess water from the soil and acting as a medium to increase the shear strength of the soil. In some places, the removal of trees and vegetation on the Nilgiris slope may increase soil erosion or the frequency of slope failure.

16.3.1.2.5 Climate change

Climate change is having a significant impact in the Nilgiris district, resulting in unseasonal and high-intensity rains that cause crop damage. Despite growing water scarcity and frequent droughts during the summer months, man–animal conflict has increased. Climate change is expected to have a wide-ranging impact on global biodiversity in the 21 century (Cordellier et al., 2012). According to ATREE (Ashoka Trust for Research in Ecology and Environment) the proportion of rain falling in the form of more intense rain has been increased in the recent past (Sibi, 2018).

Krishnaswamy, an eco-hydrologist by training, has been monitoring the changes in Nilgiris Biosphere Reserve since 2012. As per the annual report by India Meteorological Department (2012), the average annual precipitation recorded in the Nilgiris has dramatically shown variations ranging from 1456 mm in 2012, 1524 mm in 2013, 1951 mm in 2014, 1721 mm in 2015, and 887 mm in 2016. This variation in precipitation indicates that it is against an expected annual rainfall of 1920 mm. They are not, however, evenly distributed across the nine months of the year; rather, they exhibit disruptive patterns of change in their frequency and intensity at certain times and nonexistence at others with no discernible pattern.

16.3.1.2.6 Urbanization and biodiversity

The Nilgiris district is endowed with vital natural resources, making it critical to halt uncontrolled and unplanned urbanization and forest denudation to preserve the fragile ecosystem. Natural resources in the Nilgiris have been under enormous strain as a result of the growing human population and increased stress on the ecosystem as a result of changing consumer behavior. The unprecedented rate of urbanization puts pressure on natural resources in Nilgiris, which is vulnerable to massive changes due to anthropogenic interventions in land use patterns, resulting in pollution, biodiversity loss, and changes in hydro-geomorphology, among other things. According to McKee et al. (2003) and Cincotta et al. (2000), the majority of urban growth will occur in areas near the world's biodiversity hotspots.

16.3.1.3 Destruction of Shola Grasslands

What is Shola Grassland Ecosystem?

Tropical wet evergreen forests known locally as sholas grasslands (derived from the Tamil word "Sholai") can be found in the upper reaches of India's Western Ghats. The Shola ecosystem is unique, spanning the higher elevations of Tamil Nadu, Kerala, and Karnataka's southern Western Ghats. It was made up of vast meadows of tropical grassland interspersed with forest. The forest contains patches of dwarf evergreen native trees, and the hilly areas are covered with native grass species (Sasmitha et al., 2020). The vegetation is double-layered, with a continuous closed tree canopy, and it has a high water retention capacity, moisture-dependent vegetation, and so on. Because of their unique climatic conditions and high isolation, Shola forests-grasslands have a high level of endemism and are the richest biomes in terms of flora and fauna.

The Sholas grassland vegetation is rich in endemic flora and fauna species that prefer tropical and temperate climates. Phytogeographical analysis of genus-wise distribution reveals that shola fragments are typically found in temperate (Rubus, Daphiphyllum, and Eurya) or subtropical (Rhododendron, Berberis, Mahonia, and Himalayan) climates (Sasmitha et al., 2020). Species found within shola fragments, on the other hand, are Indo-Malayan or Indian in origin. Members of the Rubiaceae, Lauraceae, Myrtaceae, Myrsinaceae, Symplocaceae, and Oleaceae dominate the shola's overstory, while Fabaceae, Asteraceae, and Acanthaceae dominate the dicotyledonous understory. Soil moisture (overstorey and understorey) and soil nitrogen (understorey only) were found to have a significant influence on species along the edge-interior gradients in shola fragments.

Species discovered within shola fragments, on the other hand, are Indo-Malayan or Indian in origin. Members of the Rubiaceae, Lauraceae, Myrtaceae, Myrsinaceae, Symplocaceae, and Oleaceae dominate the shola, while Fabaceae, Asteraceae, and Acanthaceae dominate the dicotyledonous understory. Soil moisture (overstorey and understorey) and soil nitrogen (understorey only) were found to have a significant influence on species in shola fragments along the edge-interior gradients. Exotic fast-growing trees such as Black wattle (Acacia mearnsii) and eucalyptus have caused havoc in these valuable landscapes. Exotic tree invasion, extensive commercial and monoculture plantations, shola species harvesting, and cattle grazing have all contributed to the extinction of the shola-grassland ecosystem. To address these issues, it is critical to identify the knowledge gap and apply the current state of knowledge.

16.3.2 The Uttarakhand, Himalayas

Uttarakhand, also known as Dev Bhoomi or the Land of Gods, is one of India's northernmost states. The pristine environment and remoteness of the land earned it the title "Land of Gods." After the 1962 Indo-China border war, this pristine land began to be encroached upon. This was accelerated further by the creation of the separate state of Uttarakhand (earlier it was a part of Uttar Pradesh, one of the northern states of India).

Uttarakhand is located in the Himalayas, on the southern slope. Vegetation characteristics change as altitude changes. The terrain is mostly ice and rock at the highest points. Between 3000 and 5000 m, alpine shrubs and meadows can be found. Rhododendron is the dominant plant in this area. Subalpine conifers come next. Coming lower from this point, at around 3000–2600 m, one can find Himalayan broadleaf forests.

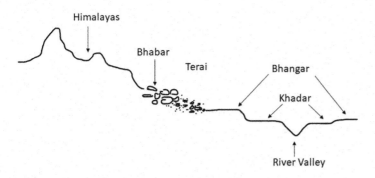

FIGURE 16.4 Types of terrain in Northern India. Source: *https://digitallylearn.com/difference-between-bhabar-terai-bhangar-and-khadar-upsc-ias/*.

TABLE 16.1 Types of terrains which are associated with mountainous regions in northern part of India.

Region	Location and extent	Climate and vegetation	Terrain and water bodies/access to water
Bhabar	Location: Shivalik Foothills Extent: 8−16 km wide	Climate: Subtropical Vegetation: Himalayan subtropical, pine forests, and Himalayan subtropical broadleaf forests Not suitable for agriculture	Terrain: It comprises pebble, rocks, debris washed from higher ranges The terrain is in the form of porous beds Water flows underground and streams disappear due to the porosity of the rocks
Terai	Location: South of the Bhabar and run parallel to it Extent: It is 20−30 km wide.	Reclaimed land used for agriculture. The speed of the Himalayan Rivers is slowed down in the Terai region and these rivers deposit fertile silt during the monsoons.	Terrain: Composed of comparatively finer alluvium and is covered by clay-rich swamps and forests The underground streams of the Bhabar reemerge on the surface here
Bhangar	Location: South of Terai region Extent: Part of Indo-Gangetic Plains and are old alluvial plains	Suitable for cultivation	Terrain: Old alluvial plains Not very fertile Contains calcareous deposits locally known as "Kankar" "Barind" and "Bhur" are other names to these plains
Khadar	Location: Floodplains of Indo Gangetic region	It is suitable for extensive cultivation/ more fertile than Bhangar	Terrain: New alluvium plains Frequent flooding help in renewal of rich soil

The *Terai* region is the lowest of these mountains, where *Duar* savanna and grasslands meet the deciduous forests of the Indo Gangetic planes to form the landscape. These belts are referred to as Bhabar. Although much of the land in this area has been cleared for agricultural purposes, only a few areas have retained their natural form (Fig. 16.4; Table 16.1).

16.3.2.1 Threats to Uttarakhand region

Some of the pressing threats in the region of Uttarakhand are as under:

4. Land use and land cover change

16.3.2.1.1 Deforestation

Uttarakhand has a significant amount of forest cover (38,000 km^2). The forests are rich in biodiversity, with numerous tree, shrub, herb, medicinal plant, and animal species. However, over the last few decades, these forests have suffered greatly due to a variety of commercial activities such as mining, hydropower plants, road construction, irrigation, and so on (Karnik, 2020). Forest loss has a variety of short- and long-term consequences, including soil loss, water loss, and disruption of ecological cycles. In extreme cases, it can force locals to flee their homes, a situation known as environmental refugees.

16.3.2.1.2 Land-use changes

Urbanization has had a significant impact on the state of Uttarakhand. Over the last few decades, there has been a surge in construction and conversion of forests to buildable areas, or they have been cleared for agricultural purposes. Unplanned constructions, encroachment on drainage areas, and drainage blockage (resulting in slope fragility) pose a serious threat to the state's environment (Nakano et al., 2016).

16.3.2.1.3 Climate change and flash floods

The phenomenon of climate change is taking its toll on Uttarakhand. The state is home to around 1266 glacial lakes (extents ranging from 500 m^2–2 L m^2) which are slowly melting due to temperature rise. Flooding in the Kedarnath Shrine in 2013 is evidence of this phenomenon (Goswami, 2017).

16.3.2.1.4 Water crisis and hydropower projects

Flash floods might be increasing in the state of Uttarakhand, yet the water woes of the state are not decreasing. Many perennial rivers are gradually dying. For example, the flow of the Kosi River has declined from 790 L/S in 2012 to just 75 L/S in 2016. A large part of the reason can be attributed to the numerous dam projects that have been built in the state. These projects have been scrutinized from the beginning. Some experts believe that the widespread construction of dams in geologically vulnerable areas is destroying traditional agriculture and other river-based livelihood activities (Goswami, 2017).

16.3.2.1.5 Illegal mining

Illegal mining has been a major problem in the state of Uttarakhand. The town of Udham Singh Nagar, which includes the towns of Khatima, Sitarganj, Kichha, Bazpur, and Kashipur, is witness to a massive amount of illegal mining. The illegal mining operations excavate and extract sand, stones, and clay, leaving these areas vulnerable to flooding and soil erosion during the rainy season (Rajput and Saxena, 2020).

16.4 Inferences from comparative studies

(Table 16.2)

TABLE 16.2 Inferences: issues and actions taken.

Parameters	Issues in Himalayan regions	Action taken	Issues in Nilgiris regions	Action taken
Deforestation	The last few decades have seen an extreme loss of these forests owing to many commercial activities like mining, hydropower plants, road construction, irrigation,	Laws restricting the mining activities from Sunrise till sunset to curb illegal mining Seed bomb campaign by Uttarakhand villages to replenish forest cover in the state. Revival of Chipko Movement to stop cutting of trees	Encroachment of forest reserves by many private resorts have led to habitat destruction, irreversible damage to the ecosystem, deforestation, and disruptions in the natural flow of water	Persons who got permission to cut the trees from the district committee were asked to plant more than 50% of native saplings procured from the forest department nursery. Moreover, a green tax of Rs 30 is levied on all vehicles entering the districts of Nilgiris, and tax money has been used for greening activities such as creating parks, public toilets, etc.
Land-use changes	Unplanned constructions, encroachment of drainage areas, blockage of drainage (leading to fragility of slopes) is a serious threats to the state's environment	Government Policies and Action plans are there but they need to be regulated and effectively strengthened	The massive influx of tourists (Tourists footfall is calculated to be 50 lakhs annually and the population of Ooty is 7.5 lakhs) creates a huge negative impact where the ecologically sensitive zones are at great risk such as pollution, encroachment in forest areas which include Indiscriminate and unregulated construction activities.	Unnadha Udhagai or Sublime Udhagai encompasses a complete ban on single-use plastic and plastic products where Tourism has an important role to play. Geological Survey of India (GSI), earmarked 283 vulnerable locations as no-construction zones or designated green zones.
Water Scarcity, Loss of ecosystem Services and indigenous species, and Climate change	Loss of Perennial rivers and drying of ponds due to climate change and increase in urbanization	Massive tree plantation drives in the catchment area of the rivers are being proposed so that it replenishes the water sources	The Nilgiris has the source of two main river systems namely- Bhavani and the Moyar. They emerge from high altitudes and flow down as cascades to the Cauvery basin in Tamil Nadu. Traditionally Nilgiris is considered to be water abundant region. Due to fluctuations in climate, this forest-grassland mosaic is facing the challenge of an extreme water stress crisis. The other alternate sources of water such as the springs and small hill wetlands have been overlooked or neglected. The crisis is rooted in the colonial history of the region which was driven by profit-oriented practices (Aditi, 2020). Invasive species in the form of exotic vegetation are found growing in this region forming potential harmful threats for these tropical ecosystems. Changes in the landscape by the British led to the complete disappearance of localized native plants. After independence, the Indian government attempted to plant more Eucalyptus and wattle hoping to boost a paper industry and as a consequence has put immense pressure on water sources.	Site-specific restoration program is the need of an hour.

(Continued)

TABLE 16.2 (Continued)

Parameters	Issues in Himalayan regions	Action taken	Issues in Nilgiris regions	Action taken
Livelihood of communities	Climate change has affected the way people are seeking their livelihood patterns	Cash crops have been improvised, delayed Sowing is done, market dependence has increased, afforestation, replacing the types of crops used, increased use of pesticides (Rautela and Karki, 2015)	As Vasamalli Kurtozen (a member of the Tamil Nadu State Wildlife Board) recent study it is evident that between 1973 and 2017, 66% of the grasslands in the Western Ghats were lost to spreading of alien invasive species (Nina, 2021). With the loss of the grasslands, Todas (an indigenous tribe of Nilgiris) found they could no longer sustain large buffalo herds. Traditionally Todas life revolves around these Buffalos and every household has at least 100 buffalos. Buffalos were a show of strength for each [Toda] clan.	Grasslands have to be restored

TABLE 16.3 Studies on landslide hazard zonation in Nilgiris district.

S. No.	Authors	Type of study on landslide hazard zonation in Nilgiris district
1.	Ramakrishnan et al. (2002)	Aerial or orthophotos were used to create thematic layers in a GIS platform. The author's intention in this study is to create a landslide hazard zonation map of Kotagiri, which is located in the northwestern part of The Nilgiris district (Ramakrishnan et al., 2002).
2.	Vasantha Kumar and Bhagavanulu (2007)	This study focuses on the effect of deforestation and increase in landslide risk in various sectors of Nilgiris (Vasantha Kumar and Bhagavanulu, 2007).
3.	Chandrasekharan (2010)	The study investigates how rainfall-induced landslides and the resulting damage occurred in the Nilgiris district in 2009. For example, the causes of structural failures caused by landslides were investigated (Chandrasekharan, 2010).
4.	Gurugnanam et al. (2013)	This study is the compilation of landslide location details from various literature using GIS within Nilgiris by Gurugnanam et al. (2013).
5.	Gobinath et al. (2015)	In the Nilgiris district, the relationship between soil roots and landslides in relation to slope stabilization was investigated. According to the literature review, landslide studies in Nilgiris can be divided into three categories. These are their names: (1) mapping landslide hazard/ susceptibility zones using remote sensing and GIS using various computational methods (2) thorough geotechnical investigation, including landslide-induced structural failure and damage; and (3) application of landslide studies to land use regulations and disaster preparedness. However, there have been no updates to the hazard zonation map since 1980 (Gobinath et al., 2015).

16.5 Recommendations

Several agencies and independent researchers have conducted landslide research in the Nilgiris district, but there is insufficient research. In other words, there is no systematic documentation or data available on landslides in the Nilgiris district. The following table shows some of the research work which has been established so far concerning the Nilgiris district (Table 16.3):

16.6 Strategies

16.6.1 Strategies to protect Nilgiris Mountains from the above issues

16.6.1.1 *Landslide hazard zonation*

For landslide risk reduction, a Landslide Risk Management Strategy' is required. Nilgiris is vulnerable to various types of landslides, which cause significant damage in terms of habitat loss and property destruction.

The strategy document will be a comprehensive road map for the need to strengthen and mainstream landslide disaster preparedness, mitigation, response, and relief mechanisms through mapping, early warning system (EWS), awareness and capacity building, and the formulation of regulations/policies specifically for mountain zones. In hazard-prone areas, vulnerable slopes should be identified.

16.6.1.2 Strategies to mitigate climate change—forests and biodiversity

The following strategies developed by Govt of Tamil Nadu taking the Green India Mission as guidance which would address various concerns related to climate change.

1. Increase the extent of forest cover, both inside and outside notified forest areas
2. Promote Conservation efforts to sustain both marine and terrestrial biodiversities
3. Creation of ethnographical records/register concerning plants and animals of Tamil Nadu forests for documentation and associated cultural knowledge
4. Undertake Lower and middle strata diversification
5. Managing forest fires and alien invasive species (AIS)
6. Emphasizing awareness and capacity-building activities that prepare communities and organizations to combat the issues of climate change.

16.6.2 Strategies to protect the Himalayas from aforementioned issues

The few strategies which the author states below are reiterated again and again by various organizations who intend to protect the Himalayas. The need of the hour is to start implementing them now:

1. Strategic strengthening various types of planning instruments like:
   a. City development plans
   b. State disaster risk reduction frameworks/ plans
   c. Climate change adaptation plans particularly the geo-hydrological disasters in rapidly urbanizing Himalayan towns
2. The collaborative effort of all societal stakeholders, namely local governing bodies, communities, and industries, to more effectively evolve and integrate conservation measures.
3. Actively working toward biodiversity conservation in association with natural resource management, sustainable development, and education (WWF, 2020). World Wildlife Fund is a pioneer in this field. Some of WWF's current projects in the Himalayan terrain include: research on vulnerable and endangered species, capacity building for local communities, promoting sustainable infrastructure development, and developing more effective poaching control networks.
4. The authorities must develop a pan-Himalayan strategy based on the contextual availability of various natural resources so that development-related threats do not result in environmental damage.
5. The issue of landslides must be addressed. Slope failure may occur frequently during heavy rains. It is necessary to bolt the rocks, divert water run-off, install proper drainage channels, and align roads properly. Slope changes should be kept to a minimum (Rawat et al., 2015).
6. Increase forest cover and biodiversity. Try to adapt local best practices. Make them known to the mainstream.

7. Check invasive species of both flora and fauna to protect the native species.
8. Management of forest fires as a lot of resources get destroyed during this phenomenon.
9. Capacity building of communities who are dependent on forest ecosystem services so that they can be better equipped in case of disaster.

16.7 Conclusion

The current study looks into the dangers of urbanization in mountainous areas. According to the study, urbanization can be directly and indirectly linked to biodiversity loss, degradation of mountainous terrains, and negative effects on locals. Urbanization not only provides the city/town (mountainous or otherwise) with a more contiguous built-up space, but it also provides the region with a more powerful impetus in terms of economic development. However, it has a negative impact, as do both sides of the coin. Unfortunately, cities and towns functionally leap into protected green space, altering the natural ecosystem. At this point, it is difficult to attribute land degradation to the underlying causes. In such cases, a complex web of causality is at work. Many researchers have attempted to comprehend how a specific stressor, such as biodiversity loss, water scarcity, land degradation, flooding, and so on, affects social and ecological systems. However, fewer studies are conducted in isolation. A comprehensive study that analyzes and evaluates how multiple stressors affect such systems at the same time is still lacking. This type of research is critical in the context of mountainous research. Because we consider land degradation to be a multicriterion, more systematic studies in a mountainous region would be required. More realistic modeling of land degradation dynamics, for example, or a better study of subterranean structures/lifeforms, could help us better estimate the impact of urbanization on mountainous regions. As a result, the study concludes by proposing additional research that thoroughly analyzes any particular area using a holistic multicriteria approach to arrive at any concrete solution rather than a fragmented approach.

References

Aditi, S., 2020. Neeru and the Nilgiris, s.l., Water Practitioners Network.

CEPF, 2021. Protecting Biodiversity by Empowering People. [Online] Available at: <https://www.cepf.net/our-work/biodiversity-hotspots/western-ghats-and-sri-lanka/threats> (accessed September 2021.).

Chandrasekharan, S., 2010. Assessment of damages induced by recent landslides in Ooty, TN. Indian Geotechnical Society 687−688.

Chandrasekaran, S., Senthil Kumar, V., Maji, V.B., 2016. Landslides in Nilgiris: causal factors and remedial measures. Indian Geotechnical Conference 15−17.

Cincotta, R., Wisnewski, J., Engelman, R., 2000. Human population in the biodiversity hotspots. Nature 404, 990−992. Available from https://doi.org/10.1038/35010105.

Conservation International, 2021. Conservation International. [Online] Available at: https://www.conservation.org/priorities/biodiversity-hotspots.

Coppin, N.J., Richards, I.G., 1990. Use of Vegetation in Civil Engineering. Butterworth, London, p. 272.

Cordellier, M., Pfenninger, A., Streit, B., Pfenninger, M., 2012. Assessing the effects of climate change on the distribution of pulmonate freshwater snail biodiversity. Marine Biology.

Das, A., et al., 2016. Managing soils of the lower Himalayas, Encyclopedia of Soil Science, *third ed.* Taylor & Francis, s.l.

Ding, Y., Peng, J., 2018. Impacts of urbanization of mountainous areas on resources and environment: based on ecological footprint model. MDPI 10 (765).

Gopal, R., 2018. Bhadra voluntary relocation. Wildlife Conservation Society 12.

Goswami, S., 2017. Uttarakhand: As Problems Pile Up, Youths See Wisdom in Migration. [Online] Available at: https://www.downtoearth.org.in/news/governance/uttarakhand-as-problems-pile-up-youths-see-wisdom-in-migration-57333.

Gurugnanam, B., Arunkumar, M., Bairavi, S., Dharanirajan, Kesavan, 2013. Assessment on landslide occurrence: a recent survey in Nilgiri, Tamil Nadu, India. International Journal of Environmental Science and Technology 1252–1256.

ICIMOD, 2009. The Changing Himalayas: Impact of Climate Change on Water Resources and Livelihoods in the Greater Himalayas. International Centre for Integrated Mountain Development, Kathmandu.

India Today, 2017. 7 Major Mountain Ranges in India: Some of the Highest Mountains in the World. [Online] Available at https://www.indiatoday.in/education-today/gk-current-affairs/story/seven-major-mountain-ranges-of-india-969214-2017-04-03.

IUCN, 2020. World Commission on Protected Areas. [Online] Available at: https://www.iucn.org/commissions/world-commission-protected-areas/our-work/mountains (accessed 05.09.22.).

Jamwal, N., 2020. Rapid Urbanization, Climate Change Increasing Water Stress in the Hindu Kush Himalayas. [Online] Available at https://en.gaonconnection.com/rapid-urbanization-and-climate-change-are-increasing-water-stress-in-the-hindu-kush-himalayan-region/.

Karnik, V., 2020. Uttarakhand Has Lost 50,000 Hectares of Forest Land in the Past 20 Years Due to Commercial Activities: Report. [Online] Available at https://weather.com/en-IN/india/environment/news/2020-12-08-uttarakhand-lost-50000-hectares-forest-past-20-commercial.

Kaur, S., Purohit, M.K., 2012. Rainfall Statistics of India - 2012. India Meteorological Department. Available from: http://hydro.imd.gov.in/hydrometweb/(S(eibinu55lyexxhqecj1iapn2))/PRODUCTS/Publications/Rainfall%20Statistics%20of%20India%20-%202012/Rainfall%20Statistics%20of%20India%20-%202012.pdf.

Khawas, V., 2009. Environmental challenges and human security in the Himalaya. Environmental Concerns and Sustainable Development: Some Perspectives from India. TERI, New Delhi, pp. 32–81.

Kilian Jornet Foundation, 2021. Kilian Jornet Foundation. [Online] Available at: https://www.kilianjornetfoundation.org/why-mountains-matter/ (accessed 18.09.21.).

Korner, C., Ohsawa, M., 2005. Millennium Ecosystem Assessment: Mountain Systems. Mountain Partnership, s.l.

Lamsal, P., Kumar, L., Aryal, A., Atreya, K., 2018. Invasive alien plant species dynamics in the Himalayan region under climate change. Ambio 697–710.

Madhu, M., et al., 2017. Bio-physical-chemical studies of swamps in the Nilgiris, Tamil Nadu. Tropical Ecology 621–635.

Mani, M., 2021. Published on End Poverty in South Asia. [Online] Available at https://blogs.worldbank.org/endpovertyinsouthasia/how-air-pollution-accelerating-himalayan-glaciers-melt.

McDougall, K.L., et al., 2011. Plant invasions in mountains: global lessons for better management. Mountain Research and Development 380–387.

McKee, J.K., Sciulli, P.W., David Fooce, C., Thomas, A. 2003. Forecasting global biodiversity threats associated with human population growth. Waite, Biological Conservation 115, 161–164.

Menon, S., Bawa, K.S., 1998. Deforestation in the tropics: reconciling disparities in estimates for India. Royal Swedish Academy of Science 576–577. *Allen Press.*

Murugesan, J., 2013. The Nilgiris District: Modifications of Hill Environment and Implications Using Geo-Spatial Techniques. Post Graduate and Research Department of Geography, Government Arts College, s.l.

Myers, N., 1988. Threatened biotas: hot spots in tropical forests (1990) The Environmentalist 1–20.

Nakano, G., Rautela, P., Shaw, R., 2016. Uttarakhand disaster and land use policy changes. Land Use Management in Disaster Risk Reduction. Springer, Tokyo, pp. 237–252.

Narain, S., 2013. Himalayas: The Agenda for Development and Environment. [Online] Available at https://www.downtoearth.org.in/news/himalayas-the-agenda-for-development-and-environment-41486.

Nina, V., 2021. Environmental Activism in the Nilgiris: Too Early to Rejoice? [Online] Available at https://india.mongabay.com/2021/07/environmental-activism-in-the-nilgiris-too-early-to-rejoice/.

Pathak, R., Negi, V.S., Rawal, R.S., Bhatt, I.D., 2019. Alien plant invasion in the Indian Himalayan Region: state of knowledge and research priorities. Biodiversity and Conservation 3073–3102.

Rajput, M., Saxena, S., 2020. Crackdown on Illegal Mining in Uttarakhand, Strict Curbs in US Nagar. [Online] Available at https://www.hindustantimes.com/dehradun/crackdown-on-illegal-mining-in-uttarakhand-strict-curbs-on-mining-in-us-nagar/story-LcdM1tIRJ5pSZaZTMBeNDM.html.

4. Land use and land cover change

Ramakrishnan, S., et al., 2002. Landslide disaster management and planning- a GIS based approach. Indian Cartographer, MFDM 192–195.

Rao, R., 2021. Floristic Diversity in Western Ghats: Documentation, Conservation and Bioprospection—A Priority Agenda for Action. ENVIS @CES, Indian Institute of Science, Bangalore.

Rautela, P., Karki, B., 2015. Impact of climate change on life and livelihood of indigenous people of higher Himalaya in Uttarakhand, India. American Journal of Environmental Protection 112–124.

Ravindran, Gobinath, Pattukandan Ganapathy, Ganapathy, et al., 2015. Studies on soil–root stabilisation for slope protection—a study using some grass varieties. The International Journal of Earth Sciences and Engineering 1526–1531.

Rawat, M., Joshi, V., Uniyal, D., Rawat, B., 2015. Investigation of hill slope stability and mitigation measures in Sikkim Himalaya. International Journal of Landslide and Environment 8–15.

Rural Transformation Project, T, 2020. District Diagnostic Survey (DDS), *The Nilgiris District*. Govt of Tamil Nadu, s.l.

Sasmitha, R., Muhammad Iqshanullah, A., Arunachalam, R., 2020. Ecosystem changes in shola forest-grassland mosaic of the Nilgiri Biosphere Reserve (NBR). Environmental Issues and Sustainable Development.

Sibi, A., 2018. India Climate Dialogue. [Online] Available at: https://indiaclimatedialogue.net/2018/12/25/nilgiris-ecosystem-threatened-climate-change-2/ (accessed 21.09.21.).

Singh, C., et al., 2021. The Himalayan natural resources: challenges and conservation for sustainable development. Journal of Pharmacognosy and Phytochemistry 1643–1648.

SPA Bhopal, 2018. Coimbatore Regional Plan 2038. SPA Bhopal, s.l.

Subburaj, A., 2008. District Groundwater Brochure, Nilgiri District, Tamil Nadu, Technical Report Series, South Eastern Coastal Region. Government of India, Ministry of Water Resources, Central Ground Water Board, Chennai.

Thennavan, E., Ganapathy, G.P., 2020. Evaluation of landslide hazard and its impacts on hilly environment of the Nilgiris District—a geospatial approach. Geoenvironmental Disasters.

Thirumalai, P., Anand, P., Murugesan, J., 2015. The Nilgiris: landslide prone zones and human influence on the modification of hill environment using geospatial technology. Archives of Applied Science Research. pp. 64–68.

Tiwari, P.C., Tiwari, A., Joshi, B., 2018. Urban growth in Himalaya: understanding the process and options for sustainable development. Journal of Urban and Regional Studies on Contemporary India 4 (2), 15–27.

UNEP, 2021. Ecosystem Restoration for People, Nature and Climate. United Nations Environment Programme (UNEP) and Food and Agriculture Organization (FAO), s.l.

Varnes, D., 1984. Slope movement types and processes—landslides, analysis and control, special report. National Academy of Sciences 11–34.

Vasantha Kumar, S., Bhagavanulu, D., 2007. Effect of deforestation on landslides in Nilgiris district—a case study. Journal of the Indian Society of Remote Sensing 105–108.

Williams, M., 1989. Deforestation: past and present. Progress in Human Geography 13 (2), 176–208.

WWF, 2020. WWF General Information Page. [Online] Available at: https://wwf.panda.org/discover/knowledge_hub/where_we_work/eastern_himalaya/solutions2/.

17

Prediction of land cover changes of Khagrachhari Hilly Upazila using artificial-neural-network-based cellular automata model

G.N. Tanjina Hasnat[1] and Rahul Bhadouria[2]

[1]Institute of Forestry and Environmental Sciences, University of Chittagong, Chattogram, Bangladesh [2]Department of Environmental Studies, Delhi College of Arts and Commerce, University of Delhi, New Delhi, India

O U T L I N E

17.1 Introduction

Land use and land cover (LULC) change is the alteration of the earth's surface that influences global warming and climate change (Mahmood et al., 2010; Thapa, 2021), which could be incorporated with changes in the natural environment such as forestry, bare land, water body, built-up area, agriculture due to natural hazards or man-made activities (Rosa et al., 2014; Anurag and Pradhan, 2018; Kafy et al., 2020; Saddique et al., 2020; Tan et al., 2020; Prijith et al., 2021). The natural calamities and human-induced changes alter the land surface processes, that is, biophysics, biogeochemistry, biogeography of the terrestrial surface, hydrology, biodiversity, and climate (Ionita et al., 2015; Hibbard et al., 2017; Wedajoa et al., 2020). LULC has a direct or indirect impact on species distribution and has become a major issue at the regional and global levels in recent decades (Hassan et al., 2016; Verma et al., 2020).

The Chittagong Hill Tracts (CHTs), which are part of the Hindu Kush Himalayas, cover more than 40% of Bangladesh's total forest area (Ahammad and Stacey, 2016). The ecosystems of living organisms in the hilly area are more complex (Primack and Morrison, 2013) and support almost 80% of the country's total biodiversity (Jashimuddin and Inoue, 2012). Because of the different altitudes and topography, the vegetation coverage and natural land use pattern here are quite different from the rest of the country, which helps to regulate climate, air quality, and water flow better than any other region (Rasul et al., 2004; Ahammad and Stacey, 2016). Khagrachhari Sadar is one of Bangladesh's hilly Upazilas that has seen changes in land use, particularly in vegetation cover, over the last several decades (Rasul et al., 2004; Kibria et al., 2015).

Models of LULC examine past changes and forecast future changes (Khwarahm et al., 2020; Leta et al., 2021; Sibanda and Ahmed, 2021). Different dynamic prediction models are used to simulate LULC change including empirical-statistical models (Mutz et al., 2021), optimization models (Sarir et al., 2021), agent-based models (Heppenstall et al., 2021), cellular automata-Markov chain models (Gharaibeh et al., 2020), multivariate spatial models (Gelfand, 2021), Markov analysis models (Ackroyd et al., 2021), cellular automata (CA) models (Lv et al., 2021), and artificial-neural-network-based cellular automata (ANN-CA) models (Saha et al., 2021). Different LULC models serve different purposes and each has advantages and disadvantages. The ANN-CA model is used in this study to predict future land-use changes using GIS and RS techniques (Tabarestani and Afzalimehr, 2021). The model deals with multiple parameters related to topography and land cover (Agatonovic-Kustrin and Beresford, 2000; Maind and Wankar, 2014; Abiodun et al., 2018). The ANN-CA technique can help with forest conservation, landscape management and development, and species distribution management in the future (Saputra and Lee, 2019; Abbas et al., 2021).

The current study sought to forecast the earth's features, specifically the vegetation coverage of Khagrachhari Sadar, a hilly Upazila in the CHTs. The study was carried out to determine the historical changes in LULC, particularly vegetation coverage, over the last 30 years and to forecast future changes over the next 30 years. Using satellite imagery, the ANN-CA model is used to characterize land-use patterns from 1990 to 2014 and to forecast for 2050. Change simulation provides the necessary information for future land cover prediction, identifies degraded land, deforested areas, and fragmented forest covers, and is very useful for future planning and decision making.

17.2 Methods and methodology

17.2.1 Study site

Khagrachhari Sadar Upazila is an Upazila of the CHTs, Bangladesh which is located at 23.0417°N and 91.9944°E (BBS, 2013). Covering a total area of 297.91 km^2, the Upazila have a 111,833 population along with 68,952 ethnic people. The major river of the Upazila is Chengi River, the maximum temperature lies from 25.6°C to 36.2°C, and minimum from 11.5°C to 13.6°C. Rainfall varied between 1824 mm and 2535 mm, and the percentage of humidity was from 62.8% to 78.0% (BBS, 2013). A total of 31,917 acres agricultural, 41,520 acres high, 8214 acres medium and 23,882 acres low land are recorded in the Upazila. Khagrachhari forest is divided into three types: tropical evergreen, semievergreen, and deciduous. The hills of this area are home to a variety of crops as well as some valuable forest timber trees. The homestead flora is diverse, with a wide range of trees, shrubs, and undergrowth. Plantations can be found alongside roads (BBS, 2013).

17.2.2 Data collection, processing, and base map preparation

Landsat images of Khagrachhari Upazila were obtained from the United States Geological Survey (USGS) (esrthexplorer.usgs.gov) for March, April, and May 1990, 2014, and 2020, respectively. The current study concentrated on the historical analysis and prediction of LULC changes, particularly changes in vegetation cover. In this regard, satellite data for the equivalent season, especially after the leaf flashing period, was downloaded with the intention of covering the entire natural vegetation. Images from March, April, and May were deliberately chosen to avoid cloud cover. The leaves begin to flash properly after February, and the rainy season begins after May. As a result, images were obtained consciously whenever they were available between March and May (Hasnat, 2021).

One Landsat 5 Thematic Mapper (TM) and two Landsat 8 Operational Land Imager (OLI) were obtained (Table 17.1) for conducting the research. The processing of the image for classification was carried out using ArcGIS. ERDAS IMAGINE, ArcGIS, and QGIS software were used for image preprocessing, atmospheric and radiometric corrections, image analysis, and map production. Tables, graphs, and charts were also created using Microsoft Excel. Because the primary goal of the research was to determine how vegetation changed over time, only three types of land use/land cover were identified: vegetation, water bodies, and bare land (Fig. 17.1).

In this study, three categories of LULC maps (Fig. 17.1) and two exploratory maps, that is, digital elevation model (DEM) and distance from road maps are used as the spatial variables. Climate, demographics, policy, regulation, and human development are not taken into account. A confusion matrix was used to calculate the percentage of correctness of the projected classification versus the true one, as well as to determine the quality information derived from remotely sensed data. Three types of kappa statistics were used to estimate the quality of the classified classes: kappa overall, kappa histogram, and kappa location.

TABLE 17.1 Images of satellites with their acquisition path, date, and resolution.

Satellite data	Path/row	Spatial resolution (m)	Date of acquired data
Landsat 5 TM	136/044	30	1990/03/05
Landsat 8 OLI	136/044	30	2014/04/24
Landsat 8 OLI	136/044	30	2020/05/10

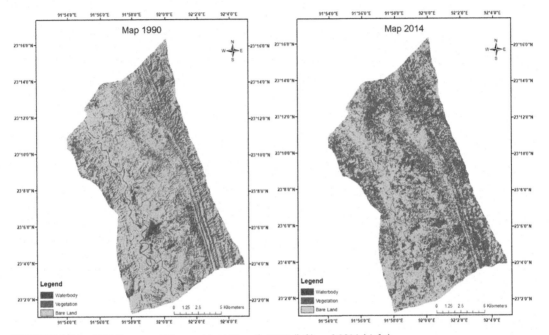

FIGURE 17.1 LULC maps of Khagrachhari Upazila 1990 (left) and 2014 (right).

17.2.3 Cellular automata model based on artificial neural networks

The ANN-CA model is used to monitor, model, and simulate changes, characterize LULC trends from one time to another, and predict LULC changes in the future of a specific area of interest (AOI). The ANN-CA model simulates and predicts LULC changes over time by providing Spatio-temporal dynamic modeling. It is the most widely used model among LULC modeling tools and techniques (Zeshan et al., 2021).

The ANN-CA model was used in this study to simulate and predict LULC changes over the study period. The ANN-CA model simulates and predicts data using an analytical process (Fig. 17.2). ANN calculates the LULC transition probability using several output neurons and simulates multiple LULC transitions. The transitional probabilities learned from the ANN learning process are used by CA to forecast LULC changes (Kafy et al., 2021).

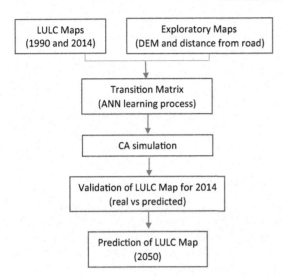

FIGURE 17.2   Structure for modeling and forecasting LULC changes using an artificial-neural-network-based cellular automaton (ANN-CA).

To begin the simulation, the inputs must be defined. The neural network is fed by a set of spatial variables in the pixel-based simulation. As the input variable in the study, two exploratory maps and three categories of LULC maps are loaded. As a result, five spatial variables are generated for each pixel and used for simulation and modeling of LULC changes. The spatial variables can be expressed through Eq. (17.1).

$$X = [x_1, x_2, x_3, \ldots, \quad x_4]T \tag{17.1}$$

where x_n is the n-th spatial variable and T is transposition.

To evaluate each correlation between the spatial variables, a two-way raster comparison was performed. The initial and final time periods are then analyzed to calculate the LULC area and changes for each variable category. The calculation generates a transition matrix as well as the area and changes, which show the proportion of pixels that change from one category to another. The transition probability is then modeled by ANN using three different layers: input, hidden, and output. The five spatial variables are placed in the input layer. Each spatial variable is linked to a neuron in this layer. The hidden layer neuron receives the signal from the input layer and computes it to produce accurate results. Training the model with the hidden layer takes very little time. Three neurons are used in the study to model. Three neurons corresponding to the three classifications of LULC maps are included in the output layer. A transitional probability is generated by the output layer. LULC changes are determined in the simulation by comparing the transition probability values.

17.2.4 Transitional potential modeling

Artificial Neural Network (ANN) deals with a large amount of input data, and in the present study, multiple parameters related to topography or land cover are entered into the analysis. The learning curve and the validation curve in Fig. 17.3 revealed that the validation curve completed a nice quicker process when the training curve needed a longer time. The learning and validation curves based on LULC maps from 1990 and 2014 show a moderately high training loss at first, then gradually decreasing and finally almost flattening. The training and validation losses were discovered to be similar, but the validation loss was slightly greater than the training loss. As a result, the overall neural network learning curve represents a well-fitting model (Fig. 17.3). The learning curve is estimated to simulate and predict future maps.

17.2.5 Cellular automata simulation, prediction, and model validation

Following the determination of the transition probability, the LULC change modeling is carried out using the CA simulation. CA simulation requires many iterations to determine the changes in the cells. CA were used in this study to forecast future land-use changes using transitional potential modeling (ANN) based on LULC between 1990 and 2014. For simplicity, the study area was only divided into a broader classification of—vegetation, bare land, and water body - because the main purpose of the study was to find out the areas that changed from vegetation coverage to other land uses and from other land uses to vegetation coverage. Based on real LULC maps of 1990 and 2014, the map of 2020 is simulated, and then validation was conducted by comparing with the real map of 2020 (Fig. 17.4). The kappa coefficients are used in the validation to evaluate and compare the real and predicted LULC maps. Validation compared the quality of the predicted map to the actual map. Following validation, the future LULC map of 2050 is projected by analyzing previous and current LULC change trends (Fig. 17.4).

The validation can be strengthened by calculating the kappa coefficient between the predicted and actual vegetation coverage maps. Because the original kappa coefficient does not distinguish between quantification and location error, its expressiveness is limited. The cause-dependent kappa indices - kappa overall, kappa histogram, and kappa location can be used to solve this problem. Kappa overall indicates the degree to which the projected and real maps agree. The spatial accuracy in the overall landscape is

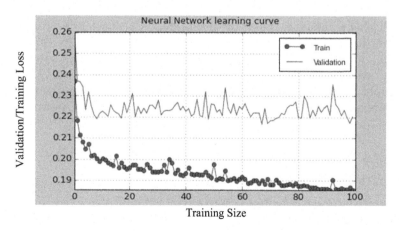

FIGURE 17.3 Neural network learning curves with training and validation datasets for Khagrachhari.

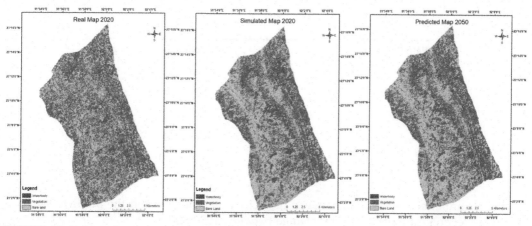

FIGURE 17.4 LULC maps—real 2020 (left), estimated 2020 (middle), and predicted 2050 (right).

computed by the kappa location. The Kappa histogram separates quantity and allocation and displays changes in inter-rater agreement between the simulated and reference maps. The generated kappa location and histogram values, as well as the percent of correctness, assess overall accuracy in terms of both location and quantity. The statistics also assess the degree of agreement between the simulated and base maps. The ranges of kappa value to measure agreement and inter-rater reliability (Viera and Garrett, 2005) are demonstrated in Table 17.2.

According to Saputra and Lee (2019), the combination of two criteria, altitude and distance from the road, yields the highest percent (%) of correctness and kappa value. This combination has the greatest influence on LULC changes of any other criterion. The transition matrix of the three LULC categories is calculated between 1990 and 2014 to determine the percentage of changes from one land use to another. The results are then used to simulate and forecast the LULC for 2020 and 2050. The transition matrix was also calculated for the 2020 and 2050 maps.

17.3 Results and discussion

Using categorize historical satellite images, changes in LULC are predicted for a given year. After determining how much land-use change has occurred between earlier and later times, the ANN-CA model estimates transitions. During the modeling process, two maps are generated: the simulated map and the predicted map. As a result, the model is used to analyze, predict, and validate land-use changes over time. The total results are divided into three broad categories. The first section examines changes between 1990 and 2014 using real 1990 and 2014 maps as well as a simulated 2020 map. This simulated map was also validated against a real 2020 map. As a result, the total change between 1990 and 2020 was investigated. In the second section, the 2050 map was predicted using real maps from 1990 and 2014, and the differences between real 2020 and 2050 were estimated.

TABLE 17.2   Kappa value ranges to measure inter-rater reliability.

Values	Strength of agreement
<0	Poor
0.01–0.40	Slight
0.41–0.60	Moderate
0.61–0.80	Substantial
0.81–1.00	Almost perfect

Finally, a comparison was made between the actual map of 1990 and the predicted map of 2050. This section estimates the total change between 1990 and 2050.

17.3.1 Changes in LULC between 1990 and 2014

The changes in the area of three categories of LULC between the years 1990 and 2014 are calculated. Fig. 17.5 shows the change in land use over time from 1990 to 2014. In comparison to 1990, bare land has decreased while vegetation has increased significantly over the time period. In previous years, there has been a significant impact on land use composition. An excessive land-use change in the Khagrachhari Sadar Upazila is recorded during 1990 and 2014 (Fig. 17.5). Table 17.3 and Fig. 17.6 depict the changes. Overall, different land uses to show the different trends in changes. In 1990, bare land covered almost three quarters (69.82%) of the total land area. An average change of (−)14.46% which is equivalent to 3,761.64 ha of the land area occurred from 1990 to 2014 that reduced the bare land to slightly more than half of the total area (55.36%) in 2014. The vegetation coverage increased from 7679.97 ha (29.51%) to 11,603.79 ha (44.59%) with a change of 15.08% land area comparable with 3923.82 ha. Along with vegetation and bare land, a significant change was observed in the water bodies. Almost (−)0.62% land area similar to 162.18 ha changes occurred from 1990 to 2014. A huge amount of land (162.18 ha) was converted to other land uses and reduced the water body from 174.33 ha (0.67%) in 1990 to 12.15 ha (0.05%) in 2014.

By comparing the percentage of area changes, the LULC transition matrix is produced. The transition matrix represents the land uses that have been replaced with others. The table below represented the transition matrix of LULC from 1990 to 2014 (Table 17.4). The transition matrix represents the land uses that have been replaced with others. Since 1990, the water body has seen the greatest percentage of land-use changes, with 0.61% converted to bare land. Almost 0.33% of the vegetation was also converted to bare land. From 1990 to 2014, nearly 0.39% of the water body and 0.35% of bare land were converted to vegetation coverage. The amount of vegetation and bare land converted to water bodies is quite small, with 0.000832% vegetation and 0.000233% bare land converted to water bodies between 1990 and 2014.

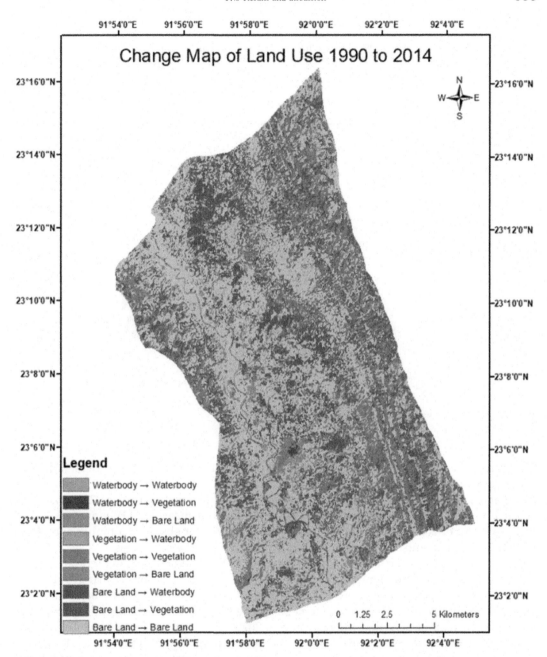

FIGURE 17.5　Change area between 1990 and 2014.

TABLE 17.3   The land use distribution and changes in 1990 and 2014.

Class	1990 (area in ha)	2014 (area in ha)	Change (area ha)	1990 (%)	2014 (%)	Change (%)
Bare land	18,167.67	14,406.03	− 3761.64	69.82	55.36	− 14.46
Vegetation	7679.97	11,603.79	3923.82	29.51	44.59	15.08
Water body	174.33	12.15	− 162.18	0.67	0.05	− 0.62

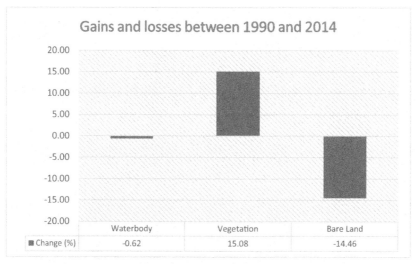

FIGURE 17.6   The changes in the LULC area in 1990 and 2014.

Accuracy is estimated and kappa statistics are analyzed for better understanding of maps agreement. Validation computes kappa statistics, i.e., kappa overall, kappa histogram, and kappa location. In the present study, when a projected map of 2020 was produced by using 2014 classified maps, a less agreement is measured. Whereas, using the classified map of 2020, when a map of 2020 was simulated, it observed a higher agreement (Table 17.5). The results revealed the actual facts as well as the correctness.

A percent of the correctness of 63.14% resulted in the validation showing that the predicted area for 2020 is moderately the same as the real area in a similar year (Table 17.5). The validation kappa values were kappa overall 0.27, kappa histogram 0.88, and kappa location 0.31, indicating a poor agreement, strong agreement, and poor agreement between the predicted map of 2020 and the real map of 2020, respectively (Table 17.5). This means that the actual map differs greatly from the predicted map. The projection was done from the map 2014, which was the main reason for the low percentage of correctness and agreement. Between 2014 and 2020, there was a significant change in land use. The lower the percent of correctness, the greater the differences between the real and predicted maps. The findings show that a number of changes in land use pattern occurred during the historical period (Fig. 17.6).

TABLE 17.4 Transition matrix (%) of LULC between 1990 and 2014.

	Water body	Vegetation	Bare land
Water body	0.008776	0.385132	0.606092
Vegetation	0.000832	0.673104	0.326064
Bare land	0.000233	0.35047	0.649297

TABLE 17.5 Validation results of simulation for 2014 and 2020.

Simulation method	ANN (simulated 2020 map from 2014 map)	ANN (simulated 2020 map from real 2020 map)
% of correctness	63.13565	85.82002
Kappa (overall)	0.27192	0.71402
Kappa (histo)	0.87689	0.71555
Kappa (loc)	0.31010	0.99787

17.3.2 Predicted future changes in LULC between 2020 and 2050

The classified maps were used to calculate the LULC difference between the actual land use in 2020 and the predicted land use in 2050. The land uses of three categories exhibit a variety of change trends. The vegetation and water body are projected to decline over time, while bare land will increase (Fig. 17.7). An average of 2297.25 ha (8.83%) bare land is estimated to be increased from 12,073.23 ha (46.40%) in 2020 to 14,370.48 ha (55.22%) in 2050. The vegetation coverage is expected to decrease from 13,713.39 ha (52.70%) in 2020 to 11,643.84 ha (44.75%) in 2050. A decrease of 0.88% which is equivalent to 227.70 ha water body is estimated to be decreased in the similar time period and it reduced from 2020 to 2050 in the amount of 235.35 ha to 7.65 ha, respectively. Table 17.6 and Fig. 17.8 depict the changes in vegetation, bare land, and water body from 2020 to 2050.

The transition matrix percentage of LULC is also examined from 2020 to 2050, and it is discovered that a significant amount of land uses is expected to be replaced with others. The greatest percentage of land (0.52%) can be converted from water bodies to bare land. A significant portion of the water body (0.46%) is also expected to be converted to vegetation, while 0.43% of the vegetation is expected to be converted to bare land and 0.000144% to the water body. When the bare land was projected to determine the converted land cover, the results revealed that a significant amount (0.31%) of the bare land will be replaced by vegetation and a minor amount (0.000082%) by the water body (Table 17.7).

This section predicts future land-use trends and changes. The study's findings indicate that most of the vegetated area will be disappeared by 2050 (Fig. 17.7) if the other regulating factors remain constant in this time frame. The future changes are calculated by analyzing the previous changes. The changes are also influenced by the spatial variables used in the study, such as the DEM and road network.

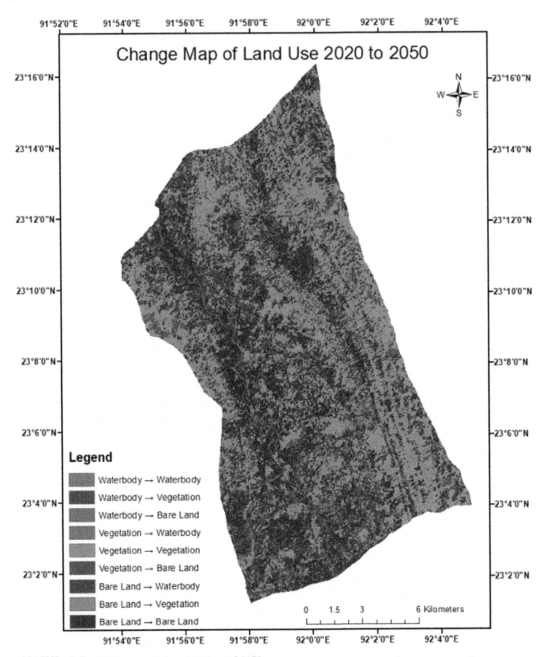

FIGURE 17.7   Change area between 2020 and 2050.

TABLE 17.6 The land use distribution and changes in 2020 and 2050.

Class	2020 (area in ha)	2050 (area in ha)	Change (area ha)	2020 (%)	2050 (%)	Change (%)
Bare land	12,073.23	14,370.48	2297.25	46.40	55.22	8.83
Vegetation	13,713.39	11,643.84	− 2069.55	52.70	44.75	− 7.95
Water body	235.35	7.65	− 227.70	0.90	0.03	− 0.88

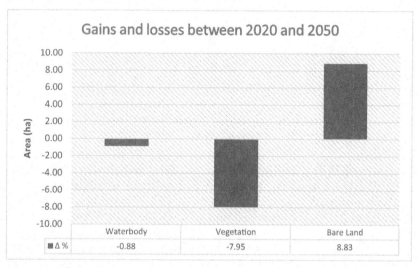

FIGURE 17.8 The changes in the LULC area in 2020 and 2050.

TABLE 17.7 Transition matrix (%) of LULC between 2020 and 2050.

	Water body	Vegetation	Bare land
Water body	0.019885	0.464245	0.51587
Vegetation	0.000144	0.572563	0.427293
Bare land	0.000082	0.305039	0.694879

17.3.3 Changes in LULC between 1990 and 2050

A comparative study was conducted among the real maps of 1990, 2014, 2020 and predicted maps of 2020 and 2050. Fig. 17.9 shows the change in land use over time from 1990 to 2050 calculated from each real and predicted map. Comparing 1990, almost 18,167.67 ha was bare in 1990. A gradual decrease in the area was observed up

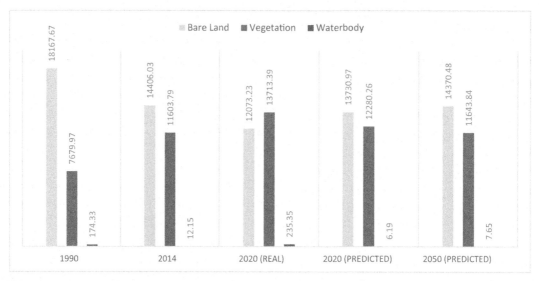

FIGURE 17.9 Historical and predicted changes of land from 1990 to 2050.

to the real scenario of 2020 (12,073.23 ha). A gradual increase was observed again in the projected map of 2050 where the bare land is predicted to cover almost 14,370.48 ha area. The analysis revealed that an average of 24.47% of bare land is decreased during the last 30 years and 8.83% will increase in the next 30 years. When comparing the natural vegetation, a gradual increase is calculated in the image analysis from 7679.97 ha in 1990 to 13,713.39 ha in the real map of 2020 (+6033.42 ha). In the simulated and predicted map, the vegetation coverage was found in a decreased amount. In the simulated map of 2020, the vegetation is calculated 12,280.26 ha and in 2050 it is 11,643.84 ha. Almost 24.24% of vegetation is increased in the last 30 years and is predicted to be decreased by 7.95% in 2050. Among the water bodies in the different time periods, a significant amount of decrease is observed in the maps from 1990 (174.33 ha) to 2014 (12.15 ha). The actual map of 2020 revealed a dramatic increase in the water body (235.35 ha). But the predicted map of 2050 shows a histrionic decrease in water body to 7.65 ha (Fig. 17.9). From 1990 to 2020, the water body increased by 0.23% and is estimated to be decreased by 0.88% between 2020 and 2050.

Comparing the total changes of land use from 1990, an area of 4,017.66 ha (48%) bare land and 159.17 ha (2%) water body was found to be reduced by 2050, whereas, 4,182.30 ha (50%) vegetation cover is predicted to be increased (Fig. 17.10).

17.3.4 Vegetation change dynamics between 1990 and 2050

The primary goal of this study was to determine changes in vegetation coverage from 1990 to 2020, with projections to 2050. The map for 2050 is created using the CA simulation. Following that, two change maps are created for the years 1990 to 2020 and 2020 to 2050 (Fig. 17.11). The change maps clearly illustrate that a considerable amount of

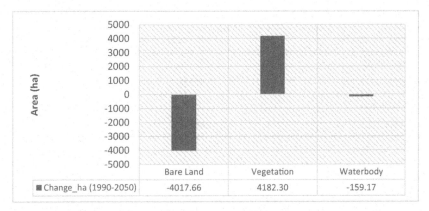

FIGURE 17.10 A total change in land use area and a percentage between 1990 and 2050.

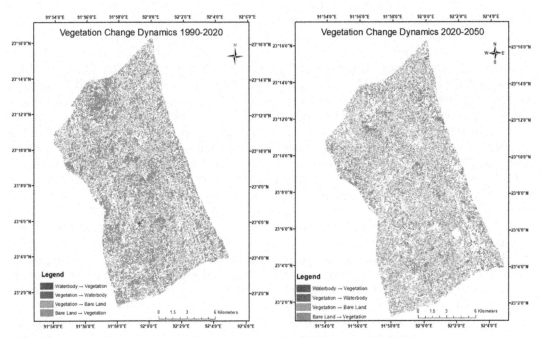

FIGURE 17.11 Vegetation cover changes between 1990−2000 (left) and 2020−50 (right).

vegetation-covered area is replaced by other categories and vice versa. In the change map of 1990−2020, 97.71 ha land is already converted to the water body and 2242.17 ha to barren land. In the similar period, a significant amount of barren land (8,364.37 ha) and water body (53.59 ha) is shifted to vegetation coverage. In the projected map of 2020−50, it is calculated that 6.33 ha and 2385.63 ha vegetation-covered area will be replaced by water body and barren land, respectively, whereas, 6266.15 ha barren land and 65.73 ha water body is expected to be replaced with vegetation cover by 2050 from 2020 (Table 17.8).

TABLE 17.8 Vegetation cover changes to and from since 1990−2020 and 2020−50.

Change (vegetation ↔ others)	1990−2020 (ha)	2020−50 (ha)
Vegetation → water body	97.71	6.33
Vegetation → barren land	2242.17	2385.63
Barren land → vegetation	8364.37	6266.15
Water body → vegetation	53.59	65.73

The results of the present study show a high impact on LULC changes both in the simulated and predicted maps (Singh et al., 2015; Saputra and Lee, 2019; Khwarahm et al., 2020). The kappa index shows low agreement between the maps of real 2014 and simulated 2020, indicating significant land-use changes over the in between 6 years. Separately, the kappa index between the real and simulated 2020 maps revealed a high level of agreement (Saputra and Lee, 2019; Khwarahm et al., 2020), which was rational. A comparison of historical and future LULC changes was performed. The findings revealed that vegetation coverage has increased over the last 30 years, while bare land has decreased (Liu et al., 2014; Sun et al., 2015). The most vegetation coverage increase is observed during the last 6 years (2014−20) in comparison to the first 24 years, based on classified and analyzed images from 1990, 2014, and 2020 (1990−2014). This is because the locals are concerned about environmental and economic development. Taking on lease bare hills and planting perennial crops on them is becoming increasingly popular among hill residents (Acharjee, 2019). According to field observations, most people are leasing the hills of Khagrachhari Sadar Upazila for at least 30 years to cultivate fruit and timber perennial trees. Even if they plant for economic development, the massive plantation over time transforms the barren hills into vegetative covered land. This scenario can be seen on the 2020 map. Furthermore, in recent days, those who were not interested in planting trees near homesteads have become interested in homestead gardening. However, over the next 30 years, it is predicted that vegetation coverage will decrease while bare land will increase. According to the research, the trend and popularity of cultivating bare land may decline in the future, or the trees that were recently cultivated may complete their rotation and harvesting will occur. Long-term effects of climatic factors, global warming, extreme weather events, ecological degradation, hydrological variation, soil erosion, and, most importantly, human activities altered historical and future landscape dynamics. The situation appears to be deteriorating in the future, with climate change and land-use change posing threats.

17.4 Conclusion

LULC are critical for resource management, sustainable development, and long-term planning. The immense importance of forest or vegetation coverage in balancing ecology and environment is undeniable. The importance of vegetation in enriching and conserving biodiversity and managing the landscape cannot be overstated. The ANN-CA model is used to simulate and predict LULC characteristics and changes in preferences with vegetation coverage in Khagrachhari, one of the major hilly Upazilas of CHTs, Bangladesh. According to historical data, vegetation coverage has increased in the last 30 years, while bare lands have decreased and water bodies have increased, indicating a fluctuating pattern in land-use changes. According to future projections, vegetation coverage and water bodies will decrease significantly, while bare land will increase. The ANN-CA model is thought to be useful for macroscale future development planning as well as sustainable forest management. Generally, LULC changes occurred as a result of disagreements over land interests. Only land cover changes and topographic variations were used as variables in this study. Future research could be enhanced by including demographic, policy, regulatory, and climatic variables.

Conflict of interest

There is no conflict of interest in any regard in this research work.

References

Abbas, Z., Yang, G., Zhong, Y., Zhao, Y., 2021. Spatiotemporal change analysis and future scenario of LULC using the CA-ANN approach: a case study of the greater bay area, China. Land 10 (6), 584. Available from: https://doi.org/10.3390/land10060584.

Abiodun, O.I., Jantan, A., Omolara, A.E., Dada, K.V., Mohamed, N.A., Arshad, H., 2018. State-of-the-art in artificial neural network applications: a survey. Heliyon 4 (11), e00938. Available from: https://doi.org/10.1016/j.heliyon.2018.e00938.

Acharjee, D. (2019, August 3). Resolution to LAND DISPUTES IN CHT: DCs seek nod to start survey, lease out land. Minister insists on preparing rules for CHT Land Dispute Resolution Commission Act. The Independent. https://m.theindependentbd.com/printversion/details/210120. Viewed on: 30.08.2021.

Ackroyd, S.A., Huang, E.S., Kurnit, K.C., Lee, N.K., 2021. Pembrolizumab and lenvatinib versus carboplatin and paclitaxel as first-line therapy for advanced or recurrent endometrial cancer: a Markov analysis. Gynecologic Oncology 162 (2), 249–255. Available from: https://doi.org/10.1016/j.ygyno.2021.05.038.

Agatonovic-Kustrin, S., Beresford, R., 2000. Basic concepts of artificial neural network (ANN) modeling and its application in pharmaceutical research. Journal of Pharmaceutical and Biomedical Analysis 22 (5), 717–727. Available from: https://doi.org/10.1016/S0731-7085(99)00272-1.

Ahammad, R., Stacey, N., 2016. Forest and agrarian change in the Chittagong Hill Tracts region of Bangladesh. Agrarian Change in Tropical Landscapes 1–323.

Anurag, Saxena, A., Pradhan, B., 2018. Land use/land cover change modelling: issues and challenges. Journal of Rural Development 37 (2), 413–424.

Bangladesh Bureau of Statistics (BBS), 2013. District Statistics 2011, Khagrachhari. Bangladesh Bureau of Statistics (BBS), Statistics and Informatics Division (SID), Ministry of Planning Government of the People's Republic of Bangladesh.

Gelfand, A.E., 2021. Multivariate spatial process models. Handbook of Regional Science 1985–2016.

Gharaibeh, A., Shaamala, A., Obeidat, R., Al-Kofahi, S., 2020. Improving land-use change modeling by integrating ANN with Cellular Automata-Markov Chain model. Heliyon 6 (9), e05092. Available from: https://doi.org/10.1016/j.heliyon.2020.e05092.

Hasnat, G.N.T., 2021. A time series analysis of forest cover and land surface temperature change over Dudpukuria-Dhopachari wildlife sanctuary using landsat imagery. Frontiers in Forests and Global Change 4, 104.

Hassan, Z., Shabbir, R., Ahmad, S.S., Malik, A.H., Aziz, N., Butt, A., et al., 2016. Dynamics of land use and land cover change (LULCC) using geospatial techniques: a case study of Islamabad Pakistan. SpringerPlus 5 (1), 1–11. Available from: https://doi.org/10.1186/s40064-016-2414-z.

Heppenstall, A., Crooks, A., Malleson, N., Manley, E., Ge, J., Batty, M., 2021. Future developments in geographical agent-based models: challenges and opportunities. Geographical Analysis 53 (1), 76–91. Available from: https://doi.org/10.1111/gean.12267.

Hibbard, K., Hoffman, F., Huntzinger, D.N., West, T., 2017. Changes in land cover and terrestrial biogeochemistry. In: Wuebbles, D.J., Fahey, D.W., Hibbard, K.A., Dokken, D.J., Stewart, B.C., Maycock, T.K. (Eds.), Climate Science Special Report: A Sustained Assessment Activity of the U.S. Global Change Research Program. U.S. Global Change Research Program, Washington, DC, pp. 405–442.

Ionita, I., Fullen, M.A., Zgłobicki, W., Poesen, J., 2015. Gully erosion as a natural and human-induced hazard. Natural Hazards 79, 1–5. Available from: https://doi.org/10.1007/s11069-015-1935-z.

Jashimuddin, M., Inoue, M., 2012. Management of village common forests in the Chittagong Hill tracts of Bangladesh: historical background and current issues in terms of sustainability. Open Journal of Forestry 2 (3), 121–137. Available from: https://doi.org/10.4236/ojf.2012.2301.

Kafy, A.A., Naim, M.N.H., Subramanyam, G., Ahmed, N.U., Al Rakib, A., Kona, M.A., et al., 2021. Cellular automata approach in dynamic modelling of land cover changes using RapidEye images in Dhaka, Bangladesh. Environmental Challenges 4, 100084. Available from: https://doi.org/10.1016/j.envc.2021.100084.

Kafy, A.A., Rahman, M.S., Hasan, M.M., Islam, M., 2020. Modelling future land use land cover changes and their impacts on land surface temperatures in Rajshahi, Bangladesh. Remote Sensing Applications: Society and Environment 18, 100314. Available from: https://doi.org/10.1016/j.rsase.2020.100314.

Khwarahm, N.R., Qader, S., Ararat, K., Al-Quraishi, A.M.F., 2020. Predicting and mapping land cover/land use changes in Erbil/Iraq using CA-Markov synergy model. Earth Science Informatics 14 (1), 393–406. Available from: https://doi.org/10.1007/s12145-020-00541-x.

Kibria, A.S.M.G., Inoue, M., Nath, T.K., 2015. Analysing the land uses of forest-dwelling indigenous people in the Chittagong Hill tracts, Bangladesh. Agroforestry Systems 89 (4), 663–676. Available from: https://doi.org/10.1007/s10457-015-9803-0.

Leta, M.K., Demissie, T.A., Tränckner, J., 2021. Modeling and prediction of land use land cover change dynamics based on land change modeler (LCM) in Nashe Watershed, Upper Blue Nile Basin, Ethiopia. Sustainability 13 (7), 3740. Available from: https://doi.org/10.3390/su13073740.

Liu, X., Zhang, J., Zhu, X., Pan, Y., Liu, Y., Zhang, D., et al., 2014. Spatiotemporal changes in vegetation coverage and its driving factors in the Three-River Headwaters Region during 2000–2011. Journal of Geographical Sciences 24 (2), 288–302. Available from: https://doi.org/10.1007/s11442-014-1088-0.

Lv, J., Wang, Y., Liang, X., Yao, Y., Ma, T., Guan, Q., 2021. Simulating urban expansion by incorporating an integrated gravitational field model into a demand-driven random forest-cellular automata model. Cities 109, 103044. Available from: https://doi.org/10.1016/j.cities.2020.103044.

Mahmood, R., Pielke Sr, R.A., Hubbard, K.G., Niyogi, D., Bonan, G., Lawrence, P., et al., 2010. Impacts of land use/land cover change on climate and future research priorities. Bulletin of the American Meteorological Society 91 (1), 37–46. Available from: https://doi.org/10.1175/2009BAMS2769.1.

Maind, S.B., Wankar, P., 2014. Research paper on basic of artificial neural network. International Journal on Recent and Innovation Trends in Computing and Communication 2 (1), 96–100.

Mutz, S.G., Scherrer, S., Muceniece, I., Ehlers, T.A., 2021. Twenty-first century regional temperature response in Chile based on empirical-statistical downscaling. Climate Dynamics 56 (9), 2881–2894. Available from: https://doi.org/10.1007/s00382-020-05620-9.

Prijith, S.S., Srinivasarao, K., Lima, C.B., Gharai, B., Rao, P.V.N., SeshaSai, M.V.R., et al., 2021. Effects of land use/land cover alterations on regional meteorology over Northwest India. Science of the Total Environment 765, 142678. Available from: https://doi.org/10.1016/j.scitotenv.2020.142678.

Primack, R.B., Morrison, R.A., 2013. Extinction, Causes of, Encyclopedia of Biodiversity, second ed.pp. 401–412.

Rasul, G., Thapa, G.B., Zoebisch, M.A., 2004. Determinants of land-use changes in the Chittagong Hill Tracts of Bangladesh. Applied Geography 24 (3), 217–240. Available from: https://doi.org/10.1016/j.apgeog.2004.03.004.

Rosa, I.M., Ahmed, S.E., Ewers, R.M., 2014. The transparency, reliability and utility of tropical rainforest land-use and land-cover change models. Global Change Biology 20 (6), 1707–1722. Available from: https://doi.org/10.1111/gcb.12523.

Saddique, N., Mahmood, T., Bernhofer, C., 2020. Quantifying the impacts of land use/land cover change on the water balance in the afforested River Basin, Pakistan. Environmental Earth Sciences 79 (19), 1–13. Available from: https://doi.org/10.1007/s12665-020-09206-w.

Saha, T.K., Pal, S., Sarkar, R., 2021. Prediction of wetland area and depth using linear regression model and artificial neural network based cellular automata. Ecological Informatics 62, 101272. Available from: https://doi.org/10.1016/j.ecoinf.2021.101272.

Saputra, M.H., Lee, H.S., 2019. Prediction of land use and land cover changes for north sumatra, indonesia, using an artificial-neural-network-based cellular automaton. Sustainability 11 (11), 3024. Available from: https://doi.org/10.3390/su11113024.

Sarir, P., Chen, J., Asteris, P.G., Armaghani, D.J., Tahir, M.M., 2021. Developing GEP tree-based, neuro-swarm, and whale optimization models for evaluation of bearing capacity of concrete-filled steel tube columns. Engineering with Computers 37 (1), 1–19. Available from: https://doi.org/10.1007/s00366-019-00808-y.

Sibanda, S., Ahmed, F., 2021. Modelling historic and future land use/land cover changes and their impact on wetland area in Shashe sub-catchment, Zimbabwe. Modeling Earth Systems and Environment 7 (1), 57–70. Available from: https://doi.org/10.1007/s40808-020-00963-y.

Singh, S.K., Mustak, S., Srivastava, P.K., Szabó, S., Islam, T., 2015. Predicting spatial and decadal LULC changes through cellular automata Markov chain models using earth observation datasets and geo-information. Environmental Processes 2 (1), 61–78. Available from: https://doi.org/10.1007/s40710-015-0062-x.

Sun, W., Song, X., Mu, X., Gao, P., Wang, F., Zhao, G., 2015. Spatiotemporal vegetation cover variations associated with climate change and ecological restoration in the Loess Plateau. Agricultural and Forest Meteorology 209, 87–99. Available from: https://doi.org/10.1016/j.agrformet.2015.05.002.

Tabarestani, E.S., Afzalimehr, H., 2021. Artificial neural network and multi-criteria decision-making models for flood simulation in GIS: Mazandaran Province, Iran. Stochastic Environmental Research and Risk Assessment 35, 2439–2457. Available from: https://doi.org/10.1007/s00477-021-01997-z.

Tan, J., Yu, D., Li, Q., Tan, X., Zhou, W., 2020. Spatial relationship between land-use/land-cover change and land surface temperature in the Dongting Lake area, China. Scientific Reports 10 (1), 1–9. Available from: https://doi.org/10.1038/s41598-020-66168-6.

Thapa, P., 2021. The relationship between land use and climate change: a case study of Nepal. doi:10.5772/intechopen.98282.

Verma, P., Singh, R., Singh, P., Raghubanshi, A.S., 2020. Chapter 1 - urban ecology – current state of research and concepts. Urban Ecology. Elsevier, pp. 3–16. Available from: https://doi.org/10.1016/B978-0-12-820730-7.00001-X.

Viera, A.J., Garrett, J.M., 2005. Understanding interobserver agreement: the kappa statistic. Family Medicine 37 (5), 360–363.

Wedajoa, G.K., Muletac, M.K., Awoke, B.G., 2020. Separate and combined impacts of land cover and climate changes on hydrological responses of Dhidhessa River Basin, Ethiopia. Preprints 2020. Available from: https://doi.org/10.20944/preprints202012.0595.v1.

Zeshan, M.T., Mustafa, M.R.U., Baig, M.F., 2021. Monitoring land use changes and their future prospects using GIS and ANN-CA for Perak River Basin, Malaysia. Water 13 (16), 2286. Available from: https://doi.org/10.3390/w13162286.

Plant functional traits and ecological sustainability

Plant functional traits: mountainous soil function and ecosystem services

Sarika[1,2] *and Hardik Manek*[3]

[1]School of Environmental Science, Jawaharlal Nehru University, New Delhi, Delhi, India
[2]Ashoka Trust for Research in Ecology and the Environment, Royal Enclave, Srirampura, Jakkur, Bengaluru, Karnataka, India [3]National Institute of Technology, Durgapur, West Bengal, India

18.1 Introduction

The classification of species based on their functions has long been a practice among ecologists. Because of the enormous diversity of animal and plant species, developing rules for ecosystem protection has always been a major challenge for researchers. It wasn't until the early 1990s that researchers became interested in understanding a species' functional traits and using them to predict community response to changing environments. Thus, studying plant functional traits can help us understand the impact of changing community composition on the ecosystem (Lavorel and Garnier, 2002). Predicting changes in

Understanding Soils of Mountainous Landscapes
DOI: https://doi.org/10.1016/B978-0-323-95925-4.00001-7

347

ecological processes by understanding a species' functional traits is regarded as the Holy Grail by modern ecologists (Funk et al., 2017). Under the stress of climate change, the world is experiencing a rapid loss of biodiversity in this postindustrial era, putting both the ecosystem and human society under immense strain. As a result, such pressure has prompted researchers to investigate the mechanism by which changing biodiversity affects ecosystem functioning. Various researchers have defined functional traits of a species based on their knowledge and research. However, a species' functional trait is one of its core morpho-physio-phenological traits (Rawat et al., 2015). These characteristics influence the growth, survival, mortality, and reproduction of the species as a result of interactions with abiotic and biotic environments, as shown in Fig. 18.1. Thus, changing climatic conditions will force species to evolve its functional trait for surviving as well as for inter and intraspecific interaction (Malézieux et al., 2007; Caruso et al., 2020).

The functional traits also include aspects such as alterations in various characters which include photosynthesis rates, biomass allocation, tissue turnover, etc. (Roscher et al., 2012). Changes in these functional traits have serious consequences for ecosystem function and feedback mechanisms (both positive and negative). This results in a change of properties of ecosystem components such as soil temperature, decomposition, carbon cycling, water availability, ecosystem services, energy balance, and soil carbon potential, etc. (Falster et al., 2011; Lavorel et al., 2013; Bu et al., 2017). All of these elements, regardless of how they interact, have an impact on regional and global climate (Bjorkman et al., 2018; Campetella et al., 2020). Numerous ecological studies, as shown in Table 18.1, investigate the prediction of ecological communities' responses to biotic and abiotic stress and their impact on ecosystem functioning by developing systematic models based on plant functional traits (Heilmeier, 2019). In an ecosystem, for example, the interaction of several flowering plant species, pollinators, and other organisms. Whereas nutrient cycling is an excellent example for understanding the interaction of organisms at different trophic levels, carbon cycling is an excellent example for understanding the interaction of organisms at different trophic levels (Roscher et al., 2012). In this chapter, we will understand and investigate the concept of functional traits, their role, and importance in the ecosystem. The chapter emphasizes the relationship between mountain soil ecosystem functions, services, and their interdependence with functional traits. Furthermore, the role of functional traits in facilitating prediction, as well as their key importance of various disrupting forces such as climate change, changing soil functions, and so on, will be covered, which can have a significant impact on conservation plans and ecosystem management.

18.2 Mountainous soil ecosystem

Mountain peoples have developed robust systems and ways to cope with their challenging conditions over millennia in order to thrive at high elevations. Soil, a finite resource that plays a critical role in supporting ecosystem services, is essential for both agricultural production and environmental stability. Soil can hold water as a natural reservoir, absorb it, filter it, and use it to build its own flood and drought resistance (Rabot et al., 2018; Nannipieri et al., 2003). This demonstrates that healthy soils play an important role in carbon storage and emission reduction, which contributes to climate change mitigation, because soils contain more carbon than the combined atmosphere and all vegetation

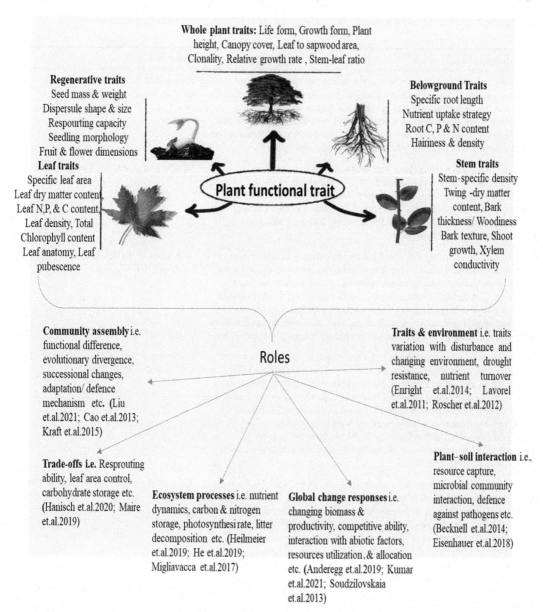

Whole plant traits: Life form, Growth form, Plant height, Canopy cover, Leaf to sapwood area, Clonality, Relative growth rate , Stem-leaf ratio

Regenerative traits
Seed mass & weight
Dispersule shape & size
Respourting capacity
Seedling morphology
Fruit & flower dimensions

Belowground Traits
Specific root length
Nutrient uptake strategy
Root C, P & N content
Hairiness & density

Leaf traits
Specific leaf area
Leaf dry matter content,
Leaf N,P, & C content,
Leaf density, Total
Chlorophyll content
Leaf anatomy, Leaf
pubescence

Stem traits
Stem-specific density
Twing -dry matter
content, Bark
thickness/ Woodiness
Bark texture, Shoot
growth, Xylem
conductivity

Plant functional trait

Roles

Community assembly i.e. functional difference, evolutionary divergence, successional changes, adaptation/ defence mechanism etc. (Liu et.al.2021; Cao et.al.2013; Kraft et.al.2015)

Traits & environment i.e. traits variation with disturbance and changing environment, drought resistance, nutrient turnover (Enright et.al.2014; Lavorel et.al.2011; Roscher et.al.2012)

Trade-offs i.e. Resprouting ability, leaf area control, carbohydrate storage etc. (Hanisch et.al.2020; Maire et.al.2019)

Ecosystem processes i.e. nutrient dynamics, carbon & nitrogen storage, photosynthesi rate, litter decomposition etc. (Heilmeier et.al.2019; He et.al.2019; Migliavacca et.al.2017)

Global change responses i.e. changing biomass & productivity, competitive ability, interaction with abiotic factors, resources utilization , & allocation etc. (Anderegg et.al.2019; Kumar et.al.2021; Soudzilovskaia et.al.2013)

Plant–soil interaction i.e., resource capture, microbial community interaction, defence against pathogens etc. (Becknell et.al.2014; Eisenhauer et.al.2018)

FIGURE 18.1 Showing different type of plant functional traits and their roles in environment.

(Blum, 2005). Mountain soils are frequently characterized as skeletal, acidic, and infertile. Mountain soils are typically shallow because extreme cold inhibits biological activity, soil genesis, and development. They are also highly variable within small areas due to exposure and slope. Elevation has a negative impact on their fertility and development (Nanda, 2015). In cold alpine regions, freeze-thaw cycles affect soil aggregation, integrity,

TABLE 18.1   Various studies showing plant functional traits in relation to changing climate and soil functions.

Description	Reference
Determine the mechanism by which plant functional characteristics influence climate change. Describes in-depth research on plant functional qualities, such as functional traits of roots and leaves and their relationships. The prediction of functional trait development and the link between soil interaction, climate change, and plant functional features were used to describe the direction of changing climatic conditions.	Zhang and Li (2019), Falster et al. (2011), Allen et al. (2011)
The growth season temperature was measured at locations within 3000–4130 m, spanning at around 5.3°C. Trait data from local populations (Leaf thickness, dry matter content of leaves, N and C isotopes in leaves, N, P and C content of leaf, leaf total area and specific area) were discovered to be linked with plant resource utilization, growth, and life cycle strategies.	Vandvik et al. (2020); Geng et al. (2017)
Climate change is putting the world's forests' resilience and tolerance to extreme weather to the test. Plant functional traits have been observed to be adaptive to climate change, with patterns of trait correlations and coordination evolving (mechanistic trade-off). According to the findings, qualities will discriminate between populations separated by climatic gradients, implying variation in adaptivity, and that other associated traits will also be revamped in response to current and future climate situations.	Ahrens et al. (2020), Woodward and Cramer (1996)
Energy, water, and carbon flows from terrestrial ecosystems have an impact on the atmosphere. Because low soil moisture reduces both transpiration and evaporation in plants while raising local temperatures, extreme climatic episodes such as heat and cold waves, as well as severe drought, can be easily exacerbated by land-atmosphere feedback. Drought intensification is influenced by a combination of environmental and plant functional characteristics, particularly those related to water transportation and maximum gas exchange rate through leaves across the plant hydraulic continuum.	Anderegg et al. (2019), Dubuis et al. (2013), Rillig et al. (2019)
The filtering effect of climatic, disturbance, and biotic variables results in a link between environment and plant features, that is, sets of plant qualities associated with different environmental situations. According to this theory, climate conditions may have a "filtering effect" on important plant functions. Filtering of vegetative traits was more prevalent than filtering of regenerative traits. Lifespan, canopy height, specific leaf area, leaf weight ratio, ramification, carbon investment in support tissue, and pollination mode were all investigated.	Diaz et al. (1998), Vandvik et al. (2020), Enright et al. (2014)

productivity, and moisture retention. Many plants, including crops, have adapted to the rocky soils of mountains, and many densely forested mountains help to reduce erosion (Barni et al., 2007). Local and global scales are critical in addressing the issue of soil spatial variability, which affects soil ecosystem function and service complexity. Soil ecosystems are extremely dependent on their parent materials, temperature, relief, the biosphere, and human activity. Soil interactions with these forming forces can be found in all ecosystems, but they are more visible in mountainous areas.

Mountain height and relief have a significant impact on soil energy balance and, as a result, soil temperature. The length of snow cover and precipitation, for example, differ between windward and leeward areas (Stockmann et al., 2013). Temperature and water availability in soils drive chemical and physical weathering. Geological uplift, varying source material hardness, and resistance to erosion and weathering all contribute to the relief. The parent material determines which plant species can grow and thus has an indirect impact on soil formation (Barni et al., 2007). Because soils can go through periods of progressive and regressive development, soil attributes and morphology cannot always be directly linked to surface age (Smith, 2008). Polygenetic soils are common in geomorphologically active areas where there is erosion or buildup. Mountain soil formation is characterized by the redistribution of soil material on mountain slopes (Kapos et al., 2000).

Climate change and intensive land use have the potential to have a significant impact on mountain soils, which are extremely dynamic and sensitive systems. In some mountainous areas, the rate of human-induced erosion far outstrips the rate of (maximum) soil formation. Carbon sequestration is harmed by rapidly degrading soil and the loss of plant-development areas (Barni et al., 2007). Environmental and site factors strongly influence soil pedogenesis, organic matter input, and turnover, resulting in soils that are thick and anisotropic under undisturbed conditions and produce distinctly distinguishable horizons (Tappeiner et al., 2008). Organic material decomposition may be slowed under certain conditions, such as colder temperatures or an abundance of water (anoxic environments). As a result, the rate of biomass generation exceeds the rate of decay (plant and soil respiration) (Rabot et al., 2018). Paludification is the waterlogging of terrestrial soils caused by organic material buildup, which eventually results in the formation of histosols, which are distinguished by thick organic horizons and play an important role in alpine flora survival (Blyth et al., 2002). The region's plant groups include rock, mountain soil, and grassland. Many plant and animal species rely on alpine soils for water and nutrients (e.g., Rendzina, Para-rendzina, Ranker, and Cambisol, among others). Soil conditions and processes generally control the composition and diversity of plant species in terrestrial ecosystems (Lal, 2004). The bare surfaces left by receding glaciers allow researchers to observe primary soil formation phases, validating current hypotheses about ecosystem evolution and determining the rate of soil formation processes (Cambardella and Elliott, 1992; Misra et al., 2008). It is realistically possible to overcome the challenges associated with sustainable mountain advancement and promote more resilient livelihoods for mountain people by gaining a solid understanding of mountain soils, their composition, dynamics, and value. Addressing mountain soils is critical when dealing with mountains today because most soil survey projects around the world are geared toward arable agricultural production, leaving little information on mountain soils. As a result, there is a significant lack of coordinated ecosystem management.

18.3 Plant functional traits

Plant functional types are the result of ecologists' long-term efforts to discover simplified patterns for understanding flora diversity as well as ecosystem complexity (Díaz et al., 2013). A variety of factors are being studied, including a more empirical investigation of ecosystem function and an improved assessment of ecosystem susceptibility to environmental changes (Kraft et al., 2015). The filtering impact of climatic disturbance and biotic variables both act as a bridge between the environment and plant features, that is, the association of specific plant traits with specific environmental situations. These filters determine how a group of species is organized into a local community (Meng et al., 2007, Diaz et al., 1998). Several studies have been conducted to investigate the action of "filters" at various scales. Climate, local disturbance, stress, and intra- and interinteractions are among the filters. All have a proclivity to act on smaller spatial scales (Díaz and Cabido, 1997). At any given site, a hierarchy of filters can be observed; only traits that are sustainable under the existing environmental conditions as well as the dominant disturbance pattern have a chance of not being filtered out upon interaction with various other organisms (Faucon et al., 2017; Van Der Valk, 1981). Plant interaction with topographical-edaphic environments, disturbance, and vital attributes, as well as the relationship between changing climate and vegetation on a regional to global scale, are all topics of emerging research in this field (He et al., 2019). Environmental filters such as regime disruption, climate, and biotic interaction among species may be altered at any scale in an unusual pattern, allowing the outcome of the species shift to be predicted for the next century. Plant stability, recolonization, and any key disturbance such as local stress due to anthropogenic activities, changing climate, and various species mobility across the landscape are all influenced by regeneration traits, whereas vegetative traits are linked to various in situ ecosystem activities (Diaz et al., 1998). The consistency in the relationship between trait, environment, and upcoming climatic circumstance, which results in the distribution of dominant plant traits along difficult terrains, can be used as a benchmark for changing climate conditions (Funk et al., 2017). Under various climatic conditions, the use of functional types or plant traits rather than species is more advantageous because it provides more opportunities to assess and anticipate reactions of distinct floras or species that are no longer in the regional pool (Fornara and Tilman, 2008).

Plant functional types have become an important tool for reducing the various complexity of species diversity in relation to their structure and function, as well as for the behavior of species assemblages in future habitats (Funk et al., 2017). Historically, functional forms were associated with plant physiognomy, on the premise that architecture is more or less related to survival, such as a plant's ability to withstand changing climatic conditions (Yang et al., 2015). Plant functional characteristics influence various ecosystem activities, particularly carbon storage in the soil, which is a key component of the carbon cycle (Xi et al., 2021). Plant traits influence phenomena such as net carbon influx, transmission, and underlying biomass storage, as well as different soil carbon and nutrient releasing processes such as breathing, fire, and leaching from the soil (Faucon et al., 2017). Plant functional traits are an important tool in climate change research because they can bridge the gap between ecosystem activities and plant physiology and community (Schweiger et al., 2017). Functional traits are morphological, physiological, and phenological characteristics of a species that influence fitness by affecting the species' three basic individual performance components, namely reproduction, growth, and survival (Faucon et al., 2017; Wood et al., 2015). Functional traits can reveal more about a species' interaction and response to its surroundings, which can be used to make predictions and provide a powerful tool for dealing with ecological issues (Garnier and Navas, 2012). The conditions of abiotic sites have a direct impact on ecosystem processes because they can alter the flux rate of matter and energy in the ecosystem, as well as indirectly by altering the plants physiological rate, which is dictated by the functional features (Niu et al., 2018).

According to Díaz and Cabido (1997), plant functional type terminology is interconnected and can only be understood in conjunction with a number of other essential terminologies. Trait is one of the most commonly used terms for plant functions. Functional traits are characteristics that a plant may have that are strategic or adaptive to the plant.Various terminologies are presented in Table 18.2 are used by the researcher as per their convenience, such as "Plant Characters" (McIntyre et al., 1999), "Plant Characteristics" and attributes (Grime, 1979; Lavorel et al., 1997) etc. According to Woodward and Cramer (1996), a trait can appear in the form of various states, expressions and might vary in size and quantity. However, most researchers are now using the term "vital Attribute" coined by Nobel et al (1980). According to Pillar (1999), the most appropriate terminology is "Trait State," which defines the species' functional trait. The second phrase is "trade-off," which acts as a negative link between plant features that are not optimized at the same time, such as nutrient productivity and mean residence time for nutrients above ground (De Long et al., 2019). Fundamental tradeoffs that define a species' ecological role are represented by functional characteristics (Westoby and Wright, 2006). For example, the tradeoff between the number of seeds produced and the size of each seed (Yang et al., 2015). The large seeds produce strong seedlings that can thrive in a resource-limited environment with fierce competition for survival; however, they are produced in very small quantities (Malézieux et al., 2007). Tiny seeds, on the other hand, fail in high-competition environments but proliferate in large quantities, increasing the likelihood of dispersal to ephemeral locations with few competitors and abundant resources. As a result, seed size is one of the easiest functional features to quantify, and it can reveal a lot about a plant's relative dispersal and competitive abilities (Liu et al., 2021). Thus, when combined with other major functionality features, these features contribute to identifying and understanding the ecological responsibilities of various plant species. Photosynthesis (nutrient

TABLE 18.2 Mountain soil and its key ecosystem services.

Ecosystem services of mountain soils (Popescu, 2015; Veith and Shaw, 2011)	
Life supporting	• Supports plants in their physical growth. • Renewing, retaining, and dispersing nutrients aids in the process of nutrient cycle. • Biodiversity and productivity are maintained. • Provides a place for the spread of genetic material.
Regulating	• Carbon and oxygen cycles are regulated by the hydrologic cycle, and also balancing all the plant nutrients, that is, nitrogen, phosphorus, potassium, calcium, and sulfur etc. • Climate change mitigation and adaptation through resilience building and carbon storage. • Biodiversity activities and functions are supported by 25% of the world's biodiversity. • Pest, disease, and pollution control and suppression are supported by microorganism processes.
Provisioning	• For the distribution of food, fiber, bioenergy, and pharmaceuticals provides the foundation. • Stores and releases water for the plant's growth and human's use.
Cultural	• Spiritual or historical significance might be found. • Provides the foundation for recreational and ecotourism environments. • Supports urban settlement and infrastructure. • Contains raw materials for building.

concentration in leaf), light capturing (through specific leaf area, also defined as leaf area per unit dry mass of leaf), pest resistance (toughness of leaves, density of wood), and various other competitive ability aspects are thought to yield similar insights (plant size) (Yadav et al., 2020; Malézieux et al., 2007) (Table 18.3).

In the context of flora, the term function refers to the mechanisms by which the plant performs its functions. Some notable functions include photosynthetic CO_2-fixation steps C3, C4, and CAM (Crassulacean Acid Metabolism) (Niu et al., 2018). Furthermore, the presence of one of these functions suggests the presence of other plant processes and

TABLE 18.3 Illustrates definitions of the plant functional traits and its essential related terms.

Key definitions in the functional trait and its related terms	
1 Traits (Díaz et al., 2013; Reiss et al., 2009)	Trait refers to those attributes, morphological, physiological, or phenological, measured at the level of the individual. This could be the integrated expression of an organism's fundamental biochemical and biophysical features as well as activities; if a trait is a mixture of these characteristics or even one of them in and of itself is a question of viewpoint and study aim. For example, Toughness of leaf could be considered as a trait which depends on the anatomical characters such as density, architecture of venation, texture and specific area etc. and chemical characteristics like lignin concentration, or it can be thought of as a trait within itself.
2 Functional traits (McGill et al., 2006; Violle et al., 2007; Reiss et al., 2009)	Functional traits are phenotypic characteristics that determine an organism's effect on processes and response to multiple environmental conditions. Individual species' phenotypes express biochemical, morphological, structural, phenological or physiological or behavioral features that are thought to be significant to their response to the environment and/or their influence on ecosystem attributes.
3 Effect traits (Minden and Kleyer, 2011; Suding et al., 2008; Bu et al., 2017)	The influence of a species on ecosystem characteristics and the benefits or losses that human civilization gets from it is based on its functional traits, independent as to whether or not such traits constitute an evolutionary advantage for the individual. Water-holding capacity in bryophytes (maintaining hydrology of ecosystem), nitrogen content of leaves in vascular plants (nutrient cycling rate), and burrowing habit (changing soil structure) or gut digesting characteristics in organisms are examples of impact attributes.
4 Response traits (Liu et al., 2021; De Deyn et al., 2008; Suding et al., 2008)	Species ability to colonize and prosper in any habitat, as well as to ensure environmental change, and is influenced by response qualities. Seed size (which affects recruitment capability under various disturbance), thickness of bark (which affects fire resistance), and size of leaf are all examples of plant characteristics (leading to different heat balances). In some circumstances, the same trait can function as both a response and an effect trait: for example, nitrogen content of leaf in case of plants and body size in case of animals both influence many environmental responses and ecosystem features.

characteristics, such as temperature, drought, and light sensitivity (Yang et al., 2015). Plant function is characterized through plant properties, which are now referred to as traits, as the environment is treated in two ways: broadly, in terms of global change (particularly in climate and atmospheric CO_2), or narrowly, in terms of local stress (Smith et al., 1993). Lavorel et al. (1997), citing Grime (1979), establishes an evident link: "determining adaptive strategies entails relating species features to resource availability variation." While alluding to Raunkir, Woodward and Cramer (1996) associates the concept of functional types to the "capabilities of plants to survive harsh situations." In reference to the Plant functional type's "core term," few researchers use it in reference to functional species classifications for use in functional ecosystem research and assessments of 'ecosystem sensitivity to environmental changes. Lavorel et al. (1997) proposed the following definition: "nonphylogenetic groupings of species that behave similarly in an ecosystem-based on a set of common biological attributes." Plant functional categories enable us to describe vegetation types structurally and functionally without having to understand regional characteristics ranging from plant communities to formations. It also helps us understand the ecological limits of floral communities.

The terminologies used to differentiate various Plant Functional Traits, which are defined in terms of environmental adaptation and involvement in ecological processes (Wood et al., 2015). It was observed by Geng et al. (2012) that greatly productive plant species prevail in the area with abundant resource availability. Plant characteristics such as specific leaf area and tree heights are more developed and visible in nutrient-rich areas. However, research should be conducted to determine whether nondominant species that require fewer resources can be found in such an environment, thereby contributing to overall ecosystem productivity (Palma et al., 2021). It is necessary to clarify and unify the lexicon used to describe plant functional categories, characteristics, and strategies. The underlying premise is that plants have characteristics that allow them to functionally adapt to specific environmental features, such as difficulties, which are commonly referred to as restrictions (Bordin and Müller, 2019). As a result, the term "plant functional type" is widely used in this anthropocene epoch. For comparing and understanding various types of plant functional both hierarchically and articulately, a comprehensive framework of environmental factors is required. However, our understanding of all of these processes is still limited, and more research is required.

18.4 Soil functions

Soil is an important and mostly nonrenewable component of the natural world. Global economies rely on the natural environment's goods and services. As a natural resource, soils perform a variety of critical environmental, social, and economic functions. Some of the services provided by soil symbiotic networks include soil coagulation and consolidation, aeration and water retention ability, nutrient cycling, deposition, and storage, abiotic or biotic stress tolerance, and mitigating the impact of external conditions on plants (Rabot et al., 2018). There are many ecological functions, such as the production of biomass (such as raw materials, fodder, sustainable sources, and food), the protection of living entities (such as the creation of a buffer zone between groundwater, vegetation, and the atmosphere cover, which has a significant impact on the water cycle), and the preservation of genetic diversity (Blum, 2005). There are three nonecological functions that soils do in addition to their ecological roles. These three nonecological activities are: (Benedetti et al., 2013). Among the many functions that comprise the natural landscape are geological and cultural heritage (i.e., safeguarding and concealing valuable paleontological or archeological assets) (Vasenev et al., 2018). Using basic sensors, it is impossible to assess the soil's ability to retain carbon dioxide, filter out pollutants, generate biomass, convert materials, or serve as a home for organisms. The functions in the soil are developing proprietors through extensive interaction with physical, chemical, and biological processes (Nannipieri et al., 2003). These processes interact with one another in a variety of nonlinear ways, and soil characteristics change as a result of external disturbances (i.e., farming). As a result, soils are thought to be complex (Drobnik et al., 2018). By the 1990s, determining the soil's fertility and suitability for various crops had taken precedence. Since then, scientists' perspectives on soil functions have shifted, reflecting the idea that soils are critical to the health of ecosystems in a variety of ways, including climate regulation, water volume and quality, nitrogen cycling, and an enormous variety of ecosystems (Blum, 2005). Soil ecosystem preservation is an essential component of sustainable development priorities such as food, water, climate, and biodiversity. Thus, soil ecosystem services are inextricably linked to the tangible benefits that human civilizations derive from healthy soil (Blum, 2014). As a result, soil, which is an important "natural capital," was considered. The role of soil in sustaining life on Earth is described by ecosystem services, which are defined as interacting soil processes with no social-economic value (Baveye et al., 2016). As a result, having a well-functioning biophysical soil is critical for supporting biodiversity and habitat conservation. Despite the fact that there is no single description of soil functions, the World Soil Information Network summary states that "soil is our life support system" (Dominati et al., 2010). Soils help plants by anchoring roots, retaining water, and providing nutrients. In the soil, there are numerous bacteria, earthworms, and termites that fix and decompose organic matter. Soils are essential for a variety of ecosystem services, including climate and water control, carbon storage, and nutrient cycling. Because these services may be jeopardized in damaged ecosystems, the international community has set specific goals for ecosystem restoration (Benedetti et al., 2013). As illustrated in Fig. 18.2, numerous soil ecosystem functions may be inferred from quantifiable soil variables such as physical, chemical, and biological features, which can be found in a broad range of soil samples. In the soil, nutrients and other components are stored and released in a controlled fashion (Bouma, 2014).

According to Blum (2005), soil functions include biomass production, human and environmental protection, gene storage, human activity as its physical basis, raw material supply, and geogenic and cultural legacy. The European Commission added a seventh factor to emphasize the soil's ability to act as a carbon sink. Soil functions are closely related to soil quality, which is defined by an American Soil Science Working Group as the "ability of a particular type of soil to function within natural or managed ecosystem borders" (Vogel et al., 2018). Soil assessment emphasizes the various roles that soils play. Healthy soil provides us with clean air and water, lush forests and crops, fertile farmland, wildlife diversity, and breathtaking scenery (Rillig et al., 2019). Everything soil accomplishes through below summaries key functions:

Nutrient Cycling—In the soil, nutrients and other minerals are stored, released, and cycled. Throughout those biogeochemical cycles, which are similar to the water cycle,

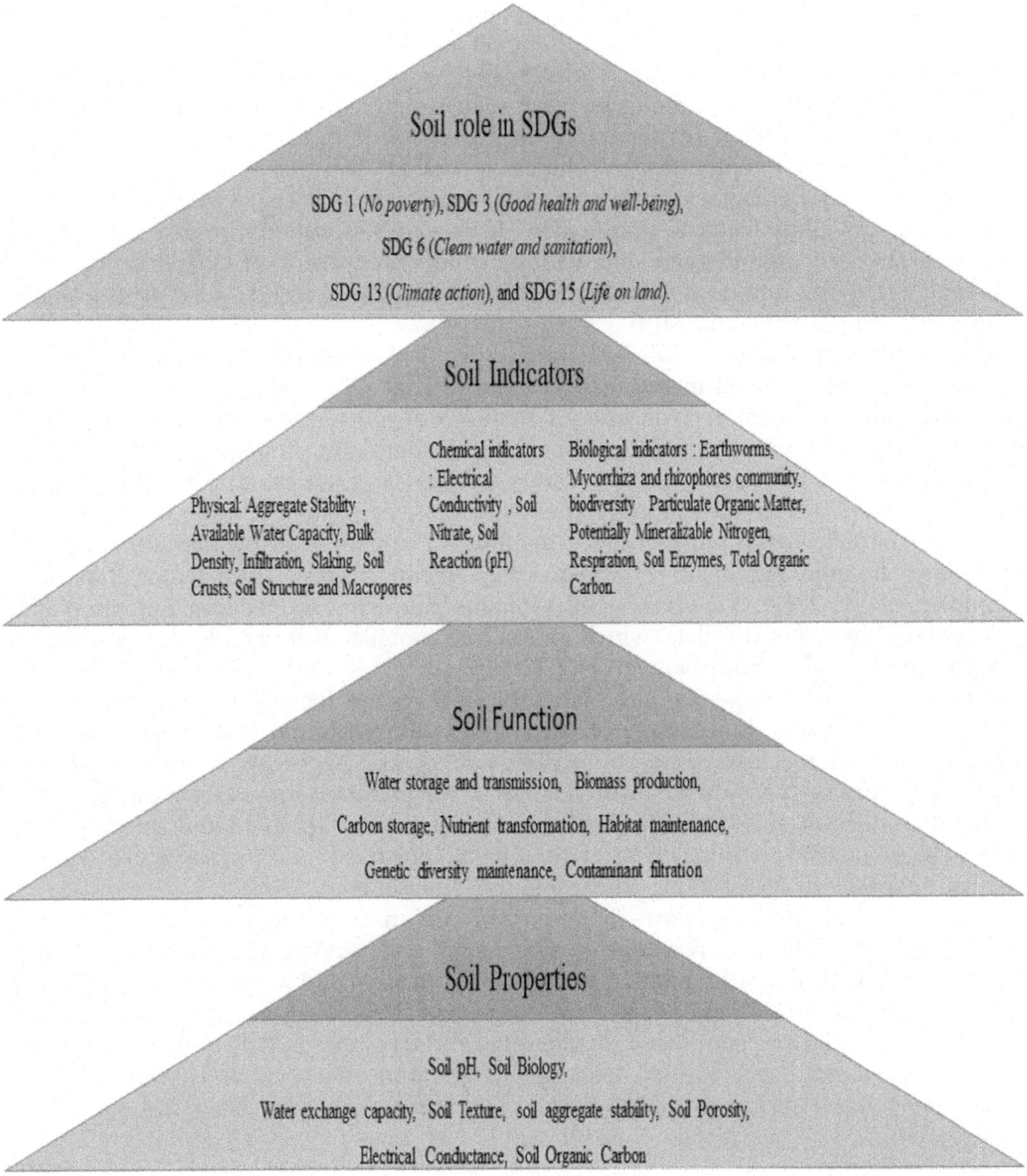

Soil role in SDGs

SDG 1 (*No poverty*), SDG 3 (*Good health and well-being*),

SDG 6 (*Clean water and sanitation*),

SDG 13 (*Climate action*), and SDG 15 (*Life on land*).

Soil Indicators

Physical: Aggregate Stability, Available Water Capacity, Bulk Density, Infiltration, Slaking, Soil Crusts, Soil Structure and Macropores

Chemical indicators : Electrical Conductivity , Soil Nitrate, Soil Reaction (pH)

Biological indicators : Earthworms, Mycorrhiza and rhizophores community, biodiversity Particulate Organic Matter, Potentially Mineralizable Nitrogen, Respiration, Soil Enzymes, Total Organic Carbon.

Soil Function

Water storage and transmission, Biomass production,

Carbon storage, Nutrient transformation, Habitat maintenance,

Genetic diversity maintenance, Contaminant filtration

Soil Properties

Soil pH, Soil Biology,

Water exchange capacity, Soil Texture, soil aggregate stability, Soil Porosity,

Electrical Conductance, Soil Organic Carbon

FIGURE 18.2 Representation of various soil properties, its functions and various soil indicators along with SDGs (Sustainable development goals) which are directly and indirectly related to soil.

nutrients can be converted into plant-available forms, held in the soil, or even transferred to water or air (Briones, 2014). The admission of oxygen into interparticle pores maintains an oxygenated environment, allowing nitrification and other aerobic microbial nutrient changes to occur. When O_2 ingress to intraparticle pores is restricted, anoxic conditions are created in macroaggregates, allowing denitrification and other anaerobic nutritional changes to occur (Nannipieri et al., 2017; Rabot et al., 2018).

Water Relations—Water and other solutes, such as phosphorus, pesticides, nitrogen, as well as other nutrients and chemicals mixed with water, can be controlled through flow, soil drainage, and retention (Kirkham, 2014). When soil is functioning properly, it separates water for aquifer replenishment, which is used by soil microbiota and plants. The formation of macroaggregates creates massive interparticle holes, which improve fluid permeability and allow water to drain easily. Increased permeability improves absorption and recharging for groundwater and vadose zone water retention (Vogel et al., 2018). Macroaggregate tiny intraparticle holes boost water-holding potential and plant-available storage of water (Passot et al., 2019).

Biodiversity and Habitat—Soil is a unique physical, chemical, and biological habitat for animals, plants, and soil microorganisms (Rillig et al., 2019). Through macroaggregate formation, anoxic exterior to oxic interior micropores are formed. Fine root and fungal hyphae penetration and respiration are enabled by external macroaggregate pores (Upton et al., 2019). Better bulk soil cohesion, water stability, and adhesion to roots and fungal hyphae are associated with microaggregate and macroaggregate forms. In general, the greater the particle mass and resistance to physical mobilization by rainfall, wind, and water flow, the more aggregate soil adheres to roots and hyphae (Eisenhauer et al., 2018; Nannipieri et al., 2017). Vegetation cover retains droplets and reduces the mechanical forces exerted by water droplets, which loosen and mobilize soil particles for erosion and transport in overland runoff (Banwart et al., 2017).

Filtering and Buffering—Soil acts as a filtration system for air and water quality, as well as other resources. In a variety of ways, excess nutrients and toxic chemicals can be destroyed or rendered inaccessible to animals and plants. Aggregate layers with mineral and organic surfaces provide adsorptive surfaces for pollutant immobilization in flowing fluids (McBratney et al., 2019). The heterogeneity of the environment and selection pressures help to sustain a distinct type of functional consortium, as well as those with the metabolic capacity to digest living contaminants (Drobnik et al., 2018).

Physical Stability and Support—The porous structure of soil that allows air and water to pass through while also resisting erosive forces and serving as a medium for plant roots. Soils also serve as a foundation for human structures and a barrier to ancient treasures (Rabot et al., 2018). Plant biomass decomposes in macroaggregates, which improves soil fertility and plant development. O_2 can enter through interparticle pore gaps, facilitating root respiration and aerobic bacteria proliferation (Herrick and Wander, 2018). Organo-mineral complexes shield adsorbate organic materials from biological and chemical degradation.

18.5 Relationship of plant functional traits with soil function

The keyword phrase "plant*" AND "trait*" AND "service*" was chosen to encompass a wide range of research that focused on the relationships between plant functional traits and ecosystem services (Hanisch et al., 2020). The effects of functional plant traits on ecosystem functioning are process-dependent, and include soil characteristics that influence carbon and nutrient stock in the soil. Although functional diversity influences a wide range of ecological processes and provides a wide range of ecosystem services, its effects on the intensity of ecological processes (i.e., synergetic effects) are debatable (Perez-Harguindeguy et al., 2016). The "mass ratio hypothesis," proposed by Grime (1979), states that the impact of a functional plant characteristic on ecosystem attributes is heavily influenced by the relative contribution of those plants to overall biomass or abundance. As a result, aggregated characteristics of dominant species would have an impact on ecological processes such as carbon and nutrient cycling (Cornelissen et al., 2003). The essential traits demonstrated several stages of variation, including stature, leaf economics ranging from productivity to carbon and nitrogen retention, and linked or dependent changes in root and stem traits (Kraft et al., 2015). Only a few studies have found that functional divergence has a positive effect on ecological processes (Hanisch et al., 2020).

Functional diversity was found to have a significant positive effect on carbon cycling, soil carbon sequestration, and net aboveground primary productivity in a herbaceous habitat (Kunstler et al., 2016). The presence of highly complementary functional groups explains increased nitrogen and carbon inputs via thick root biomass, as well as higher accumulation of long roots into deeper soil, indicating a positive effect of functional diversity (Roscher et al., 2012). These studies supported the notion that different trophic levels can be distinguished by species abundance, such as insect abundance, functional group composition, such as the relative abundance of soil fungus and bacteria, and continuous functional features, such as the relative abundance of soil biodiversity, that is, its potential activities, and so on. All of this leads to the design of soil functions in relation to its interaction (both positive and negative) with biotic and abiotic factors, thereby shaping functional traits (Perez-Harguindeguy et al., 2016; Lavorel et al., 2011). Plant performance is linked to soil microbial populations through abiotic factors such as pH and nutrient availability (König et al., 2010). Community succession may be facilitated by early plant-induced soil and nutrient changes that encourage the growth and establishment of subsequent plant species (Enright et al., 2014) Plant functional characteristics provide feedback by influencing community dynamics through their effects on soil microbial populations, which are also influenced by soil condition during community succession (Dubuis et al., 2013; Xiao Juan and Ke Ping, 2015). Significant research field investigations have shown links between plant, soil abiotic property, and microbial community interactions that suggest plant traits may be the main determinants (Peco et al., 2017). Recent research has focused on the role of plant characteristics in driving ecological processes. There is a direct relationship between soil processes such as decomposition and nutrient cycling and the proportion of a plant community made up of species with "slow" vs. "rapid" functional qualities (Zhang et al., 2021; de Bello et al., 2010). Understanding the increased carbon allocation and underground biomass storage via the variety of plant functional traits has been addressed (Dubuis et al., 2013).

The characteristics of roots, shoots, and leaves have been linked to processes such as belowground carbon inputs, decomposition rates, soil carbon storage, and soil physical features, among others, all of which have implications for soil function (Lavorel et al., 2007). There are various soil features that play a vital role in maintaining relationship with plant functional traits few of them are:

1. **Stability of soil Aggregation:** It represents the reaction of soil aggregates to precipitation and plant growth and is a common method for determining soil stability. When wet, unstable aggregates tend to form a layer of slacked soil, which reduces infiltration, increases surface runoff, and inhibits plant growth (De Deyn et al., 2008; Ali et al., 2017).

2. **Soil penetration resistance** is a composite property that is influenced by fundamental soil properties such as soil-metal friction, soil cohesion, and soil compressibility. Penetration resistance is related to several other important factors, including root elongation rate (Burylo et al., 2012; Erktan and Rey, 2013; Kervroëdan et al., 2018).

3. **Soil shear strength** is a soil property that reflects the soil's cohesiveness and resistance to shear forces such as gravity, flowing fluids, and mechanical stresses. It demonstrates how the soil root matrix supports the terrain, making it far more robust than the soil or roots alone, and how this reinforced earth can withstand environmental and human effects (Demenois et al., 2018; Pohl, 2010).

4. A variety of factors affect soil stability, both directly and indirectly. For example, abiotic soil qualities such as clay content and soil texture, biotic vegetation features such as vegetation cover, species richness, and variety of floral species, and functional features such as soil fauna, plant diversity, and plant root system are all critical for maintaining soil stability (Pohl et al., 2009; Garcia et al., 2019).

Functional diversity is a key indicator of biodiversity that acknowledges the attributes of the species in a group (Ali et al., 2017). These are the characteristics that set species apart in terms of their ecological roles or interactions with other species and the environment. The size of the plant body, its sensitivity and tolerance to environmental variation, motility, morphology, ability to fix nitrogen, and nutrient requirements are just a few examples of these phytoplankton characteristics (Rawat et al., 2020). In terrestrial plant communities, more complex characteristics such as growth rates, nutritional requirements, and water uptake have been incorporated (Henn et al., 2019).

The field of determining whether roots have an economic spectrum similar to leaves, and whether belowground characteristics can be used in conjunction with aboveground features to more accurately forecast soil parameters and ecosystem functions, has seen a massive surge (Zhang et al., 2018; Adler et al., 2014). Roumet et al. (2016) demonstrated that root respiration rate is directly proportional to SRL and nitrogen concentration in the root, whereas lignin to nitrogen concentration ratio and RDMC (Root Dry Matter Content) are inversely proportional to the mass remaining after decomposition. In a Mediterranean rangeland study, it was discovered that leaf and root features responded identically to nitrogen scarcity, indicating coordination between above- and below-ground resource acquisition processes and validating the presence of a root economic spectrum (Fort et al., 2017). Meanwhile, an investigation by Kramer-Walter et al. (2016) on 66 tree species in New Zealand indicated a considerable link between growth rate and the plants' aboveground traits, but little or no correlation with the plants' below-ground attributes. A study comparing above- and below-ground characteristics of 34 tree species found little evidence for a root economic spectrum; rather, root attributes were more strongly influenced by phylogenetic relatedness. These reported correlations between above- and below-ground features appear to change with environment, emphasizing the gaps in our current understanding of shoot—root trait interaction (Prieto et al., 2016; Defrenet et al., 2016).

Different traits can be used to define various elements of ecological functioning as demonstrated in Fig. 18.3; some are directly linked to certain functions, while others just serve as indirect indications. As a result, approaches in Community ecology that combine theories of community assembly, ecological strategies, and functional diversity have the ability to harmonize functional and demographic differences among coinhibiting species in terms of the coexistence process (Adler et al., 2014; Xiao Juan and Ke Ping, 2015). Feeding strategies, mobility, body size, and longevity have all been linked to changes in species distribution in populations that have been subjected to stressors such as sewage pollution and fishing (Meng et al., 2007). There are numerous characteristics that can be used to describe ecological functioning, but not all of them are equally important. Various characteristics can be used to define different aspects of ecological functioning, and some of them are intertwined. Despite this progress, it is still unclear which plant functional traits (aboveground and/or belowground) demonstrate soil quality and ecosystem activity accuracy (Cao et al., 2013; Geng et al., 2017; Bu et al., 2017). It's also unclear whether these plant trait-soil linkages apply to mixed plant communities in the physical world. Indeed, it is unclear whether plant traits-soil interactions will ever contribute to understanding the mechanisms that drive processes such as net ecosystem exchange and ecosystem respiration, which determine whether an ecosystem acquires or loses carbon. Many ecological processes and habitats must be studied in order to better understand the contentious implications of functional diversity.

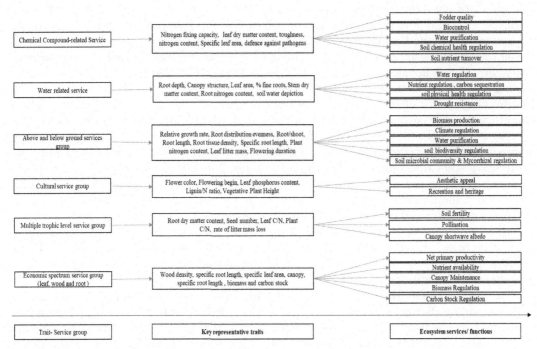

FIGURE 18.3 Demonstrates linkage between various Trait-service groups represented by key traits with ecosystem services.

18.6 Carbon dynamics

Plant traits that control soil carbon sequestration are divided into two categories: those that control soil carbon input via below-ground carbon allocation and primary productivity, and those that control soil carbon loss via VOC (Volatilization of organic compounds), respiration, leaching, and fire, among others (De Deyn et al., 2008; Conti and Díaz, 2013). Because of interactions between plant traits and soil diversity, all of these systems have an effect on the balance of soil carbon output and intake (Faucon et al., 2017). Plant growth rate is related to the amount and form of carbon returned to the soil, as well as their fate in the soil (Lavorel et al., 2007).

Plants with faster intrinsic growth rates have higher photosynthetic efficiency than slower intrinsic growth rates, but at the expense of lower dry matter accumulation, lower carbon quantities in above- and below-ground plant organs, and a shorter life span (Fornara and Tilman, 2008). Soil carbon is primarily derived from decaying plant tissues both below and above ground, but root exudates (5%–33% of daily photo assimilation) are also a significant carbon input source in soil, particularly in dynamically developing flora (Bardgett et al., 2005). Slow-growing and fast-growing flora contribute to the soil carbon pool in two ways: by providing low-quality plant materials and by adding large amounts of carbon to the soil pool (Werner and Máguas, 2010). A species must have enough light and nutrients, as well as favorable environmental conditions, to become dominant through rapid growth. All of these factors have an impact on the various plant strategy attributes for adaptation and survival (Kumar et al., 2021). As a result, low-quality litter will be the primary source of soil carbon input in biomes with a short growing season and limited nutrient availability, while primary productivity will be the source of soil carbon sequestration in biomes with a longer growing season and greater nutrient availability (Soudzilovskaia et al., 2013; Anderegg et al., 2019). Plants and other soil heterotrophs, as a result, contribute to both the input and output carbon fixation in soils. The amount and quality of plant and heterotrophic carbon pools, which affect carbon utilization efficiency and carbon storage time in the soil, determine net soil carbon sequestration (Kumar et al., 2021). Slower growth rates are associated with more resistant carbon pools in general, whereas faster growth rates are associated with massive, rapid fluxes and relatively small soil carbon pools (Becknell and Powers, 2014). Rhizo deposits and ecosystem engineers typically aid in the occlusion of carbon in soil minerals, allowing the organism to stay for a longer period of time. The ability of a plant to respond to mutualistic symbionts is critical for soil carbon absorption because symbiotic relationships can help plants absorb limited resources more effectively, increasing plant output. Across all plant functional categories, growth forms, and biomes, there appears to be a significant correlation between plant respiration rates and nitrogen content of plant tissues (Liu et al., 2021). Under current conditions, carbon loss in forest soils may have a negative impact on soil biodiversity, productivity, and climatic feedbacks (Soudzilovskaia et al., 2013). In order to mitigate climate change, it is critical to gain a better understanding of the underlying processes and mechanisms of soil carbon sequestration (De Deyn et al., 2008). As a result, understanding how plant functional traits affect carbon sequestration and soil stability has emerged as a critical environmental concern. In terms of land area, an ecologically valuable temperate forest sequesters 50 times more CO_2 than a monoculture farm (Gregorich et al., 2001). As a result, the carbon cycle, in particular, is dependent on forests to function properly. Natural forests absorb more CO_2 in temperate climate zones, indicating that the rate of CO_2 assimilation can be used to estimate carbon removal efficiency from the atmosphere and fix it as carbon stock (Migliavacca et al., 2017). Organic soil physiological parameters such as photosynthetic rate, transpiration rate, total chlorophyll, and carotenoid predict carbon and sequestration. It's because *Pinus roxburghii* forests absorb more carbon and fix more CO_2 molecules in the atmosphere (Kumar et al., 2021). CO_2 is stored in wood, which is then consumed by the plant. Carbon is released into the environment when a plant or tree dies. As a result, this is not the only way for carbon to be recycled into the atmosphere (Ghimire, 2020).

In a variety of scenarios, including differences in leaf area, carbon partitioning patterns, and variation in wood and root respiration rates, net photosynthesis has been found to be poorly related to growth rate (Singh et al., 2014). Plants are the primary source of organic carbon in the soil, either through the breakdown of aerial plant parts or through root casualty, exudates, and root respiration from underground plant parts like roots (Ahrens et al., 2020). The root system loses approximately 40% of the photosynthates produced by plant components to the rhizosphere. Many factors, such as plant age, various biotic and abiotic stimuli, and so on, influence the rate of loss (Malézieux et al., 2007). Global CO_2 increases and temperature rises may have a variety of effects on soil carbon imports via photosynthetic controls, as well as carbon losses via respiration and breakdown (Mayfield et al., 2013). Previous research has shown that adverse weather conditions reduce photosynthetic rates due to stomatal and nonstomatal factors, resulting in lower carbon sequestration (Jin et al., 2014).

Strong radiation and cold temperatures reduce photosynthetic rate by decreasing stomatal conductance, decreasing RuBP carboxylase activity, harming photosynthetic rate, and increasing respiration rate. The highest photosynthetic rate was discovered in low-temperature native plants with abundant soil moisture ecosystems. Plants' water efficiency improves due to increased photosynthesis and decreased water loss. As a result, more carbon dioxide in the atmosphere appears to be beneficial to plants, as they grow faster, use less water, and sequester more carbon (Rawat et al., 2015). Chlorophyll, carotenoids, and other light-sensitive pigments absorb solar energy in photosynthetic cells. In the presence of carbon dioxide, these cells can convert solar energy into energy-dense organic molecules like glucose. All these cells regulate the global carbon cycle and thus produce a significant portion of the oxygen in the earth's atmosphere. These photosynthetic pigments regulate carbon stock and sequestration as a result of their effect on photosynthesis rate (Gauslaa et al., 2017; He et al., 2019). Changes in the soil's abiotic conditions produced by plant characteristics may have a significant and long-term influence on SOC sequestration (Ahrens et al., 2020). Thus, plant canopy characteristics that influence surface albedo, evapotranspiration, wind speed and direction, and ice cover may compensate for the effects of resource volume and quality on carbon inflow and outflow (Cheng et al., 2021).

Plant physical features, such as deep and branching roots, can also improve soil carbon stabilization by occlusion and erosion reduction (Kervroëdan et al., 2018). Plants with specific ecophysiological features are selected by the climate, which results in communities with comparable response traits (Garcia et al., 2019). Because many plant traits are interdependent, climate influences groups of plant qualities that function as carbon-cycling impact traits (Soudzilovskaia et al., 2013). Despite vast differences in plant species richness, similar SOC concentrations in tundra, boreal, and tropical forests show that biomes with greater species diversity do not always have higher SOC pools than biomes with fewer species (Heilmeier, 2019).

The composition of plant traits, on the other hand, has a significant impact on soil carbon sequestration within biomes. Plant characteristics, such as flammability and combustibility, have a significant impact on fire regimes, resulting in sudden and massive carbon losses (Wellstein et al., 2013). While the relative abundances and productivities of the major plant functional categories, as well as their characteristics, are most likely the most important factors influencing soil carbon dynamics, interactions between plant species, as well as their avoidance, may also be important (Faucon et al., 2017). To increase soil carbon input relative to soil carbon loss, coexistence, or the development of plant features that improve community-level carbon storage, could be used (Bu et al., 2017). Plants frequently have characteristics that increase SOC sequestration while others reduce it. Shrubs, for example, raise soil temperature or moisture regimes for decomposition during the winter and spring, but the low decomposability of their litter has a negative feedback effect on carbon cycling (Pausas et al., 2004). Similarly, while increased root and rhizome biomass may aid in SOC sequestration, poor litter quality, rapid root exudation, and airways in the roots and stem can all contribute to SOC loss (Fornara and Tilman, 2008).

The rate at which primary producers use carbon and the rate at which organisms release carbon are both critical for terrestrial ecosystems to successfully regulate climate. The net balance of these intake, release, and storage processes at any given time determines the potential for carbon sequestration. Because plant species differ in their ability to absorb, store, and release carbon, the functional composition of plant communities must play a significant role in determining the carbon sinks in ecosystems subject to a specific regional environmental control (De Deyn et al., 2008; Migliavacca et al., 2017). Plant characteristics are intimately linked to the carbon cycling characteristics of soil microorganisms via host adaptation (symbionts) and, to a lesser extent, through plant litter quality and abiotic soil condition change (temperature, moisture, aeration and pH). Optimizing the balance of soil carbon intake and release improves soil carbon storage by improving the interaction of mineral and organic soil components across the soil profile, including increasing plant decomposer and mineralized nutrient use efficiency (Singh et al., 2014; Rawat et al., 2015).

18.7 Nutrient dynamics

Plants and different trophic levels collaborate to provide a wide range of environmental services (de Bello et al., 2010; Cardinale et al., 2012). Carbon and nitrogen cycling, as well as a wide range of soil activities, necessitate complex interactions among plants, herbivores, predators, and soil organisms (Nannipieri et al., 2003). Together with abiotic soil variables such as pH and nutrient availability, soil microbial populations are critical regulators of plant performance (Nannipieri et al., 2017). The bacterial community composition was substantially related with all soil abiotic variables (pH, ammonium nitrate and nitrite, total Nitrogen, total Carbon etc.) (Briones, 2014). Furthermore, soil pH, ammonium nitrate, total Carbon, and total Nitrogen were all found to be significant predictors of fungal community composition (Rabot et al., 2018).

Total soil quantity The most consistent indicator of fungal community composition is soil pH, while nitrogen concentration is the most reliable indicator of bacterial community composition. Plant functional traits such as height, specific leaf area, stem and leaf nitrogen concentrations, root carbon nitrogen concentrations, and seed mass are all strongly correlated with bacterial community composition, whereas height, leaf N:P, seed mass, root carbon concentration, and specific leaf area are most significantly associated with fungal community composition (Passot et al., 2019; Kirkham, 2014). Whole-plant anatomy with carbon and nutrient allocation amongst plant organs have an impact on nutrient cycle and soil carbon sequestration (Kumar et al., 2021).

Plant community succession occurs when the influence of a pioneer plant on the allocation of soil nutrients favors the development and survival of subsequent plants. As a result, soil conditions may have an impact on the ambient sorting of functional features of plants during community succession (Perez-Harguindeguy et al., 2016). Plant characteristics may influence community dynamics by influencing the soil microbial community. Because scarcity of resources is a significant determinant in the formation, evolution, and functionality of plant communities, physiological variables that directly govern nutrient intake may be preferable (McGill et al., 2006). Maire et al. (2009) discovered evidence for a trade-off between nitrate (NO_3), ammonia (NH^{4+}), and other nitrogen compounds in coexisting grass species where root structure was not a concern (Bu et al., 2017). Root phosphatase activity, a critical component of phosphorus acquisition, was found to be positively associated with root specific surface area and negatively associated with root diameter in tropical tree species (Ushio et al., 2015). Changes in root physical attributes may represent a species' physical response to its habitat in terms of nutrient absorption. Many species have drought tolerance, shallow soil, and seasonal waterlogging, all of which increase carbon input. These characteristics, along with others such as thermotolerance and low mineral and nutritional value, are extremely important in areas where permafrost persists. Coniferous and peatland vegetation, for example, is known for its ability to withstand cold temperatures, low mineral fertilizer requirements, and a wide range of extreme weather conditions (Shemesh et al., 2010). There is a considerable impact on carbon sequestration by soil ecosystem engineers, but probably much more by physical structure alterations (Conti and Díaz, 2013). Plant residue chemical properties associated with functional plant diversity will indeed influence nutrient availability; for example, cereal residues contain relatively low levels of N and P and may thus take longer to mineralize than residues containing relatively high levels of P and N, which promote soil nutrient recycling (Kraft et al., 2015). The ability of earthworms and termites to create soil ecosystems is aided by the quality of plant litter, most notably an adequate C: N ratio with ideal size. Furthermore, N and P concentrations improve the rate of aboveground litter decomposition (Cao et al., 2013). In comparison to root dry matter content and lignin: nitrogen ratio, root breakdown rate is more favorably impacted by root nitrogen content, root respiration, and specific root length. Nitrogen and phosphorus levels, as well as tissue density, all have an effect on the pace of stem decay (Roumet et al., 2016; Grigulis et al., 2013) quantified the relative contributions of the plant community and microbial functional variables to nitrogen cycling and related ecosystem processes.

They discovered a gradient extending from standing green biomass and litter, which are primarily determined by plant characteristics, to potential nitrogen mineralization and soil inorganic nitrogen leaching, which are primarily determined by microbial characteristics. Increased fodder production has been linked to more predatory species, longer grass swards, bacteria-dominated soil microbial communities, and rapid microbial activity (Cantarel et al., 2015). Conservative and soil microbial communities characterized by fungi and slow-acting bacteria, on the other hand, are frequently associated with low production, but they promote carbon and nitrogen storage in the soil (Briones, 2014; Nannipieri et al., 2017). Because multiple ecosystem services rely on the same or similar traits, trait-based analyses of ecosystem service provision can be combined with prior knowledge of plant functional syndromes and trade-offs to improve our understanding of versatility (the simultaneous provision of multiple services) and ecosystem service trade-offs.

18.8 Miscellaneous

Pollination is dependent on interactions between pollinators, plants, and the species with which they engage (e.g., Predators; Kremen et al., 2007). As a result, the next step in biodiversity-ecosystem functionality research has been proposed: combining a multi-trophic perspective with an attribute-based approach. Two important plant characteristics, leaf economics and size, have been linked to a number of environmental aspects related to the biogeochemical cycle and resource flow through food webs (Roscher et al., 2012). Above-ground gross primary output, litter degradation in the field and under controlled conditions, digestion, nitrification, soil moisture, and water uptake have all been shown to be influenced by plant functional characteristics at the community level (Xiao Juan and Ke Ping, 2015). The size axis represents a variation of vegetation organization diversity. This gradient would result in a trade-off between higher overall resting biomass in tall and dense communities (Falster et al., 2011) and short vegetation, which is associated with water conservation owing to low transpiration fluxes and a reduced risk of fire initiation due to fuel constraints (Lavorel and Garnier, 2002). There are numerous heat and humidity effects on SOC mineralization and carbon leaching; as a result, plant canopy characteristics such as surface albedo, evapotranspiration, wind speed, and snow cover may overcome the effects of resource quantity and quality on carbon in- and output (Falster et al., 2011).

According to studies, the various functional characteristics at the community level in the tropical rainforest responded to physical environmental differences, affecting above ground carbon stores directly at each spatial scale (Kunstler et al., 2016). Abiotic site conditions had no direct effect on above-ground carbon stock, but they did have an indirect effect via changes in plant functional features. Plant functional characteristics are more important than environmental influences in causing variations in above-ground carbon stock (McIntyre et al., 1999; Geng et al., 2017). All of these findings imply that plant functional characteristics act as mediators in the regulation of ecosystem effects caused by abiotic site conditions.

18.9 Conclusion

Indeed, recent research has highlighted the significance of functional features in trophic level interactions and their implications for ecosystem function. Functional characteristics allow us to characterize how the functional composition of communities responds to environmental gradients and influences ecological processes and service delivery.

Furthermore, in a rapidly changing world, the differential effects of environmental change on ecosystem function must be addressed. For several fundamental plant functional traits, such as height, the leaf economics spectrum, and associated or independent variations in root or stem properties, there are numerous areas of diversity. Other trophic levels can be classified based on their abundance (e.g., insect abundance), functional group dynamics (e.g., soil fungus and bacteria abundance), or continuous functional qualities (e.g., relative abundance of soil fungi and bacteria) (e.g., bacterial activities). The scale-dependent relationship between community functions and habitat heterogeneity reveals that species' functional qualities associated with resource exploitation and fitness adaptation can vary significantly. Service-oriented ecosystem management remains a critical concern in an era of major disruption, such as ecological restoration, but trait-based understanding opens up new avenues for more comprehensive, integrated solutions.

References

Adler, P.B., Salguero-Gómez, R., Compagnoni, A., Hsu, J.S., Ray-Mukherjee, J., Mbeau-Ache, C., et al., 2014. Functional traits explain variation in plant life history strategies. Proceedings of the National Academy of Sciences 111 (2), 740–745.

Ahrens, C.W., Andrew, M.E., Mazanec, R.A., Ruthrof, K.X., Challis, A., Hardy, G., et al., 2020. Plant functional traits differ in adaptability and are predicted to be differentially affected by climate change. Ecology and Evolution 10 (1), 232–248.

Ali, H.E., Reineking, B., Münkemüller, T., 2017. Effects of plant functional traits on soil stability: intraspecific variability matters. Plant and Soil 411 (1–2), 359–375.

Allen, D.E., Singh, B.P., Dalal, R.C., 2011. Soil health indicators under climate change: a review of current knowledge. Soil Health and Climate Change 25–45.

Anderegg, W.R., Trugman, A.T., Bowling, D.R., Salvucci, G., Tuttle, S.E., 2019. Plant functional traits and climate influence drought intensification and land–atmosphere feedbacks. Proceedings of the National Academy of Sciences 116 (28), 14071–14076.

Banwart, S.A., Bernasconi, S.M., Blum, W.E., de Souza, D.M., Chabaux, F., Duffy, C., et al., 2017. Soil functions in Earth's critical zone: key results and conclusions. Advances in Agronomy 142, 1–27.

Bardgett, R.D., Bowman, W.D., Kaufmann, R., Schmidt, S.K., 2005. A temporal approach to linking aboveground and belowground ecology. Trends in Ecology & Evolution 20 (11), 634–641.

Barni, E., Freppaz, M., Siniscalco, C., 2007. Interactions between vegetation, roots, and soil stability in restored high-altitude ski runs in the Alps. Arctic, Antarctic, and Alpine Research 39 (1), 25–33.

Baveye, P.C., Baveye, J., Gowdy, J., 2016. Soil "ecosystem" services and natural capital: critical appraisal of research on uncertain ground. Frontiers in Environmental Science 4, 41.

Becknell, J.M., Powers, J.S., 2014. Stand age and soils as drivers of plant functional traits and aboveground biomass in secondary tropical dry forest. Canadian Journal of Forest Research 44 (6), 604–613.

Benedetti, A., Dell'Abate, M.T., Napoli, R., 2013. Soil functions and ecological services. The Soils of Italy. Springer, Dordrecht, pp. 179–203.

Bjorkman, A.D., Myers-Smith, I.H., Elmendorf, S.C., Normand, S., Rüger, N., Beck, P.S., et al., 2018. Plant functional trait change across a warming tundra biome. Nature 562 (7725), 57–62.

Blum, W.E., 2005. Functions of soil for society and the environment. Reviews in Environmental Science and Bio/Technology 4 (3), 75–79.

Blum, W.E., 2014. Soil and food security under global change. Otosclerosis: A Misnomer 15.

Blyth, S., Groombridge, B., Lysenko, I., Miles, L., Newton, A., 2002. UNEP–WCMC (United Nations Environment Programme–World Conservation Monitoring Centre) Report, Mountain Watch: Environmental Change and Sustainable Development in Mountains. UNEP World Conservation Monitoring Centre, Cambridge, UK, p. 19.

Bordin, K.M., Müller, S.C., 2019. Drivers of subtropical forest dynamics: the role of functional traits, forest structure and soil variables. Journal of Vegetation Science 30 (6), 1164–1174.

Bouma, J., 2014. Soil science contributions towards sustainable development goals and their implementation: linking soil functions with ecosystem services. Journal of Plant Nutrition and Soil Science 177 (2), 111–120.

Briones, M.J.I., 2014. Soil fauna and soil functions: a jigsaw puzzle. Frontiers in Environmental Science 2, 7.

Bu, W., Schmid, B., Liu, X., Li, Y., Härdtle, W., von Oheimb, G., et al., 2017. Interspecific and intraspecific variation in specific root length drives aboveground biodiversity effects in young experimental forest stands. Journal of Plant Ecology 10 (1), 158–169.

Burylo, M., Rey, F., Bochet, E., Dutoit, T., 2012. Plant functional traits and species ability for sediment retention during concentrated flow erosion. Plant and Soil 353 (1), 135–144.

Cambardella, C.A., Elliott, E.T., 1992. Particulate soil organic-matter changes across a grassland cultivation sequence. Soil Science Society of America Journal 56 (3), 777–783.

Campetella, G., Chelli, S., Simonetti, E., Damiani, C., Bartha, S., Wellstein, C., et al., 2020. Plant functional traits are correlated with species persistence in the herb layer of old-growth beech forests. Scientific Reports 10 (1), 1–13.

Cantarel, A.A., Pommier, T., Desclos-Theveniau, M., Diquélou, S., Dumont, M., Grassein, F., et al., 2015. Using plant traits to explain plant–microbe relationships involved in nitrogen acquisition. Ecology 96 (3), 788–799.

Cao, K., Rao, M., Yu, J., Liu, X., Mi, X., Chen, J., 2013. The phylogenetic signal of functional traits and their effects on community structure in an evergreen broad-leaved forest. Biodiversity Science 21 (5), 564.

Cardinale, B.J., Duffy, J.E., Gonzalez, A., Hooper, D.U., Perrings, C., Venail, P., et al., 2012. Biodiversity loss and its impact on humanity. Nature 486 (7401), 59–67.

Caruso, C.M., Mason, C.M., Medeiros, J.S., 2020. The evolution of functional traits in plants: is the giant still sleeping? International Journal of Plant Sciences 181 (1), 1–8.

Cheng, X., Ping, T., Li, Z., Wang, T., Han, H., Epstein, H.E., 2021. Effects of environmental factors on plant functional traits across different plant life forms in a temperate forest ecosystem. New Forests 1–18.

Conti, G., Díaz, S., 2013. Plant functional diversity and carbon storage–an empirical test in semi-arid forest ecosystems. Journal of Ecology 101 (1), 18–28.

Cornelissen, J.H.C., Lavorel, S., Garnier, E., Díaz, S., Buchmann, N., Gurvich, D.E., et al., 2003. A handbook of protocols for standardised and easy measurement of plant functional traits worldwide. Australian Journal of Botany 51 (4), 335–380.

de Bello, F., Lavorel, S., Díaz, S., Harrington, R., Cornelissen, J.H., Bardgett, R.D., et al., 2010. Towards an assessment of multiple ecosystem processes and services via functional traits. Biodiversity and Conservation 19 (10), 2873–2893.

De Deyn, G.B., Cornelissen, J.H., Bardgett, R.D., 2008. Plant functional traits and soil carbon sequestration in contrasting biomes. Ecology Letters 11 (5), 516–531.

De Long, J.R., Jackson, B.G., Wilkinson, A., Pritchard, W.J., Oakley, S., Mason, K.E., et al., 2019. Relationships between plant traits, soil properties and carbon fluxes differ between monocultures and mixed communities in temperate grassland. Journal of Ecology 107 (4), 1704–1719.

Defrenet, E., Roupsard, O., Van den Meersche, K., Charbonnier, F., Pastor Pérez-Molina, J., Khac, E., et al., 2016. Root biomass, turnover and net primary productivity of a coffee agroforestry system in Costa Rica: effects of soil depth, shade trees, distance to row and coffee age. Annals of Botany 118 (4), 833–851.

Demenois, J., Carriconde, F., Bonaventure, P., Maeght, J.L., Stokes, A., Rey, F., 2018. Impact of plant root functional traits and associated mycorrhizas on the aggregate stability of a tropical Ferralsol. Geoderma 312, 6–16.

Díaz, S., Cabido, M., 1997. Plant functional types and ecosystem function in relation to global change. Journal of Vegetation Science 8 (4), 463–474.

Diaz, S., Cabido, M., Casanoves, F., 1998. Plant functional traits and environmental filters at a regional scale. Journal of vegetation science 9 (1), 113–122.

Díaz, S., Purvis, A., Cornelissen, J.H., Mace, G.M., Donoghue, M.J., Ewers, R.M., et al., 2013. Functional traits, the phylogeny of function, and ecosystem service vulnerability. Ecology and Evolution 3 (9), 2958–2975.

Dominati, E., Patterson, M., Mackay, A., 2010. A framework for classifying and quantifying the natural capital and ecosystem services of soils. Ecological Economics 69 (9), 1858–1868.

Drobnik, T., Greiner, L., Keller, A., Grêt-Regamey, A., 2018. Soil quality indicators—from soil functions to ecosystem services. Ecological Indicators 94, 151–169.

Dubuis, A., Rossier, L., Pottier, J., Pellissier, L., Vittoz, P., Guisan, A., 2013. Predicting current and future spatial community patterns of plant functional traits. Ecography 36 (11), 1158–1168.

Eisenhauer, N., Hines, J., Isbell, F., van der Plas, F., Hobbie, S.E., Kazanski, C.E., et al., 2018. Plant diversity maintains multiple soil functions in future environments. eLife 7, e41228.

Enright, N.J., Fontaine, J.B., Lamont, B.B., Miller, B.P., Westcott, V.C., 2014. Resistance and resilience to changing climate and fire regime depend on plant functional traits. Journal of Ecology 102 (6), 1572–1581.

Erktan, A., Rey, F., 2013. Linking sediment trapping efficiency with morphological traits of Salix tiller barriers on marly gully floors under ecological rehabilitation. Ecological Engineering 51, 212–220.

Falster, D.S., Brännström, Å., Dieckmann, U., Westoby, M., 2011. Influence of four major plant traits on average height, leaf-area cover, net primary productivity, and biomass density in single-species forests: a theoretical investigation. Journal of Ecology 99 (1), 148–164.

Faucon, M.P., Houben, D., Lambers, H., 2017. Plant functional traits: soil and ecosystem services. Trends in Plant Science 22 (5), 385–394.

Fornara, D.A., Tilman, D., 2008. Plant functional composition influences rates of soil carbon and nitrogen accumulation. Journal of Ecology 96 (2), 314–322.

Fort, F., Volaire, F., Guilioni, L., Barkaoui, K., Navas, M.L., Roumet, C., 2017. Root traits are related to plant water-use among rangeland Mediterranean species. Functional Ecology 31 (9), 1700–1709.

Funk, J.L., Larson, J.E., Ames, G.M., Butterfield, B.J., Cavender-Bares, J., Firn, J., et al., 2017. Revisiting the Holy Grail: using plant functional traits to understand ecological processes. Biological Reviews 92 (2), 1156–1173.

Garcia, L., Damour, G., Gary, C., Follain, S., Le Bissonnais, Y., Metay, A., 2019. Trait-based approach for agroecology: contribution of service crop root traits to explain soil aggregate stability in vineyards. Plant and Soil 435 (1), 1–14.

Garnier, E., Navas, M.L., 2012. A trait-based approach to comparative functional plant ecology: concepts, methods and applications for agroecology. A review. Agronomy for Sustainable Development 32 (2), 365–399.

Gauslaa, Y., Solhaug, K.A., Longinotti, S., 2017. Functional traits prolonging photosynthetically active periods in epiphytic cephalolichens during desiccation. Environmental and Experimental Botany 141, 83–91.

Geng, Y., Ma, W., Wang, L., Baumann, F., Kühn, P., Scholten, T., et al., 2017. Linking above-and belowground traits to soil and climate variables: an integrated database on China's grassland species.

Geng, Y., Wang, Z., Liang, C., Fang, J., Baumann, F., Kühn, P., et al., 2012. Effect of geographical range size on plant functional traits and the relationships between plant, soil and climate in Chinese grasslands. Global Ecology and Biogeography 21 (4), 416–427.

Ghimire, P., 2020. Carbon stocks in Chir Pine (*Pinus roxburghii*) forests on two different aspects in the Mahabharat Region of Makawanpur, Nepal. Journal of Tropical Forestry and Environment 10 (1).

Gregorich, E.G., Drury, C.F., Baldock, J.A., 2001. Changes in soil carbon under long-term maize in monoculture and legume-based rotation. Canadian Journal of Soil Science 81 (1), 21–31.

Grigulis, K., Lavorel, S., Krainer, U., Legay, N., Baxendale, C., Dumont, M., et al., 2013. Relative contributions of plant traits and soil microbial properties to mountain grassland ecosystem services. Journal of Ecology 101 (1), 47–57.

Grime, J.P., 1979. Primary strategies in plants. Transactions of the Botanical Society of Edinburgh. Taylor & Francis Group, pp. 151–160, Vol. 43, No. 2.

Hanisch, M., Schweiger, O., Cord, A.F., Volk, M., Knapp, S., 2020. Plant functional traits shape multiple ecosystem services, their trade-offs and synergies in grasslands. Journal of Applied Ecology 57 (8), 1535–1550.

He, N., Liu, C., Piao, S., Sack, L., Xu, L., Luo, Y., et al., 2019. Ecosystem traits linking functional traits to macroecology. Trends in Ecology & Evolution 34 (3), 200–210.

Heilmeier, H., 2019. Functional traits explaining plant responses to past and future climate changes. Flora 254, 1–11.

Henn, J.J., Yelenik, S., Damschen, E.I., 2019. Environmental gradients influence differences in leaf functional traits between native and non-native plants. Oecologia 191 (2), 397–409.

Herrick, J.E., Wander, M.M., 2018. Relationships between soil organic carbon and soil quality in cropped and rangeland soils: the importance of distribution, composition, and soil biological activity. Soil Processes and the Carbon Cycle. CRC Press, pp. 405–425.

Jin, D., Cao, X., Ma, K., 2014. Leaf functional traits vary with the adult height of plant species in forest communities. Journal of Plant Ecology 7 (1), 68–76.

Kapos, V., Rhind, J., Edwards, M., Price, M.F., Ravilious, C., 2000. Developing a map of the world's mountain forests. Forests in sustainable mountain development: a state of knowledge report for 2000. Task Force on Forests in Sustainable Mountain Development 4–19.

Kervroëdan, L., Armand, R., Saunier, M., Ouvry, J.F., Faucon, M.P., 2018. Plant functional trait effects on runoff to design herbaceous hedges for soil erosion control. Ecological Engineering 118, 143–151.

Kirkham, M.B., 2014. Principles of Soil and Plant Water Relations. Academic Press.

König, S., Wubet, T., Dormann, C.F., Hempel, S., Renker, C., Buscot, F., 2010. TaqMan real-time PCR assays to assess arbuscular mycorrhizal responses to field manipulation of grassland biodiversity: effects of soil characteristics, plant species richness, and functional traits. Applied and Environmental Microbiology 76 (12), 3765–3775.

Kraft, N.J., Godoy, O., Levine, J.M., 2015. Plant functional traits and the multidimensional nature of species coexistence. Proceedings of the National Academy of Sciences 112 (3), 797–802.

Kramer-Walter, K.R., Bellingham, P.J., Millar, T.R., Smissen, R.D., Richardson, S.J., Laughlin, D.C., 2016. Root traits are multidimensional: specific root length is independent from root tissue density and the plant economic spectrum. Journal of Ecology 104 (5), 1299–1310.

Kremen, C., Williams, N.M., Aizen, M.A., Gemmill-Herren, B., LeBuhn, G., Minckley, R., et al., 2007. Pollination and other ecosystem services produced by mobile organisms: a conceptual framework for the effects of land-use change. Ecology Letters 10 (4), 299–314.

Kumar, A., Kumar, P., Singh, H., Bisht, S., Kumar, N., 2021. Relationship of physiological plant functional traits with soil carbon stock in the temperate forest of Garhwal Himalaya. Current Science 120 (8), 1–6.

Kunstler, G., Falster, D., Coomes, D.A., Hui, F., Kooyman, R.M., Laughlin, D.C., et al., 2016. Plant functional traits have globally consistent effects on competition. Nature 529 (7585), 204–207.

Lal, R., 2004. Soil carbon sequestration to mitigate climate change. Geoderma 123 (1–2), 1–22.

Lavorel, S., Garnier, E., 2002. Predicting changes in community composition and ecosystem functioning from plant traits: revisiting the Holy Grail. Functional Ecology 16 (5), 545–556.

Lavorel, S., Díaz, S., Cornelissen, J.H.C., Garnier, E., Harrison, S.P., McIntyre, S., et al., 2007. Plant functional types: are we getting any closer to the Holy Grail? Terrestrial Ecosystems in a Changing World. Springer, Berlin, Heidelberg, pp. 149–164.

Lavorel, S., Grigulis, K., Lamarque, P., Colace, M.P., Garden, D., Girel, J., et al., 2011. Using plant functional traits to understand the landscape distribution of multiple ecosystem services. Journal of Ecology 99 (1), 135–147.

Lavorel, S., McIntyre, S., Landsberg, J., Forbes, T.D.A., 1997. Plant functional classifications: from general groups to specific groups based on response to disturbance. Trends in Ecology & Evolution 12 (12), 474–478.

Lavorel, S., Storkey, J., Bardgett, R.D., De Bello, F., Berg, M.P., Le Roux, X., et al., 2013. A novel framework for linking functional diversity of plants with other trophic levels for the quantification of ecosystem services. Journal of Vegetation Science 24 (5), 942–948.

Liu, C., Li, Y., Yan, P., He, N., 2021. How to improve the predictions of plant functional traits on ecosystem functioning? Frontiers in Plant Science 12.

Maire, V., Gross, N., da Silveira Pontes, L., Picon-Cochard, C., Soussana, J.F., 2009. Trade-off between root nitrogen acquisition and shoot nitrogen utilization across 13 co-occurring pasture grass species. Functional Ecology 23 (4), 668–679.

Malézieux, E., Lamanda, N., Laurans, M., Tassin, J., Gourlet-Fleury, S., 2007. Plant functional traits and types: their relevance for a better understanding of the functioning and properties of agroforestry systems. CATIE .

Mayfield, M.M., Dwyer, J.M., Chalmandrier, L., Wells, J.A., Bonser, S.P., Catterall, C.P., et al., 2013. Differences in forest plant functional trait distributions across land-use and productivity gradients. American Journal of Botany 100 (7), 1356–1368.

McBratney, A., Field, D., Morgan, C.L., Huang, J., 2019. On soil capability, capacity, and condition. Sustainability 11 (12), 3350.

McGill, B.J., Enquist, B.J., Weiher, E., Westoby, M., 2006. Rebuilding community ecology from functional traits. Trends in Ecology & Evolution 21 (4), 178−185.

McIntyre, S., Díaz, S., Lavorel, S., Cramer, W., 1999. Plant functional types and disturbance dynamics−introduction. Journal of Vegetation Science 10 (5), 603−608.

Meng, T.T., Ni, J., Wang, G.H., 2007. Plant functional traits, environments and ecosystem functioning. Chinese Journal of Plant Ecology 31 (1), 150.

Migliavacca, M., Perez-Priego, O., Rossini, M., El-Madany, T.S., Moreno, G., VanderTol, C., et al., 2017. Plant functional traits and canopy structure control the relationship between photosynthetic CO_2 uptake and far-red sun-induced fluorescence in a Mediterranean grassland under different nutrient availability. New Phytologist 214 (3), 1078−1091.

Minden, V., Kleyer, M., 2011. Testing the effect−response framework: key response and effect traits determining above-ground biomass of salt marshes. Journal of Vegetation Science 22 (3), 387−401.

Misra, S., Dhyani, D., Maikhuri, R.K., 2008. Sequestering carbon through indigenous agriculture practices. Leisa India 10 (4), 21−22.

Nanda, V.P., 2015. The journey from the millennium development goals to the sustainable development goals. Denver Journal of International Law & Policy 44, 389.

Nannipieri, P., Ascher, J., Ceccherini, M., Landi, L., Pietramellara, G., Renella, G., 2003. Microbial diversity and soil functions. European Journal of Soil Science 54 (4), 655−670.

Nannipieri, P., Ascher, J., Ceccherini, M.T., Landi, L., Pietramellara, G., Renella, G., 2017. Microbial diversity and soil functions. European Journal of Soil Science 68 (1), 12−26.

Niu, S., Classen, A.T., Luo, Y., 2018. Functional traits along a transect.

Noble, I.R., Slatyer, R.O., 1980. The use of vital attributes to predict successional changes in plant communities subject to recurrent disturbances. Vegetatio 43 (1), 5−21.

Palma, E., Vesk, P.A., White, M., Baumgartner, J.B., Catford, J.A., 2021. Plant functional traits reflect different dimensions of species invasiveness. Ecology 102 (5), e03317.

Passot, S., Couvreur, V., Meunier, F., Draye, X., Javaux, M., Leitner, D., et al., 2019. Connecting the dots between computational tools to analyse soil−root water relations. Journal of Experimental Botany 70 (9), 2345−2357.

Pausas, J.G., Bradstock, R.A., Keith, D.A., Keeley, J.E., 2004. Plant functional traits in relation to fire in crown-fire ecosystems. Ecology 85 (4), 1085−1100.

Peco, B., Navarro, E., Carmona, C.P., Medina, N.G., Marques, M.J., 2017. Effects of grazing abandonment on soil multifunctionality: the role of plant functional traits. Agriculture, Ecosystems & Environment 249, 215−225.

Perez-Harguindeguy, N., Diaz, S., Garnier, E., Lavorel, S., Poorter, H., Jaureguiberry, P., et al., 2016. Corrigendum to: new handbook for standardised measurement of plant functional traits worldwide. Australian Journal of Botany 64 (8), 715−716.

Pohl, M., 2010. How does plant diversity affect soil aggregate stability and erosion processes in disturbed alpine ecosystems? (No. THESIS). EPFL.

Pillar, V.D., 1999. On the identification of optimal plant functional types. Journal of Vegetation Science 10 (5), 631−640.

Pohl, M., Alig, D., Körner, C., Rixen, C., 2009. Higher plant diversity enhances soil stability in disturbed alpine ecosystems. Plant and Soil 324 (1), 91−102.

Popescu, O.C., 2015. United Nations decade on biodiversity: strategies, targets and action plans. Urbanism Arhitectură Construcții 6 (2), 37−50.

Prieto, I., Stokes, A., Roumet, C., 2016. Root functional parameters predict fine root decomposability at the community level. Journal of Ecology 104 (3), 725−733.

Rabot, E., Wiesmeier, M., Schlüter, S., Vogel, H.J., 2018. Soil structure as an indicator of soil functions: a review. Geoderma 314, 122−137.

Rawat, M., Arunachalam, K., Arunachalam, A., 2015. Plant functional traits and carbon accumulation in forest. Climate Change and Environmental Sustainability 3 (1), 1−12.

Rawat, M., Arunachalam, K., Arunachalam, A., Alatalo, J.M., Kumar, U., Simon, B., et al., 2020. Relative contribution of plant traits and soil properties to the functioning of a temperate forest ecosystem in the Indian Himalayas. Catena 194104671.

Reiss, J., Bridle, J.R., Montoya, J.M., Woodward, G., 2009. Emerging horizons in biodiversity and ecosystem functioning research. Trends in Ecology & Evolution 24 (9), 505−514.

Rillig, M.C., Ryo, M., Lehmann, A., Aguilar-Trigueros, C.A., Buchert, S., Wulf, A., et al., 2019. The role of multiple global change factors in driving soil functions and microbial biodiversity. Science 366 (6467), 886–890.

Roscher, C., Schumacher, J., Gubsch, M., Lipowsky, A., Weigelt, A., Buchmann, N., et al., 2012. Using plant functional traits to explain diversity–productivity relationships. PLoS One 7 (5), e36760.

Roumet, C., Birouste, M., Picon-Cochard, C., Ghestem, M., Osman, N., Vrignon-Brenas, S., et al., 2016. Root structure–function relationships in 74 species: evidence of a root economics spectrum related to carbon economy. New Phytologist 210 (3), 815–826.

Schweiger, A.K., Schütz, M., Risch, A.C., Kneubühler, M., Haller, R., Schaepman, M.E., 2017. How to predict plant functional types using imaging spectroscopy: linking vegetation community traits, plant functional types and spectral response. Methods in Ecology and Evolution 8 (1), 86–95.

Shemesh, H., Arbiv, A., Gersani, M., Ovadia, O., Novoplansky, A., 2010. The effects of nutrient dynamics on root patch choice. PLoS One 5 (5), e10824.

Singh, N., Tamta, K., Tewari, A., Ram, J., 2014. Studies on vegetational analysis and regeneration status of *Pinus roxburghii*, Roxb. and *Quercus leucotrichophora* forests of Nainital Forest Division. Global Journal of Science Frontier Research 14 (3), 41–47.

Smith, P., 2008. Land use change and soil organic carbon dynamics. Nutrient Cycling in Agroecosystems 81 (2), 169–178.

Smith, T.M., Shugart, H.H., Woodward, F.I., Burton, P.J., 1993. Plant functional types. Vegetation Dynamics & Global Change. Springer, Boston, MA, pp. 272–292.

Soudzilovskaia, N.A., Elumeeva, T.G., Onipchenko, V.G., Shidakov, I.I., Salpagarova, F.S., Khubiev, A.B., et al., 2013. Functional traits predict relationship between plant abundance dynamic and long-term climate warming. Proceedings of the National Academy of Sciences 110 (45), 18180–18184.

Stockmann, U., Adams, M.A., Crawford, J.W., Field, D.J., Henakaarchchi, N., Jenkins, M., et al., 2013. The knowns, known unknowns and unknowns of sequestration of soil organic carbon. Agriculture, Ecosystems & Environment 164, 80–99.

Suding, K.N., Lavorel, S., Chapin Iii, F.S., Cornelissen, J.H., Díaz, S., Garnier, E., et al., 2008. Scaling environmental change through the community-level: a trait-based response-and-effect framework for plants. Global Change Biology 14 (5), 1125–1140.

Tappeiner, U., Tasser, E., Leitinger, G., Cernusca, A., Tappeiner, G., 2008. Effects of historical and likely future scenarios of land use on above-and belowground vegetation carbon stocks of an alpine valley. Ecosystems 11 (8), 1383–1400.

Upton, R.N., Bach, E.M., Hofmockel, K.S., 2019. Spatio-temporal microbial community dynamics within soil aggregates. Soil Biology and Biochemistry 132, 58–68.

Ushio, M., Fujiki, Y., Hidaka, A., Kitayama, K., 2015. Linkage of root physiology and morphology as an adaptation to soil phosphorus impoverishment in tropical montane forests. Functional Ecology 29 (9), 1235–1245.

Van Der Valk, A.G., 1981. Succession in wetlands: a gleasonian approach. Ecology 62 (3), 688–696.

Vandvik, V., Halbritter, A.H., Yang, Y., He, H., Zhang, L., Brummer, A.B., et al., 2020. Plant traits and vegetation data from climate warming experiments along an 1100 m elevation gradient in Gongga Mountains, China. Scientific Data 7 (1), 1–15.

Vasenev, V.I., Van Oudenhoven, A.P.E., Romzaykina, O.N., Hajiaghaeva, R.A., 2018. The ecological functions and ecosystem services of urban and technogenic soils: from theory to practice (a review). Eurasian Soil Science 51 (10), 1119–1132.

Veith, C., Shaw, J., 2011. Why invest in sustainable mountain development?

Violle, C., Navas, M.L., Vile, D., Kazakou, E., Fortunel, C., Hummel, I., et al., 2007. Let the concept of trait be functional!. Oikos 116 (5), 882–892.

Vogel, H.J., Bartke, S., Daedlow, K., Helming, K., Kögel-Knabner, I., Lang, B., et al., 2018. A systemic approach for modeling soil functions. Soil 4 (1), 83–92.

Wellstein, C., Chelli, S., Campetella, G., Bartha, S., Galiè, M., Spada, F., et al., 2013. Intraspecific phenotypic variability of plant functional traits in contrasting mountain grasslands habitats. Biodiversity and Conservation 22 (10), 2353–2374.

Werner, C., Máguas, C., 2010. Carbon isotope discrimination as a tracer of functional traits in a Mediterranean macchia plant community. Functional Plant Biology 37 (5), 467–477.

Westoby, M., Wright, I.J., 2006. Land-plant ecology on the basis of functional traits. Trends in Ecology & Evolution 21 (5), 261−268.

Wood, S.A., Karp, D.S., DeClerck, F., Kremen, C., Naeem, S., Palm, C.A., 2015. Functional traits in agriculture: agrobiodiversity and ecosystem services. Trends in Ecology & Evolution 30 (9), 531−539.

Woodward, F.I., Cramer, W., 1996. Plant functional types and climatic change: introduction. Journal of Vegetation Science 7 (3), 306−308.

Xi, N., Adler, P.B., Chen, D., Wu, H., Catford, J.A., van Bodegom, P.M., et al., 2021. Relationships between plant−soil feedbacks and functional traits. Journal of Ecology 109 (9), 3411−3423.

Xiao Juan, L.I.U., Ke Ping, M.A., 2015. Plant functional traits—concepts, applications and future directions. Scientia Sinica Vitae 45 (4), 325−339.

Yadav, A., Verma, P., Raghubanshi, A.S., 2020. An overview of the role of plant functional traits in tropical dry forests. Handbook of Research on the Conservation and Restoration of Tropical Dry Forests 89−113.

Yang, Y., Zhu, Q., Peng, C., Wang, H., Chen, H., 2015. From plant functional types to plant functional traits: a new paradigm in modelling global vegetation dynamics. Progress in Physical Geography 39 (4), 514−535.

Zhang, D., Peng, Y., Li, F., Yang, G., Wang, J., Yu, J., et al., 2021. Changes in above-/below-ground biodiversity and plant functional composition mediate soil respiration response to nitrogen input. Functional Ecology 35 (5), 1171−1182.

Zhang, H., Li, W., Adams, H.D., Wang, A., Wu, J., Jin, C., et al., 2018. Responses of woody plant functional traits to nitrogen addition: a meta-analysis of leaf economics, gas exchange, and hydraulic traits. Frontiers in Plant Science 9, 683.

Zhang, J., Li, P., 2019. Response of plant functional traits to climate change. IOP Conference Series: Earth and Environmental Science. IOP Publishing, p. 032078, Vol. 300, No. 3.

Further reading

Box, E.O., 1996. Plant functional types and climate at the global scale. Journal of Vegetation Science 7 (3), 309−320.

Karlen, D.L., Stott, D.E., 1994. A framework for evaluating physical and chemical indicators of soil quality. Defining Soil Quality for a Sustainable Environment 35, 53−72.

Schulte, R.P., Bampa, F., Bardy, M., Coyle, C., Creamer, R.E., Fealy, R., et al., 2015. Making the most of our land: managing soil functions from local to continental scale. Frontiers in Environmental Science 3, 81.

Smith, C.M., 1999. Plant resistance to insects. Biological and Biotechnological Control of Insects. LLC, Boca Raton, FL, pp. 171−205.

Weigelt, A., Bol, R., Bardgett, R.D., 2005. Preferential uptake of soil nitrogen forms by grassland plant species. Oecologia 142 (4), 627−635.

Socioeconomic and ecological sustainability of agroforestry in mountain regions

Mushtaq Ahmad Dar[1], *Rishikesh Singh*[1], *Mustaqeem Ahmad*[2], *Shalinder Kaur*[1], *Harminder P. Singh*[2] and *Daizy R. Batish*[1]

[1]Department of Botany, Panjab University, Chandigarh, India [2]Department of Environment Studies, Panjab University, Chandigarh, India

19.1 Introduction

Almost one-fourth of the earth's surface is covered by the mountains, which cover highlands and plateaus above 1500 m mean sea level. It directly (12%) and/or indirectly (40%)

covers a significant proportion of global biodiversity and cultural diversity (Chaudhary et al., 2017; Han et al., 2021). These ecosystems are dominated primarily by forests and pastures, and 22% of the world's population relies directly and indirectly on their diverse ecological benefits. As a result, mountain ecosystems have vital resources that benefit a large number of people. Mountain ecosystems are biodiversity hotspots, and they are highly vulnerable to various environmental factors under the climate change scenario (Elsen and Tingley, 2015). Natural habitat loss results in massive species extinctions due to low-level adaptations and high-altitude extinctions caused by rising temperatures (Steinbauer et al., 2018; Knoke et al., 2020). Studies on mountain-based biodiversity changes are increasingly documented (Peters et al., 2019) but the impact of environmental changes on ecosystems is still challenging due to its rapidly changing abiotic conditions over short distances (Rahbek et al., 2019).

Mountain ecosystems are typically fragile and vulnerable due to their low resilience caused by steepness, low temperatures, and insulation. Soils are thin, young, and easily eroded. Low temperatures cause vegetation to grow slowly and soil to form slowly. Furthermore, higher altitudes' extreme diurnal temperature fluctuations necessitate special adaptations for survival. In this harsh biological environment, ecosystem recovery can take centuries (Ives and Messerli, 1990). Mountain regions face numerous threats and challenges, the majority of which are related to invasive species, land use change, overuse of land, and climate change (Egarter and Vigl, 2021). Mountain ecosystems are not only threatened by natural hazards, but they are also more likely to be harmed by humans than other types of landscapes (Poore, 1992). For example, shifting cultivation is one of the major land uses in the mountain ecosystems of Northeast India (Nath et al., 2020). The cultivation of steep slopes using lowland farming methods has resulted in disasters, resulting not only in the loss of life and property, but also in damage to the production base that may take centuries to recover (Byers and Sainju, 1994). Furthermore, the rapid growth of industrialization and population has increased the demand for natural resources such as land to sustain livelihoods; exploitation and extraction of natural resources has rendered the ecosystems fragile and unsustainable (Sundriyal and Sharma, 1996).

Overexploitation of land resources is one of the most serious challenges to sustainable development. As a result, millions of people who rely on forests and forest resources are becoming increasingly vulnerable. Such ecosystem pressure is hastening environmental degradation and necessitating the development of ecological systems for better adaptation and long-term management of mountain ecosystems (Rodríguez et al., 2006). Several attempts have been made to develop new socioecological system integrated modeling approaches in order to improve the evaluation and understanding of the interactions of social-ecological systems and associated ecosystem services (Linders et al., 2020; Mao et al., 2021). For example, progress has been made in developing a definition of climate-smart forests to mitigate the effects of climate change and to develop sustainable management strategies to improve the ecosystem's sustainability (Bowditch et al., 2020). In this connection, agroforestry systems have often been proposed as the best management practices for the socioeconomic and ecological development in the mountain ecosystems (Nath et al., 2020). This chapter provides a brief overview of the socioeconomic benefits and ecological sustainability of agroforestry systems, with a focus on mountain ecosystems.

19.1.1 Agroforestry: definition and importance

Agroforestry is a broad term for land use systems that combine woody perennials (such as trees and shrubs) with herbaceous plants such as crops and grasses, as well as livestock, to provide a variety of ecosystem services (Shin et al., 2020; Paudel et al., 2022). The basic idea behind agroforestry practice underlines that trees are an intrinsic part of the natural ecosystems, with a range of benefits provided to the soil, other plant species, and to the overall biodiversity of the ecosystem. Thus, agroforestry is an integrated (land use) system based on the coherent interactions of social and ecological aspects (Batish et al., 2008a; Nair, 1993). Agroforestry results in a unique ecological interaction and maximized economic returns by increasing the number of trees and farming plants on one piece of land (Muchane et al., 2020). Agroforestry practices are widely used in mountains to cope with disasters (e.g., eruptions, landslides, etc.) and to maximize natural resource utilization (Bachri et al., 2015; Utami et al., 2018). These systems are purposefully planned and managed to increase the positive links between trees and non-trees, and include a variety of practices such as contour cultivation, intercropping, established shelter belts, river areas, woody pastures, orchards, and so on (Mosquera-Losada et al., 2011).

Agroforestry systems provide food, fodder, fiber, fuel wood and timber, medicinal resources, improve the livelihoods of smallholder farmers, increase tree cover, and reduce pressure on forests, all of which aid in socioeconomic and ecological development (Bijalwan et al., 2011; Ahmad et al., 2017; Atreya et al., 2021; Singh et al., 2021). Furthermore, ecosystem services provided by agroforestry systems such as agro-silviculture, silvo-pasture, agro-silvo-pasture, forest farming, and others in the form of increased soil productivity, soil conservation from erosion, nutrient cycling, microclimate improvement, carbon sequestration, bio-energy and biofuel sources contribute to mitigating the effects of climate change (Fanish and Priya, 2013; Kay et al., 2019; Reppin et al., 2019; Singh et al., 2021; Paudel et al., 2022). The Intergovernmental Panel on Climate Change (IPCC) proposed agroforestry integration as a potential and cost-effective strategy to address climate change in its fifth assessment report (IPCC, 2014). Furthermore, agroforestry practices improve biodiversity, reduce crop failure risks, reduce fertilizer requirements, and prevent land degradation (UNDP, 2018), garnering policymakers' attention for integrated ecosystem management (Garrity, 2004). As a result, the agroforestry system has dramatically evolved, with an emphasis on how it works, the diversity of local, landscape, and regional knowledge systems, economic valuation, and environmental services provided (de Souza et al., 2012). Agroforestry has been recognized as a viable option for socioeconomic and ecological development in hilly regions of Timor-Leste, which is a highly vulnerable country to climate change and disaster risks such as floods and erosions (EU, 2007; Ismail et al., 2019). Depending on the degree of steepness of the hills, four types of agroforestry models are commonly used in Temor-Leste: alley cropping (i.e., planting along contour lines to reduce landslides and facilitate water flows), trees-along borders (i.e., hedgerow planting along steep slopes to conserve soil), random mixers (i.e., irregular planting of trees along with annual crops, particularly in low (flat and wide) land areas), and alternate row planting (UNDP, 2018; Paudel et al., 2022). Agroforestry systems are practiced on 27.36 million hectares (9.37% of total geographical area) in India, and this figure is expected to rise in the near future (Forest Survey of India FSI, 2017).

Fruit and fodder trees are commonly used in agroforestry systems to generate income (Ahmad et al., 2017). Different forage plants (trees and crops) such as *Acacia leucocephala*,

Albizia lebbek, Hibiscus tiliaceus, and *Prosopis* spp. are used in dryland areas; *Gliricidia sepium, Leucaena leucocephala, Samanea saman,* and *Sesbania grandiflora* are used in moderate rainfall areas; and *Albizia chinensis, Calliandra calothyrsus* and *Paraserianthes falcataria* are generally used in different agroforestry systems in Timor-Leste (Paudel et al., 2022). Furthermore, *Albizia procera, Alnus nepalensis,* and *Bauhinia* spp. have been widely used in Nepal's hills (Atreya et al., 2021). Similarly, pineapple (*Ananas comosus*)-based agroforestry systems (PAFS) adapted as post-fire sedentary systems developed after shifting cultivation by the Hmar tribes of Southern Assam in India's Sub-Himalayan region have been identified as potential methods for land restoration, biodiversity conservation, household income and food production enhancement, and thus leading to socioeconomic and ecological development of the area (Reang et al., 2022). PAFS incorporates the planting of multipurpose trees such as *Aquilaria malaccensis* (a critically endangered plant) and *Parkia timoriana,* which provide farmers with additional economic benefits (Reang et al., 2022). As a result, strengthening links between knowledge systems through participatory community management approaches is regarded as critical for sustainable forestry and agroforestry systems (Adhikari et al., 2020; Dhakal and Rai, 2020). Overall, the benefits of agroforestry can be grouped into two major components viz. socioeconomic benefits and ecological benefits (Fig. 19.1), which have been elaborated in the following sections:

19.2 Social and economic benefits of agroforestry

Because of its low-cost and diverse tree-crop produces, an important feature of the agroforestry system is the social and economic benefits to society's most vulnerable groups, particularly women, children, and marginalized sections (Blare and Useche, 2015; Kiptot, 2015). Agroforestry evolved from vast stores of traditional ecological knowledge for maintaining and properly utilizing an area's landscape (Reang et al., 2022). These traditional knowledge stores incorporate an in-depth understanding of the landscape that is in sync with plant selection, climatic conditions, and socioecological conditions in the area (Reang et al., 2022). Traditional agroforestry systems help in fulfilling the basic needs (e.g., fencing, food, fuel wood, handicrafts, housing, medicine, etc.) and income generation by the local society, particularly poor people (Nair et al., 2017; Horst and Hovorka, 2019) via providing raw-materials for the small-scale enterprises such as basket-making, bio-briquette, carpentry, essential oils, handmade-paper, medicinal plants, saw milling and wood carving (Atreya et al., 2021). Farmers (also known as agroforestry adopters) cultivate crops and vegetables, as well as trees and other livestock, in mountain ecosystems to help improve their livelihood (Rozaki et al., 2021). Furthermore, farmers benefited from agroforestry systems in the form of fruit, fodder, livestock, timber, and non-timber forest products (NTFPs), which improved their socioeconomic status by increasing job opportunities and livelihood security (Atreya et al., 2021). However, the adoption of agroforestry systems varies according to market dynamics, farmers' socioeconomic conditions (or cultures), and landscape edaphic conditions such as aspect, elevation, soil type, and slope, as well as climatic conditions such as temperature and precipitation, among other factors (Subedi et al., 2018; Atreya et al., 2021). According to Bertsch-Hoermann et al. (2021), implementing

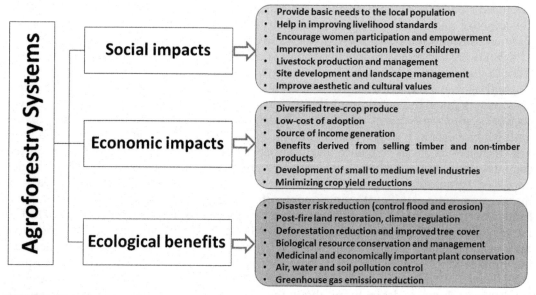

FIGURE 19.1 Impact of agroforestry systems on different social, economic and ecological components of an ecosystem.

agroforestry systems on marginal lands would be more beneficial in minimizing yield reductions than on high-yielding lands.

Men and women are equally involved in agroforestry systems (Dey et al., 2014), though women have been reported to face significant barriers to its adoption (Catacutan and Naz, 2015). The level of education, lack of access to technological advancements and markets, gender bias (including age, life-cycle stage, marital status), and labor availability all play a significant role in women's participation (Gelinas et al., 2015). However, women are reported to be more likely to use agroforestry systems than men, who prefer monoculture (Blare and Useche, 2015). A study conducted over a decade in Banswara (Rajasthan) and Dahod (Gujarat), semiarid regions in western India, discovered that agroforestry provided numerous benefits to women in the area (Bose, 2015). The study discovered that institutional arrangements were required if the excluded segments of these societies were to rise further. Chakraborty et al. (2015) examined the socioeconomic conditions of farmers in the Manirampur and Baghepura districts of Jessore, Bangladesh, and discovered that farmers who practice agroforestry fare better socially and economically than farmers who do not practice agroforestry. According to Rozaki et al. (2021), the division of labor involves both men and women in agroforestry systems. Men are generally involved in physical labor, whereas women are involved in maintenance and product sales in the disaster-prone area of Mt. Merapi, Indonesia. In a study conducted in the Philippines, Bugayong (2003) found that nearly 80% of those involved reported medium to high income changes as a result of agroforestry. Similarly, in the Attappady Kerala Block, the Nilgiri Biosphere Reserve agroforestry study found significant improvements in the socioeconomic security, diet, and livelihood standards of those involved (Kumar et al., 2006). According to Saha et al. (2010), the economic and social benefits of agroforestry systems are critical in determining whether a farmer considers it a viable alternative to conventional "modern" farming practices.

Agroforestry systems are investments of some kind, and timber planting, such as teak (*Tectona grandis*) and silver oak (*Grevilea robusta*), may not yield many benefits in the early stages (Dagar et al., 2014). Compared to the exclusive forestry land use, agroforestry practices have shown increased profitability for silvo-arable (Yates et al., 2007) and silvo-pastoral systems in terms of initial costs (Rodríguez et al., 2006). Agroforestry can provide multiple harvests in a single year, removing both work and income throughout the year. This increases financial resilience and reduces crop failure vulnerability, which is all too common with single cultivation or monoculture practices (Murthy et al., 2016). Under improved agroforestry systems, the National Research Centre for Agroforestry (NRCAF), Jhansi estimated a significant employment potential of 943 million people per year, down from 25.4 million ha (National Research Centre for Agroforestry NRCAF, 2007). The farmer's earning potential is increased (Murthy et al., 2016). Aside from increased agroforestry yield potential, product diversification increases economic yield potential by generating multiproduct annual and consistent revenues over time and lowering agricultural risks (Benjamin and Sauer, 2018). Murthy et al. (2016) studied agricultural growth systems in Northeast India to learn more about their potential, costs, and benefits. This study found that *Psidium* spp. in agricultural systems produced 2.96 times as much as similar treeline-free systems. Similarly, Singh and Pandey (2011) found that the benefit-to-cost ratio for traditional agroforestry systems combining *Acacia* and *Oryza* species was 21.47. Iskandar et al. (2016) conducted a study of the incomes of tribal and nontribal communities in Tripura West district and found that tree-crop yields from agroforestry were significant in tribal communities (at INR 24,075).

Traditional agroforestry (e.g., agri-horticulture, agri-silviculture, and agri-horti-silviculture) can be seen as the dominant land-use systems in the Garhwal Himalaya region of Uttarakhand, India (Singh et al., 2021). Because the region is rich in biodiversity, particularly medicinal plant diversity (68 plant species belonging to 38 families), agroforestry systems in this hilly area provide local people with livelihood options (Bijalwan et al., 2011; Parihaar et al., 2014). For example, in the region, child education rates have improved by 86, 98, and 100% in marginal, miniature, and medium-large landholding families, respectively (Singh et al., 2021). *Grewia optiva* had been widely used as fodder trees in the central and north-western Himalayan regions. Most of the average fuel wood consumption (84.41−538.45 kg/day/village) and fodder requirements (305.02−2015.52 kg/day/village) in lean periods were derived from the traditional (*Grewia optiva*) agroforestry systems (Bijalwan et al., 2014; Singh et al., 2021). Similarly, *G. optiva* and *Morus alba* as fodder trees, and *Azadiracta indica*, and *Dalbergia sissoo* as timber trees, have been widely used in Kumaun Himalayan traditional agroforestry systems (Bijalwan et al., 2014). Furthermore, the integration of agroforestry systems that are in synchronization with market demands increases farmers' income. Farmers in the eastern hilly areas of Nepal, for example, began cardamom-based agroforestry practices in 1964 after receiving increased demand and premium prices. Himalayan Alder (Utis - *Alnus* spp.) was used as the shade providing tree for cardamom (*Amomum subulatum*) growth (Atreya et al., 2021). Both plant species are compatible enough to coexist in moist, degraded hilly terrain. Cardamom accounts for 7% of Nepal's agricultural exports and employs approximately 67000 Nepalese households (DiCarlo et al., 2018).

Similarly, coffee-based agroforestry is a significant land-use system in Nepal, providing improved livelihood options to impoverished families (Atreya et al., 2021). Fruit trees, leguminous species, and vegetables intercropped within coffee agroforestry systems provide additional benefits to farmers. Most agroforestry systems in the Nepalese Himalaya generate more wealth from tree species than neighboring natural forests (Hurley et al., 2015). Overall, if properly managed, agroforestry systems provide significant benefits to marginal to small landholding farmers, particularly in mountainous regions.

19.2.1 Agroforestry for aesthetics and culture

Agroforestry systems not only provide direct socioeconomic benefits, but they also contribute to the aesthetic and cultural development of regions. The incorporation of trees into landscapes, for example, can make the landscape more attractive and beautiful; as a result, traditional agroforestry systems such as parks, orchard pastures, and wooden pastures are valued for visual decoration and beautification (McAdam et al., 2009). Franco et al. (2003) noted that the values that society places on non-market aspects like beauty need to be taken into consideration. This tends to make resource allocation in a landscape management setup more efficient. Despite a long history of woodland and orchard pastures, alpine woodland pastures, parks, and pannage, the cultural aspects of traditional agroforestry systems, particularly in temperate regions, are frequently overlooked (McAdam et al., 2009). Such systems are integrated with regional lifestyles, which include nomadism and traditional practices such as hedge laying and pollarding, among others. Local practice, law, and customs shape such landscapes' symbolic and cultural perceptions (Ispikoudis and Sioliou, 2005). Although only a few remnants of these traditional land use patterns remain today, they are internationally recognized for the significance and value of such cultural landscapes by the United Nations Educational, Scientific, and Cultural Organization (UNESCO). Aside from various socioeconomic benefits, agroforestry systems provide a variety of ecological benefits, which are highlighted in the following section.

19.3 Impact of agroforestry on sustainable ecosystem management

Land degradation as a result of intensive agricultural practices is a major issue that has been reported globally, particularly in Asian countries (Singh et al., 2019; Atreya et al., 2021). Practices that include the ability to restore degraded land are critical. Aside from various socioeconomic benefits, agroforestry systems provide several ecological benefits to local communities, including flood and erosion control in mountain regions, biological conservation, climate regulation, post-fire ecosystem restoration, carbon sequestration, air and water quality maintenance, and so on (Jose, 2009; Santoro et al., 2020; Scheper et al., 2021; Reang et al., 2022). Plantation of native and/or naturalized plant species such as *Choerospondias axillaris*, *Melia azedarach*, and *Prunus cerasoides*, for example, has the potential to restore degraded lands in Nepal's hilly regions (Jha et al., 2016). Similarly, agroforestry systems based on *Acacia nilotica*, *Butea monosperma*, and *Terminalia arjuna* have been recognized for land restoration by improving soil nutrient status and cycling (Jhariya et al., 2013; Singh et al., 2021). The following subsections highlight various ecological aspects of agroforestry:

19.3.1 Agroforestry for biodiversity

Agroforestry systems have been identified as a type of integrated land use that increases biodiversity while decreasing habitat loss (Suryanto et al., 2011; Guillerme et al., 2020). However, several socioeconomic, environmental, and spatial factors influence the agroforestry systems' biodiversity conservation potential (Whitney et al., 2018). Agroforestry contributes to biodiversity conservation by extending natural habitats, creating corridors between reserves, buffering reserves, and increasing landscape heterogeneity in a variety of ways (Nath et al., 2020). Further, in these systems, the pressures on formally protected reserves are reduced by the native and endangered trees species (Sharma and Vetaas, 2015), medicinal plants species (Rao and Arora, 2004) and fuel wood species (Kumar et al., 2011). Thus, agroforestry improves biodiversity by reducing deforestation and pressure on forest resources from local communities and fauna that rely on the surrounding forests and agricultural systems directly or indirectly (Zinngrebe et al., 2020; Kadigi et al., 2021). Furthermore, the use of farmers' traditional knowledge aids in the conservation and prioritization of high-value tree species plantation (Reang et al., 2022). Furthermore, because of ecological niche differentiation, complementary use of resources (e.g., nutrients, solar radiation, and water) encourages different species to coexist in agroforestry systems (Smith et al., 2013). In the Tunisian mountain oases (e.g., Chebika, Midès, and Tamaghza), agroforestry systems with three vegetation layers (first (upper) layer of date palms, second layer of smaller fruit trees (e.g., almonds, figs, olives, and pomegranates), and third (lowest) layer of annual crops (e.g., alfalfa, barley, sorghum, wheat, and vegetables)) provide the best examples (Dosskey, 2001). In such composite systems, each layer helps other crops by making the best use of resources such as air humidity, shade, sun radiation, and so on (Santoro et al., 2020). Furthermore, agroforestry and beekeeping have been identified as a practice for sustainable resource use, biodiversity conservation, and income generation by rural communities in Tanzania's Uluguru mountain ecosystems (Degu and Megerssa, 2020; Kadigi et al., 2021). Agroforestry systems can also help to reduce species extinction outside of formal conservation areas. Murthy et al. (2016) discovered that in Eastern and West Africa, agroforestry systems typically contain more than half of the trees in nearby primary forests. Despite increased competition for habitat and unsustainable land use patterns, many agroforestry systems have been observed to maintain ecological heterogeneity in the ecosystem (Bhagwat et al., 2008).

19.3.2 Agroforestry for pollution management

Agroforestry helps to reduce air pollution and improve air quality in a variety of ways by utilizing agroforestry features such as wind breaks and shelters. They protect buildings and roads from drifting snow in colder countries. Agroforestry systems have been proposed as a potential solution for lowering greenhouse gas (GHG) emissions and increasing O_2 levels (Nair et al., 2010). Further, microclimate changes (viz., 2–5°C reduction in soil and air temperature and 5%–10% increase in humidity) have been reported in the tea-crop agroforestry systems as compared to the nearby tree-less agricultural systems (Mukherjee et al., 2015; Singh et al., 2021). Wind chills are reduced, crops are protected, wildlife habitats are supplied, carbon dioxide in the air is removed, oxygen circulation is improved, wind velocity is reduced, airborne wind erosion and particulate matter are reduced, and animal odor is reduced, to name a few benefits of the agroforestry system (Tyndall and Colletti, 2011). Shelter belts have gained popularity as a means of reducing cattle odors (Tyndall and Colletti, 2007). Another time-tested benefit is that the agroforestry system filters particles in air streams, removes aerosol odors, and provides clean water. Agriculture has a wide range of effects on water quality, including changes in water chemistry caused by eutrophication, pesticide pollution, increased erosion and sediment load, and so on (Moss, 2008). River buffers, in particular, can be extremely beneficial in reducing pollution. The remaining elements are washed away by a flushing surface (Cassman, 1999). Further agroforestry has shown number of reductions in non-point source pollution of agricultural land (Dosskey, 2001) that includes reduction of field surface rinse, filters of groundwater rinse, elimination of surface rushing, filtering of stream water and reduction in bank erosion, etc. Agroforestry vegetative buffers have been shown to reduce pollution from non-point sources associated with row crop farming (Anderson et al., 2009). Deep-rooted trees provide a safety net for certain types by reusing excess nutrients and making efficient use of nutrients (van Noordwijk and van de Geijn, 1996). More research into the role of agroforestry in pollutant management in mountainous ecosystems is required.

19.3.3 Agroforestry for soil fertility improvement

Agroforestry has proven to be an effective method for improving soil quality and conserving water resources (Kumar et al., 2006; Murthy et al., 2013). The physicochemical properties of the soil, such as pH, moisture, water holding capacity, and organic matter, revealed that land use pattern influenced the soil's properties and fertility. For example, the shadow of a tree can reduce evapo-transpiration from plants, increasing soil water content in open pastures where trees are absent (Joffre et al., 1999). Reang et al. (2022) discovered that plant litter falls improve soil water and nutrient quality by reducing evaporation and improving soil structure in the PAFS developed at slash-and-burn landscapes in India's Sub-Himalayan region. Planting trees and cultivating crops that are beneficial to humans while also increasing soil nutrients are common in tropical agroforestry systems. The use of nitrogen-fixing plants (herbs, shrubs, or trees) in agroforestry systems has the potential to restore degraded land soil fertility (Ahmad et al., 2021). The integration of both nitrogen and nonnitrogen fixer species in tropical agroforestry systems benefits the soil environment by improving N-content by nitrogen-fixer species and soil organic C (SOC) content and other soil physicochemical properties by litter inputs from nonnitrogen fixer plant species (Jose, 2010). Litter inputs, for example, act as a potential source of slow release soil nutrients both above and below the surface in a *Ficus benghalensis*-based agroforestry system in the southern dry-climatic region of Karnataka (Dhanya et al., 2013). Neem plantation along with black gram (*Phaseolus mungo*) as ground crop-based agroforestry system at NRCAF, Jhansi region of Uttar Pradesh revealed sufficient return of soil macronutrients viz. N, P, K, and Ca at the rate of 98, 2.25, 32, and 131 kg/ha, respectively, to the soil by the neem plants (Pandey et al., 2012). Similarly, studies conducted on various agroforestry systems at the Central Research Institute for Dryland Agriculture (CRIDA) in Hyderabad, India, revealed an improvement in soil pH and SOC content (Sharma et al., 2019). In the North-eastern Himalayan region of India, improved soil hydrophysical properties were observed in multipurpose tree species-based agroforestry systems (Saha et al., 2010). Furthermore, in the Palani Hills of southern India, well-managed invasive leguminous tree (*Acacia* sp.)-based agroforestry systems were observed to improve soil productivity while reducing the negative impacts of invasion (Tassin et al., 2012).

Microorganisms in the soil play an important role in nutrient cycling and litter decomposition. In agroforestry systems, microbial diversity and activity have been reported to increase, which improves nutrient cycling and soil productivity (Raj et al., 2014). A variety of microbial diversity, including symbiotic and non-symbiotic N-fixers, disease-preventing endophytic microbes, and phosphate-solubilizing organisms, has been observed in agroforestry systems, depending on the tree and crop species used (Sridhar and Bagyaraj, 2018). In comparison to sole plantation systems, a *Prosopis*-based silvo-pastoral agroforestry system had higher microbial biomass C, improved soil pH, SOC, available macronutrients (NPK), physical properties, and alkali nutrient exchange (Singh and Dhayani, 2014). Furthermore, agroforestry systems have shown that they can recover polluted soils while also reducing soil acidification and salinization (Murthy et al., 2013). As a result, agroforestry systems can be recommended for restoring alkali soil productivity. Agroforestry is also a key sustainable practice for ground resource management because it restores the environment and maintains land resources (Dhyani and Chauhan, 1995). According to studies, agroforestry systems can be used in phytoremediation by using short rotation woody plants to remediate contaminated soil and groundwater systems (Rockwood et al., 2004). Overall, agroforestry systems can be used wisely to increase soil productivity in mountain ecosystems. Benefits of agroforestry systems on different soil properties have been outlined in Fig. 19.2.

19.3.4 Agroforestry for improving carbon sequestration potential

Carbon sequestration is the removal and storage of carbon from the atmosphere via physical or biological processes in carbon sinks such as the ocean, vegetation, and soil. Agroforestry's primary role in climate change mitigation may be to reduce carbon emissions through productive atmospheric carbon sequestration. The role of forest vegetation in soil carbon sequestration has received a lot of attention (Sharma et al., 2016; Pandit et al., 2017). Studies on the carbon sequestration potential of agroforestry systems, on the other hand, are limited in comparison to those on natural forest systems, but they are emerging rapidly nowadays (Bajracharya et al., 2015; Feliciano et al., 2018; Lawson et al., 2019; Atreya et al., 2021). Several studies have highlighted the ability of agroforestry systems to act as effective carbon sinks (Montagnini and Nair, 2004; Mutuo et al., 2005). In India, with an average carbon sequestration potential of 25 t C/ha (Pandey, 2011), agroforestry systems approximately store 280 Mt C in different climatic zones of the country (Forest Survey of India FSI, 2013). In an extensive review on carbon sequestration potential of agroforestry systems, Feliciano et al. (2018) reported highest carbon sequestration rate (4.38 t C/ha) for silvo-pastoral system

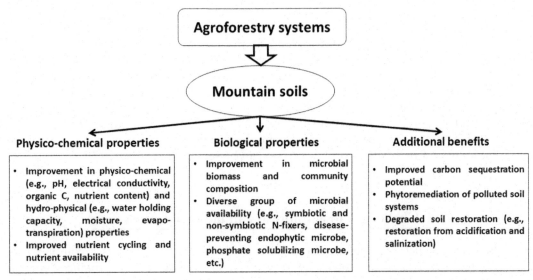

FIGURE 19.2   Impact of agroforestry systems on physicochemical and biological components of mountain soils.

developed from grassland, followed by home-gardens (3.8 t C/ha) developed from underutilized land. Different factors such as geographical locality, climate, previous land uses, species selection and their characteristics, species age, system types, management practices, etc. play major role in deciding the carbon sequestration potential of agroforestry systems (Dhyani et al., 2016; Atreya et al., 2021).

The agroforestry system can limit carbon emissions in a variety of ways, including conserving existing carbon pools, avoiding deforestation, and slashing and burning alternatives. Increased soil carbon pools, for example, and reduced use of synthetic fertilizers, for example, provide additional opportunities for carbon sequestration in the PAFS (Reang et al., 2021). Other methods of reducing carbon emissions include improved fallows and tree integration, biofuel replacement, and biological energy planting in place of fossil fuel use (Montagnini and Nair, 2004). According to an estimate by Reang et al. (2021), the PAFS having 180–225 mg/ha carbon storage capacity may attained the carbon sequestration potential to that of natural forests in a time span of ~ 15 years. However, there is a trade-off of biomass for various purposes at different time intervals in agroforestry systems, which affects their actual sequestration potential and is frequently underestimated (Bertsch-Hoermann et al., 2021). In addition to the significant amount of carbon stored in aboveground biomass, agroforestry systems can store carbon belowground. Carbon sequestration potential of aboveground components, for example, is generally estimated for agroforestry systems (Le Noë et al., 2021), but belowground components such as litter, root biomass, and SOC also contribute significantly in temperate agroforestry systems, which is often overlooked (Cardinael et al., 2017; Lim et al., 2018; Bertsch-Hoermann et al., 2021).

The tree component of agroforestry systems can be a significant carbon sink in agricultural areas. Because of their widespread applicability in existing agricultural systems, agroforestry systems are important carbon reservoirs. However, when averaged above-ground carbon is considered, significant amounts of carbon are stored in comparison to degraded land, croplands, or pastures (Albrecht and Kandji, 2003). Although agroforestry systems are well established and recognized for their ability to store additional carbon, the potential trade-off between carbon storage and profitability must be considered when promoting such systems (Nair et al., 2010). Arborescence in agroforestry can increase the amount of carbon sequestered when compared to a single cropping or pasture crop field (Sharrow and Ismail, 2004; Kirby and Potvin, 2007). Furthermore, by increasing the rotation ages of trees and/or shrubs and producing long-lasting products after harvesting the most abundant and permanent form of carbon can be sequestered (Jose, 2009).

19.4 Challenges to agroforestry in mountain ecosystems

Agroforestry, which includes tree-crop-livestock integrated systems, is a complex system that requires a balanced blend of traditional and technological advancements to ensure a sustainable future (Atreya et al., 2021). Farmers can increase their income by several folds by integrating traditional knowledge and proper tree/crop selection (Bayala et al., 2018). Aside from the various positive ecological aspects highlighted in the preceding subsections, agroforestry systems may have some negative effects in the area such as shading, resource competition, allelopathy, and exotic species invasion if not properly designed and managed (Kohli et al., 2008). Furthermore, the presence of grasses in tree-crop-grass combined systems may have facilitative or inhibitory effects on trees and crops, affecting the overall productivity of the system (Bhadouria et al., 2020). Thus, the success of these systems is determined by the appropriate selection of technology in agroforestry. Planting many non-native tree species in agroforestry systems, for example, has been reported for aesthetics, erosion control, fodder, shade, and timber production while undermining traditional knowledge and native tree species (Guillerme et al., 2020). Among several non-native species, the risk of some becoming invasive always exists (Richardson and Rejmánek, 2011). According to Udawatta et al. (2019), agroforestry is a sustainable practice that may promote invasive species if proper care is not taken. Because of the abandonment of agricultural sites, particularly in mountain ecosystems, the risk of invasion has increased. For example, abandonment of land uses resulted in a decline in agroforestry systems, which allowed invasive species to spread in France's Oussouet valley at the foothills of the Pyrenees Mountains (Guillerme et al., 2020). Furthermore, the allelopathic effects of invasive species (e.g., *Eucalyptus* sp., *Populus deltoids*, *L. leucocephala*, *Juglans nigra*, *G. sepium*, etc.) may result in decreased agroforestry system performance (Batish et al., 2008b). However, studies conducted in Brazil revealed that exotic species have less favorable habitat conditions in agroforestry systems, and the majority of the species have been filtered out in the understory (Ramos et al., 2015; Cordeiro et al., 2018). Although the mechanism is still unclear and requires data support at larger scales (Ramos et al., 2015; Guillerme et al., 2020). Furthermore, in most countries, a lack of market support and infrastructure, the involvement of middlemen in the sale of agroforestry products, poor transportation facilities in mountainous landscapes, poor research, and cumbersome legislative procedures are major barriers to wider adoption of agroforestry practices by farmers (Lundahl and Sjöholm, 2013; Sharma et al., 2017).

Though livestock are intrinsic components of agroforestry and provide numerous benefits to farmers in the form of food and income, their free grazing causes significant disturbances in agroforestry systems, which in most cases leads to their failure (Bettencourt et al., 2015). Furthermore, farmers in several regions face major challenges such as a lack of human resources, poor education and limited understanding levels of farmers in some areas, poor scientific understanding, higher initial adoption costs, and so on (da Silva et al., 2019; Atreya et al., 2021; Paudel et al., 2022). However, in some regions, such as Nepal, agroforestry systems are being revisited in light of market and infrastructure developments, resulting in increased farmer income (Aryal et al., 2019; Pandit et al., 2019). Temor-Leste should diversify its agroforestry systems by integrating agriculture with high-value trees such as sandalwood and teak, as well as bamboo, coffee, cocoa, coconuts, nuts, and so on (Paudel et al., 2022). It would aid in improving rural communities' livelihoods and building resilience to climate change (Fanzo et al., 2017). The inclusion of nitrogen-fixing trees and shrubs alongside cereal crops has been seen to help stabilize crop production in South African countries such as Temor-Leste (Sileshi et al., 2011). Increasing monetary compensation for ecosystem services, such as carbon sequestration provided by agroforestry systems, as well as appropriate land use policy documentation, would be a critical step toward climate change mitigation and wider adoption of these systems (Bertsch-Hoermann et al., 2021). Overall, different agroforestry system challenges can be managed by integrating traditional knowledge with technological advancements and government support in the form of incentives and infrastructural development required during the initial establishment stages.

19.5 Conclusion and future recommendations

Agroforestry is widely promoted as a sustainable practice that combines the best aspects of agriculture and forestry. The practice is now widely recognized as an applied science that helps thousands of tropical smallholders with food safety, poverty reduction, and ecosystem resilience. Agroforestry systems have the potential to improve farmers' socioeconomic conditions by providing a diverse range of livelihood benefits. Agroforestry has the potential to improve farmers' livelihoods by providing a plethora of alternatives and opportunities to increase farms and revenues, thereby assisting in poverty reduction. The systems also contribute to the ecological integrity of mountain ecosystems. Agroforestry systems provide significant ecological benefits such as improved soil fertility, nutrient cycling, biodiversity enrichment, microbial activities, water and air purification, and carbon sequestration. There are also a few barriers to agroforestry system adoption that can be overcome by integrating an appropriate knowledge base, management support, and government initiatives.

Acknowledgments

MAD and MA are thankful to CSIR, New Delhi, India for providing financial assistance as Senior Research Fellowship. RS gratefully acknowledges the financial support from DST-SERB, New Delhi as National Post-Doctoral Fellowship Scheme (Grant number: PDF/2020/001607).

References

Adhikari, B., Lodhiyal, N., Lodhiyal, L.S., 2020. Assessment of crop yield, productivity and carbon sequestration in agroforestry systems in Central Himalaya, India. Agroforestry Systems 94 (1), 281–296.

Ahmad, F., Goparaju, L., Qayum, A., 2017. Agroforestry suitability analysis based upon nutrient availability mapping: a GIS-based suitability mapping. AIMS Agriculture and Food 2 (2), 201–220. Available from: https://doi.org/10.3934/agrfood.2017.2.201.

Ahmad, F., Uddin, M.M., Goparaju, L., Dhyani, S.K., Oli, B.N., Rizvi, J., 2021. Tree suitability modeling and mapping in Nepal: a geospatial approach to scaling agroforestry. Modeling Earth Systems and Environment 7 (1), 169–179.

Albrecht, A., Kandji, S.T., 2003. Carbon sequestration in tropical agroforestry systems. Agriculture, Ecosystems & Environment 99 (1–3), 15–27.

Anderson, S.H., Udawatta, R.P., Seobi, T., Garrett, H.E., 2009. Soil water content and infiltration in agroforestry buffer strips. Agroforestry Systems 75 (1), 5–16.

Aryal, K., Thapa, P.S., Lamichhane, D., 2019. Revisiting agroforestry for building climate resilient communities: a case of package-based integrated agroforestry practices in Nepal. Emerging Science Journal 3 (5), 303–311. Available from: https://doi.org/10.28991/esj-2019-01193.

Atreya, K., Subedi, B.P., Ghimire, P.L., Khanal, S.C., Charmakar, S., Adhikari, R., 2021. Agroforestry for mountain development: prospects, challenges and ways forward in Nepal. Archives of Agriculture and Environmental Science 6 (1), 87–99.

Bachri, S., Stötter, J., Monreal, M., Sartohadi, J., 2015. The calamity of eruptions, or an eruption of benefits? Mt. Bromo human-volcano system a case study of an open-risk perception. Natural Hazards and Earth Systems Science 15 (2), 277–290. Available from: https://doi.org/10.5194/nhess-15-277-2015.

Bajracharya, R.M., Shrestha, H.L., Shakya, R., Sitaula, B.K., 2015. Agro-forestry systems as a means to achieve carbon co-benefits in Nepal. Journal of Forest and Livelihood 13 (1), 58–70.

Batish, D.R., Kohli, R.K., Jose, S., Singh, H.P. (Eds.), 2008a. Ecological Basis of Agroforestry. CRC Press, Florida.

Batish, D.R., Singh, H.P., Kohli, R.K., 2008b. Allelopathic tree–crop interactions under agroforestry systems. In: Batish, D.R., Kohli, R.K., Jose, S., Singh, H.P. (Eds.), Ecological Basis of Agroforestry. CRC Press, Florida, pp. 37–50.

Bayala, J., Dayamba, S.D., Ayantunde, A.A., Somda, J., Ky-Dembele, C., Bationo, et al., 2018. Methodological guide: community participatory inventory and prioritization of climate smart crop-livestock-agroforestry technologies/practices. ICRAF Technical Manual World Agroforestry Centre, Nairobi.

Benjamin, E.O., Sauer, J., 2018. The cost effectiveness of payments for ecosystem services—smallholders and agroforestry in Africa. Land Use Policy 71, 293–302.

Bertsch-Hoermann, B., Egger, C., Gaube, V., Gingrich, S., 2021. Agroforestry trade-offs between biomass provision and aboveground carbon sequestration in the alpine Eisenwurzen region, Austria. Regional Environmental Change 21 (3), 1–15.

Bettencourt, E.M.V., Tilman, M., Narciso, V., Carvalho, M.L.D.S., Henriques, P.D.D.S., 2015. The livestock roles in the wellbeing of rural communities of Timor-Leste. Revista de Economia e Sociologia Rural 53, 63–80.

Bhadouria, R., Srivastava, P., Singh, R., Tripathi, S., Verma, P., Raghubanshi, A.S., 2020. Effects of grass competition on tree seedlings growth under different light and nutrient availability conditions in tropical dry forests in India. Ecological Research 35 (5), 807–818.

Bhagwat, S.A., Willis, K.J., Birks, H.J.B., Whittaker, R.J., 2008. Agroforestry: a refuge for tropical biodiversity? Trends in Ecology & Evolution 23 (5), 261–267.

Bijalwan, A., Manmohan, J., Dobriyal, R., 2014. Productivity of wheat as intercrop in *Grewia optiva* based traditional agroforestry system along altitudinal gradient and aspect in mid hills of Garhwal Himalaya, India. American Journal of Environmental Protection 2 (5), 89–94.

Bijalwan, A., Sharma, C.M., Kediyal, V.K., 2011. Socioeconomic status and livelihood support through traditional agroforestry systems in hill and mountain agro-ecosystems of Garhwal Himalaya. Indian Forester 137 (12), 1423–1431.

Blare, T., Useche, P., 2015. Is there a choice? Choice experiment to determine the value men and women place on cacao agroforests in coastal Ecuador. International Forestry Reviews 17 (4), 46–60. Available from: https://doi.org/10.1505/146554815816086390.

Bose, P., 2015. India's drylands agroforestry: a ten-year analysis of gender and social diversity, tenure and climate variability. International Forestry Review 17 (4), 85−98.

Bowditch, E., Santopuoli, G., Binder, F., Del Rio, M., La Porta, N., Kluvankova, T., et al., 2020. What is climate-smart forestry? A definition from a multinational collaborative process focused on mountain regions of Europe. Ecosystem Services 43, 101113.

Bugayong, L.A., 2003. Socioeconomic and environmental benefits of agroforestry practices in a community-based forest management site in the Philippines. International Conference on Rural Livelihoods, Forests and Biodiversity 19−23.

Byers, E., Sainju, M., 1994. Mountain ecosystems and women: opportunities for sustainable development and conservation. Mountain Research and Development 14 (3), 213−228.

Cardinael, R., Chevallier, T., Cambou, A., Béral, C., Barthès, B.G., et al., 2017. Increased soil organic carbon stocks under agroforestry: a survey of six different sites in France. Agriculture, Ecosystems & Environment 236, 243−255.

Cassman, K.G., 1999. Ecological intensification of cereal production systems: yield potential, soil quality, and precision agriculture. Proceedings of the National Academy of Sciences 96 (11), 5952−5959.

Catacutan, D., Naz, F., 2015. Gender roles, decision-making and challenges to agroforestry adoption in Northwest Vietnam. International Forestry Reviews 17 (4), 22−32. Available from: https://doi.org/10.1505/146554815816086381.

Chakraborty, M., Haider, M.Z., Rahaman, M.M., 2015. Socio-economic impact of cropland agroforestry: evidence from Jessore district of Bangladesh. International Journal of Research in Agriculture and Forestry 2 (1), 11−20.

Chaudhary, S., Tshering, D., Phuntsho, T., Uddin, K., Shakya, B., Chettri, N., 2017. Impact of land cover change on a mountain ecosystem and its services: case study from the Phobjikha valley, Bhutan. Ecosystem Health and Sustainability 3 (9), 1393314.

Cordeiro, A.D.A.C., Coelho, S.D., Ramos, N.C., Meira-Neto, J.A.A., 2018. Agroforestry systems reduce invasive species richness and diversity in the surroundings of protected areas. Agroforestry Systems 92 (6), 1495−1505.

Dagar, J.C., Pandey, C.B., Chaturvedi, C.S., 2014. Agroforestry: a way forward for sustaining fragile coastal and island agro-ecosystems. Agroforestry Systems in India: Livelihood Security & Ecosystem Services. Springer, New Delhi, pp. 185−232.

Degu, T.K., Megerssa, G.R., 2020. Role of beekeeping in the community forest conservation: evidence from Ethiopia. Bee World 97, 98−104.

Dey, S., Resurreccion, B.P., Doneys, P., 2014. Gender and environmental struggles: voices from Adivasi Garo community in Bangladesh. Gender Place Culture 21 (8), 945−962. Available from: https://doi.org/10.1080/0966369X.2013.832662.

de Souza, H.N., de Goede, R.G., Brussaard, L., Cardoso, I.M., Duarte, E.M., Fernandes, R.B., et al., 2012. Protective shade, tree diversity and soil properties in coffee agroforestry systems in the Atlantic Rainforest biome. Agriculture, Ecosystems & Environment 146 (1), 179−196.

da Silva, F., Dassir, M., Mujetahid, M., Nadirah, S., 2019. Collaboration based agroforestry development strategy in Laubonu Village, Atsabe Subdistrict, District Ermera, Timor Leste. Advances in Environmental Biology 13, 10−16.

Dhakal, A., Rai, R.K., 2020. Who adopts agroforestry in a subsistence economy? Lessons from the Terai of Nepal. Forests 11 (5), 565.

Dhanya, B., Viswanath, S., Purushothaman, S., 2013. Crop yield reduction in ficus agroforestry systems of Karnataka, Southern India: perceptions and realities. Agroecology and Sustainable Food Systems 37 (6), 727−735.

Dhyani, S.K., Chauhan, D.S., 1995. Agroforestry interventions for sustained productivity in north-eastern hill region of India. Range Management & Agroforestry 16 (1), 79−85.

Dhyani, S.K., Ram, A., Dev, I., 2016. Potential of agroforestry systems in carbon sequestration in India. Indian Journal of Agricultural Science 86 (9), 1103−1112.

DiCarlo, J., Epstein, K., Marsh, R., Måren, I., 2018. Post-disaster agricultural transitions in Nepal. Ambio 47 (7), 794−805.

Dosskey, M.G., 2001. Toward quantifying water pollution abatement in response to installing buffers on crop land. Environmental Management 28 (5), 577−598.

Egarter Vigl, L., Marsoner, T., Giombini, V., Pecher, C., Simion, H., Stemle, E., et al., 2021. Harnessing artificial intelligence technology and social media data to support Cultural Ecosystem Service assessments. People and Nature 3 (3), 673−685.

Elsen, P.R., Tingley, M.W., 2015. Global mountain topography and the fate of montane species under climate change. Nature Climate Change 5 (8), 772–776.

European Union (EU), 2007. Disaster Preparedness in Timor-Leste. Disaster Preparedness, Humanitarian Aid and Civil Protection Directorate General (DIPECHO), and Timor Leste Living Standards Survey. Available online: <https://ec.europa.eu/echo/files/policies/dipecho/presentations/est_timor.pdf?fbclid=IwAR26DqR3-5WlKmBhmEs1jVVXfJAEJfDwn1Mk78PcU_Vw-U6PUrXf8o4uYv4> (accessed on 14 February 2021).

Fanish, S.A., Priya, R.S., 2013. Review on benefits of agroforestry system. International Journal of Educational Research 1 (1), 1–12.

Fanzo, J., Boavida, J., Bonis-Profumo, G., McLaren, R., Davis, C., 2017. Timor Leste strategic Review: Progress and Success in Achieving the Sustainable Development Goal 2. Centre of Studies for Peace and Development (CEPAD) Timor-Leste and John Hopskin University. Available online: <https://www.interpeace.org/wp-content/uploads/2017/06/TL-SR-May-15-V2-21_-May24-FINAL.pdf> (accessed on 20 September 2021).

Feliciano, D., Ledo, A., Hillier, J., Nayak, D.R., 2018. Which agroforestry options give the greatest soil and above ground carbon benefits in different world regions? Agriculture, Ecosystems and Environment 254, 117–129.

Forest Survey of India (FSI), 2013. State of forest report. FSI, Ministry of Environment and Dehradun, India.

Forest Survey of India (FSI), 2017. India State of Forest Report. Ministry of Environment and Forests, Dehradun, India.

Franco, D., Franco, D., Mannino, I., Zanetto, G., 2003. The impact of agroforestry networks on scenic beauty estimation: the role of a landscape ecological network on a socio-cultural process. Landscape and Urban Planning 62 (3), 119–138.

Garrity, D.P., 2004. Agroforestry and the achievement of the Millennium Development Goals. Agroforestry Systems 61, 5–17.

Gelinas, N., Lavoie, A., Labrecque, M.F., Olivier, A., 2015. Linking women, trees and sheep in Mali. International Forestry Reviews 17 (4), 76–84. Available from: https://doi.org/10.1505/146554815816086462.

Guillerme, S., Barcet, H., de Munnik, N., Maire, E., Marais-Sicre, C., 2020. Evolution of traditional agroforestry landscapes and development of invasive species: lessons from the Pyrenees (France). Sustainability Science 15 (5), 1285–1299.

Han, C., Liu, Y., Zhang, C., Li, Y., Zhou, T., Khan, S., et al., 2021. Effects of three coniferous plantation species on plant-soil feedbacks and soil physical and chemical properties in semi-arid mountain ecosystems. Forest Ecosystems 8 (1), 1–13.

Horst, G., Hovorka, A.J., 2019. Fuelwood: the other renewable energy source for Africa? Biomass & Bioenergy 33 (11), 1605–1616.

Hurley, P.T., Emery, M.R., McLain, R., Poe, M., Grabbatin, B., Goetcheus, C.L., 2015. Whose urban forest? The political ecology of foraging urban nontimber forest products. In: Isenhour, C., McDonagh, G., Checker, M. (Eds.), Sustainability in the Global City: Myth and Practice. pp. 187–212.

Iskandar, J., Iskandar, B.S., Partasasmita, R., 2016. Responses to environmental and socio-economic changes in the Karangwangi traditional agroforestry system, South Cianjur, West Java. Biodiversitas Journal of Biological Diversity 17 (1), 332–341.

Intergovernmental Panel on Climate Change (IPCC), 2014. Summary for policy makers. In Climate Change 2014: Impacts, Adaptation, and Vulnerability; In IPCC Working Group II contribution to AR5; IPCC, Geneva, Switzerland. Available online: <http://ipcc-wg2.gov/AR5/images/uploads/IPCC_WG2AR5_SPM_Approved.pdf> (accessed on 3 March 2021).

Ismail, C.J., Takama, T., Budiman, I., Knight, M., 2019. Comparative study on agriculture and forestry climate change adaptation projects in Mongolia, the Philippines, and timor leste. In: Castro, P., Azul, A., Leal Filho, W., Azeiteiro, U. (Eds.), Climate Change-Resilient Agriculture and Agroforestry. Springer, Cham, Switzerland, pp. 413–430.

Ispikoudis, I., Sioliou, K.M., 2005. Cultural aspects of silvopastoral systems. In: Silvopastoralism and sustainable land management. Proceedings of an international congress on silvopastoralism and sustainable management held in Lugo, Spain, April 2004 (pp. 319–323). CABI Publishing.

Ives, J.D., Messerli, B., 1990. Progress in theoretical and applied mountain research, 1973–1989, and major future needs. Mountain Research and Development 10 (2), 101–127.

Jhariya, M.K., Raj, A., Sahu, K.P., Paikra, P.R., 2013. Neema tree for solving global problem. Indian Journal of Applied Research 3 (10), 66–68.

Jha, R.K., Baral, S.K., Aryal, R., Thapa, H.B., 2016. Restoration of degraded sites with suitable tree species in the Mid-hills of Nepal. Banko Janakari 23 (2), 3–13.

Joffre, R., Rambal, S., Ratte, J.P., 1999. The dehesa system of southern Spain and Portugal as a natural ecosystem mimic. Agroforestry Systems 45 (1), 57–79.

Jose, S., 2009. Agroforestry for ecosystem services and environmental benefits: an overview. Agroforestry Systems 76 (1), 1–10.

Jose, S., 2010. Agroforestry for ecosystem services and environmental benefits: an overview. Agroforestry Systems 76, 1–10.

Kadigi, W.R., Ngaga, Y.M., Kadigi, R.M., 2021. Economic viability of smallholder agroforestry and beekeeping projects in Uluguru Mountains, Tanzania: a cost-benefit analysis. Open Journal of Forestry 11 (02), 83.

Kay, S., Rega, C., Moreno, G., den Herder, M., Palma, J.H.N., et al., 2019. Agroforestry creates carbon sinks whilst enhancing the environment in agricultural landscapes in Europe. Land Use Policy 83, 581–593.

Kiptot, E., 2015. Gender roles, responsibilities, and spaces: implications for agroforestry research and development in Africa. International Forestry Reviews 17 (4), 11–21.

Kirby, K.R., Potvin, C., 2007. Variation in carbon storage among tree species: implications for the management of a small-scale carbon sink project. Forest Ecology and Management 246 (2–3), 208–221.

Knoke, T., Paul, C., Rammig, A., Gosling, E., Hildebrandt, P., Härtl, F., et al., 2020. Accounting for multiple ecosystem services in a simulation of land-use decisions: does it reduce tropical deforestation? Global Change Biology 26 (4), 2403–2420.

Kohli, R.K., Singh, H.P., Batish, D.R., Jose, S., 2008. Ecological interactions in agroforestry: an overview. In: Batish, D.R., Kohli, R.K., Jose, S., Singh, H.P. (Eds.), Ecological Basis of Agroforestry. CRC Press, Florida, pp. 3–14.

Kumar, M., Lakiang, J.J., Gopichand, B., 2006. Phytotoxic effects of agroforestry tree crops on germination and radicle growth of some food crops of Mizoram. Lyonia 11 (2), 83–89.

Kumar, J.N., Patel, K., Kumar, R.N., Bhoi, R.K., 2011. An evaluation of fuelwood properties of some Aravally mountain tree and shrub species of Western India. Biomass and Bioenergy 35 (1), 411–414.

Lawson, G., Dupraz, C., Watté, J., 2019. Can silvoarable systems maintain yield, resilience, and diversity in the face of changing environments? Agroecosystem Diversity. Elsevier, pp. 145–168.

Le Noë, J., Erb, K.H., Matej, S., Magerl, A., Bhan, M., et al., 2021. Socioecological drivers of long-term ecosystem carbon stock trend: an assessment with the LUCCA model of the French case. Anthropocene 33, 100275.

Lim, S.S., Baah-Acheamfour, M., Choi, W.J., Arshad, M.A., Fatemi, F., et al., 2018. Soil organic carbon stocks in three Canadian agroforestry systems: from surface organic to deeper mineral soils. Forest Ecology and Management 417, 103–109.

Linders, T.E., Bekele, K., Schaffner, U., Allan, E., Alamirew, T., Choge, S.K., et al., 2020. The impact of invasive species on social-ecological systems: relating supply and use of selected provisioning ecosystem services. Ecosystem Services 41, 101055.

Lundahl, M., Sjöholm, F., 2013. Improving the Lot of the Farmer: Development Challenges in Timor-Leste during the Second Decade of Independence. Asian Economic Papers 12, 71–96.

Mao, Z., Centanni, J., Pommereau, F., Stokes, A., Gaucherel, C., 2021. Maintaining biodiversity promotes the multi-functionality of social-ecological systems: holistic modelling of a mountain system. Ecosystem Services 47, 101220.

McAdam, J.H., Burgess, P.J., Graves, A.R., Rigueiro-Rodríguez, A., Mosquera-Losada, M.R., 2009. Classifications and functions of agroforestry systems in Europe. Agroforestry in Europe. Springer, Dordrecht, pp. 21–41.

Montagnini, F., Nair, P.K.R., 2004. Carbon sequestration: an underexploited environmental benefit of agroforestry systems. New Vistas in Agroforestry. Springer, Dordrecht, pp. 281–295.

Mosquera-Losada, M.R., Freese, D., Rigueiro-Rodríguez, A., 2011. Carbon sequestration in European agroforestry systems. Carbon Sequestration Potential of Agroforestry Systems. Springer, Dordrecht, pp. 43–59.

Moss, B., 2008. Water pollution by agriculture. Philosophical Transactions of the Royal Society B: Biological Sciences 363 (1491), 659–666.

Muchane, M.N., Sileshi, G.W., Gripenberg, S., Jonsson, M., Pumariño, L., Barrios, E., 2020. Agroforestry boosts soil health in the humid and sub-humid tropics: a meta-analysis. Agriculture, Ecosystems & Environment 295, 106899.

Mukherjee, A., Banerjee, S., Nanda, M.K., Sarkar, S., 2015. Microclimate study under agroforestry system and its impact on performance of tea. Journal of Agrometeorology 10 (1), 99–105.

Murthy, I.K., Dutta, S., Varghese, V., Joshi, P.P., Kumar, P., 2016. Impact of Agroforestry systems on Ecological and socio-economic systems: a review. Global Journal of Science Frontiers in Research: Health, Environment and Earth Science 16 (5), 15–27.

Murthy, I.K., Gupta, M., Tomar, S., Munsi, M., Tiwari, R., Hegde, G.T., et al., 2013. Carbon sequestration potential of agroforestry systems in India. Journal of Earth Science and Climate Change 4 (1), 1–7.

Mutuo, P.K., Cadisch, G., Albrecht, A., Palm, C.A., Verchot, L., 2005. Potential of agroforestry for carbon sequestration and mitigation of greenhouse gas emissions from soils in the tropics. Nutrient Cycling in Agroecosystems 71 (1), 43–54.

Nair, P.R., Nair, V.D., Kumar, B.M., Showalter, J.M., 2010. Carbon sequestration in agroforestry systems. Advances in Agronomy 108, 237–307.

Nair, P.K.R., 1993. An Introduction to Agroforestry. Springer, Dordrecht, ISBN: 978-0-7923-2134-7.

Nair, P.K.R., Viswanath, S., Lubina, P.A., 2017. Cinderella agroforestry systems. Agroforestry Systems 91 (5), 901–917.

Nath, A.J., Sahoo, U.K., Giri, K., Sileshi, G.W., Das, A.K., 2020. Incentivizing hill farmers for promoting agroforestry as an alternative to shifting cultivation in Northeast India. Agroforestry for Degraded Landscapes. Springer, Singapore, pp. 425–444.

National Research Centre for Agroforestry (NRCAF), 2007. National Research Centre for Agroforestry (NRCAF): Vision-2025: NRCAF Perspective Plan Jhansi, India .

Pandey, A.K., Gupta, V.K., Solanki, K.R., 2012. Productivity of neem-based agroforestry system in semi-arid region of India. Range Management & Agroforestry 31 (2), 144–149.

Pandey, D.N., 2011. Multifunctional agroforestry systems. Current Science 92 (4), 1–8.

Pandit, R., Neupane, P.R., Wagle, B.H., 2017. Economics of carbon sequestration in community forests: evidence from REDD + piloting in Nepal. Journal of Forest Economics 26, 9–29.

Pandit, B.H., Nuberg, I., Shrestha, K.K., Cedamon, E., Amatya, S.M., Dhakal, B., et al., 2019. Impacts of market-oriented agroforestry on farm income and food security: insights from Kavre and Lamjung districts of Nepal. Agroforestry Systems 93 (4), 1593–1604.

Parihaar, R.S., Bargali, K., Bargali, S.S., 2014. Diversity and uses of ethno-medicinal plants associated with traditional agroforestry systems in Kumaun Himalaya. Indian Journal of Agricultural Sciences 84 (12), 1470–1476.

Paudel, S., Baral, H., Rojario, A., Bhatta, K.P., Artati, Y., 2022. Agroforestry: opportunities and challenges in Timor-Leste. Forests 13 (1), 41.

Peters, M.K., Hemp, A., Appelhans, T., Becker, J.N., Behler, C., Classen, A., et al., 2019. Climate–land-use interactions shape tropical mountain biodiversity and ecosystem functions. Nature 568 (7750), 88–92.

Poore, D. (Ed.), 1992. Guidelines for Mountain Protected Areas. IUCN-The World Conservation Union.

Rahbek, C., Borregaard, M.K., Colwell, R.K., Dalsgaard, B.O., Holt, B.G., Morueta-Holme, N., et al., 2019. Humboldt's enigma: what causes global patterns of mountain biodiversity? Science 365 (6458), 1108–1113.

Raj, A., Jhariya, M.K., Pithoura, F., 2014. Need of agroforestry and impact on ecosystem. Journal of Plant Development and Science 6 (4), 577–581.

Ramos, N.C., Gastauer, M., de Almeida Campos Cordeiro, A., et al., 2015. Environmental filtering of agroforestry systems reduces the risk of biological invasion. Agroforestry Systems 89, 279–289.

Rao, V.R., Arora, R.K., 2004. Rationale for conservation of medicinal plants. Medicinal Plants Research in Asia 1, 7–22.

Reang, D., Hazarika, A., Sileshi, G.W., Pandey, R., Das, A.K., Nath, A.J., 2021. Assessing tree diversity and carbon storage during land use transitioning from shifting cultivation to indigenous agroforestry systems: implications for REDD + initiatives. Journal of Environmental Management 298, 113470.

Reang, D., Nath, A.J., Sileshi, G.W., Hazarika, A., Das, A.K., 2022. Post-fire restoration of land under shifting cultivation: a case study of pineapple agroforestry in the Sub-Himalayan region. Journal of Environmental Management 305, 114372.

Reppin, S., Kuyah, S., de Neergaard, A., Oelofse, M., Rosenstock, T.S., 2019. Contribution of agroforestry to climate change mitigation and livelihoods in Western Kenya. Agroforestry Systems 94, 203–220. Available from: https://doi.org/10.1007/s10457-019-00383-7.

Richardson, D.M., Rejmánek, M., 2011. Trees and shrubs as invasive alien species—a global review. Diversity & Distributions 17 (5), 788–809.

Rockwood, D.L., Naidu, C.V., Carter, D.R., Rahmani, M., Spriggs, T.A., Lin, C., et al., 2004. Short-rotation woody crops and phytoremediation: opportunities for agroforestry? New Vistas in Agroforestry. Springer, Dordrecht, pp. 51–63.

Rodríguez, J.P., Beard Jr, T.D., Bennett, E.M., Cumming, G.S., Cork, S.J., Agard, J., et al., 2006. Trade-offs across space, time, and ecosystem services. Ecology and Society 11 (1), 28.

Rozaki, Z., Rahmawati, N., Wijaya, O., Safitri, F., Senge, M., Kamarudin, M.F., 2021. Gender perspectives on agro-forestry practices in Mt. Merapi hazards and risks prone area of Indonesia. Biodiversitas Journal of Biological Diversity 22 (7), 2980–2987.

Saha, R., Ghosh, P.K., Mishra, V.K., Majumdar, B., Tomar, J.M.S., 2010. Can agroforestry be a resource conserva-tion tool to maintain soil health in the fragile ecosystem of north-east India? Outlook on Agriculture 39 (3), 191–196.

Santoro, A., Venturi, M., Ben Maachia, S., Benyahia, F., Corrieri, F., Piras, F., et al., 2020. Agroforestry heritage sys-tems as agrobiodiversity hotspots. The case of the mountain oases of Tunisia. Sustainability 12 (10), 4054.

Scheper, A.C., Verweij, P.A., Kuijk, M., 2021. Post-fire forest restoration in the humid tropics: a synthesis of avail-able strategies and knowledge gaps for effective restoration. Science of the Total Environment 771, 144647. Available from: https://doi.org/10.1016/j.scitotenv.2020.144647.

Sharma, R., Chauhan, S.K., Tripathi, A.M., 2016. Carbon sequestration potential in agroforestry system in India: an analysis for carbon project. Agroforestry Systems 90 (4), 631–644.

Sharma, K.L., Ramachandra, Raju, K., Das, S.K., Prasadrao, B.R.C., Kulkarni, B.S., et al., 2019. Soil fertility and quality assessment under tree-, crop-, and pasture-based land-use systems in a rainfed environment. Communications in Soil Science & Plant Analysis 40, 9–10.

Sharma, P., Singh, M.K., Tiwari, P., Verma, K., 2017. Agroforestry systems: opportunities and challenges in India. Journal of Pharmacoglogy & Phytochemistry, SP-1 953–957.

Sharma, L.N., Vetaas, O.R., 2015. Does agroforestry conserve trees? A comparison of tree species diversity between farmland and forest in mid-hills of central Himalaya. Biodiversity and Conservation 24 (8), 2047–2061.

Sharrow, S.H., Ismail, S., 2004. Carbon and nitrogen storage in agroforests, tree plantations, and pastures in west-ern Oregon, USA. Agroforestry Systems 60 (2), 123–130.

Sileshi, G.W., Akinnifesi, F.K., Ajayi, O.C., Muys, B., 2011. Integration of legume trees in maize-based cropping systems improves rain use efficiency and yield stability under rain-fed agriculture. Agriculture Water Management 98, 1364–1372.

Shin, S., Soe, K.T., Lee, H., Kim, T.H., Lee, S., Park, M.S., 2020. A systematic map of agroforestry research focusing on ecosystem services in the Asia-Pacific Region. Forests 11, 368.

Singh, A.K., Dhayani, S.K., 2014. Agroforestry policy issues and challenges. Agroforestry systems in India: Livelihood security & ecosystem services 367–372 (eds.), India. *Advances in Agronomy.*

Singh, V., Johar, V., Kumar, R., Chaudhary, M., 2021. Socio-economic and environmental assets sustainability by agroforestry systems: a review. International Journal of Agriculture, Environment and Biotechnology (IJAEB) 14 (4), 521–533. Available from: https://doi.org/10.30954/0974-1712.04.2021.6.

Singh, V.S., Pandey, D.N., 2011. Multifunctional Agroforestry Systems in India: Science-based Policy Options. Climate Change and CDM Cell, Rajasthan State Pollution Control Board.

Singh, R., Singh, H., Raghubanshi, A.S., 2019. Challenges and opportunities for agricultural sustainability in changing climate scenarios: a perspective on Indian agriculture. Tropical Ecology 60 (2), 167–185.

Smith, J., Pearce, B.D., Wolfe, M.S., 2013. Reconciling productivity with protection of the environment: is temper-ate agroforestry the answer? Renewable Agriculture & Food Systems 28, 80–92.

Sridhar, K.R., Bagyaraj, D.J., 2018. Microbial biodiversity in agroforestry systems. In: Dagar, J., Tewari, V. (Eds.), Agroforestry. Springer, Singapore, pp. 645–667. Available from: https://doi.org/10.1007/978-981-10-7650-3_25.

Steinbauer, M.J., Grytnes, J.-A., Jurasinski, G., Kulonen, A., Lenoir, J., Pauli, H., et al., 2018. Accelerated increase in plant species richness on mountain summits is linked to warming. Nature 556 (7700), 231–234.

Subedi, Y., Mulia, R., Cedamon, E., Lusiana, B., Shrestha, K., Nuberg, I., 2018. Local knowledge on factors leading to agroforestry diversification in Mid-hills of Nepal. Journal of Forest and Livelihood 15 (2), 32–51.

Sundriyal, R.C., Sharma, E., 1996. Anthropogenic pressure on tree structure and biomass in the temperate forest of Mamlay watershed in Sikkim. Forest Ecology and Management 81 (1-3), 113–134.

Suryanto, P., Hamzah, M.Z., Mohamed, A., Alias, M.A., 2011. Silviculture agroforestry regime: compatible man-agement in Southern Gunung Mt. Merapi National Park, Java, Indonesia. International Journal of Biology 3 (2), 115–126.

Tassin, J., Rangan, H., Kull, C.A., 2012. Hybrid improved tree fallows: harnessing invasive woody legumes for agroforestry. Agroforestry Systems 84 (3), 417–428.

Tyndall, J., Colletti, J., 2007. Mitigating swine odor with strategically designed shelterbelt systems: a review. Agroforestry Systems 69 (1), 45–65.

Tyndall, J., Colletti, J., 2011. Mitigating swine odor with strategically designed shelterbelt systems: a review. Agroforestry Systems 69 (1), 45–65.

Udawatta, P.R., Rankoth, L., Jose, S., 2019. Agroforestry and biodiversity. Sustainability 11 (10), 2879.

United Nations Development Programme (UNDP), 2018. Community Agroforestry Guide. UNDP-DARDC Project. Available online: <https://www.tl.undp.org/content/timor_leste/en/home/library/resilience/community-agroforestry-guide.html> (accessed on 12 January 2021).

Utami, S.N.H., Purwanto, B.H., Marwasta, D., 2018. Land management for agriculture after The 2010 Mt. Merapi eruption. Planta Tropika Journal of Agro Science 6 (1), 32–38. Available from: https://doi.org/10.18196/pt.2018.078.32-38.

van Noordwijk, M., van de Geijn, S.C., 1996. Root, shoot and soil parameters required for process-oriented models of crop growth limited by water or nutrients. Plant and Soil 183 (1), 1–25.

Whitney, C.W., Luedeling, E., Tabuti, J.R.S., Nyamukuru, A., Hensel, O., Gebauer, J., et al., 2018. Crop diversity in home gardens of southwest Uganda and its importance for rural livelihoods. Agriculture and Human Values 35, 399–424. Available from: https://doi.org/10.1007/s10460-017-9835-3.

Yates, C., Dorward, P., Hemery, G., Cook, P., 2007. The economic viability and potential of a novel poultry agroforestry system. Agroforestry Systems 69 (1), 13–28.

Zinngrebe, Y., Borasino, E., Chiputwa, B., Dobie, P., Barcia, E., Gassner, A., et al., 2020. Agroforestry governance for operationalizing the landscape approach: connecting conservation and farming actors. Sustainability Science 15, 1417–1434.

Index

Note: Page numbers followed by "*f*" and "*t*" refer to figures and tables, respectively.

Printed in the United States
by Baker & Taylor Publisher Services